Introduction to Linear and Nonlinear Finite Element Analysis

Bo Liu · Cuiyun Liu

Introduction to Linear and Nonlinear Finite Element Analysis

北京航空航天大学出版社
BEIHANG UNIVERSITY PRESS

Springer

Bo Liu
School of Aeronautics Science
and Engineering
Beihang University
Beijing, China

Cuiyun Liu
School of Materials Science
and Engineering
Beihang University
Beijing, China

ISBN 978-981-95-9669-0 ISBN 978-981-95-9670-6 (eBook)
https://doi.org/10.1007/978-981-95-9670-6

Jointly published with Beihang University Press Co., Ltd
The print edition is not for sale in China (Mainland). Customers from China (Mainland) please order the print book from: Beihang University Press Co., Ltd.

The original submitted manuscript has been translated into English. The translation was done using artificial intelligence. A subsequent revision was performed by the author(s) to further refine the work and to ensure that the translation is appropriate concerning content and scientific correctness. It may, however, read stylistically different from a conventional translation.

This Springer imprint is published by the registered company Springer Nature Singapore Pte Ltd.
The registered company address is: 152 Beach Road, #21-01/04 Gateway East, Singapore 189721, Singapore

Preface

Computational solid mechanics permeates through all branches of the solid mechanics. By applying specific theories from solid mechanics and utilizing modern electronic computers and various numerical methods, it addresses practical problems in the field. Computational solid mechanics plays a vital role in structural strength analysis for modern manufacturing industries, national defense technology, and the research and development of weapon systems. The foundation of computational solid mechanics is the finite element method, which is further divided into linear and nonlinear finite element methods. Currently, most textbooks in the field focus on fundamental concepts, primarily covering linear finite elements, while works on nonlinear finite elements often provide limited foundational explanations. Therefore, drawing on years of experience in teaching, research, and software development, the author has attempted to write a comprehensive, practical, and foundational book on computational solid mechanics. This book incorporates many of the latest international research achievements from the past two decades, as well as the author's own research findings from the past ten years.

Computational mechanics software is a crucial component of industrial software for research and design, and its theoretical foundation lies in the finite element method. Compared to international computational mechanics software, there is a significant gap in both the scale and level of development of such software in China. In terms of overall functionality and performance, domestic products are still unable to compete with advanced foreign counterparts. Not only are major industrial projects almost entirely reliant on foreign software for engineering mechanics calculations, but even the engineering mechanics computations in most small and medium-sized design institutes are dominated by foreign programs. Over the past decade, the author has been dedicated to the development of computational mechanics software with independent intellectual property rights, and this book is a partial summary of the author's explorations in this field.

This book is divided into eight chapters, including:

Chapter 1 addresses general problems and nonlinearities in solid mechanics, providing a detailed summary of the main principles for formulating finite element equations for solid and structural mechanics problems, as well as the fundamental steps required to solve general small-strain solid mechanics problems. The equations are presented in both the strong form of partial differential equations and the weak form in integral representation. The chapter also offers a detailed introduction to variational principles and explains how certain problems become nonlinear. As this chapter provides a broad overview, beginners are advised to focus on understanding the basic concepts rather than delving deeply into the equations. After reading the subsequent chapters, revisiting this chapter will help achieve a more comprehensive understanding of its content.

Chapter 2 covers Galerkin approximations: irreducible and mixed formulations. It introduces the use of weak formulations to construct finite element-based approximate solutions, leading to the well-known Galerkin method. Assuming that readers may not be familiar with the finite element method for small deformation linear problems, the chapter provides a complete overview of the fundamental steps for solving transient problems and the solution of nonlinear algebraic systems, considering both irreducible and mixed approximation approaches. The chapter concludes with a thermal analysis example based on nonlinear quasi-harmonic equations.

Chapter 3 focuses on the finite element analysis of trusses and beams. It first introduces the energy functionals for truss (bar) elements, torsional shaft elements, beam bending elements, and shear beam bending elements. The chapter then discusses h- and p-type finite elements for trusses and beams. Trusses and beams are among the simplest and most widely used structural components, and their corresponding finite element formulations are two typical one-dimensional finite elements. Therefore, this chapter provides a detailed introduction along with multiple examples.

Chapter 4 focuses on the finite element analysis of two-dimensional and three-dimensional solids, both of which utilize C^0 elements and therefore share certain similarities. This chapter provides a detailed discussion of linear and higher-order triangular elements, as these procedures form the foundation for constructing other two- and three-dimensional elements. Building on this, the chapter introduces linear and higher-order quadrilateral, tetrahedral, prismatic, hexahedral, and pyramidal elements. Quadrilateral and hexahedral elements are simple, accurate, and efficient, but their geometric representation capabilities are not as versatile as those of triangular and tetrahedral elements.

Chapter 5 covers the finite element analysis of plates and solid shells, including thin plate elements, thick plate elements, and three-dimensional degenerated shell elements. This chapter presents conforming thin plate elements based on the hierarchical finite element method, with lower-order elements as special cases. The thick plate elements are analogous to those used in plane problems. They are generally similar, so the main discussion focuses on reduced integration, mixed methods, and the hierarchical quadrature element method. Three-dimensional degenerate shell elements are almost identical to thick plate elements, and similar analogies exist for general curved shells.

Chapter 6 covers Cartesian tensors, including coordinate axis rotation, Cartesian tensors, pseudo-tensors, tensor algebra, and the applications of second-order and higher-order Cartesian tensors in physics and engineering.

Chapter 7 covers finite element analysis of elastoplastic problems, including one-dimensional and multidimensional elastoplasticity. It analyzes two-dimensional and three-dimensional elastoplastic problems based on the J_2 flow theory and higher-order spectral element methods.

Chapter 8 focuses on the finite element analysis of geometrically nonlinear problems. It first introduces the fundamental kinematic relationships used in finite deformation solid mechanics, followed by a discussion of the variational formulation for finite deformations and the higher-order spectral quadrature element analysis of Mindlin plates and shells. Finally, numerical examples are provided for two-dimensional, three-dimensional, and plate and shell problems based on the higher-order spectral quadrature element method.

To ensure that this book comprehensively covers the essential foundational knowledge and theories while also providing depth, practicality, and innovation, a wide range of classic textbooks, monographs, and the latest research findings were consulted during its preparation. The book also incorporates numerous original research results, as detailed below:

Chapter 1, "General Problems and Nonlinearity in Solid Mechanics," mainly refers to *The Finite Element Method for Solid and Structural Mechanics* by Zienkiewicz et al., published in 2014, and *The Finite Element Method: A Practical Course* by G. R. Liu and S. S. Quek, published in 2003.

Chapter 2, "Galerkin Approximations: Irreducible and Mixed Forms," is primarily based on the 2014 publication by Zienkiewicz et al., *The Finite Element Method for Solid and Structural Mechanics.*

Chapter 3, "Finite Element Analysis of Trusses and Beams," mainly refers to *The Finite Element Method: A Practical Course* by G. R. Liu and S. S. Quek, published in 2003, and *Introduction to Finite Element Vibration Analysis* by M. Petyt, published in 2010.

Chapter 4, "Finite Element Analysis of Two- and Three-Dimensional Solids," mainly refers to the following works: "*A Differential Quadrature Hierarchical Finite Element Method*" by Bo Liu et al., published in 2021; and "*The Finite Element Method: A Practical Course*" by G. R. Liu and S. S. Quek, published in 2003.

Chapter 5, "Finite Element Analysis of Plates and Solid Shells," mainly refers to *The Finite Element Method: Its Basis and Fundamentals* by Zienkiewicz et al., published in 2013, and *A Differential Quadrature Hierarchical Finite Element Method* by Liu Bo et al., published in 2021.

Chapter 6, "Cartesian Tensor," is primarily based on *Higher Mathematics for Physics and Engineering* by H. Shima and T. Nakayama, published in 2010.

Chapter 7, "Finite Element Analysis of Elastoplastic Problems," is primarily based on N. H. Kim's 2015 publication, *Introduction to Nonlinear Finite Element Analysis*, as well as the master's thesis of the author's graduate student, Song Jiajia.

Chapter 8, "Finite Element Analysis of Geometric Nonlinear Problems," mainly refers to *Non-linear Finite Element Analysis of Solids and Structures* by R. Borst et al., published in 2012, as well as the dissertation of the author's graduate student, Shi Tao.

The first five chapters of this book cover the most fundamental knowledge in computational solid mechanics, encompassing the core principles of the linear finite element method. Compared to most related works, this book includes additional topics such as hierarchical quadrature element method, pyramid elements, and three-dimensional degenerated shell elements. Chapters 7 and 8 provide a systematic summary of the foundational knowledge on the topics. Apart from the basic knowledge presented in Chaps. 1–3 and 6, the book consistently incorporates the author's original research achievements in hierarchical quadrature element method. This book serves as a valuable reference for scholars and engineering professionals in computational solid mechanics and related fields and can also be used as a graduate-level textbook in computational solid mechanics. Drawing on the author's extensive experience in computational mechanics software development, the book offers a comprehensive summary of the relevant theoretical knowledge, making it an important reference for researchers in the field.

The completion of this book would not have been possible without the invaluable support of Prof. Xing Yufeng. Graduate students Lan Yingying, Wu Jiaojiao, Xie Pan, Xu Pai, Shi Tao, Song Jiajia, and Guo Mao contributed to the preparation of the initial draft. The author has made multiple rounds of additions, deletions, proofreading, and revisions to the manuscript. Special thanks are also due to the National Natural Science Foundation of China for financial support (Project Grant Numbers: 12472194, 11972004, 12002018, 11772031, 11402015).

This book took more than five years to complete, during which it was used multiple times as lecture notes for graduate courses. The content has been continuously revised and improved to ensure correctness. However, due to the limitations of the author's knowledge and time, errors or inadequacies may remain. Readers are sincerely invited to provide corrections and suggestions.

Beijing, China
August 2023

Bo Liu

Contents

Chapter 1
General Problems and Nonlinearity in Solid Mechanics

The human mind has inherent limitations and cannot fully grasp the behavior of complex entities in a single operation. Therefore, a common approach for engineers, scientists, and even economists is to decompose systems into individual components or "elements" whose behaviors are more easily understood [1–3]. The original system is then reconstructed from these components to naturally study its behavioral processes.

In many cases, we use a finite number of "elements" to explicitly define a model, a process known as discretization. In other cases, the subdivision continues indefinitely, and the problem can only be defined using infinitesimal mathematical assumptions, leading to differential equations or equivalent definitions with an infinite number of elements. Such systems are regarded as continuous.

With the advent of electronic computers, discrete problems can generally be solved with ease, even when the number of elements is very large. However, since all computers have limited capacity, continuous problems can only be solved exactly through mathematical operations, and exact solutions are typically applicable only to very simple geometric models.

To address the challenges of solving continuous problems in practice, engineers and mathematicians have developed various discretization methods, all of which involve an approximation process. As the number of discrete variables increases, these approximations are expected to converge to the true continuous solution in the limit.

Mathematicians and engineers approach the discretization of continuous problems differently. Mathematicians have developed general techniques that can be directly applied to the control of differential equations, such as finite difference approximations, various weighted residual methods, or approximation techniques defined by the stationary value of a functional. On the other hand, engineers often address problems more intuitively by drawing analogies between actual discrete elements and finite portions of a continuum. For example, M. J. Turner and colleagues demonstrated that

B. Liu and C. Liu, *Introduction to Linear and Nonlinear Finite Element Analysis*,
https://doi.org/10.1007/978-981-95-9670-6_1

by substituting properties and simplifying a continuum into a series of small parts or "elements," one can effectively capture its overall behavior.

The term "finite element" originated from the perspective of direct analogy in engineering. R. W. Clough was the first to use this term, and its adoption marked the emergence of a standardized method applicable to discrete systems. This is of paramount importance both conceptually and computationally: it is, first and foremost, easy to understand, and it also provides a unified framework for the development of various problems and standard computational procedures.

Since the early 1960s, the finite element method has made significant progress. Today, the purely mathematical approach and the "direct analogy" approach have been unified. The finite element method is now regarded as a general procedure for discretizing problems in continuum mechanics defined by mathematics.

In analyzing discrete problems, a set of standard methods has been developed over the years. Civil structural engineers first calculate the force–displacement relationship for each element in the structure, then assemble the structure by applying local equilibrium conditions at each "node" or connection point to obtain the global structural equations. Finally, the unknown displacements are determined by solving these equations.

The existence of a unified approach to the "standard discrete problem" allows us to preliminarily define the finite element method as an approach for approximating the solution of continuous problems. The specific procedure is as follows:

- Discretize the problem domain into numerous small elements, then use simple polynomials to represent the displacement within each element and obtain the element equations.
- Assemble the equations of each element with those of adjacent elements to form the global finite element equations for the entire problem. The complete displacement field is then obtained by solving these global equations.

We find that many classical mathematical or engineering approximation processes share this characteristic. Therefore, regarding the finite element concept, the origin and exact time of its invention are difficult to determine.

The finite element procedure described above may not appear difficult. However, upon closer examination, a series of questions naturally arise: How should the displacement function be assumed, and in what simple form should it be expressed? How can we ensure that the assumed displacement function satisfies the governing differential equations? How can the final displacement field be obtained from the assumed displacement function? Indeed, we can simply assume a displacement function, but certain principles must be followed in order to derive a discretized system of equations, from which the final displacement field can be obtained by solving the system. By adhering to these principles, the governing equations can be satisfied in an optimal (though not exact) manner. This chapter will provide a detailed introduction to the main principles for formulating finite element equations in solid and structural mechanics problems.

Many introductory articles on the finite element method discuss approaches for solving linear elasticity and field equations. However, in practical applications, the presence of nonlinear effects and geometries with "thin" dimensions in one or more directions often limit the applicability of linear elasticity or, more generally, linear behavior, making problem-solving more challenging. In this book, we will present methods for addressing both types of problems.

There are two types of nonlinear behavior in solids: material nonlinearity and geometric nonlinearity. The simplest form of material nonlinear behavior is nonlinear elastic behavior, in which stress and strain are no longer linearly proportional but remain reversible. In more general cases, the material response during loading and unloading may differ, with elastoplastic behavior being a typical example.

When the deformation of a solid reaches a state fundamentally different from its undeformed configuration, a finite deformation state arises. In such cases, the linear strain–displacement relations or equilibrium equations cannot be applied to the undeformed geometry. Even before the onset of finite deformation, phenomena such as buckling or load bifurcation can be observed in solids, necessitating consideration of nonlinear equilibrium effects. The classical Euler beam buckling equilibrium equation, which incorporates the influence of axial loads, is a typical example of such problems. When deformations are large, boundary conditions may also become nonlinear. For instance, pressure loads may remain perpendicular to the deformed body, or the deformed boundary may interact with another body. The latter defines what is commonly referred to as a contact problem, which is also a widely studied area.

General two-dimensional or three-dimensional finite element formulations are often ineffective for solving solid mechanics problems involving one (or more) small dimensions. The primary reason is that the resulting algebraic systems are numerically ill-conditioned. Therefore, it is necessary to employ alternative classical structural mechanics theories to address such problems. For example, a plate is a flat structure characterized by a thin (small) dimension known as its thickness; a shell is a spatially curved structure with a similarly small thickness; beams, frames, or rods are structures with two small dimensions. In this book, we integrate classical structural mechanics approaches more closely with three-dimensional solid mechanics theory, thereby deriving standard finite element formulations that are easier to solve.

Today, computational solid mechanics and structural mechanics are applied across a wide range of problem-solving scenarios, from remote-controlled toy cars to full-scale aircraft. In addressing such problems, it is often necessary to use both structural elements (beams and shells) and solid elements to achieve accurate modeling and solution.

Aerospace engineering problems played a significant role in driving early research activities in the finite element method. For example, in the 1960s, manned spaceflight programs were a national priority in the United States, and many researchers around the world were actively engaged in studies related to space structures. Thin shells and flexible beams were often used to model such structures.

This book addresses problems in solid mechanics and structural mechanics, as well as practical and efficient finite element formulations for solving such problems. The content of this chapter is highly comprehensive, and beginners may find it challenging to fully grasp the details of the equations presented. Therefore, it is recommended that beginners focus on understanding the fundamental concepts rather than delving too deeply into the equations at this stage; after reading the subsequent chapters, revisiting this chapter will facilitate a more thorough understanding. This chapter also introduces the notation and methods adopted throughout the book.

1.1 Small Deformation Solid Mechanics Problems

1.1.1 Strong Form Equation: Index Notation

In this section, we will discuss how various equations in solid mechanics become nonlinear under certain circumstances, particularly when nonlinear stress–strain relationships are employed. Through this introduction, readers will observe that the formal transition between linear and nonlinear problems is, in fact, quite straightforward.

The field equations of solid mechanics are defined by the strain–displacement relations, equilibrium equations (momentum balance), boundary conditions, initial conditions, and constitutive equations.

In this book, when expressing various equations and relationships, we will use two types of notation: the first is the index notation of Cartesian tensors, and the second is the matrix notation. Generally, we find that both are helpful for describing specific aspects of the formulas. For example, when addressing large deformation problems, the most straightforward way to derive the so-called "geometric" or "initial stress" stiffness is by using index notation; in most other cases, matrix notation proves to be more convenient. We will also provide the requirements for converting between these two forms.

When using index notation, an index that appears only once in any equation is called a free index, while an index that appears more than once is called a dummy index. A dummy index may appear only twice in any equation and implies summation over its range. For example, if two vectors, $\boldsymbol{a}_i$ and $\boldsymbol{b}_i$, each have three components ($i = 1, 2, 3$), the expression $\boldsymbol{a}_i\boldsymbol{b}_i$ represents the following relationship:

$$\boldsymbol{a}_i\boldsymbol{b}_i = \boldsymbol{a}_1\boldsymbol{b}_1 + \boldsymbol{a}_2\boldsymbol{b}_2 + \boldsymbol{a}_3\boldsymbol{b}_3$$

Note that the dummy index can be replaced by any other index without changing its meaning, i.e.,

$$\boldsymbol{a}_i\boldsymbol{b}_i = \boldsymbol{a}_j\boldsymbol{b}_j$$

1.1.1.1 Coordinates and Displacement

For a fixed Cartesian coordinate system, we denote the coordinates as x, y, z, or in index notation as x_1, x_2, x_3. Thus, the coordinate vector is given by

$$\mathbf{x} = x_1\boldsymbol{e}_1 + x_2\boldsymbol{e}_2 + x_3\boldsymbol{e}_3 = x_i\boldsymbol{e}_i$$

Here, $\boldsymbol{e}_i$ denotes the unit basis vectors in the Cartesian coordinate system, and the summation convention is employed.

Similarly, displacement can be represented as u_1, u_2, u_3 (or alternatively as u, v, w), and the displacement vector

$$\boldsymbol{u} = u_1\boldsymbol{e}_1 + u_2\boldsymbol{e}_2 + u_3\boldsymbol{e}_3 = u_i\boldsymbol{e}_i$$

Physical quantities are typically represented using components. In general, coordinates and displacements are denoted by x_i and u_i, respectively, where the index i ranges from 1 to 3 in three dimensions (and from 1 to 2 in two dimensions).

1.1.1.2 Strain–Displacement Relationship

Strain can be represented by a Cartesian tensor as follows

$$\varepsilon_{ij} = \frac{1}{2}\left(\frac{\partial u_i}{\partial x_j} + \frac{\partial u_j}{\partial x_i}\right)$$

As long as the deformation is very small, this form of representation is valid. What we refer to as the small deformation problem is

$$\left|\varepsilon_{ij}\right| \ll 1 \quad \text{and} \quad \left|\omega_{ij}\right| \ll \left\|\varepsilon_{ij}\right\|$$

Here, | • | denotes the absolute value, || • || represents a certain norm, and ω_{ij} indicates an infinitesimal rotation, that is,

$$\omega_{ij} = \frac{1}{2}\left(\frac{\partial u_i}{\partial x_j} - \frac{\partial u_j}{\partial x_i}\right)$$

Therefore, the displacement gradient can be expressed as

$$\frac{\partial u_i}{\partial x_j} = \varepsilon_{ij} + \omega_{ij}$$

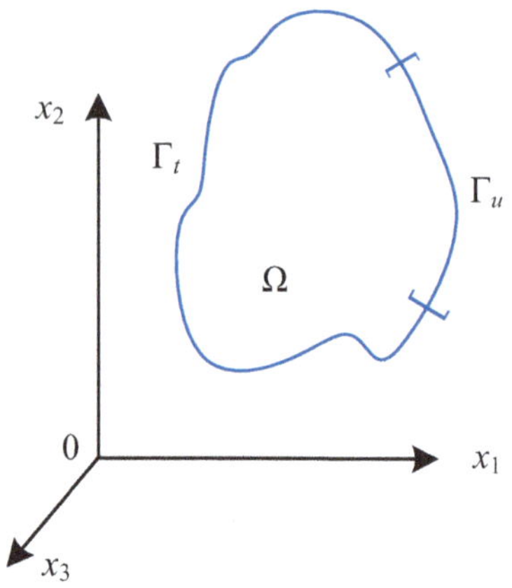

Fig. 1.1 Computational domain (Ω) and boundaries: traction boundary (Γ_t), and displacement boundary (Γ_u)

1.1.1.3 Balance Equation (Momentum Balance)

The index form of the balance equation (linear momentum balance) is expressed as

$$\sigma_{ji,j} + b_i = \rho \ddot{u}_i, \quad i, j = 1, 2, 3 \tag{1.1}$$

Here, σ_{ij} denotes the components of the Cauchy stress, ρ is the mass density, and b_i represents the components of the body force. The partial derivatives with respect to coordinates and time are expressed as follows:

$$f_{,i} = \frac{\partial f}{\partial x_i} \quad \text{and} \quad \dot{f} = \frac{\partial f}{\partial t}$$

Then in Eq. (1.1), $\ddot{u}_i$ represents acceleration, and the right-hand side of Eq. (1.1) corresponds to the inertial force.

Similarly, the stress symmetry resulting from moment equilibrium (angular momentum balance) can be expressed in index notation as

$$\sigma_{ij} = \sigma_{ji} \tag{1.2}$$

Equations (1.1) and (1.2) hold at all points x_i in the computational domain Ω (see Fig. 1.1).

1.1.1.4 Boundary Conditions

The stress boundary condition is specified by the traction condition, namely,

$$t_i = \sigma_{ji} n_j = \bar{t}_i$$

Note: The overline "–" above a variable denotes the given function.

Let Γ_t denote the set of all points on this boundary.

Similarly, the displacement boundary condition is given by

$$u_i = \overline{u}_i$$

Similarly, let Γ_u denote the set of all points where this boundary condition is applied.

A special type of displacement boundary condition is the periodic boundary condition. In this case, the boundary conditions are given by

$$u_i(x_j + \Delta x_j) = u_i(x_j) + \overline{g}_i$$

Here, Δx_j represents the period length, and $\overline{g}_i$ denotes the displacement variation between two points. This type of boundary condition is commonly used in the Representative Volume Element (RVE) to derive approximate constitutive models for composites and other types of structured materials.

1.1.1.5 Initial Conditions

For transient problems in which the inertial term $\rho\,\ddot{u}_i$ in Eq. (1.1) cannot be neglected, initial conditions for displacement and velocity are required. These conditions, defined at the initial time denoted by "0" within the domain Ω, are given as follows:

$$u_i(x_j, 0) = \overline{d}_i(x_j)$$
$$\dot{u}_i(x_j, 0) = \overline{v}_i(x_j)$$

In certain problems, it is also necessary to specify the initial stress state.

1.1.1.6 Constitutive Relations

As long as the deformation remains small, all of the above equations are applicable to any material. The specific behavior of a material is described by its constitutive equation, which relates the stress to the applied strain history and other factors that cause deformation, such as temperature.

The simplest material constitutive model is the linear elastic model, which, according to this model, holds at any time t

$$\sigma_{ij}(t) = \mathbb{C}_{ijkl}\left(\varepsilon_{kl}(t) - \varepsilon_{kl}^{(0)}(t)\right) \tag{1.3}$$

Here, $\mathbb{C}_{ijkl}$ denotes the elastic modulus, and $\varepsilon^{(0)}{}_{kl}$ represents the strain induced by factors other than displacement. For example, in thermo-elasticity problems, the strain is caused by temperature variation and can be expressed as

$$\varepsilon_{kl}^{(0)}(t) = \alpha_{kl}[T(t) - T_0] \tag{1.4}$$

where α_{kl} denotes the coefficient of linear thermal expansion, T represents the temperature at time t, and T_0 is the reference temperature at which the thermal strain is zero. In the following discussion, the time variable (i.e., time t) will no longer be explicitly included, and it is assumed that the elastic properties of the material depend only on the current value of the strain.

For linear isotropic materials, Eqs. (1.3) and (1.4) can be simplified as follows:

$$\sigma_{ij} = \lambda\delta_{ij}\left(\varepsilon_{kk} - \varepsilon_{kk}^{(0)}\right) + 2\mu\left(\varepsilon_{ij} - \varepsilon_{ij}^{(0)}\right)$$

$$\varepsilon_{ij}^{(0)} = \delta_{ij}\alpha[T - T_0]$$

Here, λ and μ are the Lamé elastic constants, and α is the coefficient of linear thermal expansion. In addition, δ_{ij} is the Kronecker function defined by

$$\delta_{ij} = \begin{cases} 1, & \text{when } i = j \\ 0, & \text{when } i \neq j \end{cases}$$

Many materials are neither linear nor elastic. Therefore, developing appropriate constitutive models to represent the experimentally observed mechanical behavior of materials is highly complex. In this book, we will introduce several classical models of mechanical behavior and demonstrate how to incorporate them into a general solution framework. Here, we simply point out how nonlinear material mechanical behavior affects the equations. To this end, we use the strain energy density function W to represent nonlinear elastic behavior, so that the stress can be expressed as

$$\sigma_{ij} = \frac{\partial W}{\partial \varepsilon_{ij}}$$

Materials that satisfy the above relationship are called hyper-elastic materials. When the strain energy is given by a quadratic form

$$W = \frac{1}{2}\varepsilon_{ij}\mathbb{C}_{ijkl}\varepsilon_{kl} - \varepsilon_{ij}\mathbb{C}_{ijkl}\varepsilon_{kl}^{(0)} \tag{1.5}$$

the linear elastic model described by Eq. (1.3) can be obtained. For hyper-elastic materials, the moduli must satisfy the symmetry

$$C_{ijkl} = C_{klij}$$

Furthermore, since both stress and strain are symmetric, there is additional symmetry:

$$C_{ijkl} = C_{jikl} = C_{ijlk} = C_{jilk}$$

More general forms are also permissible, but this will result in nonlinear elastic behavior.

1.1.2 Matrix Representation

In this book, we will frequently use matrix notation to present equations. In such cases, we represent the coordinates as

$$\boldsymbol{x} = \begin{Bmatrix} x \\ y \\ z \end{Bmatrix} = \begin{Bmatrix} x_1 \\ x_2 \\ x_3 \end{Bmatrix}$$

Express the displacement as

$$\boldsymbol{u} = \begin{Bmatrix} u \\ v \\ w \end{Bmatrix} = \begin{Bmatrix} u_1 \\ u_2 \\ u_3 \end{Bmatrix}$$

For two-dimensional problems, the third term is typically neglected.

The matrix representation of stress is given by

$$\begin{aligned} \boldsymbol{\sigma} &= \begin{bmatrix} \sigma_{11} & \sigma_{22} & \sigma_{33} & \tau_{12} & \tau_{23} & \tau_{31} \end{bmatrix}^{\mathrm{T}} \\ &= \begin{bmatrix} \sigma_{xx} & \sigma_{yy} & \sigma_{zz} & \tau_{xy} & \tau_{yz} & \tau_{zx} \end{bmatrix}^{\mathrm{T}} \end{aligned}$$

In a sequential table, the force is denoted as $\tau_{12} = \tau_{21}$, and so on.

The matrix representation of strain is

$$\begin{aligned} \boldsymbol{\varepsilon} &= \begin{bmatrix} \varepsilon_{11} & \varepsilon_{22} & \varepsilon_{33} & \gamma_{12} & \gamma_{23} & \gamma_{31} \end{bmatrix}^{\mathrm{T}} \\ &= \begin{bmatrix} \varepsilon_{xx} & \varepsilon_{yy} & \varepsilon_{zz} & \gamma_{xy} & \gamma_{yz} & \gamma_{zx} \end{bmatrix}^{\mathrm{T}} \end{aligned}$$

where the tensor is assumed to be symmetric by default, and the "engineering" shear strain is introduced.

$$\gamma_{ij} = \varepsilon_{ij} + \varepsilon_{ji} = 2\varepsilon_{ij}, \quad i \neq j$$

The introduction of this variable will simplify the subsequent expressions of matrix relationships.

The conversion between the tensor and matrix forms of the six independent components of stress and strain follows the index order listed in Table 1.1. This ordering is also used in many subsequent derivations. By simply omitting the last two terms and considering the third term as applicable, the above index relationship reduces to the two-dimensional case for plane or axisymmetric conditions.

The strain–displacement relationship can be expressed in matrix form as

$$\boldsymbol{\varepsilon} = \boldsymbol{B}\boldsymbol{u}$$

Here, the three-dimensional strain differential operator matrix is

$$\boldsymbol{B}^{\mathrm{T}} = \begin{bmatrix} \frac{\partial}{\partial x_1} & 0 & 0 & \frac{\partial}{\partial x_2} & 0 & \frac{\partial}{\partial x_3} \\ 0 & \frac{\partial}{\partial x_2} & 0 & \frac{\partial}{\partial x_1} & \frac{\partial}{\partial x_3} & 0 \\ 0 & 0 & \frac{\partial}{\partial x_3} & 0 & \frac{\partial}{\partial x_2} & \frac{\partial}{\partial x_1} \end{bmatrix}$$

In Cartesian coordinates, the equilibrium Eq. (1.4) can be written using the same differential operator, namely

$$\boldsymbol{B}^{\mathrm{T}}\sigma + \mathbf{b} = \rho\ddot{\mathbf{u}}$$

The displacement boundary condition and traction boundary condition are as follows

$$\mathbf{u} = \bar{\mathbf{u}} \qquad \text{on } \Gamma_u$$

$$\mathbf{t} = \mathbf{G}^{\mathrm{T}}\sigma = \bar{t} \text{ on } \Gamma_{\mathbf{t}}$$

where

Table 1.1 Index correspondence between tensor and matrix forms

Form	Index value					
Matrix	1	2	3	4	5	6
Tensor (1, 2, 3)	11	22	33	12	23	31
				21	32	13
Cartesian (x, y, z)	xx	yy	zz	xy	yz	zx
				yx	zy	xz
Cylindrical coordinates (r, z, θ)	rr	zz	$\theta\theta$	rz	$z\theta$	θr
				zr	θz	$r\theta$

$$G^{\mathrm{T}} = \begin{bmatrix} n_1 & 0 & 0 & n_2 & 0 & n_3 \\ 0 & n_2 & 0 & n_1 & n_3 & 0 \\ 0 & 0 & n_3 & 0 & n_2 & n_1 \end{bmatrix}$$

Here, $\boldsymbol{n} = (n_1, n_2, n_3)$ represents the direction cosines of the outward normal vector to the boundary Γ. Further observation reveals that the nonzero entries of $\boldsymbol{B}$ and $\boldsymbol{G}$ are identical.

For transient problems, the initial conditions can be expressed as

In Ω: $u(x, 0) = \overline{d}(x)$ and $\dot{u}(x, 0) = \overline{v}(x)$.

The matrix representation of the constitutive equation for linear elastic materials is

$$\boldsymbol{\sigma} = \boldsymbol{D}(\boldsymbol{\varepsilon} - \boldsymbol{\varepsilon}_0) \tag{1.6}$$

In this context, the indices ij and kl in Eq. (1.3) have been converted into a 6×6 symmetric matrix $\boldsymbol{D}$ for C_{ijkl} using Table 1.1. For general hyper-elastic materials,

$$\boldsymbol{\sigma} = \frac{\partial W}{\partial \boldsymbol{\varepsilon}}$$

1.1.3 Two-Dimensional Problems

The simplest two-dimensional problems are plane stress and plane strain problems. The former assumes that the deformation plane (for example, the x_1-x_2 plane) is very thin, with stress $\sigma_{33} = \tau_{13} = \tau_{23} = 0$; the latter assumes that on the deformation plane (for example, the x_1-x_2 plane), $\varepsilon_{33} = \gamma_{13} = \gamma_{23} = 0$. Another type is a two-dimensional function of r and z.

1.1.3.1 Plane Stress and Plane Strain Problems

For plane stress and plane strain problems with the x_1-x_2 plane as the deformation plane, assume the displacement is

$$\boldsymbol{u} = \begin{Bmatrix} u_1(x_1, x_2, t) \\ u_2(x_1, x_2, t) \end{Bmatrix}$$

Strain can thus be defined as

$$\boldsymbol{\varepsilon} = \begin{Bmatrix} \varepsilon_{11} \\ \varepsilon_{22} \\ \varepsilon_{33} \\ \gamma_{12} \end{Bmatrix} = \boldsymbol{B}\boldsymbol{u} + \boldsymbol{\varepsilon}_3 = \begin{bmatrix} \frac{\partial}{\partial x_1} & 0 \\ 0 & \frac{\partial}{\partial x_2} \\ 0 & 0 \\ \frac{\partial}{\partial x_2} & \frac{\partial}{\partial x_1} \end{bmatrix} \begin{Bmatrix} u_1 \\ u_2 \end{Bmatrix} + \begin{Bmatrix} 0 \\ 0 \\ \varepsilon_{33} \\ 0 \end{Bmatrix} \tag{1.7}$$

In Eq. (1.7), either ε_{33} is set to zero (plane strain), or σ_{33} is assumed to be zero (plane stress) and determined by the material constitutive relations. The force components are expressed in matrix form as follows:

$$\boldsymbol{\sigma}^{\mathrm{T}} = \left\{ \sigma_{11}\ \sigma_{22}\ \sigma_{33}\ \tau_{12} \right\} \tag{1.8}$$

Here, σ_{33} is determined by the material constitutive relation (plane strain) or set to 0 (plane stress).

Note: Whether in plane stress or plane strain problems, although the local "energy"

$$E = \boldsymbol{\sigma}^{\mathrm{T}}\boldsymbol{\varepsilon}$$

neither includes ε_{33}, but the stress state used for energy calculations must accurately account for the corresponding two-dimensional behavior.

Traction vector in planar problems

$$\boldsymbol{t} = \boldsymbol{G}^{\mathrm{T}}\boldsymbol{\sigma} \quad \text{where} \quad \boldsymbol{G}^{\mathrm{T}} = \begin{bmatrix} n_1 & 0 & 0 & n_2 \\ 0 & n_2 & 0 & n_1 \end{bmatrix} \tag{1.9}$$

Thus, we once again observe that $\boldsymbol{B}$ and $\boldsymbol{G}$ share the same structure of nonzero elements.

1.1.3.2 Axisymmetric Problems

In axisymmetric problems, we use the (curvilinear) cylindrical coordinate system.

$$\boldsymbol{x} = \begin{Bmatrix} x_1 \\ x_2 \\ x_3 \end{Bmatrix} = \begin{Bmatrix} r \\ z \\ \theta \end{Bmatrix}$$

This ordering allows the equations for two-dimensional axisymmetric and planar problems to be written in a very similar manner. The object is three-dimensional, but due to the definition of the surface of revolution, the properties and boundaries are independent of the θ coordinate. Therefore, in this case, the displacement field can be expressed as

$$\boldsymbol{u} = \begin{Bmatrix} u_1(x_1, x_2, t) \\ u_2(x_1, x_2, t) \\ u_3(x_1, x_2, t) \end{Bmatrix} = \begin{Bmatrix} u_r(r, z, t) \\ u_z(r, z, t) \\ u_\theta(r, z, t) \end{Bmatrix}$$

It can also be regarded as independent of θ.
The strain in the case of axial symmetry can be expressed as ($x_1 = r$ and $x_2 = z$)

$$\boldsymbol{\varepsilon} = \begin{Bmatrix} \varepsilon_{rr} \\ \varepsilon_{zz} \\ \varepsilon_{\theta\theta} \\ \gamma_{rz} \\ \gamma_{z\theta} \\ \gamma_{\theta r} \end{Bmatrix} = \begin{Bmatrix} \varepsilon_{11} \\ \varepsilon_{22} \\ \varepsilon_{33} \\ \gamma_{12} \\ \gamma_{23} \\ \gamma_{31} \end{Bmatrix} = \boldsymbol{B}^{\mathrm{T}}\boldsymbol{u} = \begin{bmatrix} \frac{\partial}{\partial x_1} & 0 & 0 \\ 0 & \frac{\partial}{\partial x_2} & 0 \\ \frac{1}{x_1} & 0 & 0 \\ \frac{\partial}{\partial x_2} & \frac{\partial}{\partial x_1} & 0 \\ 0 & 0 & \frac{\partial}{\partial x_2} \\ 0 & 0 & \frac{\partial}{\partial x_1} - \frac{1}{x_1} \end{bmatrix} \begin{Bmatrix} u_1 \\ u_2 \\ u_3 \end{Bmatrix}$$

The components of stress can be expressed as

$$\begin{aligned} \boldsymbol{\sigma} &= \begin{bmatrix} \sigma_{11} & \sigma_{22} & \sigma_{33} & \tau_{12} & \tau_{23} & \tau_{31} \end{bmatrix}^{\mathrm{T}} \\ &= \begin{bmatrix} \sigma_{rr} & \sigma_{zz} & \sigma_{\theta\theta} & \tau_{rz} & \tau_{z\theta} & \tau_{\theta r} \end{bmatrix}^{\mathrm{T}} \end{aligned}$$

Similar to the three-dimensional case, the traction vector

$$\boldsymbol{t} = \boldsymbol{G}^{\mathrm{T}}\boldsymbol{\sigma}$$

where $\boldsymbol{G}^{\mathrm{T}} = \begin{bmatrix} n_1 & 0 & 0 & n_2 & 0 & 0 \\ 0 & n_2 & 0 & n_1 & 0 & 0 \\ 0 & 0 & 0 & 0 & n_2 & n_1 \end{bmatrix}$.

Note: For a complete body of revolution, n_3 cannot exist; $\boldsymbol{B}$ and $\boldsymbol{G}$ share the same nonzero structure, except for the σ_{33} component, which differs due to the use of curvilinear coordinates.

We note that the strain–displacement relationships among the u_1, u_2, and u_3 components are decoupled. If, in the material constitutive relations expressed in cylindrical coordinates, the first four and the last two strain components are also decoupled (i.e., the first four stresses depend only on the first four strains), then the axisymmetric problem can be divided into two categories: the first type depends only on the part involving the first four strains, represented by u_1 and u_2, and is sometimes referred to as the axisymmetric problem without torsion; the second type depends only on the last two shear strains and the u_3 component, and is called the torsion problem. However, when the constitutive relations couple these two parts, as in the classical elastoplastic analysis of stretching and twisting rods, it is necessary to consider the more general case.

The strain in the case of axisymmetric problems without torsion can be expressed as

$$\boldsymbol{\varepsilon} = \begin{Bmatrix} \varepsilon_{rr} \\ \varepsilon_{zz} \\ \varepsilon_{\theta\theta} \\ \varepsilon_{rz} \end{Bmatrix} = \begin{Bmatrix} \varepsilon_{11} \\ \varepsilon_{22} \\ \varepsilon_{33} \\ \varepsilon_{12} \end{Bmatrix} = \mathbb{B}^a \boldsymbol{u} = \begin{bmatrix} \frac{\partial}{\partial x_1} & 0 \\ 0 & \frac{\partial}{\partial x_2} \\ \frac{1}{x_1} & 0 \\ \frac{\partial}{\partial x_2} & \frac{\partial}{\partial x_1} \end{bmatrix} \begin{Bmatrix} u_1 \\ u_2 \end{Bmatrix}$$

The stress is given by Eq. (1.8), and the traction force is given by Eq. (1.9).

Therefore, the only distinction between these two types of problems is the presence of u_1/x_1 in the third strain component under axisymmetric conditions (of course, there are also differences in the description of the solution domains for the two problems).

1.2 The Calculus of Variations and the Variational Formulation of Nonlinear Elasticity Problems

1.2.1 Gauss–Green Theorem

The Gauss–Green theorem (commonly referred to as Green's theorem) is a fundamental identity that relates the integral of the derivative of a function over a domain Ω to the integral of the function itself over the boundary Γ. The domain can be either two-dimensional or three-dimensional. For simplicity, the following derivation presents this relationship in the two-dimensional case.

Let Ω be a planar domain bounded by the curve Γ. First, consider the derivative of the function $f = f(x, y)$ with respect to x. The integral over Ω can be written as

$$\int_{\Omega} \frac{\partial f}{\partial x} \mathrm{d}\Omega = \int_{y_1}^{y_2} \left(\int_{x_1}^{x_2} \frac{\partial f}{\partial x} \mathrm{d}x \right) \mathrm{d}y = \int_{y_1}^{y_2} \left[f(x_2, y) - f(x_1, y) \right] \mathrm{d}y \tag{1.10}$$

where

$$x_1 = x_1(y) \quad \text{and} \quad x_2 = x_2(y)$$

According to the details in Fig. 1.2, there is

$$\frac{\mathrm{d}y}{\mathrm{d}s} = \cos\alpha = n_x \quad \Rightarrow \quad \mathrm{d}y = n_x \mathrm{d}s \tag{1.11}$$

$$-\frac{\mathrm{d}x}{\mathrm{d}s} = \sin\alpha = n_y \quad \Rightarrow \quad \mathrm{d}x = -n_y \mathrm{d}s \tag{1.12}$$

Here, n_x and n_y are the components of the unit vector $\boldsymbol{n}$, and $\boldsymbol{n}$ is perpendicular to the boundary Γ. The negative sign in Eq. (1.12) arises because, when the angle α is measured counter-clockwise from the positive x-axis, the signs of dx and sinα are opposite (see Fig. 1.2).

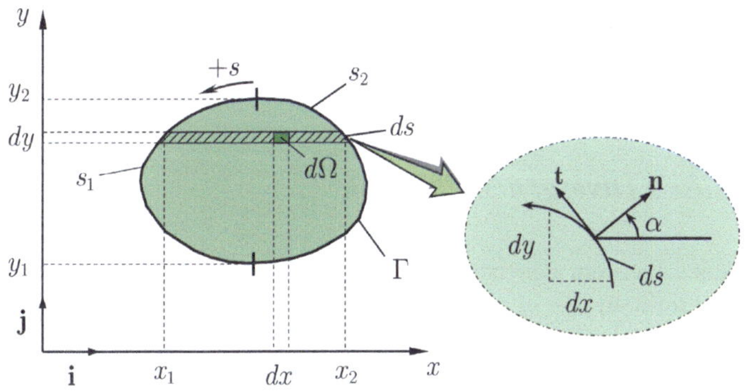

Fig. 1.2 Integral over the planar domain Ω bounded by the curve Γ

Therefore, Eq. (1.10) becomes

$$\int_{y_1}^{y_2} \left[f(x_2, y) - f(x_1, y)\right] \mathrm{d}y = \int_{s_2} f(x_2, y) n_x \mathrm{d}s - \int_{s_1} f(x_1, y) n_x \mathrm{d}s \tag{1.13}$$

In Eq. (1.13), when y changes from y_1 to y_2, the integration over s_1 is performed in the negative (clockwise) direction. If a consistent orientation is used for integration over s, the two terms on the right-hand side of Eq. (1.13) can be combined into a single expression, namely

$$\int_\Omega \frac{\partial f}{\partial x} \mathrm{d}\Omega = \int_\Gamma f n_x \mathrm{d}s \tag{1.14}$$

By interchanging x and y in Eq. (1.14), we obtain

$$\int_\Omega \frac{\partial f}{\partial y} \mathrm{d}\Omega = \int_\Gamma f n_y \mathrm{d}s \tag{1.15}$$

If g is another function of x and y, then the results of Eqs. (1.14) and (1.15) are

$$\begin{aligned} &\int_\Omega \frac{\partial (fg)}{\partial x} \mathrm{d}\Omega = \int_\Gamma fg n_x \mathrm{d}s = \int_\Omega g \frac{\partial f}{\partial x} \mathrm{d}\Omega + \int_\Omega f \frac{\partial g}{\partial x} \mathrm{d}\Omega \quad \Rightarrow \\ &\int_\Omega g \frac{\partial f}{\partial x} \mathrm{d}\Omega = -\int_\Omega f \frac{\partial g}{\partial x} \mathrm{d}\Omega + \int_\Gamma fg n_x \mathrm{d}s \end{aligned} \tag{1.16}$$

$$\begin{aligned} &\int_\Omega \frac{\partial (fg)}{\partial y} \mathrm{d}\Omega = \int_\Gamma fg n_y \mathrm{d}s = \int_\Omega g \frac{\partial f}{\partial y} \mathrm{d}\Omega + \int_\Omega f \frac{\partial g}{\partial y} \mathrm{d}\Omega \quad \Rightarrow \\ &\int_\Omega g \frac{\partial f}{\partial y} \mathrm{d}\Omega = -\int_\Omega f \frac{\partial g}{\partial y} \mathrm{d}\Omega + \int_\Gamma fg n_y \mathrm{d}s \end{aligned} \tag{1.17}$$

Equations (1.16) and (1.17) represent integration by parts in two dimensions, also known as the Gauss–Green theorem.

1.2.2 Gauss's Divergence Theorem

The divergence theorem can be readily derived from the Gauss–Green theorem. Consider a vector field $\mathbf{u} = u\mathbf{i} + v\mathbf{j}$, where $\mathbf{i}, \mathbf{j}$ denote the unit vectors along the x and y-axes, and $u = u(x, y)$ and $v = v(x, y)$ are the corresponding components. Taking $f = u$ and applying Eq. (1.14), then taking $f = v$ and applying Eq. (1.15) again, and finally adding the two results, we obtain

$$\int_\Omega \left(\frac{\partial u}{\partial x} + \frac{\partial v}{\partial y}\right)\mathrm{d}\Omega = \int_\Gamma \left(un_x + vn_y\right)\mathrm{d}s \tag{1.18}$$

If the coordinates x and y are denoted by x_1 and x_2, respectively, u_i ($i = 1, 2$) are the components of the vector field $\boldsymbol{u}$, and n_i ($i = 1, 2$) are the components of the normal vector $\boldsymbol{n}$. Thus, Eq. (1.18) can be rewritten as

$$\int_\Omega \left(\frac{\partial u_1}{\partial x_1} + \frac{\partial u_2}{\partial x_2}\right)\mathrm{d}\Omega = \int_\Gamma (u_1n_1 + u_2n_2)\mathrm{d}s \tag{1.19}$$

or using the summation convention

$$\int_\Omega \frac{\partial u_i}{\partial x_i}\mathrm{d}\Omega = \int_\Gamma u_in_i\mathrm{d}s \quad (i = 1, 2) \tag{1.20}$$

Equations (1.18) and (1.20) can also be expressed using vector notation.

$$\int_\Omega \nabla \cdot \mathbf{u}\mathrm{d}\Omega = \int_\Gamma \mathbf{u} \cdot \mathbf{n}\mathrm{d}s \tag{1.21}$$

Here, the vector symbol ∇ is defined as

$$\nabla \equiv \mathbf{i}\frac{\partial}{\partial x} + \mathbf{j}\frac{\partial}{\partial y} = \mathbf{i}_1\frac{\partial}{\partial x_1} + \mathbf{i}_2\frac{\partial}{\partial x_2}$$

Represents the differential operator that generates the gradient of a scalar field.

$\nabla \cdot$u denotes the dot product of the vector ∇ and $\boldsymbol{u}$, referred to as the divergence of the vector field $\boldsymbol{u}$ at a point within the domain Ω. The quantity $\boldsymbol{u} \cdot \boldsymbol{n}$ is called the flux of the vector field at a point on the boundary Γ. Geometrically, $\boldsymbol{u} \cdot \boldsymbol{n}$ represents the projection of $\boldsymbol{u}$ in the direction of $\boldsymbol{n}$. Equation (1.21) relates the total divergence of the vector field to the total flux, and is known as the Gauss (divergence) theorem. It is one of the most important theorems in integral calculus.

1.2.3 Green's Second Identity

Consider the functions $u = u(x, y)$ and $v = v(x, y)$, which are twice continuously differentiable in the domain Ω and continuously differentiable on the boundary Γ. Substituting $g = u$ and $f = \partial u/\partial x$ into Eq. (1.16), $g = u$ and $f = \partial v/\partial y$ into Eq. (1.17), then adding the results of the two equations yields:

$$\int_\Omega v\left(\frac{\partial^2 u}{\partial x^2}+\frac{\partial^2 u}{\partial y^2}\right)\mathrm{d}\Omega = -\int_\Omega \left(\frac{\partial u}{\partial x}\frac{\partial v}{\partial x}+\frac{\partial u}{\partial y}\frac{\partial v}{\partial y}\right)\mathrm{d}\Omega + \int_\Gamma v\left(\frac{\partial u}{\partial x}n_x+\frac{\partial u}{\partial y}n_y\right)\mathrm{d}s \tag{1.22}$$

Similarly, substituting $g = u, f = \partial v/\partial x$ into Eq. (1.16), and substituting $g = u, f = \partial v/\partial x$ into Eq. (1.17), then combining the two equations

$$\int_\Omega u\left(\frac{\partial^2 v}{\partial x^2}+\frac{\partial^2 v}{\partial y^2}\right)\mathrm{d}\Omega = -\int_\Omega \left(\frac{\partial u}{\partial x}\frac{\partial v}{\partial x}+\frac{\partial u}{\partial y}\frac{\partial v}{\partial y}\right)\mathrm{d}\Omega + \int_\Gamma u\left(\frac{\partial v}{\partial x}n_x+\frac{\partial v}{\partial y}n_y\right)\mathrm{d}s \tag{1.23}$$

Subtracting Eqs. (1.23) from (1.22) yields

$$\int_\Omega (v\nabla^2 u - u\nabla^2 v)\mathrm{d}\Omega = \int_\Gamma \left(v\frac{\partial u}{\partial n} - u\frac{\partial v}{\partial n}\right)\mathrm{d}s \tag{1.24}$$

Here, ∇^2 is called the Laplace operator or harmonic operator, defined as

$$\nabla^2 \equiv \nabla\cdot\nabla = \left(\boldsymbol{i}\frac{\partial}{\partial x}+\boldsymbol{j}\frac{\partial}{\partial y}\right)\cdot\left(\boldsymbol{i}\frac{\partial}{\partial x}+\boldsymbol{j}\frac{\partial}{\partial y}\right) = \frac{\partial^2}{\partial x^2}+\frac{\partial^2}{\partial y^2}$$

and

$$\frac{\partial}{\partial n} \equiv \boldsymbol{n}\cdot\nabla = (n_x\boldsymbol{i}+n_y\boldsymbol{j})\cdot\left(\boldsymbol{i}\frac{\partial}{\partial x}+\boldsymbol{j}\frac{\partial}{\partial y}\right) = n_x\frac{\partial}{\partial x}+n_y\frac{\partial}{\partial y}$$

is the operator for the derivative of a scalar function in the $\boldsymbol{n}$ direction. Equation (1.24) is known as the second Green's identity for the Laplacian operator or Green's reciprocity identity.

1.2.4 Integral Theorems

Many equations in solid mechanics are expressed as differential equations. For example, the equilibrium of an infinitesimal element can be described by partial differential equations. Since these differential equations are satisfied at every point within the domain, integrating them over the entire region yields an integral equation. In this section, several theorems that can be used to derive integral equations in solid mechanics will be introduced. Although the intermediate transformations are carried out in Cartesian coordinates during the derivation, the resulting integral equations are also valid in curvilinear coordinates, as the final equations are presented in a coordinate-independent form.

Divergence Theorem: The divergence theorem is a special case of Green's theorem for tensor fields. The divergence theorem relates the volume integral over a domain to the surface integral over its boundary. Let Ω be a region bounded by Γ. If tensor $\boldsymbol{A}$ has continuous partial derivatives within the domain, then the volume integral of the divergence of $\boldsymbol{A}$ over the domain can be converted to a surface integral over the boundary, that is,

$$\iint_{\Omega} \nabla \cdot \boldsymbol{A} \mathrm{d}\Omega = \int_{\Gamma} \boldsymbol{n} \cdot \boldsymbol{A} \mathrm{d}\Gamma \tag{1.25}$$

Here, $\boldsymbol{n}$ is the outward unit normal vector of the boundary Γ. A variant of the divergence theorem is the gradient theorem, in which the inner product is replaced by a dyadic (or tensor) product, that is,

$$\iint_{\Omega} \nabla \boldsymbol{A} \mathrm{d}\Omega = \int_{\Gamma} \boldsymbol{n} \otimes \boldsymbol{A} \mathrm{d}\Gamma$$

Reynolds Transport Theorem: The Reynolds transport theorem relates the time derivative of an integral equation over a domain, within which the integrand varies with time. And the region varies with time. Consider an integral of $f = f(\boldsymbol{x}, t)$ over a time-dependent domain $\Omega(t)$ bounded by $\Gamma(t)$. The time derivative of the integral of $f(\boldsymbol{x}, t)$ over the domain $\Omega(t)$ can then be expressed as

$$\frac{\mathrm{d}}{\mathrm{d}t} \iint_{\Omega} f \mathrm{d}\Omega = \iint_{\Omega} \frac{\partial f}{\partial t} \mathrm{d}\Omega + \int_{\Gamma} (\boldsymbol{n} \cdot \boldsymbol{v}) f \mathrm{d}\Gamma \tag{1.26}$$

Here, $\boldsymbol{n}(\boldsymbol{x}, t)$ denotes the outward unit normal vector of the boundary, and $\boldsymbol{v}(\boldsymbol{x}, t)$ represents the velocity of the boundary. On the right-hand side of Eq. (1.26), the first term is referred to as the partial derivative, while the second term is known as the convection term.

Note: The integral on the left side of the equation is solely a function of time, so the total derivative is used.

Integration by parts theorem: Integration by parts relates the integral of a product of functions to the integrals of their derivatives and antiderivatives. In the one-dimensional case, if $u(x)$ and $v(x)$ are two continuously differentiable functions on the interval (a, b), then integration by parts can be expressed as

$$\int_a^b u(x)v'(x)\mathrm{d}x = [u(x)v(x)]_a^b - \int_a^b u'(x)v(x)\mathrm{d}x \tag{1.27}$$

Equation (1.27) can be extended to two- or three-dimensional cases. Let Ω be the integration domain with boundary Γ. By applying integration by parts, it can be written as

$$\iint_\Omega \frac{\partial u}{\partial x_i} v\mathrm{d}\Omega = \int_\Gamma uvn_i\mathrm{d}\Gamma - \iint_\Omega u\frac{\partial v}{\partial x_i}\mathrm{d}\Omega \tag{1.28}$$

Here, n_i denotes the components of the unit normal vector to the boundary Γ. By replacing the scalar function v in Eq. (1.28) with the vector v_i and summing over i, we obtain the vector form of the equation.

$$\iint_\Omega \nabla u\cdot \boldsymbol{v}\mathrm{d}\Omega = \int_\Gamma u(\boldsymbol{v}\cdot\boldsymbol{n})\mathrm{d}\Gamma - \iint_\Omega u\nabla\cdot\boldsymbol{v}\mathrm{d}\Omega \tag{1.29}$$

By substituting the constant 1 for u in Eq. (1.29), the divergence theorem in Eq. (1.25) can be obtained. In the context of continuum mechanics, The objective of mechanics allows us to replace $\boldsymbol{v}$ with ∇v in Eq. (1.29) to obtain the following Green's identity.

$$\iint_\Omega \nabla u\cdot\nabla v\mathrm{d}\Omega = \int_\Gamma u(\nabla v\cdot\mathbf{n})\mathrm{d}\Gamma - \iint_\Omega u\nabla^2 v\mathrm{d}\Omega \tag{1.30}$$

One of the main reasons for using integration by parts is to relax the requirements for differentiability. For example, in Eq. (1.30), the right-hand side requires that $v(x)$ be twice differentiable, whereas the left-hand side is well-defined using only the first derivative of $v(x)$ transferring the additional differentiability requirement to $u(x)$.

Example (Divergence Theorem) Compute the integral $\int_S \boldsymbol{F}\cdot\boldsymbol{n}\mathrm{d}S$, where $\boldsymbol{F}$ is the given vector field $\boldsymbol{F} = 2x\boldsymbol{e}_1 + y^2\boldsymbol{e}_2 + z^2\boldsymbol{e}_3$; For the surface $S = (x^2 + y^2 + z^2 = 1)$, its unit normal vector is $\boldsymbol{n}$.

Solution. Using the divergence theorem,

$$\begin{aligned}\int_S \boldsymbol{F}\cdot\boldsymbol{n}\mathrm{d}S &= \iint_\Omega \nabla\cdot\boldsymbol{F}\mathrm{d}\Omega = 2\iint_\Omega (1+y+z)\mathrm{d}\Omega \\ &= 2\iint_\Omega \mathrm{d}\Omega = \frac{8\pi}{3}\end{aligned} \tag{1.31}$$

In the solution of Eq. (1.31), the integral of the odd function is zero because the region is symmetric.

1.2.5 Adjoint Operators

Consider the general second-order differential equation with variable coefficients

$$L(u) = A\frac{\partial^2 u}{\partial x^2} + 2B\frac{\partial^2 u}{\partial x \partial y} + C\frac{\partial^2 u}{\partial y^2} + D\frac{\partial u}{\partial x} + E\frac{\partial u}{\partial y} + Fu = 0 \tag{1.32}$$

where A, B, …, F are defined in Ω as functions of x and y. Multiplying Eq. (1.32) by a function $v = v(x, y)$ and then integrating over the domain yields

$$\int_\Omega vL(u)\mathrm{d}\Omega = 0 \tag{1.33}$$

Assume that v is twice continuously differentiable in the domain Ω and once continuously differentiable on the boundary Γ. By repeatedly applying integration by parts to Eq. (1.33) until all derivatives of u are eliminated, and combining with Eqs. (1.16) and (1.17), the reciprocity identity can be obtained as

$$\int_\Omega \left[vL(u) - uL^*(v)\right]\mathrm{d}\Omega = \int_\Gamma \left(Xn_x + Yn_y\right)\mathrm{d}s \tag{1.34}$$

where

$$L^*(v) = \frac{\partial^2 (Av)}{\partial x^2} + 2\frac{\partial^2 (Bv)}{\partial x \partial y} + \frac{\partial^2 (Cv)}{\partial y^2} - \frac{\partial (Dv)}{\partial x} - \frac{\partial (Ev)}{\partial y} + Fv = 0 \tag{1.35}$$

$$X = A\left(v\frac{\partial u}{\partial x} - u\frac{\partial v}{\partial x}\right) + B\left(v\frac{\partial u}{\partial y} - u\frac{\partial v}{\partial y}\right) + \left(D - \frac{\partial A}{\partial x} - \frac{\partial B}{\partial y}\right)uv \tag{1.36}$$

$$Y = B\left(v\frac{\partial u}{\partial x} - u\frac{\partial v}{\partial x}\right) + C\left(v\frac{\partial u}{\partial y} - u\frac{\partial v}{\partial y}\right) + \left(E - \frac{\partial B}{\partial x} - \frac{\partial C}{\partial y}\right)uv \tag{1.37}$$

The differential operator $L^*()$ defined in Eq. (1.35) is called the adjoint operator of $L()$. Equation (1.34) is a general form of Green's second identity. Setting $A = C = 1$ and $B = D = E = 0$ shows that Eq. (1.24) is a special case of Eq. (1.34). The case $F \neq 0$ does not affect Eq. (1.24), because it is equivalent to adding and subtracting the term Fuv in the integral on the left-hand side of that equation.

It can be shown that if the coefficients of operator (1.32) satisfy the following conditions

$$\begin{aligned} \frac{\partial A}{\partial x} + \frac{\partial B}{\partial y} &= D \\ \frac{\partial B}{\partial x} + \frac{\partial C}{\partial y} &= E \end{aligned} \tag{1.38}$$

then Eq. (1.35) becomes

$$L^*(v) = A\frac{\partial^2 v}{\partial x^2} + 2B\frac{\partial^2 v}{\partial x \partial y} + C\frac{\partial^2 v}{\partial y^2} + Fv$$

That is, the differential operator $L^*()$ is identical to $L()$. In this case, $L()$ is called self-adjoint; Eq. (1.38) is referred to as the self-adjoint condition.

The properties of Eq. (1.32) and the types of problems to be solved depend on $\Delta = B^2\text{-}AC$. According to the value of Δ, they can be classified as follows:

a. Elliptic type, corresponding to $\Delta < 0$.
b. Parabolic type, corresponding to $\Delta = 0$.
c. Hyperbolic type, corresponding to $\Delta > 0$.

1.2.6 Calculus of Variations and the Euler–Lagrange Equation

1.2.6.1 Calculus of Variations

An important application of the variational method in mechanics is the derivation of differential equations and associated boundary conditions that govern the response of complex structural systems, especially when other methods are not feasible. This section introduces some fundamental concepts of the variational method. The variational method is a branch of mathematical analysis used to address problems involving the maxima or minima of functionals. A functional is a mapping from a set of functions to the real numbers, typically expressed in the form of a definite integral, where the integrand depends on one or more unknown functions and their derivatives. The fundamental problem of the variational method is to find the extremal function, that is, the function for which the functional attains a stationary value (maximum or minimum), meaning the function that makes the variation of the functional zero. The most basic part of this theory is the necessary condition that the function must satisfy: typically, a differential equation with boundary conditions, namely the Euler–Lagrange equation. Solving this equation yields the required function.

A simple example of this type of problem is to find a curve $u(x)$ that passes through points $A(x_1, u_1)$ and $B(x_2, u_2)$, and generates a surface of revolution with minimal area about the x-axis, as shown in Fig. 1.3. Therefore, the function $u(x)$ can be determined by minimizing the integral representing the surface area, that is,

$$I = 2\pi \int_{x_1}^{x_2} u\left(1 + u'^2\right)^{1/2} \mathrm{d}x \tag{1.39}$$

Fig. 1.3 Function $u(x)$ and its admissible function $\tilde{u}(x)$

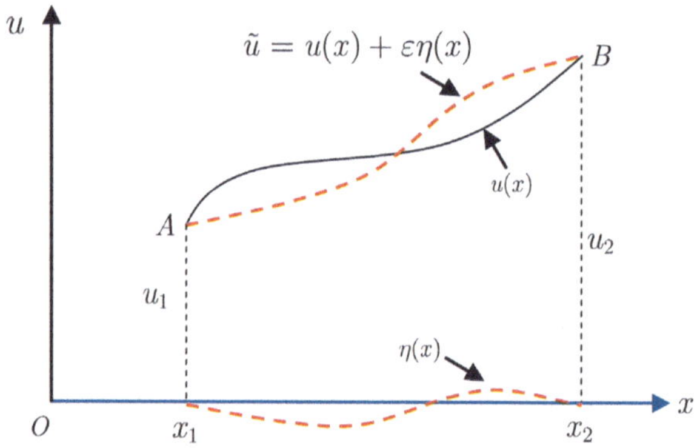

and $u(x_1) = u_1$ and $u(x_2) = u_2$. Here, it is assumed that u_1 and u_2 are non-negative. We consider all continuously differentiable functions that satisfy the geometric or essential boundary conditions (i.e., passing through points A and B). These functions form the set of admissible functions, among which we seek the function $u(x)$ that minimizes the integral (1.39). Assuming that $u(x)$ is the actual minimizing function, we choose an arbitrary continuously differentiable function $\eta(x)$ with $\eta(x_1) = \eta(x_2) = 0$, as shown in Fig. 1.3. Then for any constant ε, the function $\tilde{u}(x) = u(x) + \varepsilon\eta(x)$ is an admissible function. The variation $\varepsilon\eta(x)$ is called the variation of $u(x)$ and is usually denoted by δu, i.e.,

$$\delta u = \varepsilon\eta(x)$$

1.2.6.2 Euler–Lagrange Equation Considering Integrals

Consider the integral

$$I(u) = \int_{x_1}^{x_2} F\left(x, u, u'\right)\mathrm{d}x \tag{1.40}$$

Here, it is assumed that the integrand $F = F(x, u, u')$ possesses continuous second-order derivatives with respect to its three arguments. The integral

$$I(\varepsilon) = \int_{x_1}^{x_2} \tilde{F}\left(x, u + \varepsilon\eta, u' + \varepsilon\eta'\right)\mathrm{d}x \tag{1.41}$$

By substituting $u(x)$ in Eq. (1.40) with $u(x) + \varepsilon\eta(x)$, we obtain that $I(\varepsilon)$ becomes a function of ε. Once the functions $u(x)$ and $\eta(x)$ are given, $I(\varepsilon)$ attains its minimum when $\varepsilon = 0$. That is, we require

$$\left.\frac{\mathrm{d}I(\varepsilon)}{\mathrm{d}\varepsilon}\right|_{\varepsilon=0} = 0 \tag{1.42}$$

Differentiating Eq. (1.41) with respect to ε, and noting that the derivative outside the integral can be moved inside, we obtain

$$\frac{\mathrm{d}I(\varepsilon)}{\mathrm{d}\varepsilon} = \int_{x_1}^{x_2} \left(\frac{\partial \tilde{F}}{\partial u}\eta + \frac{\partial \tilde{F}}{\partial u'}\eta' \right) \mathrm{d}x \tag{1.43}$$

After integrating the second term of Eq. (1.43) by parts, we obtain

$$\frac{\mathrm{d}I(\varepsilon)}{\mathrm{d}\varepsilon} = \int_{x_1}^{x_2} \left(\frac{\partial \tilde{F}}{\partial u} - \frac{\mathrm{d}}{\mathrm{d}x}\frac{\partial \tilde{F}}{\partial u'} \right) \eta \mathrm{d}x + \left[\frac{\partial \tilde{F}}{\partial u'}\eta' \right]_{x_1}^{x_2}$$

It should be noted that as $\varepsilon \to 0$, $\eta(x_1) = \eta(x_2) = 0$ and $\tilde{F} \to F$, so the necessary condition (1.42) becomes

$$\int_{x_1}^{x_2} \left(\frac{\partial F}{\partial u} - \frac{\mathrm{d}}{\mathrm{d}x}\frac{\partial F}{\partial u'} \right) \eta \mathrm{d}x = 0 \tag{1.44}$$

Since $\eta(x)$ is arbitrary, the coefficient of $\eta(x)$ in Eq. (1.44) must be zero within the interval (x_1, x_2), that is,

$$\frac{\partial F}{\partial u} - \frac{\mathrm{d}}{\mathrm{d}x}\frac{\partial F}{\partial u'} = 0 \tag{1.45}$$

Therefore, if $u(x)$ minimizes (or maximizes) the integral, it must satisfy condition (1.45), which is called the Eule-Lagrange equation of the functional. Given F, the solution to the differential equation given by condition (1.45) over the interval (x_1, x_2) is the extremizing function.

The variation in F caused by δu is

$$\Delta F = F\left(x, u + \delta u, u' + \delta u'\right) - F\left(x, u, u'\right) \tag{1.46}$$

Using the Taylor series expansion on the right-hand side of Eq. (1.46), we obtain

$$\Delta \mathrm{F} = \frac{\partial \mathrm{F}}{\partial \mathrm{u}}\delta \mathrm{u} + \frac{\partial \mathrm{F}}{\partial \mathrm{u}'}\delta \mathrm{u}' + (\text{higher - order terms of } \delta \mathrm{u} \text{ and} \delta \mathrm{u}') \tag{1.47}$$

Similar to differentiation, the first two terms on the right-hand side of Eq. (1.47) are defined as the variation of F, that is,

$$\delta F = \frac{\partial F}{\partial u}\delta u + \frac{\partial F}{\partial u'}\delta u' \tag{1.48}$$

The variational operator δ is analogous to the differential operator d. For a function $F(x, y)$ with two independent variables, we have

$$\mathrm{d}F = \frac{\partial F}{\partial x}\mathrm{d}y + \frac{\partial F}{\partial y}\mathrm{d}y \tag{1.49}$$

However, the difference between the operators in Eqs. (1.48) and (1.49) is as follows: the differentiation of a function represents the first-order approximation of the function's change along a specific curve, whereas the variation of a function is the first-order approximation of the function's change from one curve to another. It should be noted that variation is similar to differentiation, namely,

$$\delta F = \frac{\partial F}{\partial x}\delta x + \frac{\partial F}{\partial u}\delta u + \frac{\partial F}{\partial u'}\delta u'$$

However, since x and $\delta x = 0$ are both complete and regular. The situation is entirely analogous to differentiation. For example,

$$\begin{aligned}
&\delta(FG) = G\delta F + F\delta G \\
&\delta F^n = n\delta F^{n-1} \\
&\delta\left(\frac{F}{G}\right) = \frac{G\delta F - F\delta G}{G^2}
\end{aligned}$$

Furthermore, since $\delta u = \varepsilon\eta(x)$, we have

$$\frac{\mathrm{d}}{\mathrm{d}x}(\delta u) = \varepsilon\frac{\mathrm{d}\eta}{\mathrm{d}x} = \delta\frac{\mathrm{d}u}{\mathrm{d}x} \tag{1.50}$$

Equation (1.50) indicates that if x is the independent variable, then the operators d/dx and δ are commutative.

According to Eq. (1.48), the variation of the integral Eq. (1.40) can be written as

$$\delta I = \int_{x_1}^{x_2} \delta F(x, u, u')\mathrm{d}x = \int_{x_1}^{x_2} \left(\frac{\partial F}{\partial u}\delta u + \frac{\partial F}{\partial u'}\delta u'\right)\mathrm{d}x \tag{1.51}$$

From Eq. (1.50), we have $\delta u\prime = (\delta u)\prime$. Thus, applying integration by parts to the second term of the integrand in Eq. (1.51), we obtain

$$\delta I = \int_{x_1}^{x_2} \left(\frac{\partial F}{\partial u} - \frac{\mathrm{d}}{\mathrm{d}x}\frac{\partial F}{\partial u'}\right)\delta u\mathrm{d}x + \left[\frac{\partial F}{\partial u'}\delta u\right]_{x_1}^{x_2} \tag{1.52}$$

Given that $\delta u(x_1) = \delta u(x_2) = 0$, i.e.,

$$\delta I = \int_{x_1}^{x_2} \left(\frac{\partial F}{\partial u} - \frac{\mathrm{d}}{\mathrm{d}x}\frac{\partial F}{\partial u'}\right)\delta u\mathrm{d}x$$

From Eq. (1.45), we know that $\delta I = 0$. Therefore, a necessary condition for the functional to attain a minimum (or maximum) is that its variation equals zero.

1.2.6.3 Natural Boundary Conditions

When the value of the unknown function $u(x)$ is not specified at one or both endpoints $x = x_1, x_2$, the variation $\delta u(x)$ is not necessarily zero at those points. However, if $u(x)$ is the function that minimizes (or maximizes) the functional among all admissible variations $\delta u(x)$, then the right-hand side of Eq. (1.52) must be zero. Therefore, the second term on the right-hand side of Eq. (1.52) must also be zero, that is,

$$\left[\frac{\partial F}{\partial u'}\delta u(x)\right]_{x_1}^{x_2} = \left[\frac{\partial F}{\partial u'}\delta u(x)\right]_{x=x_2} - \left[\frac{\partial F}{\partial u'}\delta u(x)\right]_{x=x_1} = 0 \tag{1.53}$$

If $u(x)$ is not specified at either endpoint, then $\delta u(x_1) \neq 0$ or $\delta u(x_2) \neq 0$. Therefore, for Eq. (1.53) to hold, it follows that

$$\left[\frac{\partial F}{\partial u'}\right]_{x=x_1} = 0 \tag{1.54}$$

$$\left[\frac{\partial F}{\partial u'}\right]_{x=x_2} = 0 \tag{1.55}$$

When $u(x_1)$ is not specified, Eq. (1.54) must be satisfied; when $u(x_2)$ is not specified, Eq. (1.55) must be satisfied. Both are referred to as natural boundary conditions.

Based on the above analysis, the function $u(x)$ that minimizes (or maximizes) the given functional can be obtained by solving the following boundary value problem.

$$\frac{\partial F}{\partial u} - \frac{\mathrm{d}}{\mathrm{d}x}\frac{\partial F}{\partial u'} = 0 \quad x_1 \le x \le x_2 \tag{1.56}$$

$$\begin{aligned} u(x_1) = u_1 \quad &\text{or} \quad \left[\frac{\partial F}{\partial u'}\right]_{x=x_1} = 0 \\ u(x_2) = u_2 \quad &\text{or} \quad \left[\frac{\partial F}{\partial u'}\right]_{x=x_2} = 0 \end{aligned} \tag{1.57}$$

1.2.6.4 Bivariate Functionals

Consider the functional

$$I = \int_{\Omega} F\left(x, y, u, u_{,x}, u_{,y}\right) \mathrm{d}x\mathrm{d}y \tag{1.58}$$

In this equation, $u = u(x, y)$ is the unknown continuously differentiable function associated with the stationary value of the functional; Ω is a two-dimensional domain in the x–y plane with boundary Γ (Fig. 1.4). The process of deriving this necessary condition is more complex. However, it becomes simpler if the required condition is obtained through the requirement of variation $\delta I = 0$. We can then obtain

$$\begin{aligned} \delta I &= \int_{\Omega} \delta F\left(x, y, u, u_{,x}, u_{,y}\right) \mathrm{d}x\mathrm{d}y \\ &= \int_{\Omega} \left(\frac{\partial F}{\partial u} \delta u + \frac{\partial F}{\partial u_{,x}} \delta u_{,x} + \frac{\partial F}{\partial u_{,y}} \delta u_{,y} \right) \mathrm{d}x\mathrm{d}y \end{aligned} \tag{1.59}$$

By applying integration by parts, the terms $\delta u_{,x}$ and $\delta u_{,y}$ can be eliminated from the integrand. Therefore, using the Gauss–Green theorem together with Eqs. (1.16) and (1.17), we obtain

$$\begin{aligned} \int_{\Omega} \left(\frac{\partial F}{\partial u_{,x}} \delta u_{,x} \right) \mathrm{d}x\mathrm{d}y &= \int_{\Omega} \frac{\partial F}{\partial u_{,x}} (\delta u)_{,x} \mathrm{d}x\mathrm{d}y \\ &= -\int_{\Omega} \frac{\partial}{\partial x} \frac{\partial F}{\partial u_{,x}} \delta u \mathrm{d}x\mathrm{d}y + \int_{\Gamma} \frac{\partial F}{\partial u_{,x}} n_x \delta u \mathrm{d}s \\ \int_{\Omega} \left(\frac{\partial F}{\partial u_{,y}} \delta u_{,y} \right) \mathrm{d}x\mathrm{d}y &= \int_{\Omega} \frac{\partial F}{\partial u_{,y}} (\delta u)_{,y} \mathrm{d}x\mathrm{d}y \\ &= -\int_{\Omega} \frac{\partial}{\partial y} \frac{\partial F}{\partial u_{,y}} \delta u \mathrm{d}x\mathrm{d}y + \int_{\Gamma} \frac{\partial F}{\partial u_{,y}} n_y \delta u \mathrm{d}s \end{aligned} \tag{1.60}$$

Substituting Eqs. (1.60) into (1.59), we obtain

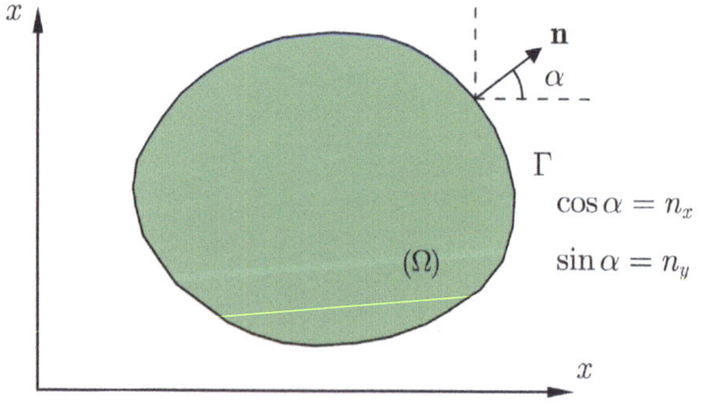

Fig. 1.4 Domain Ω bounded by the curve Γ

$$\delta I = \int_{\Omega} \delta F(x, y, u, u_{,x}, u_{,y}) \mathrm{d}x\mathrm{d}y$$
$$= \int_{\Omega} \left(\frac{\partial F}{\partial u} - \frac{\partial}{\partial x}\frac{\partial F}{\partial u_{,x}} - \frac{\partial}{\partial y}\frac{\partial F}{\partial u_{,y}} \right) \delta u \mathrm{d}x\mathrm{d}y + \int_{\Gamma} \left(\frac{\partial F}{\partial u_{,x}} n_x + \frac{\partial F}{\partial u_{,y}} n_y \right) \delta u \mathrm{d}s \tag{1.61}$$

When the function $u(x, y)$ is specified on the boundary Γ, we have $\delta u = 0$, and thus the boundary integral in Eq. (1.61) is zero. Furthermore, since

$$\frac{\partial F}{\partial u} - \frac{\partial}{\partial x}\frac{\partial F}{\partial u_{,x}} - \frac{\partial}{\partial y}\frac{\partial F}{\partial u_{,y}} = 0 \tag{1.62}$$

This is the Euler–Lagrange equation of the functional.

When the function $u(x, y)$ is not specified on the boundary Γ, we have $\delta u \neq 0$. Only when the natural boundary conditions are satisfied does the boundary integral vanish, that is,

$$\frac{\partial F}{\partial u_{,x}} n_x + \frac{\partial F}{\partial u_{,y}} n_y = 0 \quad \text{(on the boundary } \Gamma\text{)} \tag{1.63}$$

Therefore, the boundary condition on $\Gamma_1 \subseteq \Gamma$ is

$$u = \overline{u} \quad \text{or} \quad \frac{\partial F}{\partial u_{,x}} n_x + \frac{\partial F}{\partial u_{,y}} n_y = 0 \tag{1.64}$$

Here, the overline on u denotes a given quantity.

1.2.7 Variational Formulation of Nonlinear Elasticity Problems

For the strain energy of elastic materials satisfying,

$$\boldsymbol{\sigma} = \frac{\partial \boldsymbol{W}}{\partial \boldsymbol{\varepsilon}}$$

when inertial effects are not considered, the equations in Sect. 1.1 can be expressed in a variational form. The simplest form is the principle of minimum potential energy, which can be expressed as:

$$\Pi_{PE} = \int_{\Omega} W(\boldsymbol{Bu}) \mathrm{d}\Omega - \int_{\Omega} \boldsymbol{u}^{\mathrm{T}} \boldsymbol{b} \mathrm{d}\Omega - \int_{\Gamma_t} \boldsymbol{u}^{\mathrm{T}} \bar{\boldsymbol{t}} \mathrm{d}\Gamma \tag{1.65}$$

The first-order variation yields the governing equation for the functional as follows:

$$\delta\Pi_{PE} = \int_{\Omega} \delta(\boldsymbol{Bu})^{\mathrm{T}} \frac{\partial W(\boldsymbol{Bu})}{\partial \boldsymbol{Bu}} \mathrm{d}\Omega - \int_{\Omega} \delta\boldsymbol{u}^{\mathrm{T}}\boldsymbol{b}\mathrm{d}\Omega - \int_{\Gamma_t} \delta\boldsymbol{u}^{\mathrm{T}}\bar{\boldsymbol{t}}\mathrm{d}\Gamma = 0 \tag{1.66}$$

Performing integration by parts on Eq. (1.66) and rearranging yields:

$$\delta\Pi_{PE} = -\int_{\Omega} \delta\boldsymbol{u}^{\mathrm{T}}\left(\boldsymbol{B}^{\mathrm{T}}\boldsymbol{\sigma} + \boldsymbol{b}\right)\mathrm{d}\Omega + \int_{\Gamma_t} \delta\boldsymbol{u}^{\mathrm{T}}\left(\boldsymbol{G}^{\mathrm{T}}\boldsymbol{\sigma} - \bar{\boldsymbol{t}}\right)\mathrm{d}\Gamma = 0 \tag{1.67}$$

where

$$\boldsymbol{\sigma} = \frac{\partial W(\boldsymbol{Bu})}{\partial \boldsymbol{Bu}} = \frac{\partial W}{\partial \boldsymbol{\varepsilon}}$$

When the strain energy W is given in quadratic form, the linear problem described by Eq. (1.6) is obtained. In this case, the formulation becomes the principle of minimum potential energy, and the displacement field that renders W an absolute minimum is the exact solution to the problem.

Note: The principle of potential energy includes the strain–displacement equation and the elastic model expressed in terms of strain based on displacement. Therefore, in addition to this theorem, displacement boundary conditions must also be specified. This theorem represents the simplest variational form and requires only knowledge of the displacement field. This form serves as the foundation for irreducible (or displacement-based) approximate solution methods.

Hu-Washizu variational principle provides a general variational theorem that encompasses all equations and boundary conditions. The principle is expressed as follows:

$$\begin{aligned}\Pi_{HW}(\boldsymbol{u},\boldsymbol{\varepsilon},\boldsymbol{\sigma}) &= \int_{\Omega} \left[W(\boldsymbol{\varepsilon}) + \boldsymbol{\sigma}^{\mathrm{T}}(\boldsymbol{Bu} - \boldsymbol{\varepsilon})\right]\mathrm{d}\Omega \\ &\quad - \int_{\Omega} \boldsymbol{u}^{\mathrm{T}}\boldsymbol{b}\mathrm{d}\Omega - \int_{\Gamma_t} \boldsymbol{u}^{\mathrm{T}}\bar{\boldsymbol{t}}\mathrm{d}\Gamma - \int_{\Gamma_u} \boldsymbol{t}^{\mathrm{T}}(\boldsymbol{u} - \bar{\boldsymbol{u}})\mathrm{d}\Gamma \end{aligned} \tag{1.68}$$

Here, $\boldsymbol{t} = \boldsymbol{G}^{\mathrm{T}}\boldsymbol{\sigma}$. By taking the derivatives of $\boldsymbol{u}$, ε, and $\boldsymbol{\sigma}$ in Eq. (1.68), it can be shown that this principle encompasses all governing equations. Specifically, by taking the variation of Eq. (1.68) and performing integration by parts on $\delta(\boldsymbol{Bu})$, we obtain

$$\begin{aligned}\delta\Pi_{HW}(\boldsymbol{u},\boldsymbol{\varepsilon},\boldsymbol{\sigma}) &= \int_{\Omega} \delta\boldsymbol{\varepsilon}^{\mathrm{T}}\left[\frac{\partial W(\boldsymbol{\varepsilon})}{\partial \boldsymbol{\varepsilon}} - \boldsymbol{\sigma}\right]\mathrm{d}\Omega + \int_{\Omega} \delta\boldsymbol{\sigma}^{\mathrm{T}}[\boldsymbol{Bu} - \boldsymbol{\varepsilon}]\mathrm{d}\Omega - \int_{\Omega} \delta\boldsymbol{u}^{\mathrm{T}}\left[\boldsymbol{B}^{\mathrm{T}}\boldsymbol{\sigma} + \boldsymbol{b}\right]\mathrm{d}\Omega \\ &\quad - \int_{\Gamma_u} \delta\boldsymbol{\sigma}^{\mathrm{T}}\boldsymbol{G}(\boldsymbol{u} - \bar{\boldsymbol{u}})\mathrm{d}\Gamma + \int_{\Gamma_t} \delta\boldsymbol{u}^{\mathrm{T}}\left(\boldsymbol{G}^{\mathrm{T}}\boldsymbol{\sigma} - \bar{\boldsymbol{t}}\right)\mathrm{d}\Gamma = 0 \end{aligned} \tag{1.69}$$

It is evident that Hu-Washizu variational principle provides all the equations for nonlinear elastic statics problems.

We can also establish a direct connection between the Hu-Washizu principle and other variational principles. If we use the Laurent transformation to express strain $\boldsymbol{\varepsilon}$ in terms of stress, we obtain

$$U(\boldsymbol{\sigma}) + W(\boldsymbol{\varepsilon}) = \boldsymbol{\sigma}^{\mathrm{T}}\boldsymbol{\varepsilon} \tag{1.70}$$

Thus, the Hellinger–Reissner variational principle can be expressed as follows:

$$\Pi_{HR}(\boldsymbol{u},\boldsymbol{\sigma}) = \int_{\Omega}\left[\boldsymbol{\sigma}^{\mathrm{T}}\boldsymbol{B}\boldsymbol{u} - U(\boldsymbol{\sigma})\right]\mathrm{d}\Omega - \int_{\Omega}\boldsymbol{u}^{\mathrm{T}}\boldsymbol{b}\mathrm{d}\Omega - \int_{\Gamma_t}\boldsymbol{u}^{\mathrm{T}}\bar{\boldsymbol{t}}\mathrm{d}\Gamma - \int_{\Gamma_u}\boldsymbol{t}^{\mathrm{T}}(\boldsymbol{u} - \bar{\boldsymbol{u}})\mathrm{d}\Gamma = 0 \tag{1.71}$$

In the case of linear elasticity, neglecting initial strain and stress effects, we obtain

$$U(\boldsymbol{\sigma}) = \frac{1}{2}\sigma_{ij}S_{ijkl}\sigma_{kl} \tag{1.72}$$

Here, S_{ijkl} denotes the elastic compliance. Although this form is applicable to general elasticity problems, in nonlinear cases it is not always possible to establish a unique constitutive relationship based on the stress formulation. Therefore, we will frequently use the Hu-Washizu functional as the foundation for the mixed formulation.

We can also establish a direct connection between the principle of minimum potential energy and the Hu-Washizu principle. If the displacement boundary conditions are known to be satisfied, the integral term concerning Γ_u can be eliminated. Typically, in finite element approximations based on the Hu-Washizu principle (or its equivalent forms), the displacement boundary conditions are explicitly satisfied, thereby avoiding the need to approximate the Γ_u terms.

If we further assume that the strain–displacement relationship is known, then the Hu-Washizu principle becomes equivalent to the principle of minimum potential energy. In the development of finite element formulations, the principle of minimum potential energy serves as the foundation for deriving displacement-based formulations (also known as irreducible formulations), while the Hu-Washizu principle forms the basis for deriving mixed formulations. Mixed formulations offer significant advantages in constructing robust finite element methods. However, finite element formulations that express the global problem in terms of displacement also have their own advantages.

Note: The Hu-Washizu principle can be simplified to the principle of minimum potential energy, from which the correspondence between their approximate forms can be derived.

The advantage of the variational theorem is that it can automatically yield symmetry conditions. However, its obvious drawback is that it can only address elastic behavior and static problems. The weak form method introduced in the next section is applicable to both elastic and inelastic materials, and can directly account for inertial effects. We will observe that, for elastic statics problems, the weak form is equivalent to the principle of minimum potential energy.

1.3 Weak Formulation of the Governing Equations

The variational (weak) form of any governing equation is a scalar relation that can be constructed by multiplying the governing equation by an appropriate arbitrary function. This arbitrary function possesses the same free indices as the governing equation, thus becoming a dummy index to be summed over its range. Finally, the result is integrated over the problem domain and set to zero.

1.3.1 Weak Form of the Equilibrium Equation

The index notation of the equilibrium Eq. (1.1) contains a free index i. To construct a weak form, one can multiply it by an arbitrary vector with index i and integrate the result over the domain Ω. The principle of virtual work is a type of weak form, in which the arbitrary function is the virtual displacement δu_i. Thus, using the virtual displacement function yields the following weak form:

$$\delta\Pi_{eq} = \int_{\Omega} \delta u_i \left[\rho\ddot{u}_i - \sigma_{ji,j} - b_i\right] \mathrm{d}\Omega = 0 \tag{1.73}$$

Note: Virtual work is a type of Galerkin method, defined as follows

$$\delta\Pi = G = 0 \tag{1.74}$$

Here, G represents the Galerkin expression. In the next chapter, the finite element approximation will be used to solve the variational method and the Galerkin method.

In general, stress depends on strain, which is defined by the derivatives of displacement. Therefore, in Eq. (1.73), it is necessary to compute the second derivatives of displacement to obtain the integrand. To reduce (i.e., "weaken") the need to calculate second derivatives, the method of integration by parts can be employed, and by considering the symmetry of stress, we obtain

$$G_{eq} = \int_{\Omega} \delta u_i \rho \ddot{u}_i \mathrm{d}\Omega + \int_{\Omega} \delta\varepsilon_{ij} \sigma_{ij} \mathrm{d}\Omega - \int_{\Omega} \delta u_i b_i \mathrm{d}\Omega - \int_{\Gamma} \delta u_i t_i \mathrm{d}\Omega = 0 \tag{1.75}$$

The relationship between virtual strain and virtual displacement is given by

$$\delta\varepsilon_{ij} = \frac{1}{2}\left(\delta u_{i,j} + \delta u_{j,i}\right) \tag{1.76}$$

By dividing the boundary into the portion with prescribed external forces, Γ_t, and the portion with prescribed displacements, Γ_u, Eq. (1.75) can be further simplified. If all displacement boundary conditions are strictly enforced pointwise and the constraint $\delta u_i = 0$ is imposed on Γ_u, the final result can be obtained.

$$G_{eq} = \int_\Omega \delta u_i \rho \ddot{u}_i \mathrm{d}\Omega + \int_\Omega \delta \varepsilon_{ij} \sigma_{ij} \mathrm{d}\Omega - \int_\Omega \delta u_i b_i \mathrm{d}\Omega - \int_{\Gamma_t} \delta u_i \bar{t}_i \mathrm{d}\Gamma = 0 \qquad (1.77)$$

Equation (1.77) can be expressed in matrix form as

$$G_{eq} = \int_\Omega \delta \boldsymbol{u}^\mathrm{T} \rho \ddot{\boldsymbol{u}} \mathrm{d}\Omega + \int_\Omega \delta (\boldsymbol{B}\boldsymbol{u})^\mathrm{T} \sigma \mathrm{d}\Omega - \int_\Omega \delta \boldsymbol{u}^\mathrm{T} \boldsymbol{b} \mathrm{d}\Omega - \int_{\Gamma_t} \delta \boldsymbol{u}^\mathrm{T} \bar{\boldsymbol{t}} \mathrm{d}\Gamma = 0 \qquad (1.78)$$

Here, the first term represents the virtual work of internal inertial forces; the second term is the virtual work of internal stresses; the last two terms correspond to the virtual work of body forces and traction forces, respectively. In the following sections, the virtual work of inertial forces will be expressed as

$$G_{dyn} = \int_\Omega \delta \boldsymbol{u}^\mathrm{T} \rho \ddot{\boldsymbol{u}} \mathrm{d}\Omega \qquad (1.79)$$

The virtual work of internal forces is expressed as

$$G_{int} = \int_\Omega \delta (\boldsymbol{B}\boldsymbol{u})^\mathrm{T} \sigma \mathrm{d}\Omega \qquad (1.80)$$

The virtual work of external forces is expressed as

$$G_{ext} = \int_\Omega \delta \boldsymbol{u}^\mathrm{T} \boldsymbol{b} \mathrm{d}\Omega + \int_{\Gamma_t} \delta \boldsymbol{u}^\mathrm{T} \bar{\boldsymbol{t}} \mathrm{d}\Gamma \qquad (1.81)$$

Therefore, the overall weak form or Galerkin formulation can be expressed as

$$G_{eq} = G_{dyn} + G_{int} - G_{ext}$$

The above weak form provides the foundation for deriving finite element equilibrium equations in various applications. To fully formulate the problem, it is necessary to supplement it with appropriate strain–displacement relations and constitutive equations. The weak form described above can also be obtained variationally from the Hu-Washizu principle given in Eq. (1.69).

Note: Using the matrix form for defining stress and strain allows the internal work of stress and strain to be expressed as

$$\varepsilon_{ij}\sigma_{ij} = \boldsymbol{\varepsilon}^{\mathrm{T}}\boldsymbol{\sigma} = \boldsymbol{\sigma}^{\mathrm{T}}\boldsymbol{\varepsilon} \tag{1.82}$$

Similarly, the internal virtual work per unit volume can be expressed as

$$\delta W = \delta\varepsilon_{ij}\sigma_{ij} = \delta\boldsymbol{\varepsilon}^{\mathrm{T}}\boldsymbol{\sigma} \tag{1.83}$$

This will be discussed in greater detail later. In the degenerate form, the constitutive equations constructed from the six stress and strain components must be appropriately derived from the original nine tensor components.

1.3.2 Relationship Between the Strong and Weak Forms

Systems of partial differential equations constitute the strong form of governing equations in solid mechanics. Compared to the weak form, the strong form requires the relevant field variables (here referring to displacements u, v, and w) to possess strong continuity. Any function defining these field variables must be differentiable, with the order of differentiability matching the highest order of derivatives in the strong form partial differential equations. For practical engineering problems, obtaining exact solutions to strong form differential equations is typically very challenging. The finite difference method can be used to obtain approximate solutions to strong form equations and is generally suitable for problems with simple, regular geometries and boundary conditions. The weak form equations are commonly constructed using the following two methods:

- Energy principle.
- Weighted residual method.

The energy principle can be regarded as a special form of the variational principle, and this method is particularly well-suited for problems in solid mechanics and structural mechanics. The weighted residual method is a more general mathematical tool, in principle applicable to solving various systems of partial differential equations. Both methods are straightforward to understand and apply. This book will introduce the formulation of finite element equations using both approaches. The energy principle is used to address problems in solid and structural mechanics, while the weighted residual method is employed for heat transfer problems. The energy principle can also be applied to heat transfer problems, and the weighted residual method can likewise be used to solve problems in solid mechanics, with only minor procedural differences.

The weak form is typically expressed in integral form and requires the field variables to possess only limited continuity. Because it imposes less stringent requirements on the field variables and the integration operations, formulations based on the weak form generally yield a set of discretized equations, resulting in higher accuracy, especially for problems involving complex geometric models. As a result, many prefer to use the weak form to obtain approximate solutions. When the problem domain is appropriately discretized into elements, the weak form usually leads to a well-conditioned system of algebraic equations. The finite element method is a classic example of the successful application of the weak form. Since the problem domain can be discretized into various types of elements, the finite element method is applicable to most practical engineering problems with complex geometries and boundary conditions.

1.4 Finite Element Procedure

This section provides an overview of the standard finite element procedure, including domain discretization, displacement interpolation, the standard process for constructing shape functions, properties of shape functions, finite element equations in the local coordinate system, coordinate transformation, assembly of the global finite element equations, application of displacement constraints, and solution of the global finite element equations.

1.4.1 Domain Discretization

The solid is divided into N_e elements, a process known as meshing, which is typically performed using a preprocessor. This is especially necessary for complex geometries. As shown in Fig. 1.5, an example of a two-dimensional solid mesh with element and node numbering is provided.

The preprocessor generates unique numbers for all elements and nodes of the entity or structure in an appropriate manner. Elements are formed by connecting nodes in a predefined, consistent way, and continuity between elements is achieved through the nodes. All elements together constitute the model. The entire domain is divided into elements with no gaps or overlaps between them. The region can be composed of different types of elements with varying numbers of nodes, as long as adjacent elements are compatible at their boundaries (i.e., there are no gaps or overlaps). The mesh density depends on the required accuracy of the analysis and the available computational resources. Generally, refining the mesh yields more accurate results but increases computational cost. Therefore, the mesh is usually non-uniform, with finer elements used in regions where displacement gradients are large or higher accuracy is required. After discretizing the domain, it becomes easier to specify the displacement field patterns for the elements.

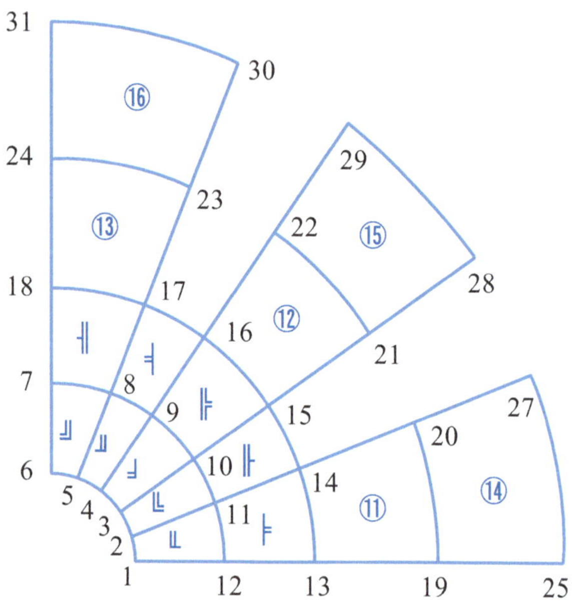

Fig. 1.5 Example of a two-dimensional solid mesh with element and node numbering

1.4.2 Displacement Interpolation

The finite element formulation must be based on a specific coordinate system. When constructing the finite element equations for an element, a local coordinate system is often used for computational convenience. The local coordinate system is defined with reference to the global coordinate system, while the global coordinate system is typically defined for the entire structure. Based on the local coordinate system defined on the element, the displacement within the element can be expressed solely in terms of the nodal displacements, obtained through polynomial interpolation, that is,

$$\boldsymbol{U}^h(x, y, z) = \sum_{i=1}^{n_d} N_i(x, y, z)\boldsymbol{d}_i = \boldsymbol{N}(x, y, z)\boldsymbol{d}_e \tag{1.84}$$

Here, the superscript h on $\boldsymbol{U}$ denotes the approximate value; n_d is the number of nodes in the element; $\mathbf{d}_i$ represents the nodal displacement at the i-th node, which is the unknown to be solved for and can be generally expressed as

$$\mathbf{d}_i = \begin{Bmatrix} d_1 \\ d_2 \\ \vdots \\ d_{n_f} \end{Bmatrix} \begin{matrix} \rightarrow \text{displacement component } 1 \\ \rightarrow \text{displacement component } 2 \\ \vdots \\ \rightarrow \text{displacement component } n_f \end{matrix}$$

Here, n_f denotes the degrees of freedom (DOF) at a node. For a three-dimensional solid, $n_f = 3$, and

$$\mathrm{d}_i = \begin{Bmatrix} u_i \\ v_i \\ w_i \end{Bmatrix} \begin{matrix} \rightarrow \text{dispacement in } x - \text{direction} \\ \rightarrow \text{dispacement in } y - \text{direction} \\ \rightarrow \text{dispacement in } z - \text{direction} \end{matrix}$$

Note: Displacement components may also include the rotations of beam and plate structures. The vector $\mathbf{d}_e$ in Eq. (1.85) represents the displacement vector of the entire element, and is given by

$$d_e = \begin{Bmatrix} d_1 \\ d_2 \\ \vdots \\ d_{n_d} \end{Bmatrix} \begin{matrix} \rightarrow \text{Displacement of Node 1} \\ \rightarrow \text{Displacement of Node 1} \\ \vdots \\ \rightarrow \text{Displacement of Node } n_d \end{matrix}$$

Therefore, the total degrees of freedom for the entire element are $n_d \times n_f$.

In Eq. (1.85), $\boldsymbol{N}$ is the matrix of nodal shape functions for the element. These functions are predefined to represent the variation of displacement with respect to coordinates, and their general form is given

$$\begin{matrix} \boldsymbol{N}(x, y, z) = [N_1(x, y, z) & N_2(x, y, z) & \dots & N_{n_d}(x, y, z)] \\ \downarrow & \downarrow & \dots & \downarrow \\ \text{Node 1} & \text{Node 2} & \dots & \text{Node } n_d \end{matrix}$$

Here, $\boldsymbol{N}_i$ is the submatrix of the shape functions for displacement components, arranged as follows:

$$\mathbf{N}_i = \begin{bmatrix} N_{i1} & 0 & 0 & 0 \\ 0 & N_{i2} & 0 & 0 \\ 0 & 0 & \ddots & 0 \\ 0 & 0 & 0 & N_{i\,n_f} \end{bmatrix}$$

where N_{ik} is the shape function for the k-th displacement component at the i-th node. For a three-dimensional solid, $n_f = 3$, and it is common to have $N_{i1} = N_{i2} = N_{i3} = N_i$. It should be noted that all displacement components at a node do not necessarily use the same shape function. For instance, different shape functions are often used for translational and rotational displacements.

Please note that this approach of assuming displacements is commonly referred to as the displacement method. There are some finite element methods where stresses are assumed as unknowns, as exemplified in Chap. 2.

1.4.3 Standard Procedure for Constructing Shape Functions

Consider an element with n_d nodes, whose nodal coordinates are $\boldsymbol{x}_i$ ($i = 1, 2, \ldots, n_d$). When $\mathbf{x}^{\mathrm{T}} = \{x\}$, it represents a one-dimensional problem, when $\mathbf{x}^{\mathrm{T}} = \{x, y\}$, it represents a two-dimensional problem, and when $\boldsymbol{x}^{\mathrm{T}} = \{x, y, z\}$, it represents a three-dimensional problem. For each displacement component of an element, there should be n_d shape functions. In the following description of the standard procedure for constructing shape functions, only one displacement component is considered. This standard procedure is applicable to any displacement component. First, the displacement components are approximated as a linear combination of n_d linearly independent basis functions $p_i(\boldsymbol{x})$, that is,

$$u^h(\mathbf{x}) = \sum_{i=1}^{n_d} p_i(\mathbf{x})\alpha_i = \mathbf{p}^{\mathrm{T}}(\mathbf{x})\boldsymbol{\alpha} \tag{1.85}$$

Here, u^h is the approximate value of the displacement component; $p_i(\boldsymbol{x})$ is a monomial basis function with respect to the spatial coordinate $\boldsymbol{x}$; α_i is the monomial $p_i(\boldsymbol{x})$; the coefficient; the vector $\boldsymbol{\alpha}$ is defined as

$$\boldsymbol{\alpha}^T = \left\{\alpha_1,\ \alpha_2,\ \alpha_3,\ \ldots,\ \alpha_{n_d}\right\}$$

Note: In Eq. (1.86), $p_i(\boldsymbol{x})$ consists of n_d one-dimensional monomials (one-dimensional problem); for two-dimensional problems, monomials can be selected based on the Pascal triangle shown in Fig. 1.6; for three-dimensional problems, monomials can be selected based on the Pascal pyramid shown in Fig. 1.7. The p-th order complete basis functions in a one-dimensional domain are

$$\boldsymbol{p}^{\mathrm{T}}(\boldsymbol{x}) = \left\{1,\ x,\ x^2,\ x^3,\ x^4,\ \ldots,\ x^p\right\}$$

The p-th order complete basis functions in a two-dimensional domain are

$$\mathbf{p}^{\mathrm{T}}(\mathbf{x}) = \mathbf{p}^{\mathrm{T}}(x, y) = \left\{1, x, y, xy, x^2, y^2, \ldots, x^p, y^p\right\}$$

The p-th order complete basis functions in a three-dimensional domain are

$$\mathbf{p}^{\mathrm{T}}(\mathbf{x}) = \mathbf{p}^{\mathrm{T}}(x, y, z) = \left\{1, x, y, z, xy, yz, zx, x^2, y^2, z^2, \ldots, x^p, y^p, z^p\right\}$$

Typically, the n_d term in $p_i(\boldsymbol{x})$ should be selected from the monomial Pascal triangle or pyramid as shown in Figures. 1.6 or 1.7, following a symmetric selection from the constant term to higher-order terms. However, in certain cases, it is possible to include only selected higher-order terms in the polynomial basis as needed.

The coefficients α_i in Eq. (1.86) can be determined by requiring that the displacement calculated using Eq. (1.86) is equal to the nodal displacement at each of the n_d

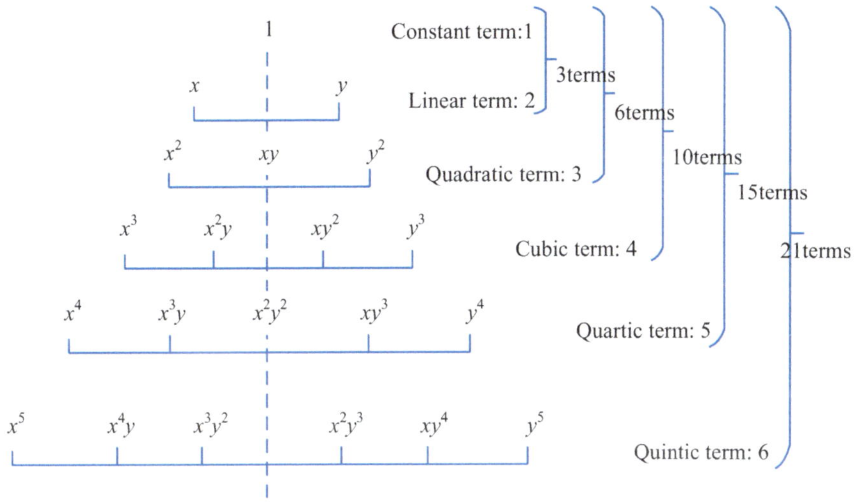

Fig. 1.6 Monomial Pascal Triangle (Two-Dimensional Case)

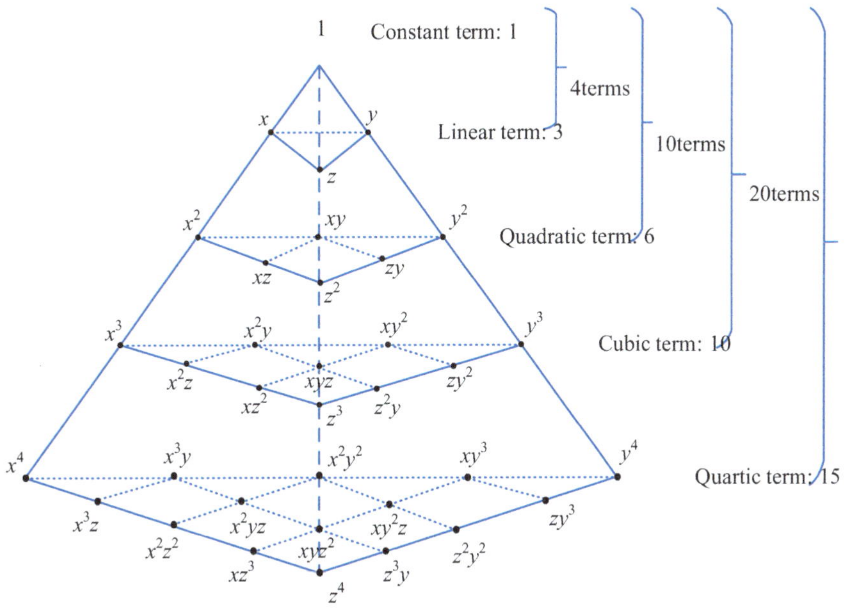

Fig. 1.7 Monomial Pascal Pyramid (Three-Dimensional Case)

nodes of the element. At node i, this yields

$$d_i = \boldsymbol{p}^{\mathrm{T}}(\boldsymbol{x})\boldsymbol{\alpha} \quad i = 1,\ 2,\ 3,\ \ldots,\ n_d \tag{1.86}$$

Here, $\boldsymbol{d}_i$ is the nodal value of u^h at $\boldsymbol{x} = \boldsymbol{x}_i$. Equation (1.87) can be expressed in matrix form as

$$\boldsymbol{d}_e = \boldsymbol{P\alpha} \tag{1.87}$$

where $\boldsymbol{d}_e$ is the vector of displacement components at all n_d nodes within the element, that is,

$$\mathbf{d}_e = \begin{Bmatrix} d_1 \\ d_2 \\ \vdots \\ d_{n_d} \end{Bmatrix}$$

Moreover, $\boldsymbol{P}$ is given by the following expression

$$\boldsymbol{P} = \begin{bmatrix} \boldsymbol{p}^{\mathrm{T}}(\boldsymbol{x}_1) \\ \boldsymbol{p}^{\mathrm{T}}(\boldsymbol{x}_2) \\ \vdots \\ \boldsymbol{p}^{\mathrm{T}}(\boldsymbol{x}_{n_d}) \end{bmatrix}$$

where $\boldsymbol{P}$ is called a moment matrix. The expanded form of $\boldsymbol{P}$ is

$$\boldsymbol{P} = \begin{bmatrix} p_1(\boldsymbol{x}_1) & p_2(\boldsymbol{x}_1) & \cdots & p_{n_d}(\boldsymbol{x}_1) \\ p_1(\boldsymbol{x}_2) & p_2(\boldsymbol{x}_2) & \cdots & p_{n_d}(\boldsymbol{x}_2) \\ \vdots & \vdots & \ddots & \vdots \\ p_1(\boldsymbol{x}_{n_d}) & p_2(\boldsymbol{x}_{n_d}) & \cdots & p_1(\boldsymbol{x}_{n_d}) \end{bmatrix} \tag{1.88}$$

The expansion of the two-dimensional polynomial basis function $\boldsymbol{P}$ is given by

$$\boldsymbol{P} = \begin{bmatrix} 1 & x_1 & y_1 & x_1y_1 & x_1^2 & y_1^2 & x_1^2y_1 & x_1y_1^2 & x_1^3 & \cdots \\ 1 & x_2 & y_2 & x_2y_2 & x_2^2 & y_2^2 & x_2^2y_2 & x_2y_2^2 & x_2^3 & \cdots \\ \vdots & \vdots & \vdots & \vdots & \vdots & \vdots & \vdots & \vdots & \vdots & \vdots \\ 1 & x_{n_d} & y_{n_d} & x_{n_d}y_{n_d} & x_{n_d}^2 & y_{n_d}^2 & x_{n_d}^2y_{n_d} & x_{n_d}y_{n_d}^2 & x_{n_d}^3 & \cdots \end{bmatrix}$$

Using Eq. (1.88) and assuming the inverse of the moment matrix $\boldsymbol{P}$ exists, we can obtain

$$\boldsymbol{\alpha} = \mathbf{P}^{-1}\mathbf{d}_e \tag{1.89}$$

Substituting Eqs. (1.90) into (1.86), we obtain

$$u^h(\mathbf{x}) = \sum_{i=1}^{n_d} N_i(\mathbf{x}) d_i \tag{1.90}$$

or expressed in matrix form as

$$u^h(\boldsymbol{x}) = \boldsymbol{N}(\boldsymbol{x}) \boldsymbol{d}_e \tag{1.91}$$

Here, $\boldsymbol{N}(\mathbf{x})$ is the matrix of shape functions $N_i(\mathbf{x})$, defined as follows:

$$\begin{aligned} \boldsymbol{N}(\boldsymbol{x}) = \boldsymbol{p}^{\mathrm{T}}(\boldsymbol{x}) \boldsymbol{P}^{-1} &= \left[\underbrace{\boldsymbol{p}^{\mathrm{T}}(\boldsymbol{x}) \boldsymbol{P}_1^{-1}}_{N_1(\boldsymbol{x})} \ \underbrace{\boldsymbol{p}^{\mathrm{T}}(\boldsymbol{x}) \boldsymbol{P}_2^{-1}}_{N_2(\boldsymbol{x})} \cdots \underbrace{\boldsymbol{p}^{\mathrm{T}}(\boldsymbol{x}) \boldsymbol{P}_n^{-1}}_{N_n(\boldsymbol{x})} \right] \\ &= \left[N_1(\boldsymbol{x}) \ N_2(\boldsymbol{x}) \cdots N_n(\boldsymbol{x}) \right] \end{aligned} \tag{1.92}$$

Here, $\boldsymbol{P}_i^{-1}$ refers to the i-th column of the matrix $\boldsymbol{P}^{-1}$, and

$$N_i(\mathbf{x}) = \mathbf{p}^{\mathrm{T}}(\mathbf{x}) \mathbf{P}_i^{-1}$$

In deriving Eq. (1.90), we assume that the inverse of $\boldsymbol{P}$ exists. If $\boldsymbol{P}^{-1}$ does not exist, the construction of the shape function fails. The existence of $\boldsymbol{P}^{-1}$ depends on the choice of basis functions and the distribution of nodes within the element. The basis functions must first be selected from a set of linearly independent bases, and then the number of basis functions is determined according to the number of nodes. The detailed discussion of this issue is rather complex, and this book will only discuss units of invertible matrices.

The derivatives of shape functions can be readily obtained, as all the basis functions involved are polynomials. The l-th derivative of a shape function can be simply expressed as

$$N_i^{(l)}(\mathbf{x}) = \left[\mathbf{p}^{(l)}(\mathbf{x})\right]^{\mathrm{T}} \mathbf{P}_i^{-1} \tag{1.93}$$

Note: There are many other methods for constructing shape functions, which do not necessarily follow the standard procedure described above. Common shortcut methods for creating shape functions for different types of elements will be discussed in later sections. These shortcut methods require the properties of shape functions introduced in the next section.

1.4.4 Properties of Shape Functions

1.4.4.1 Property 1: Consistency and Reproducibility

The consistency of shape functions within an element depends on the complete order of the monomials $p_i(\mathbf{x})$ used in Eq. (1.86). Therefore, it also depends on the number of nodes in the element. If the highest order of the monomials is k, the shape functions are said to possess C^k order consistency. For example, consider the following equation:

$$f(\mathbf{x}) = \sum_{j}^{k} p_j(\mathbf{x})\beta_j,\ k \le n_d \tag{1.94}$$

Here, $p_j(\mathbf{x})$ refers to the monomials included in Eq. (1.86). Such a physical field can be represented using all the basis function terms in Eq. (1.86), including the basis function terms in Eq. (1.95), that is,

$$f(\mathbf{x}) = \sum_{j}^{n_d} p_j(\mathbf{x})\beta_j = \mathbf{p}^{\mathrm{T}}(\mathbf{x})\boldsymbol{\alpha} \tag{1.95}$$

where

$$\boldsymbol{\alpha}^{\mathrm{T}} = [\beta_1,\ \beta_2,\ \ldots,\ \beta_k,\ 0,\ \ldots, 0]$$

Using n nodes in the computational domain of $\boldsymbol{x}$, the vector of nodal function values $\boldsymbol{d}_e$ can be obtained, that is,

$$\mathbf{d}_e = \begin{Bmatrix} f_1 \\ f_2 \\ \vdots \\ f_k \\ f_{k+1} \\ \vdots \\ f_n \end{Bmatrix} = \begin{bmatrix} p_1(\mathbf{x}_1) & p_2(\mathbf{x}_1) & \cdots & p_k(\mathbf{x}_1) & p_{k+1}(\mathbf{x}_1) & \cdots p_{n_d}(\mathbf{x}_1) \\ p_1(\mathbf{x}_2) & p_2(\mathbf{x}_2) & \cdots & p_k(\mathbf{x}_2) & p_{k+1}(\mathbf{x}_2) & \cdots p_{n_d}(\mathbf{x}_2) \\ \vdots & \vdots & \cdots & \vdots & \vdots & \vdots \\ p_1(\mathbf{x}_k) & p_2(\mathbf{x}_k) & \cdots & p_k(\mathbf{x}_k) & p_{k+1}(\mathbf{x}_k) & \cdots p_{n_d}(\mathbf{x}_k) \\ p_1(\mathbf{x}_{k+1}) & p_2(\mathbf{x}_{k+1}) & \cdots & p_k(\mathbf{x}_{k+1}) & p_{k+1}(\mathbf{x}_{k+1}) & \cdots p_{n_d}(\mathbf{x}_{k+1}) \\ \vdots & \vdots & \cdots & \vdots & \vdots & \vdots \\ p_1(\mathbf{x}_{n_d}) & p_2(\mathbf{x}_{n_d}) & \cdots & p_k(\mathbf{x}_{n_d}) & p_{k+1}(\mathbf{x}_{n_d}) & \cdots p_{n_d}(\mathbf{x}_{n_d}) \end{bmatrix} \begin{Bmatrix} \beta_1 \\ \beta_2 \\ \vdots \\ \beta_k \\ 0 \\ \vdots \\ 0 \end{Bmatrix} = \mathbf{P}\boldsymbol{\alpha} \tag{1.96}$$

Substituting Eqs. (1.97) into (1.92) yields the following approximate expression:

$$u^h(\mathbf{x}) = \mathbf{p}^{\mathrm{T}}(\mathbf{x})\mathbf{P}^{-1}\mathbf{d}_e = \mathbf{p}^{\mathrm{T}}(\mathbf{x})\mathbf{P}^{-1}\mathbf{P}\boldsymbol{\alpha} = \mathbf{p}^{\mathrm{T}}(\mathbf{x})\boldsymbol{\alpha} = \sum_{j}^{k} p_j(\mathbf{x})\alpha_j = f(\mathbf{x}) \tag{1.97}$$

This is precisely the result given by Eq. (1.96). This demonstrates that for any physical field described by Eq. (1.95), as long as the given field function is included in the basis functions used to construct the shape functions, the field function can be exactly reproduced within the element by the given shape functions. This property of shape functions is, in fact, also quite intuitive: any function of the form $f(\mathbf{x}) = \sum_{j}^{k} p_j(\mathbf{x})\beta_j$ that satisfies the conditions, it is always possible to achieve exactly by setting $\alpha_j = \beta_j$ ($j = 1, 2, \ldots, k$) and $\alpha_j = 0$ ($j = k + 1, \ldots, n_d$), as long as the moment matrix is invertible (to ensure the uniqueness of $\boldsymbol{\alpha}$).

The proof of shape function consistency implies another important property of shape functions: reproducibility, meaning that any function present in the basis can be exactly reproduced. This key property can be used to construct physical fields with specific characteristics. To ensure that shape functions possess linear consistency, it is sufficient to include constants (unity) and linear monomials in the basis functions. In this way, the properties of the shape functions can be leveraged to obtain highly accurate solutions to problems by including terms in the basis functions that closely approximate the exact solution of the problem. The distinction between consistency and reproducibility is as follows:

- Consistency depends on the full order of the basis functions.
- Reproducibility depends on the specific components included in the basis functions.

1.4.4.2 Property 2: Linear Independence

Shape functions are linearly independent because the basis functions are linearly independent and it is assumed that $\boldsymbol{P}^{-1}$ exists. The existence of $\boldsymbol{P}^{-1}$ implies that the shape functions are equivalent to the basis functions in the function space. For example, in Eq. (1.93), since the basis functions are linearly independent, the shape functions are also linearly independent. This linear independence is often overlooked; however, it forms the foundation for the delta (δ) function property of the shape functions.

1.4.4.3 Property 3: Delta (Δ) Function Properties

$$N_i(\mathbf{x}_j) = \delta_{ij} = \begin{cases} 1 & i = j, \quad j = 1,\ 2,\ \ldots,\ n_d \\ 0 & i \neq j, \quad j = 1,\ 2,\ \ldots,\ n_d \end{cases} \tag{1.98}$$

Here, δ_{ij} is the δ function. The property of the δ function implies that the shape function N_i takes a value of one at its associated node i, and is zero at all other nodes of the element (i.e., when $j \neq i$).

The delta function property of the shape function is straight forward to prove: Since the shape functions $N_i(\mathbf{x})$ are linearly independent, any vector of length n_d can be uniquely determined by a linear combination of these n_d shape functions. Suppose the displacement at node i is $\boldsymbol{d}_i$, and the displacements at all other nodes are zero, i.e.,

$$\mathbf{d}_e = \{0,\ 0,\ \ldots,\ \mathbf{d}_i,\ \ldots,\ 0\}^{\mathrm{T}}$$

Substituting the above equation into Eq. (1.91), we have at $\mathbf{x} = \mathbf{x}_j$:

$$\mathbf{u}^h(\mathbf{x}_j) = \sum_{k=1}^{n_d} N_k(\mathbf{x}_j)\mathbf{d}_k = N_i(\mathbf{x}_j)\mathbf{d}_i$$

When $i = j$, the following condition must be satisfied:

$$\mathbf{u}_i = \mathbf{d}_i = N_i(\mathbf{x}_i)\mathbf{d}_i$$

which means

$$N_i(\mathbf{x}_i) = 1$$

That is, to prove (1.99) for the case i $= j$.

$$u_j = 0 = N_i(\mathbf{x}_j)d_i$$

which requires

$$N_i(\mathbf{x}_j) = 0$$

This proves the case of $i \neq j$ in Eq. (1.99). Therefore, it can be concluded that the shape functions possess the delta function property as expressed in Eq. (1.99). Note: The shape functions of certain elements (such as beam and thin plate elements) may not possess the δ function property.

1.4.4.4 Property 4: Partition of Unity

If constants are included in the basis functions, the shape functions possess the partition of unity property, that is,

$$\sum_{i=1}^{n} N_i(\mathbf{x}) = 1 \tag{1.99}$$

This property can be easily proven using the reproducing property of shape functions. Let $u(\boldsymbol{x}) = c$, where c is a constant, then

$$\mathbf{d}_e = \begin{Bmatrix} d_1 \\ d_2 \\ \vdots \\ d_{n_d} \end{Bmatrix} = \begin{Bmatrix} c \\ c \\ \vdots \\ c \end{Bmatrix} \tag{1.100}$$

This implies that the displacements at all nodes are constant and equal. Substituting Eqs. (1.101) into (1.90) yields

$$u(\mathbf{x}) = c \underset{reappear}{=} u^h(\mathbf{x}) \underset{approximate}{=} \sum_{i=1}^{n_d} N_i(\mathbf{x}) d_i = \sum_{i=1}^{n_d} N_i(\mathbf{x}) c = c \sum_{i=1}^{n_d} N_i(\mathbf{x}) \tag{1.101}$$

Thus, Eq. (1.100) is proven. This demonstrates that the partition of unity property of shape functions within the element enables the representation of constant fields or rigid-body motions.

Note: Eq. (1.100) does not require $0 \le N_i(\boldsymbol{x}) \le 1$.

1.4.4.5 Property 5: Reproducibility of Linear Fields

If the basis functions include first-order monomials, the constructed shape functions can reproduce a linear field, that is,

$$\sum_{i=1}^{n_d} N_i(x) x_i = x \tag{1.102}$$

Here, x_i represents the nodal value of the linear field. This can be easily demonstrated using the reproducibility property of the shape functions, with a proof process similar to that of Property 4. Let $u(\boldsymbol{x}) = x$, then

$$\mathbf{d}_e = \left\{ x_1, \ x_2, \ \ldots, \ x_{n_d} \right\}^{\mathrm{T}} \tag{1.103}$$

Substituting Eqs. (1.103) into (1.91) yields

$$u^h(x) = x = \sum_{i=1}^{n_d} N_i(x) x_i$$

This completes the proof of Eq. (1.102).

Lemma 1 Conditions for the Unitary Decomposition of Shape Functions

For a general set of shape functions

$$N_i(\boldsymbol{x}) = c_{1i} + c_{2i} p_2(\boldsymbol{x}) + c_{3i} p_3(\boldsymbol{x}) + \cdots + c_{n_d i} p_{n_d}(\boldsymbol{x}) \tag{1.104}$$

Here, $p_i(\boldsymbol{x})$ ($p_1(\boldsymbol{x}) = 1$, $i = 2, \ldots, n_d$) are a set of independent basis functions. This set of shape functions can be fully decomposed into units if and only if

$$\begin{aligned} C_1 &= 1 \\ C_2 &= C_3 = \cdots = C_{n_d} = 0 \end{aligned} \tag{1.105}$$

where

$$C_k = \sum_{i=1}^{n_d} c_{ki}$$

Proof (Sufficiency) According to Eq. (1.104), the sum of the shape functions is

$$\begin{aligned} \sum_{i=1}^{n_d} N_i(\boldsymbol{x}) &= \sum_{i=1}^{n_d} c_{1i} + p_2(\boldsymbol{x}) \sum_{i=1}^{n_d} c_{2i} + p_3(\boldsymbol{x}) \sum_{i=1}^{n_d} c_{3i} + \cdots + p_{n_d}(\boldsymbol{x}) \sum_{i=1}^{n_d} c_{n_d i} \\ &= \underbrace{C_1}_{1} + \underbrace{C_2}_{0} p_2(\boldsymbol{x}) + \underbrace{C_3}_{0} p_3(\mathbf{x}) + \cdots + \underbrace{C_{n_d}}_{0} p_{n_d}(\boldsymbol{x}) = 1 \end{aligned}$$

This proves the sufficiency condition.

The necessity condition is proved as follows. To achieve partition of unity, define

$$\sum_{i=1}^{n_d} N_i(\boldsymbol{x}) = C_1 + C_2 p_2(\boldsymbol{x}) + C_3 p_3(\boldsymbol{x}) + \ldots + C_{n_d} p_{n_d}(\boldsymbol{x}) = 1 \tag{1.106}$$

or

$$(C_1 - 1) + C_2 p_2(\boldsymbol{x}) + C_3 p_3(\boldsymbol{x}) + \ldots + C_{n_d} p_{n_d}(\boldsymbol{x}) = 0 \tag{1.107}$$

If $p_i(\boldsymbol{x})$ ($p_1(\boldsymbol{x}) = 1$, $i = 2$, n_d) are a set of linearly independent basis functions, then Eqs. (1.106) or (1.107) holds.

Lemma 2 Conditions for Shape Functions to Form a Partition of Unity

Any set of n_d shape functions that satisfies the following conditions will automatically possess the partition of unity property.

Condition 1: It is expressed as a linear combination of the same set of n_d linearly independent basis functions, which include a constant term, and the matrix defined by Eq. (1.89) is of full rank.

Condition 2: Possesses the properties of the δ function.

Proof From Eq. (1.93), it can be seen that all n_d shape functions are constructed using the same set of basis functions $p_i(\boldsymbol{x})$ ($i = 1, 2, \ldots, n_d$)

$$N_i(\boldsymbol{x}) = c_{1i} + c_{2i}p_2(\boldsymbol{x}) + c_{3i}p_3(\boldsymbol{x}) + \ldots + c_{n_d i}p_{n_d}(\boldsymbol{x}) \tag{1.108}$$

This includes the constant term $p_1(\boldsymbol{x}) = 1$. The other basis functions $p_i(\boldsymbol{x})$ ($i = 2, \ldots, n_d$) in Eq. (1.108) can be monomials or any other type of basis function, provided that all basis functions (including $p_1(\boldsymbol{x})$) are linearly independent.

According to Condition 2, the shape function possesses the δ function property, that is,

$$\sum_{i=1}^{n_d} N_i(\boldsymbol{x}_j) = 1, \text{ where } j = 1, 2, \ldots, n_d \tag{1.109}$$

Substituting Eqs. (1.108) into (1.109), we obtain

$$\sum_{i=1}^{n_d} c_{1i} + p_2(\boldsymbol{x}_j)\sum_{i=1}^{n_d} c_{2i} + p_3(\boldsymbol{x}_j)\sum_{i=1}^{n_d} c_{3i} + \cdots + p_{n_d}(\boldsymbol{x}_j)\sum_{i=1}^{n_d} c_{n_d i} = 1, \text{ where } j = 1, 2, \cdots, n_d \tag{1.110}$$

or

$$\mathrm{C}_1 + \mathrm{p}_2(\mathrm{x}_j)\mathrm{C}_2 + \mathrm{p}_3(\mathrm{x}_j)\mathrm{C}_3 + \cdots + \mathrm{p}_{\mathrm{n_d}}(\mathrm{x}_j)\mathrm{C}_{\mathrm{n_d}} = 1, \quad \text{where } \mathrm{j} = 1,2,\ldots,\mathrm{n_d} \tag{1.111}$$

Equation (1.111) can be extended as follows

$$\begin{aligned} &C_1 + p_2(\boldsymbol{x}_1)C_2 + p_3(\boldsymbol{x}_1)C_3 + \ldots + p_{n_d}(\boldsymbol{x}_1)C_{n_d} = 1 \\ &C_1 + p_2(\boldsymbol{x}_2)C_2 + p_3(\boldsymbol{x}_2)C_3 + \ldots + p_{n_d}(\boldsymbol{x}_2)C_{n_d} = 1 \\ &\vdots \\ &C_1 + p_2(\boldsymbol{x}_{n_d})C_2 + p_3(\boldsymbol{x}_{n_d})C_3 + \ldots + p_{n_d}(\boldsymbol{x}_{n_d})C_{n_d} = 1 \end{aligned} \tag{1.112}$$

or, in matrix form, as

$$\begin{bmatrix} 1 & p_2(\boldsymbol{x}_2) & p_3(\boldsymbol{x}_2) & \cdots & p_{n_d}(\boldsymbol{x}_2) \\ 1 & p_2(\boldsymbol{x}_3) & p_3(\boldsymbol{x}_3) & \cdots & p_{n_d}(\boldsymbol{x}_3) \\ \vdots & \vdots & \vdots & \ddots & \vdots \\ 1 & p_2(\boldsymbol{x}_{n_d}) & p_3(\boldsymbol{x}_{n_d}) & \cdots & p_{n_d}(\boldsymbol{x}_{n_d}) \end{bmatrix} \begin{Bmatrix} C_1 - 1 \\ C_2 \\ C_3 \\ \vdots \\ C_{n_d} \end{Bmatrix} = 0 \tag{1.113}$$

Note: The coefficient matrix in Eq. (1.113) is a full-rank square matrix (Condition 1). Therefore, we have

$$C_1 = 1$$

$$C_2 = C_3 = \cdots = C_{n_d} = 0 \tag{1.114}$$

According to Lemma 1, the partition of unity property of the shape functions is thus established.

Lemma 3 Conditions for Shape Functions to Possess Linear Field Reproducibility

Any set of n_d shape functions will automatically satisfy linear field reproducibility if the following condition is met:

Condition 1: It is a linear combination of n_d linearly independent basis functions, including linear basis functions, and the matrix defined by Eq. (1.89) is of full rank.

Condition 2: It possesses the properties of the δ function.

Proof: For the general form of the shape function

$$N_i(\boldsymbol{x}) = c_{1i}p_1(\boldsymbol{x}) + c_{2i}x + c_{3i}p_3(\boldsymbol{x}) + \cdots + c_{n_d i}p_{n_d}(\boldsymbol{x}) \tag{1.115}$$

Consider the linear field $u(\boldsymbol{x}) = x$, with nodal vectors

$$\boldsymbol{d}_e = \{x_1,\ x_2,\ \ldots,\ x_{n_d}\}^{\mathrm{T}}$$

Substituting the above equation into Eq. (1.91), we obtain

$$\begin{aligned}
u^h(\boldsymbol{x}) &= \sum_{i=1}^{n_d} N_i(\boldsymbol{x})x_i \\
&= \sum_{i=1}^{n_d} [c_{1i}p_1(\boldsymbol{x}) + c_{2i}x + c_{3i}p_3(\boldsymbol{x}) + \cdots + c_{n_d i}p_{n_d}(\boldsymbol{x})]x_i \\
&= \sum_{i=1}^{n_d} c_{1i}p_1(\boldsymbol{x})x_i + \sum_{i=1}^{n_d} c_{2i}xx_i + \sum_{i=1}^{n_d} c_{3i}p_3(\boldsymbol{x})x_i + \cdots + \sum_{i=1}^{n_d} c_{n_d i}p_{n_d}(\boldsymbol{x})x_i \\
&= p_1(\boldsymbol{x})\sum_{i=1}^{n_d} c_{1i}x_i + x\sum_{i=1}^{n_d} c_{2i}x_i + p_3(\boldsymbol{x})\sum_{i=1}^{n_d} c_{3i}x_i + \cdots + p_{n_d}(\boldsymbol{x})\sum_{i=1}^{n_d} c_{n_d i}x_i \\
&= p_1(\boldsymbol{x})C_{x1} + xC_{x2} + p_3(\boldsymbol{x})C_{x3} + \cdots + p_{n_d}(\boldsymbol{x})C_{xn_d}
\end{aligned} \tag{1.116}$$

At the n_d nodes of the element, n_d equations can be derived from Eq. (1.116).

$$\begin{aligned}
u^h(\boldsymbol{x}_1) &= p_1(\boldsymbol{x}_1)C_{x1} + x_1C_{x2} + p_3(\boldsymbol{x}_1)C_{x3} + \cdots + p_{n_d}(\boldsymbol{x}_1)C_{xn_d} \\
u^h(\boldsymbol{x}_2) &= p_1(\boldsymbol{x}_2)C_{x1} + x_2C_{x2} + p_3(\boldsymbol{x}_2)C_{x3} + \cdots + p_{n_d}(\boldsymbol{x}_2)C_{xn_d} \\
&\vdots \\
u^h(\boldsymbol{x}_{n_d}) &= p_1(\boldsymbol{x}_{n_d})C_{x1} + x_{n_d}C_{x2} + p_3(\boldsymbol{x}_{n_d})C_{x3} + \cdots + p_{n_d}(\boldsymbol{x}_{n_d})C_{xn_d}
\end{aligned} \tag{1.117}$$

From the delta function property of the shape function, we have

$$\begin{aligned} u^h(\boldsymbol{x}_j) &= \sum_{i=1}^{n_d} N_i(\boldsymbol{x}_j)x_i \\ &= \underbrace{N_1(\boldsymbol{x}_j)x_1}_{0} + \underbrace{N_2(\boldsymbol{x}_j)x_2}_{0} + \cdots + \underbrace{N_j(\boldsymbol{x}_j)x_j}_{1} + \cdots + \underbrace{N_{n_d}(\boldsymbol{x}_j)x_{n_d}}_{0} \\ &= x_j \end{aligned} \tag{1.118}$$

Therefore, based on Eqs. (1.118), (1.117) can be expressed as:

$$\begin{aligned} 0 &= p_1(\mathbf{x})C_{x1} + x_1(C_{x2} - 1) + p_3(\mathbf{x}_2)C_{x3} + \cdots + p_{n_d}(\mathbf{x}_2)C_{xn_d} \\ 0 &= p_1(\mathbf{x}_3)C_{x1} + x_2(C_{x2} - 1) + p_3(\mathbf{x}_3)C_{x3} + \cdots + p_{n_d}(\mathbf{x}_3)C_{xn_d} \\ &\vdots \\ 0 &= p_1(\mathbf{x}_{n_d})C_{x1} + x_{n_d}(C_{x2} - 1) + p_3(\mathbf{x}_{n_d})C_{x3} + \cdots + p_{n_d}(\mathbf{x}_{n_d})C_{xn_d} \end{aligned} \tag{1.119}$$

Alternatively, it can be expressed in matrix form as

$$\begin{bmatrix} p_1(\boldsymbol{x}_1) & x_1 & p_3(\boldsymbol{x}_1) & \cdots & p_{n_d}(\boldsymbol{x}_1) \\ p_1(\boldsymbol{x}_1) & x_2 & p_3(\boldsymbol{x}_2) & \cdots & p_{n_d}(\boldsymbol{x}_2) \\ \vdots & \vdots & \vdots & \ddots & \vdots \\ p_1(\boldsymbol{x}_1) & x_{n_d} & p_3(\boldsymbol{x}_{n_d}) & \cdots & p_{n_d}(\boldsymbol{x}_{n_d}) \end{bmatrix} \begin{Bmatrix} C_{x1} \\ C_{x2} - 1 \\ C_{x3} \\ \vdots \\ C_{xn_d} \end{Bmatrix} = \mathbf{0} \tag{1.120}$$

Note: The coefficient matrix in Eq. (1.120) is a full-rank square matrix. Therefore, we have

$$\begin{aligned} &C_{x1} = 0 \\ &(C_{x2} - 1) = 0 \\ &C_{x3} = \cdots = C_{n_d} = 0 \end{aligned} \tag{1.121}$$

Substituting Eqs. (1.121) into (1.116), we obtain

$$u^h(x) = \sum_{i=1}^{n_d} N_i(x)x_i = x \tag{1.122}$$

The linear field reproducibility of the shape function is thus proven.

Property 3 (the δ function property) ensures that essential boundary conditions can be conveniently applied, because the displacement at one node is independent of the displacements at all other nodes. In other words, constraints are often expressed in the form of so-called single-point constraints. If the displacement at a certain node is fixed, it is sufficient to remove the corresponding row and column without affecting the others.

The proof of Property 4 provides a convenient method for verifying the partition of unity property of shape functions. As long as the basis functions include a constant (unit) basis, the constructed shape functions will possess the partition of unity property. Properties 4 and 5 are crucial for the standard patch test in finite element analysis, which has been a common method for verifying the characteristics of finite elements for decades. In a standard patch test, the patch is first divided into multiple elements, with at least one interior node; then, linear displacements are applied along the boundary (edges) of the patch. A successful patch test requires that the finite element results produce a linear displacement field (or constant strain field) at any interior node. Therefore, the ability of shape functions to reproduce linear fields forms the basis for passing the standard patch test.

Note: The linear field reproduction property of shape functions does not guarantee the success of the patch test, as there may be other sources of numerical error, such as numerical integration, that can also lead to failure.

Lemma 1 may seem redundant since Property 4 already exists. However, Lemma 1 is a very convenient property that can be used to verify the partition of unity property of shape functions constructed using alternative shortcut methods, rather than the standard procedure described in Sect. 1.4.3. By using Lemma 1, it is only necessary to check whether the shape functions satisfy Eq. (1.105).

Lemma 2 is another highly useful property that can be employed to verify the partition of unity property of shape functions. By using Lemma 2, it is only necessary to ensure that the constructed n_d shape functions possess the δ-function property and are all linear combinations of n_d linearly independent basis functions that include the constant basis function. Since it is sometimes difficult to determine whether the basis function matrix is of full rank, this book generally assumes that conventional elements are of full rank as long as the basis functions are linearly independent. In most cases, if the matrix is not of full rank, it is not possible to obtain the shape functions; if the shape functions are somehow successfully obtained, it can usually be concluded that the corresponding matrix is of full rank.

Lemma 3 is also a very useful property that can be used to verify the linear field reproduction of shape functions. By applying Lemma 3, it is only necessary to ensure that the constructed n_d shape functions possess the delta function property and are all linearly independent and are linear combinations of n_d basis functions.

1.4.5 Finite Element Equations in Local Coordinate System

After constructing the shape functions, the following procedure can be used to formulate the element finite element equations. By substituting the nodal interpolated displacement field (Eq. (1.85)) and the strain–displacement equation ($\varepsilon = \mathbf{Bu}$) into the strain energy Eq. (1.6), we obtain

$$\Pi = \frac{1}{2}\int_{V_e} \boldsymbol{\varepsilon}^{\mathrm{T}}\boldsymbol{c}\boldsymbol{\varepsilon}\mathrm{d}V = \frac{1}{2}\int_{V_e} \boldsymbol{d}_e^{\mathrm{T}}\boldsymbol{B}^{\mathrm{T}}\boldsymbol{c}\boldsymbol{B}\boldsymbol{d}_e\mathrm{d}V = \frac{1}{2}\boldsymbol{d}_e^{\mathrm{T}}\left(\int_{V_e} \boldsymbol{B}^{\mathrm{T}}\boldsymbol{c}\boldsymbol{B}\mathrm{d}V\right)\boldsymbol{d}_e \tag{1.123}$$

where the subscript e denotes the element.

Note: The volume integral over the entire domain has been replaced by a volume integral over a single element, since the assumed displacement field is required to satisfy compatibility conditions along all edges between elements.

In Eq. (1.123), $\boldsymbol{B}$ is commonly referred to as the strain matrix, defined as follows:

$$\boldsymbol{B} = \boldsymbol{L}\boldsymbol{N} \tag{1.124}$$

Here, $\boldsymbol{L}$ is the differential operator defined for different problems (using the matrix $\boldsymbol{B}$ introduced in Sect. 1.1.2). For three-dimensional solids, it is given by $\boldsymbol{\varepsilon} = \boldsymbol{B}\boldsymbol{u}$. Define

$$\boldsymbol{k}_e = \int_{V_e} \boldsymbol{B}^{\mathrm{T}}\boldsymbol{c}\boldsymbol{B}\mathrm{d}V \tag{1.125}$$

which is referred to as the element stiffness matrix, then Eq. (1.123) can be rewritten as

$$\Pi = \frac{1}{2}\mathbf{d}_e^{\mathrm{T}}\mathbf{k}_e\mathbf{d}_e \tag{1.126}$$

Note: The element stiffness matrix $\boldsymbol{k}_e$ is symmetric, because

$$[\mathbf{k}_e]^{\mathrm{T}} = \int_{V_e} \left[\mathbf{B}^{\mathrm{T}}\mathbf{c}\mathbf{B}\right]^{\mathrm{T}}\mathrm{d}V = \int_{V_e} \mathbf{B}^{\mathrm{T}}\mathbf{c}^{\mathrm{T}}\left[\mathbf{B}^{\mathrm{T}}\right]^{\mathrm{T}}\mathrm{d}V = \int_{V_e} \mathbf{B}^{\mathrm{T}}\mathbf{c}\mathbf{B}\mathrm{d}V = \mathbf{k}_e \tag{1.127}$$

This indicates that the transpose of matrix $\boldsymbol{k}_e$ is itself. When deriving Eq. (1.127), the relation $\boldsymbol{c} = \boldsymbol{c}^{\mathrm{T}}$ is used. It can be known from the symmetry of the stiffness matrix that only half of the terms in the matrix need to be calculated and stored.

Using Eq. (1.85), the kinetic energy can be expressed as

$$T = \frac{1}{2}\int_{V_e} \rho\dot{\mathbf{U}}^{\mathrm{T}}\dot{\mathbf{U}}\mathrm{d}V = \frac{1}{2}\int_{V_e} \rho\dot{\mathbf{d}}_e^{\mathrm{T}}\mathbf{N}^{\mathrm{T}}\mathbf{N}\dot{\mathbf{d}}_e\mathrm{d}V = \frac{1}{2}\dot{\mathbf{d}}_e^{\mathrm{T}}\left(\int_{V_e} \rho\mathbf{N}^{\mathrm{T}}\mathbf{N}\mathrm{d}V\right)\dot{\mathbf{d}}_e \tag{1.128}$$

Define

$$\mathbf{m}_e = \int_{V_e} \rho\mathbf{N}^{\mathrm{T}}\mathbf{N}\mathrm{d}V \tag{1.129}$$

as the mass matrix of the element, then Eq. (1.128) can be rewritten as

$$T = \frac{1}{2}\dot{\mathbf{d}}_e^{\mathrm{T}}\mathbf{m}_e\dot{\mathbf{d}}_e \tag{1.130}$$

Clearly, the element mass matrix is also symmetric.

To obtain the work done by the external force, substitute Eqs. (1.85) into (1.81) to obtain

$$W_f = \int_{V_e} \mathbf{d}_e^{\mathrm{T}} \mathbf{N}^{\mathrm{T}} \mathbf{f}_b \mathrm{d}V + \int_{S_e} \mathbf{d}_e^{\mathrm{T}} \mathbf{N}^{\mathrm{T}} \mathbf{f}_s \mathrm{d}S = \mathbf{d}_e^{\mathrm{T}} \left(\int_{V_e} \mathbf{N}^{\mathrm{T}} \mathbf{f}_b \mathrm{d}V \right) + \mathbf{d}_e^{\mathrm{T}} \left(\int_{S_e} \mathbf{N}^{\mathrm{T}} \mathbf{f}_s \mathrm{d}S \right) \tag{1.131}$$

In the above equation, the surface force only requires a surface integral over the elements located on the force boundary of the problem domain. Define

$$\mathbf{F}_b = \int_{V_e} \mathbf{N}^{\mathrm{T}} \mathbf{f}_b \mathrm{d}V \tag{1.132}$$

and

$$\mathbf{F}_s = \int_{S_e} \mathbf{N}^{\mathrm{T}} \mathbf{f}_s \mathrm{d}S \tag{1.133}$$

Thus, Eq. (1.131) can be rewritten as

$$W_f = \mathbf{d}_e^{\mathrm{T}} \mathbf{F}_b + \mathbf{d}_e^{\mathrm{T}} \mathbf{F}_s = \mathbf{d}_e^{\mathrm{T}} \mathbf{f}_e \tag{1.134}$$

Here, $\mathbf{F}_b$ and $\mathbf{F}_s$ are the nodal forces acting on the element nodes, which are equivalent to the work done by the body force and surface force applied to the element on the virtual displacement. By summing these two nodal force vectors, the total nodal force vector is obtained as

$$\mathbf{f}_e = \mathbf{F}_b + \mathbf{F}_s \tag{1.135}$$

Using the element matrix and vector, the finite element equation for a single element can be expressed as

$$\mathbf{m}_e \ddot{\mathbf{d}}_e + \mathbf{k}_e \mathbf{d}_e = \mathbf{f}_e \tag{1.136}$$

Here, $\mathbf{k}_e$ and $\mathbf{m}_e$ are the element stiffness and mass matrices, respectively, and $\mathbf{f}_e$ is the total external force vector applied to the element nodes. All of these element matrices and vectors can be obtained by integrating the given displacement shape functions.

1.4.6 Coordinate Transformations

The element equation given by Eq. (1.136) is derived based on the local coordinate system defined on the element. In general, a structure is divided into elements oriented in various directions (see Fig. 1.5). To assemble the equations of all elements into a global system of equations, it is necessary to perform a coordinate transformation for each element so as to obtain the element equations in the global coordinate system.

The coordinate transformation provides the relationship between the displacement vector $\mathbf{d}_e$ in the local coordinate system and the displacement vector $\mathbf{D}_e$ of the same element in the global coordinate system.

$$\mathbf{d}_e = \mathbf{T}\mathbf{D}_e \tag{1.137}$$

Here, $\mathbf{T}$ denotes the transformation matrix, which takes different forms depending on the type of element and will be discussed in detail in later sections. The transformation matrix is also used to relate force vectors between the local and global coordinate systems.

$$\mathbf{f}_e = \mathbf{T}\mathbf{F}_e \tag{1.138}$$

Here, $\mathbf{F}_e$ is the force vector at node i in the global coordinate system. Substituting Eqs. (1.137) and (1.138) into (1.36), the element finite element equation based on the global coordinate system can be derived as

$$\mathbf{K}_e\mathbf{D}_e + \mathbf{M}_e\ddot{\mathbf{D}}_e = \mathbf{F}_e \tag{1.139}$$

where

$$\begin{aligned} \mathbf{K}_e &= \mathbf{T}^{\mathrm{T}}\mathbf{k}_e\mathbf{T} \\ \mathbf{M}_e &= \mathbf{T}^{\mathrm{T}}\mathbf{m}_e\mathbf{T} \\ \mathbf{F}_e &= \mathbf{T}^{\mathrm{T}}\mathbf{f}_e \end{aligned} \tag{1.140}$$

1.4.7 Assembly of the Global Finite Element Equations

The finite element equations for all individual elements can be assembled to form the global finite element system of equations.

$$\mathbf{K}\mathbf{D} + \mathbf{M}\ddot{\mathbf{D}} = \mathbf{F} \tag{1.141}$$

Here, **K** and **M** represent the global stiffness and mass matrices, respectively; **D** is the displacement vector for all nodes within the problem domain; **F** is the vector of all equivalent nodal forces. The assembly process simply involves summing the contributions of all elements connected to each node. It is important to note that if the assembly process is combined with the solution of the system of equations, the assembly step can be bypassed. This means that a particular entry in the global matrix is assembled only when the equation solver operates on that entry.

1.4.8 Application of Displacement Constraints

The global stiffness matrix $\boldsymbol{K}$ in Eq. (1.141) is typically not of full rank, as displacement constraints (supports) have not yet been applied; it is non-negative definite or semi-positive definite. In physics, an unconstrained solid or structure is capable of rigid body motion. Therefore, if a solid or structure is unsupported and subjected to dynamic loading, Eq. (1.141) describes the mechanical behavior including rigid body motion. If the applied external force is static, then for any given force vector, the displacement cannot be uniquely determined by Eq. (1.141). Attempting to solve for the static displacement of a freely movable, unconstrained solid or structure is therefore meaningless.

For constrained solids and structures, constraints can be applied using the procedure described in Sect. 2.4.

1.4.9 Solving the Global Finite Element Equations

By solving the global finite element equations, the displacements at all nodes can be obtained, which in turn allows for the determination of strain and stress in any element.

1.5 Summary

This chapter summarizes the main principles for formulating finite element equations for solid and structural mechanics problems, as well as the fundamental steps required to solve general small-strain solid mechanics problems. The relevant equations are presented in both the strong form of partial differential equations and the weak form expressed as integrals. The chapter also highlights how certain problems can become nonlinear. In the next chapter, we will introduce the construction of the weak form for nonlinear transient solid mechanics problems using the finite element method and discuss their approximate solutions.

In the finite element method, the displacement field $\boldsymbol{U}$ is represented by shape functions defined over the element and expressed in terms of nodal displacements. Once the shape functions are defined, the force vector, stiffness matrix, and mass matrix can be derived. Therefore, to formulate the finite element equations for various types of structural elements, it is essential to first construct the shape function $\boldsymbol{N}$, and then obtain the strain–displacement matrix $\boldsymbol{B}$; the remaining steps are largely similar. In the following chapters on the finite element method, we will focus on the derivation and computation of shape functions and strain–displacement matrices for various solid and structural elements.

References

1. ZIENKIEWICZ O C, TAYLOR R L, ZHU J Z. The Finite Element Method: Its Basis and Fundamentals [M]. 7th ed. New York: Elsevier, 2013.
2. ZIENKIEWICZ O C, TAYLOR R L, FOX D. The Finite Element Method for Solid and Structural Mechanics [M]. 7th ed. Oxford: Butterworth-Heinemann, 2014.
3. LIU G R, QUEK SS. The Finite Element Method: A Practical Course [M]. Oxford: Elsevier Science Ltd, 2003.

Chapter 2
Galerkin Approximation: Irreducible and Mixed Forms

Chapter 1 introduced the fundamental equations for nonlinear solid mechanics problems with small strains. These equations can be expressed in strong form as a set of partial differential equations, or in variational (weak) form as integrals over the computational domain. In this chapter, we will discuss how to construct approximate solutions based on the finite element method using the weak form, leading to the Galerkin method.

Here, it is assumed that the reader is not familiar with the finite element method for small deformation linear problems. Therefore, this chapter will provide a detailed introduction to the fundamental steps for solving transient problems. It is important to note the differences between linear and nonlinear responses, and to pay close attention to the numerical procedures used to establish the final discrete form of the equations, which is the form employed in computer analysis. This chapter will present both irreducible and mixed approximation formulations. The introduction of mixed formulations aims to overcome the deficiencies that arise when using low-order elements based on the irreducible form. In particular, this chapter will discuss a mixed formulation suitable for problems exhibiting nearly incompressible behavior. Chapter 5 of this book will also address the issue of "shear locking" that may occur in plate bending.

The final section of this chapter will introduce the application of solid mechanics solution methods to thermal analysis based on nonlinear quasi-harmonic equations.

2.1 Finite Element Approximation: The Galerkin Method

The finite element approximation of a problem first requires partitioning the computational domain Ω into a set of subdomains Ω_e (called elements), that is,

B. Liu and C. Liu, *Introduction to Linear and Nonlinear Finite Element Analysis*,
https://doi.org/10.1007/978-981-95-9670-6_2

$$\Omega \approx \hat{\Omega} = \sum_{e} \Omega_e \tag{2.1}$$

Similarly, the boundary is also divided into a set of subdomains, that is,

$$\Gamma \approx \hat{\Gamma} = \sum_{e} \Gamma_e = \sum_{et} \Gamma_{t_e} + \sum_{eu} \Gamma_{u_e} \tag{2.2}$$

Here, Γ_{te} denotes the boundary segment with prescribed traction, and Γ_{ue} denotes the boundary segment with prescribed displacement.

Note: In general, the domain $\hat{\Omega}$ and boundary $\hat{\Gamma}$ used in finite element analysis are both approximations of the actual domain, and both are determined by the shape of the element boundaries.

The weak form of the governing equation is the sum of the weak forms over all elements in the computational domain $\hat{\Omega}$. Therefore, the weak form of the equilibrium equation given by Eq. (1.78) can be expressed as

$$\begin{aligned} \mathrm{G}_{eq} \approx \hat{\mathrm{G}}_{eq} &= \sum_{e} \left[\int_{\Omega_e} \delta \boldsymbol{u}^{\mathrm{T}} \rho \ddot{\boldsymbol{u}} \mathrm{d}\Omega + \int_{\Omega_e} \delta (\mathbb{B}\boldsymbol{u})^{\mathrm{T}} \boldsymbol{\sigma} \mathrm{d}\Omega - \int_{\Omega_e} \delta \boldsymbol{u}^{\mathrm{T}} \boldsymbol{b} \mathrm{d}\Omega \right] - \sum_{et} \left[\int_{\Gamma_{te}} \delta \boldsymbol{u}^{\mathrm{T}} \bar{\boldsymbol{t}} \mathrm{d}\Gamma \right] \\ &= \sum_{e} \hat{\mathrm{G}}^{e} + \sum_{et} \hat{\mathrm{G}}_{t}^{e} = 0 \end{aligned} \tag{2.3}$$

Here, $\hat{\mathrm{G}}^{e}$ represents the terms within each element domain Ω_e, and $\hat{\mathrm{G}}_t^{e}$ represents the terms on the traction boundary surface Γ_{te}. Galerkin approximation method is implemented using the dependent variable and its virtual work formulation. For irreducible (or displacement-based) finite element methods, it is only necessary to approximate $\mathbf{u}$ and $\delta\mathbf{u}$.

In order to decompose the variational or weak form into the summation form shown in Eq. (2.3), the highest-order derivatives appearing in the functional must be at least piecewise continuous so that all integrals exist, but do not cross element boundaries and affect other elements. For those containing a function of a variable whose highest derivative order is $m + 1$, the function used to approximate that variable must be m-times continuously differentiable over the entire domain $\hat{\Omega}$. Such functions are referred to as C^m continuous. For the weak form of physical problems, we typically encounter functionals involving only first-order derivatives, because here, only C^0 functions are required for approximation. In fact, some mixed-type functions lack derivatives, and these functions can be approximated by discontinuous functions within the domain $\hat{\Omega}$. Typically, attention must be paid to the order of approximation (i.e., C^m), as the exact solution may exhibit discontinuous behavior. In solid mechanics, displacement or stress discontinuities may occur at material interfaces or due to certain singular load forms (such as point loads or line loads). Material interfaces are physically real, whereas point or line loads are merely approximations of actual loads. When the exact solution is smooth, it may be beneficial to use functions with continuity higher than C^m. Therefore, recent literature has introduced some interpolation methods that enhance continuity (smoothness). In this book, we will primarily focus on problems with C^0 continuity.

Some formulas violate continuity conditions, resulting in so-called nonconforming approximations; such approximations are referred to as variational crimes. However, as long as certain requirements are met, convergence can still be achieved. Most well-behaved nonconforming schemes have ultimately been shown to belong to the class of mixed methods.

In addition to continuity requirements, a C^m function must also possess a complete polynomial of order $m + 1$ to ensure that the derivative of order $m + 1$ can be constant. Both of these requirements are discussed in standard textbooks on the finite element method and are equally applicable to nonlinear problems (including both material nonlinearity and problems involving large deformations with nonlinear kinematic conditions). Piecewise testing remains effective for evaluating the continuity and differentiability order of any approximation method.

2.1.1 Displacement Approximation

The finite element approximation of displacement is given by the following expression:

$$\boldsymbol{u}(\boldsymbol{x}, t) \approx \hat{\boldsymbol{u}} = \sum_b N_b(\boldsymbol{x})\tilde{u}_b(t) = \boldsymbol{N}(\boldsymbol{x})\tilde{\boldsymbol{u}}(t) \tag{2.4}$$

Here, N_b denotes the shape function of the element, and $\tilde{u}_b(t)$ represents the nodal displacement as a function of time. The summation is taken over the nodes associated with the element. In isoparametric elements, Eq. (2.4) is given by the following expression (taking the four-node two-dimensional quadrilateral isoparametric mapping shown in Fig. 2.1 as an example):

$$\begin{gathered}\boldsymbol{u}(\boldsymbol{\xi}, t) \approx \hat{\boldsymbol{u}}(\boldsymbol{\xi}, t) = \sum_b N_b(\boldsymbol{\xi})\tilde{\boldsymbol{u}}_b(t) = \boldsymbol{N}(\boldsymbol{\xi})\tilde{\boldsymbol{u}}(t); \\ \boldsymbol{x}(\boldsymbol{\xi}) = \sum_b N_b(\boldsymbol{\xi})\tilde{\boldsymbol{x}}_b = \boldsymbol{N}(\boldsymbol{\xi})\tilde{\boldsymbol{x}}\end{gathered} \tag{2.5}$$

Here, $\tilde{\boldsymbol{x}}$ represents the spatial coordinates of the element node, and ξ denotes the parametric coordinates of the element.

The approximate expression for virtual displacement is given by

$$\delta\boldsymbol{u}(\boldsymbol{\xi}) \approx \delta\hat{\boldsymbol{u}}(\boldsymbol{\xi}) = \sum_a N_a(\boldsymbol{\xi})\delta\tilde{\boldsymbol{u}}_a = \boldsymbol{N}(\boldsymbol{\xi})\delta\tilde{\boldsymbol{u}} \tag{2.6}$$

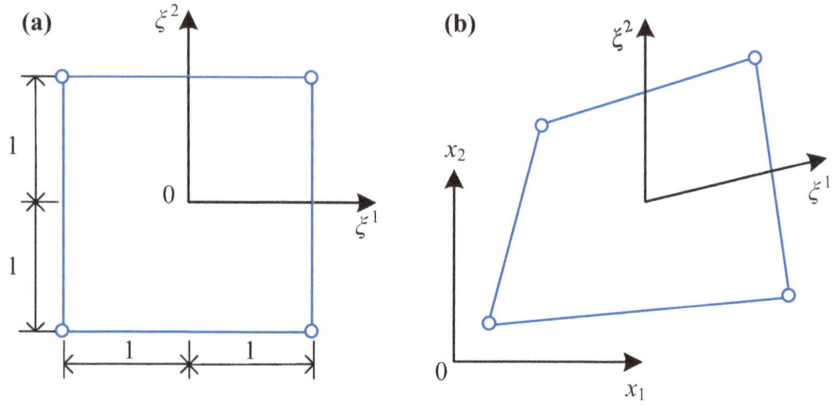

Fig. 2.1 Four-node two-dimensional quadrilateral isoparametric mapping

Chapter 4 will introduce the standard procedure for constructing certain isoparametric element shape functions.

2.1.2 *Derivatives*

The weak forms introduced in Chap. 1 all involve first-order derivatives of displacement. For the isoparametric approximation given by Eq. (2.5), the first-order derivatives of the shape functions with respect to x_j are required, which can be calculated using the chain rule as follows:

$$\frac{\partial N_a}{\partial \xi^i} = \frac{\partial x_j}{\partial \xi^i}\frac{\partial N_a}{\partial x_j} \tag{2.7}$$

Alternatively, it can be expressed in matrix form as

$$\frac{\partial N_a}{\partial \boldsymbol{\xi}} = \boldsymbol{J}\frac{\partial N_a}{\partial \boldsymbol{x}} \tag{2.8}$$

where

$$\frac{\partial N_a}{\partial \boldsymbol{\xi}} = \begin{Bmatrix} \frac{\partial N_a}{\partial \xi^1} \\ \frac{\partial N_a}{\partial \xi^2} \\ \frac{\partial N_a}{\partial \xi^3} \end{Bmatrix}, \quad \frac{\partial N_a}{\partial \boldsymbol{x}} = \begin{Bmatrix} \frac{\partial N_a}{\partial x_1} \\ \frac{\partial N_a}{\partial x_2} \\ \frac{\partial N_a}{\partial x_3} \end{Bmatrix}, \quad \boldsymbol{J} = \begin{bmatrix} \frac{\partial x_1}{\partial \xi^1} & \frac{\partial x_2}{\partial \xi^1} & \frac{\partial x_3}{\partial \xi^1} \\ \frac{\partial x_1}{\partial \xi^2} & \frac{\partial x_2}{\partial \xi^2} & \frac{\partial x_3}{\partial \xi^2} \\ \frac{\partial x_1}{\partial \xi^3} & \frac{\partial x_2}{\partial \xi^3} & \frac{\partial x_3}{\partial \xi^3} \end{bmatrix} \tag{2.9}$$

where $\mathbf{J}$ is the Jacobian transformation between $\boldsymbol{x}$ and ξ. Thus, the derivatives of the shape functions with respect to the spatial coordinates are

$$\frac{\partial N_a}{\partial \boldsymbol{x}} = \boldsymbol{J}^{-1}\frac{\partial N_a}{\partial \boldsymbol{\xi}} \tag{2.10}$$

In two-dimensional problems, only the first two coordinates are involved, so **J** reduces to a 2 × 2 matrix. We often denote differentiation as follows

$$\begin{aligned} \frac{\partial N_a}{\partial x_j} &= N_{a,x_j} \equiv N_{a,j} \\ \frac{\partial N_a}{\partial \xi^i} &= N_{a,\xi^i} \end{aligned} \tag{2.11}$$

2.1.3 Strain–Displacement Equations

The approximate expression for the strain–displacement equation is given by

$$\boldsymbol{\varepsilon} = \boldsymbol{B}\boldsymbol{u} \approx \sum_b (\boldsymbol{B}N_b)\tilde{\boldsymbol{u}}_b = \sum_b \boldsymbol{B}_b\tilde{\boldsymbol{u}}_b = \boldsymbol{B}\tilde{\boldsymbol{u}} \tag{2.12}$$

In general three-dimensional problems, the strain matrix at each node of an element is defined as follows

$$\boldsymbol{B}_b^{\mathrm{T}} = \begin{bmatrix} N_{b,x_1} & 0 & 0 & N_{b,x_2} & 0 & N_{b,x_3} \\ 0 & N_{b,x_2} & 0 & N_{b,x_1} & N_{b,x_3} & 0 \\ 0 & 0 & N_{b,x_3} & 0 & N_{b,x_2} & N_{b,x_1} \end{bmatrix} \tag{2.13}$$

For two-dimensional plane stress, plane strain, and axisymmetric problems without torsion, the strain matrix at each node of the element is defined as follows:

$$\boldsymbol{B}_b^{\mathrm{T}} = \begin{bmatrix} N_{b,x_1} & 0 & cN_b/x_1 & N_{b,x_2} \\ 0 & N_{b,x_2} & 0 & N_{b,x_1} \end{bmatrix} \tag{2.14}$$

In this context, for plane stress and strain problems, $c = 0$; for axisymmetric cases without torsion, $c = 1$.

For axisymmetric problems with torsion, the strain matrix at each node of the element is defined as follows:

$$\boldsymbol{B}_b^{\mathrm{T}} = \begin{bmatrix} N_{b,x_1} & 0 & N_b/x_1 & N_{b,x_2} & 0 & 1 \\ 0 & N_{b,x_2} & 0 & N_{b,x_1} & 0 & 0 \\ 0 & 0 & 0 & 0 & N_{b,x_2} & N_{b,x_1} - N_b/x_1 \end{bmatrix} \tag{2.15}$$

2.1.4 Weak Formulation

Substituting the above approximate expressions for displacement and strain into Eq. (2.3) yields the weak form equilibrium equation for an element, namely

$$\hat{G}^{e}_{eq} = \delta\tilde{\boldsymbol{u}}^{\mathrm{T}}\left[\int_{\Omega_e} \boldsymbol{N}^{\mathrm{T}}\rho\boldsymbol{N}\mathrm{d}\Omega\ddot{\tilde{\boldsymbol{u}}} + \int_{\Omega_e} \boldsymbol{B}^{\mathrm{T}}\boldsymbol{\sigma}\mathrm{d}\Omega - \int_{\Omega_e} \boldsymbol{N}^{\mathrm{T}}\boldsymbol{b}\mathrm{d}\Omega - \int_{\Gamma_{te}} \boldsymbol{N}^{\mathrm{T}}\bar{\boldsymbol{t}}\mathrm{d}\Gamma\right] \tag{2.16}$$

Summing over all elements and noting that the virtual displacement $\delta\tilde{\mathbf{u}}$ is arbitrary, we obtain a semi-discrete problem described by a system of ordinary differential equations.

$$\boldsymbol{M}\ddot{\tilde{\boldsymbol{u}}} + \boldsymbol{P}(\boldsymbol{\sigma}) = \boldsymbol{f} \tag{2.17}$$

where

$$\boldsymbol{M} = \sum_{e}\boldsymbol{M}^{(e)}, \boldsymbol{P} = \sum_{e}\boldsymbol{P}^{(e)}, \boldsymbol{f} = \sum_{e}\boldsymbol{f}^{(e)} \tag{2.18}$$

and the corresponding element matrices

$$\begin{aligned} \boldsymbol{M}^{(e)} &= \int_{\Omega_e} \boldsymbol{N}^{\mathrm{T}}\rho\boldsymbol{N}\mathrm{d}\Omega \\ \boldsymbol{P}^{(e)}(\boldsymbol{\sigma}) &= \int_{\Omega_e} \boldsymbol{B}^{\mathrm{T}}\boldsymbol{\sigma}\mathrm{d}\Omega \\ \boldsymbol{f}^{(e)}(\boldsymbol{\sigma}) &= \int_{\Omega_e} \boldsymbol{N}^{\mathrm{T}}\boldsymbol{b}\mathrm{d}\Omega + \int_{\Gamma_{te}} \boldsymbol{N}^{\mathrm{T}}\bar{\boldsymbol{t}}\mathrm{d}\Gamma \end{aligned} \tag{2.19}$$

where **P** is commonly referred to as the internal force term.

Although the array form given above is applicable to various types of problems, the differential volume element differs for different types of problems, as detailed in Table 2.1.

For the volume element in Table 2.1, assume that the x_3 direction corresponds to the thickness direction in the plane strain problem, and h_3 is the thickness in the plane stress problem. The thickness of the plate in the problem, and the factor 2π in axisymmetric problems, is the result of the integral $\int \mathrm{d}x_3 = \mathrm{d}\theta$.

Table 2.1 Differential volume elements for different types of problems

Problem type	Volume element
General three-dimensional problem	$\mathrm{d}\Omega = \mathrm{d}x_1\mathrm{d}x_2\mathrm{d}x_3$
Plane strain problem	$\mathrm{d}\Omega = \mathrm{d}x_1\mathrm{d}x_2$
Plane stress problem	$\mathrm{d}\Omega = h_3\mathrm{d}x_1\mathrm{d}x_2$
Axisymmetric problem	$\mathrm{d}\Omega = 2\pi x_1\mathrm{d}x_1\mathrm{d}x_2$

In the following sections, coordinates x_i, displacements u_i, and similar notations will be used to discuss the finite element formulation of solids. Unless otherwise specified, the forms $\mathbf{B}$, Ω_e, $\mathrm{d}\Omega$, etc., as described above, are assumed by default and are simplified to the problem categories already discussed (i.e., plane stress, plane strain, axisymmetric, or general three-dimensional problems).

2.1.5 *Irreducible Displacement Method*

In the case of linear elasticity, the constitutive equation is given by Eq. (1.6). According to Eq. (2.12), the result of the irreducible displacement method can be expressed as

$$\boldsymbol{P}^{(e)}(\boldsymbol{\sigma}) = \left(\int_{\Omega_e} \boldsymbol{B}^{\mathrm{T}}\boldsymbol{D}\boldsymbol{B}\mathrm{d}\Omega\right)\tilde{\boldsymbol{u}} = \boldsymbol{K}^{(e)}\tilde{\boldsymbol{u}} \tag{2.20}$$

Here, $\mathbf{K}^{(e)}$ denotes the linear stiffness matrix. However, in many cases it is necessary to use nonlinear or time-dependent stress–strain (constitutive) relationships. In such situations, it is essential to derive solution methods directly from Eqs. (2.17) to (2.19). These methods will be discussed in detail in subsequent chapters with respect to specific constitutive behaviors. At this point, it is only necessary to note

$$\boldsymbol{\sigma} = \boldsymbol{\sigma}(\boldsymbol{\varepsilon}) \tag{2.21}$$

This functional relationship may be nonlinear and, in some cases, not unique. Furthermore, when special constraints are present, such as near-incompressibility, it becomes necessary to employ mixed methods, which will be discussed in Sect. 2.5. Before addressing this approach, we first introduce the finite element formulation for the numerical solution of transient equations.

2.2 Numerical Integration

The most convenient method for calculating the integrals required in finite element formulations is numerical integration. Among the various quadrature formulas, the Gauss–Legendre quadrature, which is based on polynomial interpolation, offers the highest accuracy. Typically, tables for the Gauss–Legendre quadrature are prepared for the coordinate range $-1 \leq \xi \leq 1$ (which is also a primary reason for choosing shape functions over this interval).

The Gauss–Legendre quadrature formula for one-dimensional integration is given by

$$\int_{-1}^{1} f(\xi)\mathrm{d}\xi = \sum_{j=1}^{n} f(\xi_j)w_j + O\left(\frac{\mathrm{d}^{2n}f}{\mathrm{d}\xi_{2n}}\right) \tag{2.22}$$

Here, ξ_j denotes the evaluation points of the function, and w_j represents the weight coefficients. An n-point Gauss–Legendre quadrature formula can exactly integrate polynomials of degree up to $2n - 1$. The positions of the integration points and the corresponding weighting coefficients for the first five orders of quadrature rules are listed in Table 2.2.

Integration over multidimensional domains can be performed using the tensor product of one-dimensional formulas. Therefore, the integration formula in the two-dimensional case is

$$\int_{-1}^{1}\int_{-1}^{1} f(\xi^1, \xi^2)\mathrm{d}\xi^1\mathrm{d}\xi^2 = \sum_{j=1}^{n}\sum_{k=1}^{n} f\left(\xi_j^1, \xi_k^2\right)w_j w_k \tag{2.23}$$

The integration formula in three dimensions is

$$\int_{-1}^{1}\int_{-1}^{1}\int_{-1}^{1} f(\xi^1, \xi^2, \xi^3)\mathrm{d}\xi^1\mathrm{d}\xi^2\mathrm{d}\xi^3 = \sum_{j=1}^{n}\sum_{k=1}^{n}\sum_{l=1}^{n} f\left(\xi_j^1, \xi_k^2, \xi_l^2\right)w_j w_k w_l \tag{2.24}$$

Table 2.2 The first five-order quadrature nodes and weights of Gauss–Legendre quadrature rules ($\int_{-1}^{1} f(\xi)\mathrm{d}\xi = \sum_{j=1}^{n} f(\xi_j)w_j$)

Order	j	ξ_j	w_j	
$n = 1$	1	0	2	
$n = 2$	1	$+1/\sqrt{3}$	1	
	2	$-1/\sqrt{3}$	1	
$n = 3$	1	$+\sqrt{0.6}$	5/9	
	2	0	8/9	
	3	$-\sqrt{0.6}$	5/9	
$n = 4$	1	$+\sqrt{(3+a)/7}$	$0.5 - 1/(3a)$	$a = \sqrt{4.8}$
	2	$+\sqrt{(3-a)/7}$	$0.5 + 1/(3a)$	
	3	$-\sqrt{(3-a)/7}$	$0.5 + 1/(3a)$	
	4	$-\sqrt{(3+a)/7}$	$0.5 - 1/(3a)$	
$n = 5$	1	$+\sqrt{b}$	$(5c - 3)d/b$	$a = \sqrt{1120}$
	2	$+\sqrt{c}$	$(3 - 5b)d/c$	$b = (70 + a)/126$
	3	0	$2 - 2(w_1 + w_2)$	$c = (70 - a)/126$
	4	$-\sqrt{c}$	$(3 - 5b)d/c$	$d = 1/(15(c - b))$
	5	$-\sqrt{b}$	$(5c - 3)d/b$	

When the polynomial in any direction is of degree less than $2n$, the result is exact.

The above form is convenient for quadrilateral and hexahedral elements, but for triangular and tetrahedral elements, it can only be used after transformation.

2.2.1 Volume Integral

The volume integral problem still requires transforming the element domain Ω_e to the integration interval of the Gauss–Legendre quadrature formula, $-1 \leq \xi \leq 1$. For the Jacobian determinant **J** in Eq. (2.8), it is necessary to convert the volume element from Cartesian coordinates to natural coordinates, that is,

$$\mathrm{d}x_1\mathrm{d}x_2\mathrm{d}x_3 = \det \boldsymbol{J}\, \mathrm{d}\xi^1\mathrm{d}\xi^2\mathrm{d}\xi^3 = j(\xi^1, \xi^2, \xi^3)\mathrm{d}\xi^1\mathrm{d}\xi^2\mathrm{d}\xi^3 \tag{2.25}$$

where $\det\mathbf{J} = j$ must be positive to yield the correct volume.

Using the transformation in Eq. (2.25), the integral over a given region can be expressed as:

$$\int_{\Omega_e} f(\boldsymbol{x})\mathrm{d}\Omega = \int_{\square} \hat{f}(\boldsymbol{\xi})j(\boldsymbol{\xi})\mathrm{d}\square \tag{2.26}$$

where $\hat{f}$ is the representation of function f in the coordinate ξ, $\square$ denotes the parametric coordinate range of the considered problem's dimensionality, and $j(\xi)$ is the Jacobian transformation corresponding to the considered coordinate system. For various problems, the specific form is given by the following expression:

$$\begin{aligned}
&\textbf{ProblemTypes}\ \square \text{ — } \textbf{Area}\ j \text{ — } \textbf{Jacobiandeterminant}\\
&\text{Three - dimensional : } \mathrm{d}\xi^1\mathrm{d}\xi^2\mathrm{d}\xi^3\, j(\xi^1, \xi^2, \xi^3)\\
&\text{Plane strain: } \mathrm{d}\xi^1\mathrm{d}\xi^2\, j(\xi^1, \xi^2)\\
&\text{Plane stress: } \mathrm{d}\xi^1\mathrm{d}\xi^2\, h_3 j(\xi^1, \xi^2)\\
&\text{Axisymmetric: } \mathrm{d}\xi^1\mathrm{d}\xi^2 2\pi x_1 j(\xi^1, \xi^2)
\end{aligned}$$

For two-dimensional problems

$$j(\xi^1, \xi^2) = \det\begin{bmatrix} \frac{\partial x_1}{\partial \xi^1} & \frac{\partial x_2}{\partial \xi^1} \\ \frac{\partial x_1}{\partial \xi^2} & \frac{\partial x_2}{\partial \xi^2} \end{bmatrix} > 0 \tag{2.27}$$

Under the following criteria, the minimum number of integration points may be selected:

- When the Jacobian determinant j is constant, the integral over the element is exact.
- The resulting element stiffness matrix is of full rank.

In all cases, the patch test must be satisfied. The first criterion applies only to linear materials under small strain conditions (as discussed in this chapter). The second criterion is applicable to both linear and nonlinear problems. Numerical integration of one order lower than that required by the above criteria is referred to as reduced integration, which should generally be avoided. When some terms are integrated using "full" integration and others using "reduced" integration, this approach is called selective reduced integration; however, for nonlinear problems, this method is generally not recommended.

In the natural coordinate ξ as described above, each finite element is assumed to be a line, a quadrilateral, or a hexahedron. For other shapes, such as triangles or tetrahedra, appropriate transformations of the natural coordinates and integration formulas are required.

2.2.2 Surface Integrals

It is also necessary to perform integration over the surface of the computational element, which can be achieved by assuming dΓ as a vector oriented in the direction of the outward normal to the surface. For three-dimensional problems, this can be obtained using the vector cross product.

$$\begin{aligned} \boldsymbol{n}\mathrm{d}\Gamma = \mathrm{d}\boldsymbol{\Gamma} &= \frac{\partial \boldsymbol{x}}{\partial \xi^1} \times \frac{\partial \boldsymbol{x}}{\partial \xi^2}\mathrm{d}\xi^1\mathrm{d}\xi^2 \\ &= (\boldsymbol{v}_1 \times \boldsymbol{v}_2)\mathrm{d}\xi^1\mathrm{d}\xi^2 = \boldsymbol{v}_n\mathrm{d}\xi^1\mathrm{d}\xi^2 \end{aligned} \tag{2.28}$$

Here, ξ^1 and ξ^2 are the parametric coordinates of the surface element, and $\times$ denotes the vector cross product.

Thus, the computational formula for surface differentiation is as follows:

$$\mathrm{d}\Gamma = \left(\boldsymbol{v}_n^{\mathrm{T}}\boldsymbol{v}_n\right)^{1/2}\mathrm{d}\xi^1\mathrm{d}\xi^2 \tag{2.29}$$

The surface in a two-dimensional problem can be described solely by ξ^1, and $\partial x/\xi^2$ can be replaced by $\mathbf{e}^3$ (the unit vector perpendicular to the deformation plane).

2.3 Nonlinear Transient and Steady-State Problems

To obtain the discrete algebraic equations for transient problems, a temporal discretization method is first introduced, and the approximate solution can be expressed as

$$\tilde{\boldsymbol{u}}(t_{n+1}) \approx \boldsymbol{u}_{n+1}, \quad \dot{\tilde{\boldsymbol{u}}}(t_{n+1}) \approx \boldsymbol{v}_{n+1}, \quad \ddot{\tilde{\boldsymbol{u}}}(t_{n+1}) \approx \boldsymbol{a}_{n+1}$$

For simplicity, the tilde over discrete variables has been omitted. Therefore, at each discrete time point t_{n+1}, the equilibrium Eq. (2.17) can be expressed in the following residual form

$$\boldsymbol{\Psi}_{n+1} = \boldsymbol{f}_{n+1} - \boldsymbol{M}\boldsymbol{a}_{n+1} - \boldsymbol{P}_{n+1} \tag{2.30}$$

where

$$\boldsymbol{P}_{n+1} = \int_{\Omega} \boldsymbol{B}^{\mathrm{T}} \boldsymbol{\sigma}_{n+1} \mathrm{d}\Omega = \boldsymbol{P}(\boldsymbol{u}_{n+1}) \tag{2.31}$$

As previously noted, $\boldsymbol{P}$ can be expressed solely in terms of displacement. This holds true for elastic materials; however, for inelastic behavior, the general form of the material model depends on the variables to be determined. Nevertheless, most constitutive models can be formulated in terms of the incremental form of $\boldsymbol{u}$. Therefore, only minor modifications to the assumed form above are necessary, which will not significantly affect the following discussion.

For transient problems, the Newmark method can be used to solve equations involving second-order time derivatives. The Newmark method relates the discrete displacement, velocity, and acceleration at time t_{n+1} to their corresponding discrete values at t_n, as follows:

$$\begin{aligned} \boldsymbol{u}_{n+1} &= \boldsymbol{u}_n + \Delta t \boldsymbol{v}_n + \left(\tfrac{1}{2} - \beta\right)\Delta t^2 \boldsymbol{a}_n + \beta \Delta t^2 \boldsymbol{a}_{n+1} = \breve{\boldsymbol{u}}_{n+1} + \beta \Delta t^2 \boldsymbol{a}_{n+1} \\ \boldsymbol{v}_{n+1} &= \boldsymbol{v}_n + (1-\gamma)\Delta t \boldsymbol{a}_n + \gamma \Delta t \boldsymbol{a}_{n+1} = \breve{\boldsymbol{v}}_{n+1} + \gamma \Delta t \boldsymbol{a}_{n+1} \end{aligned} \tag{2.32}$$

where $\Delta t = t_{n+1} - t_n$ is the time increment, and $\breve{\boldsymbol{u}}_{n+1}$, $\breve{\boldsymbol{v}}_{n+1}$ depend only on the known values at t_n. This single-step method is highly desirable because it allows Δt to change from one step to the next without introducing any complexity (though excessively large time steps should always be avoided). The parameters γ and β are used to control accuracy and stability.

Thus, by solving the nonlinear system of Eq. (2.30) and the linear system with scalar coefficients (2.32), the results of the transient problem at each time step t_{n+1} can be obtained. Any one of the three variables at time t_{n+1} (i.e., $\mathbf{u}_{n+1}$, $\mathbf{v}_{n+1}$ or $\mathbf{a}_{n+1}$) can be chosen as the primary unknown. Then, by applying Eq. (2.32), a nonlinear equation in terms of this single unknown can be derived.

2.3.1 Explicit Newmark Method

A very convenient option is to set $\beta = 0$ and choose $\mathbf{a}_{n+1}$ as the primary unknown. Then, from Eq. (2.32), we obtain

$$\boldsymbol{u}_{n+1} = \breve{\boldsymbol{u}}_{n+1}$$

Therefore, according to Eq. (2.30), we have

$$\boldsymbol{M}\boldsymbol{a}_{n+1} = \boldsymbol{f}_{n+1} - \boldsymbol{P}\left(\breve{\boldsymbol{u}}\right)$$

The value of $\mathbf{a}_{n+1}$ can be directly solved, and then the velocity $\mathbf{v}_{n+1}$ can be obtained using Eq. (2.32). This is known as the explicit method, as it only requires solving linear equations.

If the matrix $\mathbf{M}$ is a diagonal matrix (or a lumped mass matrix), then solving for $\mathbf{a}_{n+1}$ is trivial and the problem can be considered solved, because

$$\boldsymbol{M}^{-1} = \begin{bmatrix} 1/M_{11} & & \\ & \ddots & \\ & & 1/M_{mm} \end{bmatrix}$$

Here, m is the total number of equations in the problem. However, explicit methods are only conditionally stable under the condition $\Delta t \leq \Delta t_{crit}$, where Δt_{crit} is related to the minimum time required for "wave propagation" to traverse any element, or to the highest "frequency" present in the finite element mesh. As a result, the explicit method may require thousands of time steps to complete a given time interval. In practice, for many transient problems and most static (steady-state) problems, it is generally more efficient to use implicit methods, which allow for larger time steps.

2.3.2 Implicit Newmark Method

In implicit methods, $\mathbf{u}_{n+1}$ is generally used as the primary variable, and Eq. (2.32) is applied to compute $\mathbf{v}_{n+1}$ and $\mathbf{a}_{n+1}$. Only consider the quasi-static case (where inertial forces can be neglected), set $\mathbf{v}_{n+1} = \mathbf{a}_{n+1} = 0$. The system of Eq. (2.30) can then be written as

$$\boldsymbol{\Psi}(\boldsymbol{u}_{n+1}) \equiv \boldsymbol{f}_{n+1} - \frac{c}{\beta \Delta t^2}\boldsymbol{M}\left[\boldsymbol{u}_{n+1} - \breve{\boldsymbol{u}}_{n+1}\right] - \boldsymbol{P}_{n+1} = \mathbf{0} \tag{2.33}$$

where $c = 1$ is used for transient problems, and $c = 0$ is used for quasi-static problems. When any term is nonlinear, the system of Eq. (2.33) requires an iterative solution. Newton's method serves as the foundation for most practical nonlinear solution techniques. In this method, a single iteration step is expressed as

$$\boldsymbol{\Psi}_{n+1}^{k+1} \approx \boldsymbol{\Psi}_{n+1}^{k} + \mathrm{d}\boldsymbol{\Psi}_{n+1}^{k} = 0 \tag{2.34}$$

Among them, for $\mathbf{f}_{n+1}$, which is independent of deformation, the increment in Eq. (2.33) can be expressed as

$$\mathrm{d}\boldsymbol{\Psi}(\boldsymbol{u}_{n+1}) \equiv -\left[\frac{c}{\beta\Delta t^2}\boldsymbol{M} + \left.\frac{\partial \boldsymbol{P}_{n+1}}{\partial \boldsymbol{u}_{n+1}}\right|_{n+1}^{k}\right]\mathrm{d}\boldsymbol{u}_{n+1}^{k} = -\boldsymbol{A}_{n+1}^{k}\mathrm{d}\boldsymbol{u}_{n+1}^{k} \tag{2.35}$$

The displacement increment is calculated as follows:

$$\boldsymbol{A}_{n+1}^{k}\mathrm{d}\boldsymbol{u}_{n+1}^{k} = \boldsymbol{\Psi}_{n+1}^{k} \tag{2.36}$$

and is updated according to the following formula

$$\begin{aligned} \boldsymbol{u}_{n+1}^{k+1} &= \boldsymbol{u}_{n+1}^{k} + \mathrm{d}\boldsymbol{u}_{n+1}^{k} \\ \boldsymbol{a}_{n+1}^{k+1} &= \frac{1}{\beta\Delta t^2}\left[\boldsymbol{u}_{n+1}^{k+1} - \breve{\boldsymbol{u}}_{n+1}\right] \\ \boldsymbol{v}_{n+1}^{k+1} &= \breve{\boldsymbol{v}}_{n+1} + \gamma\Delta t\boldsymbol{a}_{n+1}^{k+1} \end{aligned} \tag{2.37}$$

The initial iteration can be set to zero, or more appropriately, to the converged solution from the previous time step. Therefore,

$$\boldsymbol{u}_{n+1}^{1} = \boldsymbol{u}_{n} \tag{2.38}$$

Here, quantities without superscripts denote converged values. For transient problems, the initial velocity and acceleration are given by the following expressions:

$$\begin{aligned} \boldsymbol{a}_{n+1}^{1} &= \frac{1}{\beta\Delta t^2}\left[\boldsymbol{u}_{n} - \breve{\boldsymbol{u}}_{n+1}\right] \\ \boldsymbol{v}_{n+1}^{1} &= \breve{\boldsymbol{v}}_{n+1} + \gamma\Delta t\boldsymbol{a}_{n+1}^{1} \end{aligned} \tag{2.39}$$

Iterate continuously until the following convergence criterion is satisfied.

$$\left\|\boldsymbol{\Psi}_{n+1}^{k}\right\| \le \varepsilon\left\|\boldsymbol{\Psi}_{n+1}^{1}\right\| \tag{2.40}$$

or other similar criteria based on error tolerance. When all terms of Newton's method can be computed accurately, a good practice is to assume the error tolerance to be half of computer precision. Therefore, if the computer can accurately represent 16 digits, one may choose $\varepsilon = 10^{-8}$. A common choice for the parameters is $\gamma = 2$, $\beta = 1/2$, which is also referred to as the "trapezoidal integration rule"; sometimes, $\gamma = 2$, $\beta = 1/2$ is chosen.

To compute each element in Eq. (2.19), it is necessary to differentiate $\mathbf{P}$ in Eq. (2.35) with respect to it, i.e.

$$\left.\frac{\partial \boldsymbol{P}^{(e)}}{\partial \boldsymbol{u}_{n+1}}\right|_{n+1}^{k} = \int_{\Omega_e} \boldsymbol{B}^{\mathrm{T}} \boldsymbol{D}_T^k \boldsymbol{B} \mathrm{d}\Omega \equiv \boldsymbol{K}_T^k \tag{2.41}$$

Note: The above relationships are similar for both linear and nonlinear problems, but not exactly the same. In Eq. (2.41), $\mathbf{D}_T^k$ represents the tangent modulus matrix of the stress–strain relationship (which may or may not be unique, but is generally related to the nonlinear deformation mode); $\mathbf{K}_T$ denotes the tangent stiffness matrix.

This book will employ certain forms of nonlinear elasticity. Here, we present a simple approach: define the strain energy density W as a function of $\boldsymbol{\varepsilon}$, that is,

$$W = W(\boldsymbol{\varepsilon}) = W\left(\varepsilon_{ij}\right)$$

Therefore, we obtain

$$\boldsymbol{\sigma} = \frac{\partial W}{\partial \boldsymbol{\varepsilon}} \tag{2.42}$$

If the properties of the function W are known, the tangent modulus $\mathbf{D}$ can be expressed as

$$\boldsymbol{D}_T^k = \left.\frac{\partial \boldsymbol{\sigma}}{\partial \boldsymbol{\varepsilon}}\right|_{n+1}^{k} = \left.\frac{\partial^2 W}{\partial \boldsymbol{\varepsilon} \partial \boldsymbol{\varepsilon}}\right|_{n+1}^{k}$$

If the calculation of $\boldsymbol{\sigma}_{n+1}$ involves path dependency, it is necessary to track the total incremental change in the solution steps from t_n to t_{n+1}, that is,

$$\boldsymbol{u}_{n+1}^{k+1} = \boldsymbol{u}_n + \Delta \boldsymbol{u}_{n+1}^{k+1},\ \Delta \boldsymbol{u}_{n+1}^{1} = \boldsymbol{0} \tag{2.43}$$

The total increment can be obtained by summing each individual increment, that is,

$$\Delta \boldsymbol{u}_{n+1}^{k+1} = \boldsymbol{u}_{n+1}^{k+1} - \boldsymbol{u}_n = \Delta \boldsymbol{u}_{n+1}^{k} + \mathrm{d}\boldsymbol{u}_{n+1}^{k} \tag{2.44}$$

In implicit methods, it is preferable to use the displacement from the most recent iteration to compute $\mathbf{A}$ and ψ, especially when considering inelastic material behavior or large deformations. However, in some cases, using $\mathbf{A}$ from the previous iteration can help reduce computational cost.

2.3.3 Generalized Midpoint Implicit Newmark Method

The generalized midpoint implicit Newmark method enforces the momentum balance equation at the midpoint between t_n and t_{n+1}. In this method, variables are interpolated as follows:

$$\begin{aligned} \boldsymbol{u}_{n+\alpha} &= (1-\alpha)\boldsymbol{u}_n + \alpha\boldsymbol{u}_{n+1} \\ \boldsymbol{v}_{n+\alpha} &= (1-\alpha)\boldsymbol{v}_n + \alpha\boldsymbol{v}_{n+1} \\ \boldsymbol{a}_{n+\alpha} &= (1-\alpha)\boldsymbol{a}_n + \alpha\boldsymbol{a}_{n+1} \end{aligned} \tag{2.45}$$

and express the momentum balance equation as

$$\boldsymbol{\Psi}_{n+\alpha} = \boldsymbol{f}_{n+\alpha} - \boldsymbol{M}\boldsymbol{a}_{n+\alpha} - \boldsymbol{P}_{n+\alpha} = \boldsymbol{0} \tag{2.46}$$

In the above equation, there are two methods that can be used to calculate $\mathbf{P}_{n+\alpha}$. Method 1 is

$$\boldsymbol{P}_{n+\alpha} = \boldsymbol{P}(\boldsymbol{u}_{n+\alpha}) = \int_\Omega \boldsymbol{B}^{\mathrm{T}}\sigma(\boldsymbol{u}_{n+\alpha})\mathrm{d}\Omega \tag{2.47}$$

The second method interpolates $\mathbf{P}$ as

$$\boldsymbol{P}_{n+\alpha} = (1-\alpha)\boldsymbol{P}_n + \alpha\boldsymbol{P}_{n+1} = \int_\Omega \boldsymbol{B}^{\mathrm{T}}\left[(1-\alpha)\sigma_n + \alpha\sigma_{n+1}\right]\mathrm{d}\Omega \tag{2.48}$$

where $\mathbf{P}_n = \mathbf{P}(n)$.

Choosing the Newmark parameters as $\gamma = \beta$, the following simplified form can be obtained:

$$\begin{aligned} \boldsymbol{u}_{n+1} &= \boldsymbol{u}_n + \Delta t(\boldsymbol{v}_n + \boldsymbol{v}_{n+1}) \\ \frac{1}{\Delta t}(\boldsymbol{v}_{n+1} - \boldsymbol{v}_n) &= (1-\gamma)\boldsymbol{a}_n + \gamma\boldsymbol{a}_{n+1} \end{aligned} \tag{2.49}$$

Let $\alpha = \gamma$, then the momentum balance equation becomes

$$\boldsymbol{\Psi}_{n+\alpha} = \boldsymbol{f}_{n+\alpha} - \frac{1}{\Delta t}\boldsymbol{M}(\boldsymbol{v}_{n+1} - \boldsymbol{v}_n) - \boldsymbol{P}_{n+\alpha} = \boldsymbol{0} \tag{2.50}$$

This is part of the energy–momentum conservation method developed by Simo et al. For linear elasticity problems, it is easy to prove that by choosing $\alpha = \gamma = 1/2$, energy conservation can be achieved in the case of free motion (i.e., $\mathbf{f} = 0$). However, for nonlinear elasticity problems, it is necessary to modify the calculation method of $\boldsymbol{\sigma}_{n+1/2}$ for these values in order to preserve its energy conservation property.

The solution to the momentum balance equation in the above form can still use $\mathbf{u}_{n+1}$ as the primary variable. However, due to the presence of the parameter α, the linearization of $\mathbf{P}_{n+\alpha}$ requires corresponding adjustments.

2.4 Solving Systems of Nonlinear Algebraic Equations

When solving linear problems using the finite element method, it is always necessary to solve a set of simultaneous algebraic equations of the form

$$\boldsymbol{K}\boldsymbol{u} = \boldsymbol{f} \tag{2.51}$$

If the coefficient matrix is nonsingular, then the solution to these equations is unique. When solving nonlinear problems, a similar set of algebraic equations are obtained, but they are typically nonlinear. For example, in the earlier discussion of this chapter, at each discrete time point t_{n+1}, a system of equations as shown in Eq. (2.33) is obtained. Here, we consider a general problem, which can be expressed as

$$\boldsymbol{\Psi}_{n+1} = \boldsymbol{\Psi}(\boldsymbol{u}_{n+1}) = \boldsymbol{f}_{n+1} - \boldsymbol{P}(\boldsymbol{u}_{n+1}) \tag{2.52}$$

Here, $\mathbf{u}_{n+1}$ denotes the set of discretized variables, $\mathbf{f}_{n+1}$ is a vector independent of the unknowns, and $\mathbf{P}$ is a vector that depends on the unknowns. These systems of equations may have multiple solutions (i.e., there may be more than one set of $\mathbf{u}_{n+1}$ that satisfies Eq. (2.52)). Therefore, obtaining one solution does not necessarily guarantee it is the desired one. A thorough understanding of the physical nature of the problem is essential, and it is generally crucial to employ an incremental approach starting from a known solution in order to obtain meaningful results. If the problem is transient, if the constitutive relationship between stress and strain is path-dependent, or if the load–displacement path exhibits bifurcation or multiple branches at certain load levels, the incremental method must be used.

In general, the solution of nonlinear problems should always begin near a known solution. For example, suppose

$$\boldsymbol{u} = \boldsymbol{u}_{n+1}, \quad \boldsymbol{\Psi}_n = \boldsymbol{0}, \quad \boldsymbol{f} = \boldsymbol{f}_{n+1} \tag{2.53}$$

Next, the calculation begins with the variation of the load $\mathbf{f}_n$ as follows

$$\boldsymbol{f}_{n+1} = \boldsymbol{f}_n + \Delta\boldsymbol{f}_{n+1} \tag{2.54}$$

and thereby determine the increment $\Delta\mathbf{u}_{n+1}$, obtaining

$$\boldsymbol{u}_{n+1} = \boldsymbol{u}_n + \Delta\boldsymbol{u}_{n+1} \tag{2.55}$$

This is the objective of the solution. Typically, the increment $\Delta\mathbf{f}_{n+1}$ is kept small to ensure path dependency can be maintained. Moreover, such an incremental approach helps to avoid excessive iterations and to follow the physically correct path. Figure 2.2 illustrates a typical non-unique problem (the possibility of multiple solutions). If the function ψ_{n+1} first decreases and then increases as the parameter $\mathbf{u}_{n+1}$ increases

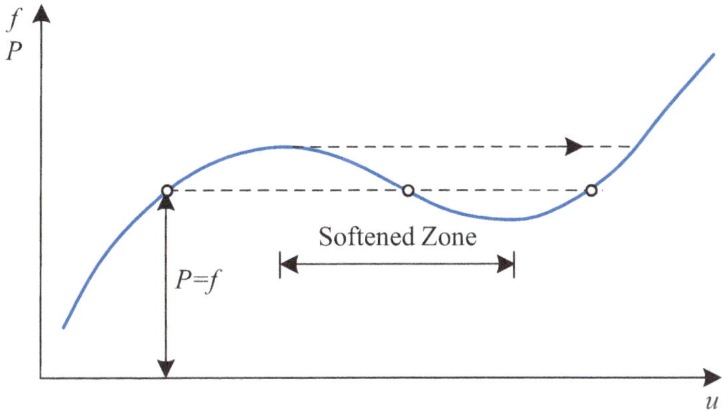

Fig. 2.2 Possibility of Multiple Solutions

uniformly, such non-uniqueness may arise. Clearly, during the complete computation process, following the path of $\Delta \mathbf{f}n + 1$ will involve both positive and negative values.

Only in cases of mild nonlinearity (and without path dependence) is it possible to obtain a solution in a single increment, that is,

$$\boldsymbol{f}_n = \mathbf{0}, \quad \Delta \boldsymbol{f}_{n+1} = \boldsymbol{f}_{n+1} \tag{2.56}$$

There is a vast body of literature on general solution methods and specific applications, making it impossible to cover all approaches comprehensively in this chapter. The next section provides an overview of the general solution procedure to present a complete outline. In the final two chapters of this book, the focus will be primarily on rate-independent material nonlinearity (plasticity) and geometric nonlinearity (large displacements).

2.4.1 Overview

The solutions to problems given by Eqs. (2.52) to (2.55) cannot be obtained directly and always require some form of iterative algorithm. Here, we will focus on algorithms that repeatedly solve linear systems of the following form (i.e., iterative methods).

$$\boldsymbol{K}^i \mathrm{d}\boldsymbol{u}_{n+1}^i = \boldsymbol{r}_{n+1}^i \tag{2.57}$$

Here, $\boldsymbol{r}$ is related to $\boldsymbol{\Psi}$, and the superscript i denotes the iteration number. In this process, it is necessary to compute the solution increment $\mathrm{d}\boldsymbol{u}^i{}_{n+1}$. Direct (Gaussian) elimination techniques or iterative methods can be used to solve the linear equations associated with each nonlinear iteration. Iterative solution methods have been shown to be the most efficient, but their full potential has yet not been realized.

Many of the iterative techniques currently used to solve nonlinear problems are derived from intuitive physical reasoning. However, these techniques are often directly related to methods in numerical analysis, and the commonly adopted nomenclature for these problems will be used below.

Although each algorithm is designed for a system of nonlinear algebraic equations, this book will use a single scalar equation to illustrate the process. While this approach is pedagogically useful, it has its limitations, as the convergence behavior of multi-degree-of-freedom problems may not be as straightforward as that of a single equation.

2.4.2 Newton's Method

The Newton method is the fastest converging approach for solving nonlinear problems, requiring only a single evaluation of $\boldsymbol{\Psi}$ per iteration. Of course, this assumes that the initial guess lies within the region of convergence, thereby avoiding divergence. The Newton method is the only algorithm with asymptotic quadratic convergence and is commonly referred to as the Newton–Raphson method.

In this iterative method, for first-order problems, Eq. (2.52) can be approximated as follows:

$$\boldsymbol{\Psi}\left(\boldsymbol{u}_{n+1}^{i+1}\right) \approx \boldsymbol{\Psi}\left(\boldsymbol{u}_{n+1}^{i}\right) + \left(\frac{\partial \boldsymbol{\Psi}}{\partial \boldsymbol{u}}\right)_{n+1}^{i} d\boldsymbol{u}_{n+1}^{i} = \mathbf{0} \tag{2.58}$$

In Eq. (2.58), the iteration count i typically starts from the following expression, assuming

$$\boldsymbol{u}_{n+1}^{1} = \boldsymbol{u}_{n} \tag{2.59}$$

Here, $\mathbf{u}_n$ refers to the converged solution at the previous load level or time step. The Jacobian matrix in the tangent direction (also known as the stiffness matrix in structural mechanics terminology) can be expressed as

$$\boldsymbol{K}_{\mathrm{T}} = \frac{\partial \boldsymbol{P}}{\partial \boldsymbol{u}} = -\frac{\partial \boldsymbol{\Psi}}{\partial \boldsymbol{u}} \tag{2.60}$$

Thus, Eq. (2.58) can be iteratively corrected as follows:

$$\begin{aligned} \boldsymbol{K}_{\mathrm{T}}^{i} d\boldsymbol{u}_{n+1}^{i} &= \boldsymbol{\Psi}_{n+1}^{i} \\ d\boldsymbol{u}_{n+1}^{i} &= \left(\boldsymbol{K}_{\mathrm{T}}^{i}\right)^{-1} \boldsymbol{\Psi}_{n+1}^{i} \end{aligned} \tag{2.61}$$

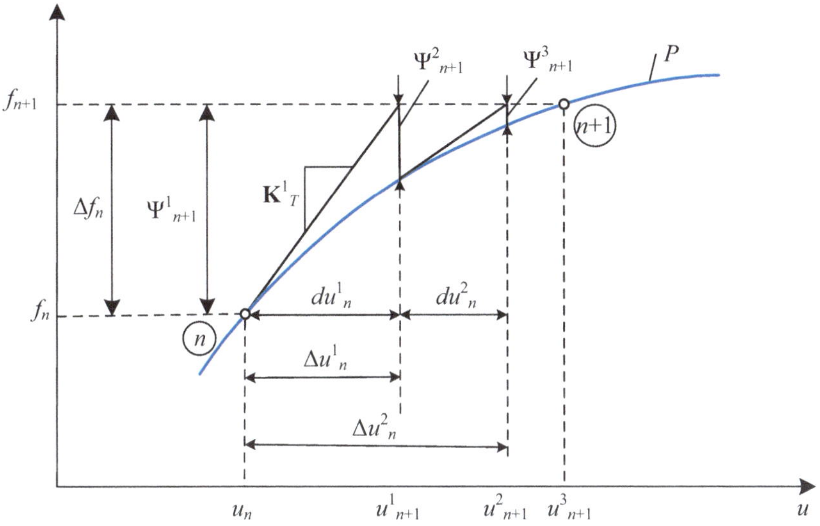

Fig. 2.3 Convergence process of the Newton method

A sequence of successive approximations yields

$$\boldsymbol{u}_{n+1}^{i+1} = \boldsymbol{u}_{n+1}^{i} + d\boldsymbol{u}_{n+1}^{i} = \boldsymbol{u}_n + \Delta\boldsymbol{u}_{n+1}^{i} \tag{2.62}$$

where

$$\Delta\boldsymbol{u}_{n+1}^{i} = \sum_{k=1}^{i} d\boldsymbol{u}_{n+1}^{k} \tag{2.63}$$

This process is illustrated in Fig. 2.3, which demonstrates that the method converges very rapidly.

The necessity of introducing the total increment $\Delta\mathbf{u}_{n+1}^{i}$ may not be immediately apparent here, but it becomes essential if the solution process is path-dependent, as is the case with certain nonlinear solid constitutive equations discussed in Chap. 7.

Although the Newton method converges rapidly, it also has some drawbacks:

- A new $\boldsymbol{K}_{\mathrm{T}}$ matrix must be computed in each iteration.
- If the direct solution method of Eq. (2.61) is used, the $\boldsymbol{K}_{\mathrm{T}}$ matrix must be decomposed in each iteration.
- In certain cases, the tangent stiffness matrix is symmetric in the solution state, but in other situations it is asymmetric (for example, in some algorithms used for large rotations or non-associated plasticity). In such cases, a non-symmetric solver is typically required.

Some alternative algorithms avoid certain drawbacks, but they often sacrifice the quadratic rate of asymptotic convergence.

2.4.3 Modified Newton's Method

This method employs essentially the same algorithm as the Newton method, but replaces the varying Jacobian matrix $\mathbf{K}_{\mathrm{T}}^{i}$ with a constant matrix approximation, that is,

$$\boldsymbol{K}_{\mathrm{T}}^{i} = \overline{\boldsymbol{K}}_{\mathrm{T}} \tag{2.64}$$

Then Eq. (2.61) can be written as

$$d\boldsymbol{u}_{n+1}^{i} = \overline{\boldsymbol{K}}_{\mathrm{T}}^{-1} \boldsymbol{\Psi}_{n+1}^{i} \tag{2.65}$$

There are many possible choices here. For example, $\overline{\mathbf{K}}_{\mathrm{T}}$ can be chosen as the $\mathbf{K}_{\mathrm{T}}^{1}$ matrix corresponding to the first iteration (as shown in Fig. 2.4a), or even as the matrix corresponding to a previous time step or load increment $\mathbf{K}^{0}$ (as shown in Fig. 2.4b). In the context of solving solid mechanics problems, this method is also known as the stress transfer method or the initial stress method. Alternatively, the approximation can be updated every few iterations, i.e., taking $\overline{\mathbf{K}}_{\mathrm{T}} = \mathbf{K}_{\mathrm{T}}^{j}$, where $j \leq i$.

Obviously, this process typically converges at a slower rate (the norm of the residual $\boldsymbol{\Psi}$ generally exhibits linear asymptotic convergence, rather than the quadratic convergence of the original Newton's method). However, it overcomes some of the drawbacks of Newton's method discussed in Sect. 2.4.2. Nevertheless, when the tangent used has a "slope" opposite to that of the current solution (as shown in regions with different slopes in Fig. 2.2), this method may also introduce some new disadvantages. The "convergence region" of the modified algorithm is generally expanded, and it can enable convergence in cases where the original algorithm would diverge. The convergence rate is quite slow. This algorithm has many variants, and if $\overline{\mathbf{K}}_{\mathrm{T}}$ is symmetric, a symmetric solver can be used.

2.4.4 Secant Method or Quasi-Newton Method

Once the first iteration described in Sect. 2.4.3 has been established,

$$d\boldsymbol{u}_{n+1}^{1} = \overline{\boldsymbol{K}}_{\mathrm{T}}^{-1} \boldsymbol{\Psi}_{n+1}^{1} \tag{2.66}$$

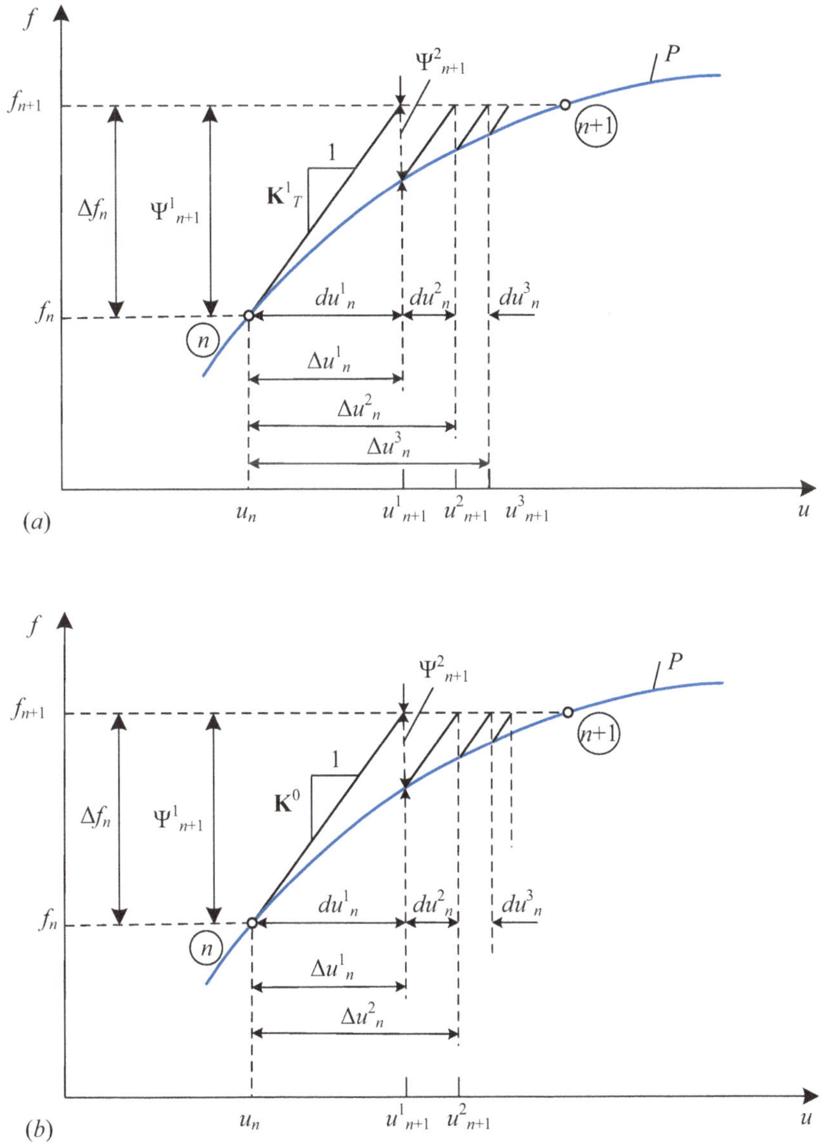

Fig. 2.4 Convergence process of the modified Newton's method

it is possible to find a secant "slope" as shown in Fig. 2.5, such that

$$d\boldsymbol{u}^1_{n+1} = \left(\overline{\boldsymbol{K}}^2_{\mathrm{T}}\right)^{-1}\left(\boldsymbol{\Psi}^1_{n+1} - \boldsymbol{\Psi}^2_{n+1}\right) \tag{2.67}$$

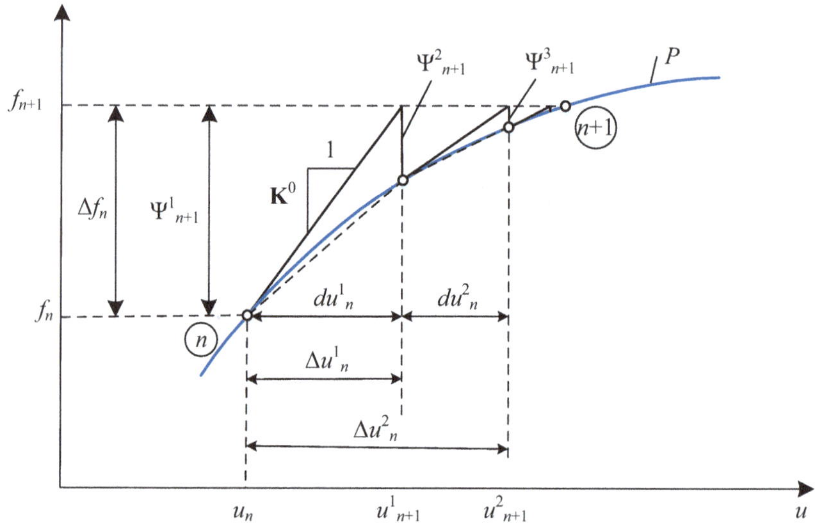

Fig. 2.5 Secant method starting from K_0

According to Eq. (2.67), the given "slope" can be used to determine **u**, that is,

$$d\boldsymbol{u}_{n+1}^{2} = \left(\overline{\boldsymbol{K}}_{s}^{2}\right)^{-1} \boldsymbol{\Psi}_{n+1}^{2} \tag{2.68}$$

More generally, when $i > 1$, one can use

$$d\boldsymbol{u}^{i} = \left(\overline{\boldsymbol{K}}_{s}^{i}\right)^{-1} \boldsymbol{\Psi}^{i} \tag{2.69}$$

to substitute Eq. (2.68). The subscript in Eq. (2.69) is omitted, where $(\boldsymbol{K}^{i}s)^{-1}$ is fixed, and

$$d\boldsymbol{u}^{i-1} = \left(\overline{\boldsymbol{K}}_{s}^{i}\right)^{-1} \left(\boldsymbol{\Psi}^{i-1} - \boldsymbol{\Psi}^{i}\right) = \left(\overline{\boldsymbol{K}}_{s}^{i}\right)^{-1} \boldsymbol{\gamma}^{i-1} \tag{2.70}$$

For the scalar system shown in Fig. 2.5, $\mathbf{K}^{i}$s can be easily determined, and its convergence rate is much faster than that of the modified Newton method (the residual norm typically achieves superlinear asymptotic convergence).

For multi-degree-of-freedom systems, determining $\mathbf{K}^{i}$s or its inverse matrix is more challenging and not unique. Many different forms of the matrix $\mathbf{K}^{i}$s can satisfy Eq. (2.70), so in practice, there are numerous alternative approaches. All of these methods utilize some form of a previously determined matrix or its inverse, updating it to satisfy Eq. (2.70). Some of these update methods preserve the symmetry of the matrix, while others do not. Any method that starts with a symmetric tangent can avoid the drawbacks of non-symmetric matrices during the Newton method's convergence process, and the convergence rate can be faster than that of the modified Newton method.

The BFGS update algorithm (named after Broyden, Fletcher, Goldfarb, and Shanno) and the DFP update algorithm (named after Davidon, Fletcher, and Powell) both preserve the symmetry and positive definiteness of the matrix, and are widely used. The inverse matrix in the BFGS update algorithm can be expressed as

$$\left(\boldsymbol{K}^i\right)^{-1} = \left(\boldsymbol{I} + \boldsymbol{w}_i\boldsymbol{v}_i^{\mathrm{T}}\right)\left(\boldsymbol{K}^{i-1}\right)^{-1}\left(\boldsymbol{I} + \boldsymbol{v}_i\boldsymbol{w}_i^{\mathrm{T}}\right) \tag{2.71}$$

Here, **I** is an identity matrix, and

$$\begin{aligned} \boldsymbol{v}_i &= \left[1 - \frac{\left(d\boldsymbol{u}^{i-1}\right)^{\mathrm{T}}\boldsymbol{\gamma}^{i-1}}{\left(d\boldsymbol{u}^i\right)^{\mathrm{T}}\boldsymbol{\Psi}^{i-1}}\right]\boldsymbol{\Psi}^{i-1} - \boldsymbol{\Psi}^i \\ \boldsymbol{w}_i &= \frac{1}{\left(d\boldsymbol{u}^{i-1}\right)^{\mathrm{T}}\boldsymbol{\gamma}^{i-1}}d\boldsymbol{u}^{i-1} \end{aligned} \tag{2.72}$$

where γ is defined by Eq. (2.70). It is easy to verify that substituting Eqs. (2.71) and (2.72) into Eq. (2.70) yields an identity. Furthermore, the form of Eq. (2.71) ensures the symmetry of the original matrix.

The update algorithm does not preserve any sparsity of the original matrix. Therefore, it is convenient to return to the original (sparse) matrix $\mathbf{K}^1$s at each iteration (as used in the first iteration) and reapply the multiplication in Eq. (2.71). The resulting algorithm is as follows:

$$\begin{aligned} \boldsymbol{b}_1 &= \prod_{j=2}^{i}\left(\boldsymbol{I} + \boldsymbol{v}_i\boldsymbol{w}_i^{\mathrm{T}}\right)\boldsymbol{\Psi}^i \\ \boldsymbol{b}_2 &= \left(\boldsymbol{K}_{\mathrm{S}}^1\right)^{-1}\boldsymbol{b}_1 \\ d\boldsymbol{u}^i &= \prod_{j=0}^{i-2}\left(\boldsymbol{I} + \boldsymbol{w}_{i-j}\boldsymbol{v}_{i-j}^{\mathrm{T}}\right)\boldsymbol{b}_2 \end{aligned} \tag{2.73}$$

This requires storing all previously iterated vectors $\mathbf{v}_j$ and $\mathbf{w}_j$ needed for subsequent multiplications.

When the number of iterations is large ($i > 15$), the efficiency of updates decreases due to initial instability. There are many ways to address this issue, among which the most effective one is to recompute and decompose the tangent matrix at the current step and restart the algorithm.

Another possible approach is to disregard all previous updates and revert to the original matrix $\mathbf{K}^1$s. This algorithm, proposed by Crisfield, is illustrated in Fig. 2.6.

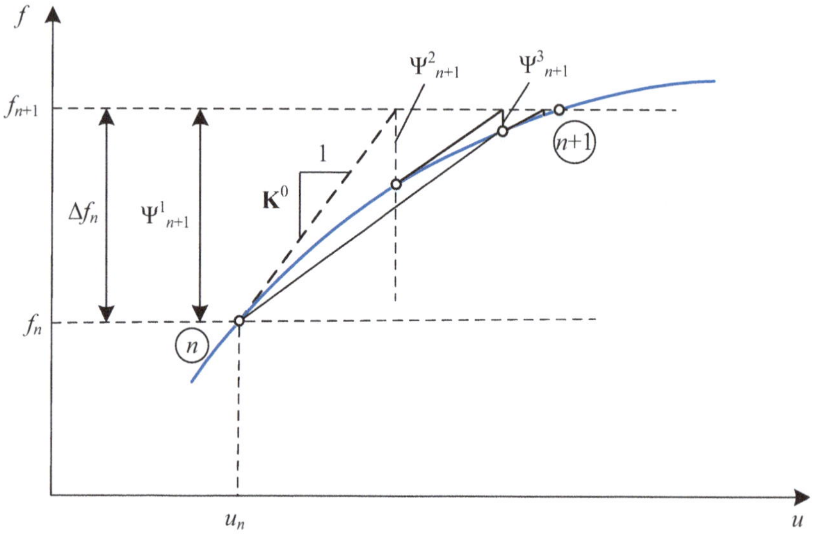

Fig. 2.6 Direct (or Picard) Iteration

As shown, it converges at a slightly slower rate but completely avoids the stability issues encountered earlier, while also reducing the required storage space and number of iterations. Clearly, this method can be applied with any secant update technique.

The algorithm in Fig. 2.6 is identical to the direct (or Picard) iteration algorithm and is particularly effective for solving nonlinear problems. It can be expressed as

$$\boldsymbol{\Psi}(\boldsymbol{u}) \equiv \boldsymbol{f} - \boldsymbol{K}(\boldsymbol{u})\boldsymbol{u} = \boldsymbol{0} \tag{2.74}$$

In this case, we can set $\mathbf{u}^1{}_{n+1} = \mathbf{u}_n$, and the iterative equation can be written as

$$\boldsymbol{u}_{n+1}^{i+1} = \left[\boldsymbol{K}\left(\boldsymbol{u}_{n+1}^{i}\right)\right]^{-1}\boldsymbol{f}_{n+1} \tag{2.75}$$

2.4.5 Line Search Algorithms: Accelerating Convergence

All the iterative methods discussed in Sect. 2.4.4 share the same structure as described in Eqs. (2.61) to (2.63), in which various approximations are made to the Newton matrix $\mathbf{K}^i$T. For all these methods, an iterative vector is determined, from which the new values of the unknowns are obtained.

$$\boldsymbol{u}_{n+1}^{i+1} = \boldsymbol{u}_{n+1}^{i} + d\boldsymbol{u}_{n+1}^{i} \tag{2.76}$$

Iteration starts from

$$\boldsymbol{u}_{n+1}^{1} = \boldsymbol{u}_{n}$$

where $\mathbf{u}n$ is the known (converged) solution from the previous time step or load step. The goal is to reduce $\psi^{i+1}n + 1$ to zero, but this is not easily achieved using any of the aforementioned algorithms, even for the scalar equation example given earlier. It may be easier to compute the value of $\psi^{i+1}n + 1$ for various choices of $\mathbf{u}n + 1$, and then use appropriate interpolation to find an approximate solution that satisfies the nonlinear equations. For multi-degree-of-freedom systems, this approach is clearly not feasible unless some scalar norm of the residual is used. One possible algorithm is

$$\boldsymbol{u}_{n+1}^{i+1,j} = \boldsymbol{u}_{n+1}^{i} + \eta_{i,j} d\boldsymbol{u}_{n+1}^{i} \tag{2.77}$$

and determine the step size $\eta_{i,j}$ so that the projection of the residual onto the search direction $d\mathbf{u}in + 1$ is zero. This projection can be defined as

$$\mathbf{G}_{i,j} \equiv \left(d\boldsymbol{u}_{n+1}^{i}\right)^{\mathrm{T}} \boldsymbol{\Psi}_{n+1}^{i+1,j} \tag{2.78}$$

where

$$\boldsymbol{\Psi}_{n+1}^{i+1,j} \equiv \boldsymbol{\Psi}\left(\boldsymbol{u}_{n+1}^{i} + \eta_{i,j} d\boldsymbol{u}_{n+1}^{i}\right), \quad \eta_{i,0} = 1$$

Of course, other norms of the residual can also be used.

This algorithm is known as line search, and $\eta_{i,j}$ can be conveniently obtained using the secant method, as illustrated in Fig. 2.7. A notable drawback of line search is that it requires multiple evaluations of $\boldsymbol{\Psi}$. However, when applied to modified or quasi-Newton methods, it can significantly accelerate overall convergence. In fact, line search is also useful in the full Newton method, as it can increase the radius of convergence. A commonly used compromise is to perform the search only under the following circumstances:

$$\mathbf{G}_{i,0} > \varepsilon \left(d\boldsymbol{u}_{n+1}^{i}\right)^{\mathrm{T}} \boldsymbol{\Psi}_{n+1}^{i+1,j} \tag{2.79}$$

Here, the tolerance ε is set between 0.5 and 0.8. This means that if the reduction in the residual during the iteration process is less than or equal to ε times its original value, line search will no longer be performed.

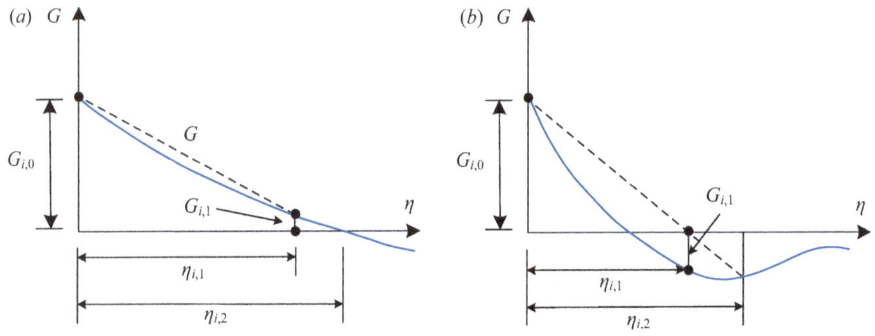

Fig. 2.7 Secant method in linear search

2.4.6 "Softening" Behavior and Displacement Control

When applying the above algorithm to load control problems, it is implicitly assumed that the iterative algorithm is associated with the positive increment of the load vector **f** in Eq. (2.54). In certain structural problems, this refers to a set of loads that can be assumed to be proportional to each other, which can be expressed as

$$\Delta \boldsymbol{f}_{n+1} = \Delta \lambda_{n+1} \boldsymbol{f}_0 \tag{2.80}$$

In many problems, there are cases where no solution exists above a certain maximum value of **f**. In such situations, the true solution corresponds to the "softening" branch, as shown in Fig. 2.2. In these cases, the problem can either be redefined as a displacement-controlled loading problem, or $\Delta\lambda_{n+1}$ should be negative. For the simple case of a single load, it is straightforward to rewrite the equations in terms of the increment of a single prescribed displacement, and considerable research has already been conducted in this area.

Among all successful methods for increasing $\Delta\lambda_{n+1}$, the original problem given by Eq. (2.52) is rewritten as the solution to the following equation:

$$\boldsymbol{\Psi}_{n+1} = \lambda_{n+1} \boldsymbol{f}_0 - \boldsymbol{P}(\boldsymbol{u}_{n+1}) = \boldsymbol{0}$$

where

$$\begin{aligned} \boldsymbol{u}_{n+1} &= \boldsymbol{u}_n + \Delta \boldsymbol{u}_{n+1} \\ \lambda_{n+1} &= \lambda_n + \Delta \lambda_{n+1} \end{aligned} \tag{2.81}$$

It is included as a variable in the incremental step. An additional constraint equation is now required to solve for the extra variable $\Delta\lambda n + 1$.

These additional constraint equations can take various forms. Riks assumed that in each increment

$$\Delta \boldsymbol{u}_{n+1}^{\mathrm{T}} \Delta \boldsymbol{u}_{n+1} + \Delta \lambda^2 \boldsymbol{f}_0^{\mathrm{T}} \boldsymbol{f}_0 = \Delta l^2 \tag{2.82}$$

Here, Δl denotes the prescribed "length" in $(n + 1)$-dimensional space. Crisfield proposed a more natural method for displacement control, requiring that

$$\Delta \boldsymbol{u}_{n+1}^{\mathrm{T}} \Delta \boldsymbol{u}_{n+1} = \Delta l^2 \tag{2.83}$$

In addition to the arc length and spherical path control methods mentioned above, there are several other approaches.

At this point, the constraint Eqs. (2.82) or (2.83) can be directly added to the system in Eq. (2.81), and the iterative methods introduced earlier can still be applied. However, the resulting "tangent" system of equations will generally lose its symmetry, so alternative approaches are often employed.

For a given i-th iteration, the solution can be expressed in the following general form:

$$\begin{aligned} \boldsymbol{\Psi}_{n+1}^{i} &= \lambda_{n+1}^{i} \boldsymbol{f}_0 - \boldsymbol{P}\left(\boldsymbol{u}_{n+1}^{i}\right) \\ \boldsymbol{\Psi}_{n+1}^{i+1} &\approx \boldsymbol{\Psi}_{n+1}^{i} + d\lambda_{n+1}^{i} \boldsymbol{f}_0 - \boldsymbol{K}_{\mathrm{T}}^{i} d\boldsymbol{u}_{n+1}^{i} \end{aligned} \tag{2.84}$$

From this, the increment of **u** can be obtained.

$$\begin{aligned} d\boldsymbol{u}_{n+1}^{i} &= \left(\boldsymbol{K}_{\mathrm{T}}^{i}\right)^{-1}\left[\boldsymbol{\Psi}_{n+1}^{i} + d\lambda_{n+1}^{i} \boldsymbol{f}_0\right] \\ d\boldsymbol{u}_{n+1}^{i} &= d\breve{\boldsymbol{u}}_{n+1}^{i} + d\lambda_{n+1}^{i} \hat{\boldsymbol{u}}_{n+1}^{i} \end{aligned} \tag{2.85}$$

where

$$\begin{aligned} d\breve{\boldsymbol{u}}_{n+1}^{i} &= \left(\boldsymbol{K}_{\mathrm{T}}^{i}\right)^{-1} \boldsymbol{\Psi}_{n+1}^{i} \\ d\hat{\boldsymbol{u}}_{n+1}^{i} &= \left(\boldsymbol{K}_{\mathrm{T}}^{i}\right)^{-1} \boldsymbol{f}_0 \end{aligned} \tag{2.86}$$

Therefore, based on the preceding constraints, an additional equation can be obtained, and together with Eq. (2.83), we have

$$\left(\Delta \boldsymbol{u}_{n+1}^{i-1} + d\boldsymbol{u}_{n+1}^{i}\right)^{\mathrm{T}}\left(\Delta \boldsymbol{u}_{n+1}^{i-1} + d\boldsymbol{u}_{n+1}^{i}\right) = \Delta l^2 \tag{2.87}$$

Here, $\Delta \mathbf{u}^{i-1}{}_{n+1}$ is defined by Eq. (2.63). Substituting Eqs. (2.85) into(2.87) yields a quadratic equation for solving the remaining unknown $d\lambda^{i}{}_{n+1}$ (the result may be negative).

The procedure proposed by Bergan differs somewhat from the one described above. This method first assumes a fixed load increment $\Delta\lambda_{n+1}$, and then uses any of the iterative algorithms introduced earlier to compute the increment $d\mathbf{u}^i n + 1$. Next, a new increment $\Delta\lambda^* n + 1$ is calculated so as to minimize the norm of the following residual:

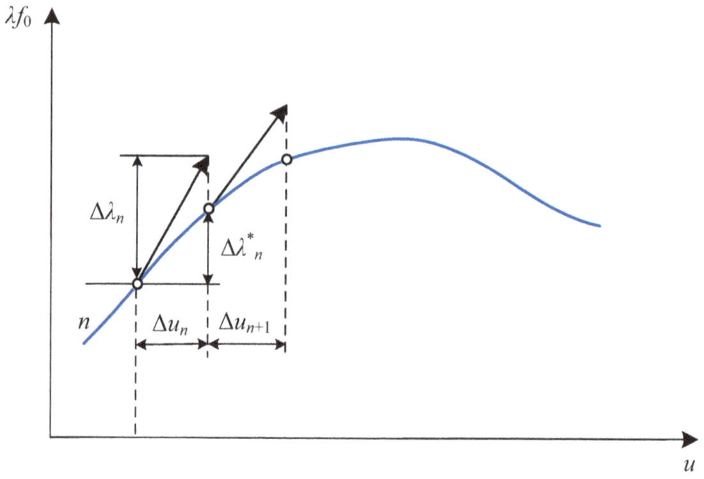

Fig. 2.8 One-dimensional illustration of the Bergan method

$$\left[\left(\Delta\lambda_{n+1}^{*}\boldsymbol{f}_0 - \boldsymbol{P}_{n+1}^{i+1}\right)^{\mathrm{T}}\left(\Delta\lambda_{n+1}^{*}\boldsymbol{f}_0 - \boldsymbol{P}_{n+1}^{i+1}\right)\right] = \Delta l^2 \tag{2.88}$$

Therefore, the result should be given by

$$\frac{\mathrm{d}\Delta l^2}{\mathrm{d}\Delta\lambda_{n+1}^{*}} = 0$$

and one can obtain the following solution

$$\Delta\lambda_{n+1}^{*} = \frac{\boldsymbol{f}_0^{\mathrm{T}}\boldsymbol{P}_{n+1}^{i+1}}{\boldsymbol{f}_0^{\mathrm{T}}\boldsymbol{f}_0} \tag{2.89}$$

This quantity may be negative, which means that the load needs to be reduced. In general, this does lead to a rapid decrease in the residual, but it also makes precise control of the displacement amplitude more difficult. Figure 2.8 illustrates the Bergan method using a one-dimensional example. This method employs displacement control, where the magnitude of $\Delta\lambda n + 1$ is determined by the slope of **KT** used in the first iteration.

2.4.7 Convergence Criteria

In all the iterative algorithms introduced above, the numerical solution can only be an approximation, so a tolerance must be set to determine when to terminate the iteration process.

Since all computer calculations are performed with finite precision, it is impossible to obtain a solution more accurate than the limit imposed by rounding errors.

Typically, the error criterion used is the norm of the displacement parameter $||d\mathbf{u}^i{}_{n+1}||$, or, more logically, the norm of the residual $||\psi^{i+1}n+1||$. In the latter case, the limit is often expressed as a certain tolerance of the load norm $||\mathbf{f}_{n+1}||$. Therefore, it can be required that

$$\left\|\boldsymbol{\Psi}_{n+1}^{i}\right\| \leq \varepsilon\left\|\boldsymbol{f}_{n+1}\right\| \tag{2.90}$$

Here, ε is a small number, and

$$\|\boldsymbol{\Psi}\| = \left(\boldsymbol{\Psi}^{\mathrm{T}}\boldsymbol{\Psi}\right)^{1/2} \tag{2.91}$$

Other norm choices also exist. For example, the residual from the first iteration can be used as the basis, that is,

$$\left\|\boldsymbol{\Psi}_{n+1}^{i}\right\| \leq \varepsilon\left\|\boldsymbol{\Psi}_{n+1}^{1}\right\| \tag{2.92}$$

The error arising from the incomplete solution of the discrete nonlinear system may accumulate in the discretized energy norm. Therefore, the same norm can be used to control the iterative process. As a third option, one may require that the error in the energy norm satisfies

$$dE^{i} = \left(d\boldsymbol{u}_{n+1}^{i,\mathrm{T}}\boldsymbol{\Psi}_{n+1}^{i}\right)^{1/2} \leq \varepsilon\left(d\boldsymbol{u}_{n+1}^{1,\mathrm{T}}\boldsymbol{\Psi}_{n+1}^{1}\right)^{1/2} \leq \varepsilon dE^{1}. \tag{2.93}$$

In each of the forms mentioned above, there may be cases where the norm on the right-hand side is zero. Therefore, the more general fourth form involves computing the norm of the element-wise residuals. If the residual of the problem is the sum over all elements, that is,

$$\boldsymbol{\Psi}_{n+1} = \sum_{e}\boldsymbol{\psi}_{n+1}^{e} \tag{2.94}$$

Here, e denotes an individual element, and ψe represents the residual of each element. Thus, the convergence criterion can be expressed as

$$\left\|\boldsymbol{\Psi}_{n+1}^{i}\right\| \leq \varepsilon\left\|\boldsymbol{\psi}_{n+1}^{e}\right\| \tag{2.95}$$

where

$$\left\|\boldsymbol{\psi}_{n+1}^{e}\right\| = \sum_{e}\left\|\left(\boldsymbol{\psi}_{n+1}^{e}\right)^{i}\right\| \tag{2.96}$$

Once the criterion is established, the question is how to choose an appropriate value for ε. When using the full Newton method (which should exhibit quadratic asymptotic convergence), the error tolerance ε can be set to half the machine precision. Therefore, if the machine precision is 16 digits, then let $\varepsilon = 10^{-8}$, quadratic convergence ensures that the residual of the next step (in the absence of rounding errors) reaches full accuracy. For modified Newton's method or quasi-Newton methods, their asymptotic convergence rates are uncertain, and more iterations are required to achieve high precision. In such cases, larger tolerance values (e.g., 0.01 to 0.001) are often used. However, for problems requiring a large number of iterations, an excessively large convergence tolerance may lead to instability in the solution. Therefore, it is recommended, whenever practically feasible, to use a tolerance equal to half the machine precision.

2.4.8 General Discussion: Incremental and Rate Methods

The various iterative methods introduced earlier provide essential tools for solving the nonlinear systems of equations arising from finite element discretization. The optimal choice of solution method depends on the specific problem. Although numerous studies have compared the computational effort of different solvers, the differences are generally minor. However, when convergence is difficult to achieve, it is undoubtedly advisable to use the exact Newton method (combined with line search). In addition, the advantage of symmetric matrix updates in quasi-Newton methods results in low computational cost, making them frequently adopted. When a nonsymmetric tangent stiffness matrix is present, it is preferable to consider nonsymmetric updates.

The direct iterative methods introduced earlier did not cover the conjugate gradient method or dynamic relaxation method, which are commonly used to solve explicit transient dynamic problems (see Sect. 2.3) in order to obtain steady-state solutions. These algorithms typically have the following characteristics:

1. The matrix used to compute the trial increment $d\mathbf{u}$ has a diagonal or highly sparse structure (thus the computational cost per iteration is very low).
2. The total number of iterations and the number of computations for the residual $\boldsymbol{\Psi}$ are relatively high.

The above features give these methods significant potential for solving large-scale problems. However, to date, these methods are generally effective only for certain types of problems. Interested readers may refer to software packages that implement the aforementioned algorithms.

On the selection of the increment size Δf or $\Delta\lambda$: First, it is clear that smaller increments reduce the total number of iterations required at each computational step, and in many applications, it is necessary to automatically specify the increment size to maintain an (almost) constant number of iterations. In such cases, methods such as the "current stiffness" introduced by Bergan may be effective.

Second, if the system exhibits path-dependent behavior (for example, in plastic constitutive models), it is recommended to use small increments to ensure the accuracy of the results. In this context, it is advisable to use the accumulated increment $\Delta \mathbf{u}^i n + 1$ rather than the incremental step $d\mathbf{u}^{\mathrm{i}}{}_{n+1}$.

Third, if only a single Newton iteration is performed for each increment of $\Delta\lambda$, the process is equivalent to directly solving the standard rate problem using forward incremental integration. In this case, it should be noted that if Eq. (2.52) is expressed as

$$\boldsymbol{P}(\boldsymbol{u}) = \lambda \boldsymbol{f}_0 \tag{2.97}$$

Differentiating with respect to λ yields

$$\frac{\mathrm{d}\boldsymbol{P}}{\mathrm{d}\boldsymbol{u}} \frac{\mathrm{d}\boldsymbol{u}}{\mathrm{d}\lambda} = \boldsymbol{f}_0 \tag{2.98}$$

Further express it as

$$\frac{\mathrm{d}\boldsymbol{u}}{\mathrm{d}\lambda} = \boldsymbol{K}_T^{-1} \boldsymbol{f}_0 \tag{2.99}$$

In incremental form, Eq. (2.99) can be expressed explicitly using the Euler method as follows:

$$\Delta \boldsymbol{u}_{n+1} = \Delta\lambda \boldsymbol{K}_{Tn}^{-1} \boldsymbol{f}_0. \tag{2.100}$$

This direct integration method, as shown in Fig. 2.9, employs the explicit Euler method and is therefore often divergent, exhibiting only conditional stability. Clearly, other methods can be used to improve accuracy and stability, such as the implicit Euler method and the Runge–Kutta method.

2.5 Boundary Conditions: Nonlinear Problems

When constructing solution schemes using the variational (weak) form, boundary conditions are classified into two categories: one is the natural boundary condition, which is inherently satisfied by the variational form and requires no special treatment; the other is the essential boundary condition, which necessitates modifications to the solution process to ensure the validity of the variational form. For example, in the irreducible displacement method, traction boundary conditions are natural boundary conditions, while displacement boundary conditions are essential boundary conditions must be applied separately.

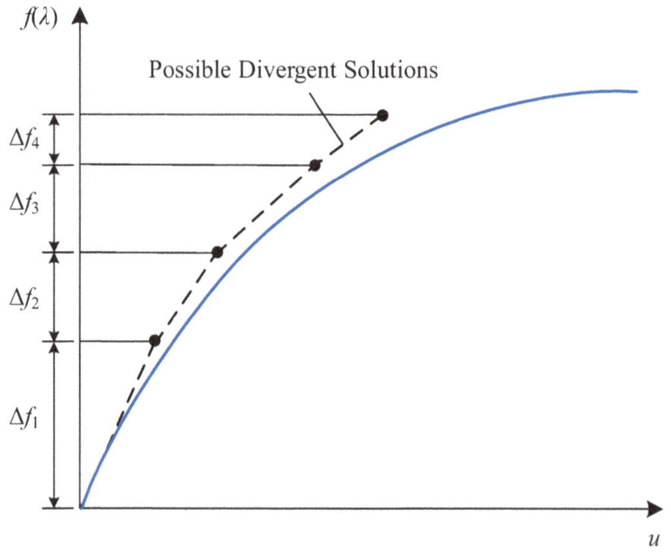

Fig. 2.9 Direct Integration Method Process

2.5.1 Displacement Boundary Conditions

The method for imposing displacement boundary conditions is given by Eq. (1.7). In finite element calculations, a common procedure for specifying displacement boundary conditions is to assign the prescribed values at the nodes, i.e.

$$(\tilde{u}_a)_i = \overline{u}_i(\boldsymbol{x}_a) \tag{2.101}$$

Here, $(\tilde{u}_a)_i$ denotes the value of node a in the i-th direction, as shown in the two-dimensional case in Fig. 2.10. It should be noted that boundary conditions are imposed on the boundary Γ_u^h of the finite element approximation, rather than on the actual boundary Γ_u. As the mesh is refined near the boundary, the two typically converge to each other at a rate equal to or greater than the error rate of other approximation methods.

The method for applying specific displacement boundary conditions is not unique; there are several approaches. Three methods are introduced below. For example, consider the linear static problem given by the following equation:

$$\begin{bmatrix} \boldsymbol{K}_{11} & \boldsymbol{K}_{12} \\ \boldsymbol{K}_{21} & \boldsymbol{K}_{22} \end{bmatrix} \begin{Bmatrix} \boldsymbol{u}_1 \\ \boldsymbol{u}_2 \end{Bmatrix} = \begin{Bmatrix} \boldsymbol{f}_1 \\ \boldsymbol{f}_2 \end{Bmatrix} \tag{2.102}$$

To impose the boundary condition $u_1 = \overline{u}_1$, we have.

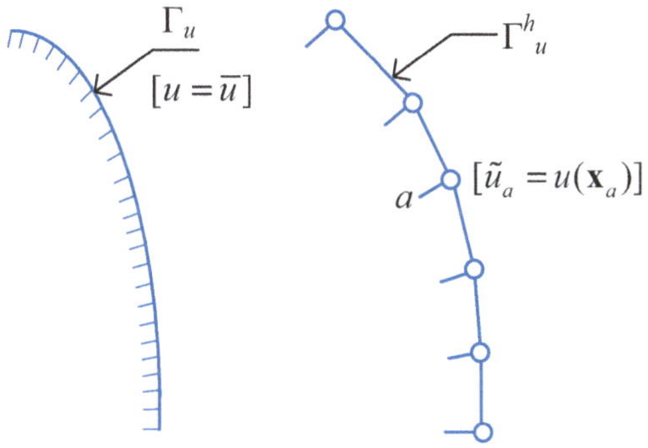

Fig. 2.10 Prescribed displacement boundary conditions

1. Substitute the first equation (associated with δu_i in the weak form) to impose the boundary condition.

$$\begin{bmatrix} \mathbf{1} & \mathbf{0} \\ \boldsymbol{K}_{21} & \boldsymbol{K}_{22} \end{bmatrix} \begin{Bmatrix} \boldsymbol{u}_1 \\ \boldsymbol{u}_2 \end{Bmatrix} = \begin{Bmatrix} \bar{\boldsymbol{u}}_1 \\ \boldsymbol{f}_2 \end{Bmatrix}$$

The desired result can thus be obtained.
If K is symmetric, this method is not efficient. However, by rewriting it as

$$\begin{bmatrix} \mathbf{1} & \mathbf{0} \\ \mathbf{0} & \boldsymbol{K}_{22} \end{bmatrix} \begin{Bmatrix} \boldsymbol{u}_1 \\ \boldsymbol{u}_2 \end{Bmatrix} = \begin{Bmatrix} \bar{\boldsymbol{u}}_1 \\ \boldsymbol{f}_2 - \boldsymbol{K}_{21}\bar{\boldsymbol{u}}_1 \end{Bmatrix} \tag{2.103}$$

The problem thus becomes symmetric again.

2. The second approach is to carry out the modifications described above and eliminate all equations corresponding to known quantities, yielding

$$\boldsymbol{K}_{22}\boldsymbol{u}_2 = \boldsymbol{f}_2 - \boldsymbol{K}_{21}\bar{\boldsymbol{u}}_1 \tag{2.104}$$

and the condition $u_1 = \bar{u}_1$. This method yields a set of final equations with the fewest unknowns, but it is not necessarily optimal for programming.

3. The third approach is the penalty factor method, whose equation is given by

$$\begin{bmatrix} \boldsymbol{K}_{11} & \boldsymbol{K}_{12} \\ \boldsymbol{K}_{21} & \boldsymbol{K}_{22} \end{bmatrix} \begin{Bmatrix} \boldsymbol{u}_1 \\ \boldsymbol{u}_2 \end{Bmatrix} = \begin{Bmatrix} \boldsymbol{k}_{11}\bar{\boldsymbol{u}}_1 \\ \boldsymbol{f}_2 \end{Bmatrix} \tag{2.105}$$

where $k_{11} = \alpha K_{11}$, and $\alpha > > 1$. This method is easy to implement but requires selecting an appropriate value for α. For simple point constraints, such as $u_1 = \bar{u}_1$, it is usually sufficient to choose $\alpha = 10^6$ to 10^8.

When dealing with transient nonlinear problems, the imposition of displacement boundary conditions becomes slightly more complex. First, boundary conditions need to be applied to the incremental equations. This requires calculating the initial values of displacement, velocity, and acceleration. As mentioned above, there are two fundamental forms to consider: explicit methods and implicit methods.

2.5.1.1 Nonlinear Explicit Methods

If the velocity term does not appear in the equilibrium equation, the explicit method is straightforward. Taking the Newmark algorithm as an example, with $\beta = 0$ and the displacement at time t_{n+1} obtained from Eq. $(2.32)_1$ and its boundary Γ_u, the new acceleration is given by

$$\boldsymbol{M}\boldsymbol{a}_{n+1} = \boldsymbol{f}_{n+1} - \boldsymbol{P}(\boldsymbol{\sigma}_{n+1}) \tag{2.106}$$

Here, **M** is given in diagonalized form. The velocity is then obtained using Eq. $(2.32)_2$ and the known acceleration. By using the diagonalized **M** matrix, there is no coupling of acceleration between boundary nodes and non-boundary nodes.

If the velocity term appears explicitly in the equilibrium equation, an iterative strategy can be employed by treating v^{n+1} as v^n, and then calculating a "predicted" value for the acceleration. The velocity is then computed using the predicted acceleration, followed by another iteration, which yields results of sufficient accuracy. It may be necessary to design an expression for updating the boundary velocity in order to maintain high precision in the final results.

2.5.1.2 Nonlinear Implicit Methods

When using the implicit method of the Newmark algorithm, both γ and β are nonzero. If the time increment Δt is 0, then neither displacement nor velocity will change (see Eq. (2.32)), and the new acceleration is determined by Eq. (2.106): only the instantaneous change of $\mathbf{f}_{n+1}$ needs to be considered. When $\Delta t > 0$, Eq. (2.36) can be used to impose the constraint $\mathbf{u}_{n+1} = \overline{\boldsymbol{u}}_{n+1}$. In the first iteration, we obtain

$$\mathrm{d}\boldsymbol{u}_{n+1}^{1} = \mathrm{d}\overline{\boldsymbol{u}}_{n+1}^{1}$$

From Eqs. $(2.37)_1$ and (2.38), we obtain $\mathbf{u}^2{}_{n+1} \equiv \overline{\boldsymbol{u}}_{n+1}$. In the first iteration, the incremental form of the displacement boundary condition is given by

$$\begin{bmatrix} \boldsymbol{A}_{11} & \boldsymbol{A}_{12} \\ \boldsymbol{A}_{21} & \boldsymbol{A}_{22} \end{bmatrix} \begin{Bmatrix} \mathrm{d}\overline{\boldsymbol{u}}_1 \\ \mathrm{d}\overline{\boldsymbol{u}}_2 \end{Bmatrix} = \begin{Bmatrix} \boldsymbol{\psi}_1 \\ \boldsymbol{\psi}_2 \end{Bmatrix} \tag{2.107}$$

In the above equation, $\overline{u}_1$ is associated with the known displacement at the boundary node, while $\mathbf{u}_2$ is set as the "unknown" displacement. The methods previously introduced for linear static problems can all be applied to solve this problem.

2.5.2 Traction Boundary Conditions

The traction boundary condition is a "natural" variational boundary condition that does not affect the displacement of active nodes on the boundary—it only influences the nodal force conditions applied. Applying a nonzero traction on the boundary requires integration over each surface element. Therefore, for a typical node a as shown in Fig. 2.11, it is necessary to compute the following integral:

$$\mathbf{f}_a = \sum_e \int_{\Gamma_t} \mathbf{N}_a \bar{\mathbf{t}} \mathrm{d}\Gamma_t \tag{2.108}$$

Here, e encompasses all elements in Γ_t that include node a (for example, in the two-dimensional case shown in Fig. 2.11, these are the two elements above and below node a). Of course, if $\bar{\mathbf{t}}$ is zero, the integral does not need to be computed.

2.5.2.1 Pressure Load

An important example is the application of normal "pressure" on a surface. In this case, the traction force is given by

$$\bar{t} = \bar{p}_n \boldsymbol{n}$$

Here, $\bar{p}_n$ denotes the prescribed normal pressure (positive in tension), and $\mathbf{n}$ is the unit normal vector to the boundary Γ_t (see Fig. 2.12). In this case, Eq. (2.108) becomes

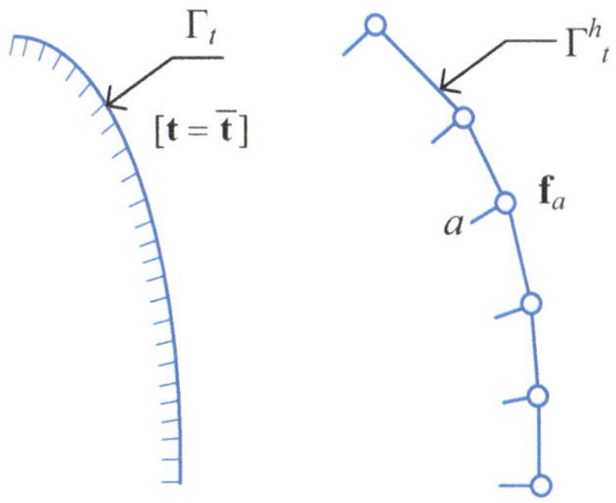

Fig. 2.11 Boundary conditions with specified traction

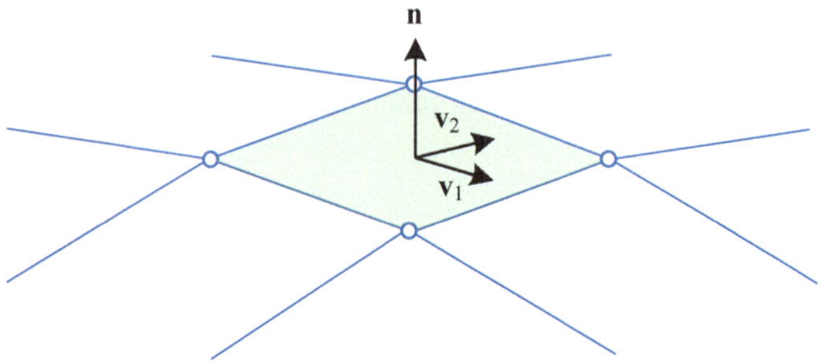

Fig. 2.12 Normal of a surface

$$f_a = \sum_e \int_{\Gamma_t} N_a \overline{p}_n \boldsymbol{n} \mathrm{d}\Gamma \tag{2.109}$$

Using Eq. (2.28), the calculation of pressure load can be expressed as

$$f_a = \sum_e \int_{\square} N_a(\boldsymbol{\xi}) \overline{p}_n(\boldsymbol{\xi}) (\boldsymbol{v}_1 \times \boldsymbol{v}_2) \mathrm{d}\square \tag{2.110}$$

where $\square = d\xi^1 d\xi^2$, and integrals are all performed directly in the natural coordinate system. For two-dimensional problems, the surface shape function is given by $N_a(\xi^1), \square = d\xi^1$, and

$$\boldsymbol{v}_2 \equiv \begin{cases} \boldsymbol{e}_3, \text{ for plane strain} \\ h_3 \boldsymbol{e}_3, \text{ for plane stress} \\ 2\pi h_3 \boldsymbol{e}_3, \text{ for axial symmetry} \end{cases}$$

Here, $\mathbf{e}_3$ is the unit normal vector for the planar problem.

2.5.3 *Mixed Displacement/Traction Boundary Conditions*

Handling mixed boundary conditions where some displacement components and some traction components are prescribed typically requires modifying the nodal parameters. For example, for a shaft with its axis in the x_3-direction and radius R rotating inside a bearing (without friction or clearance), the conditions are $u_n = u_r(R) = 0$ and $t_\vartheta(R) = t_z(R) = 0$ (with the coordinate origin at the center of the shaft). In this case, it is necessary to transform the degrees of freedom at each node on the shaft boundary so that

$$\boldsymbol{u}_a = \begin{Bmatrix} (\tilde{u}_1)_a \\ (\tilde{u}_2)_a \\ (\tilde{u}_3)_a \end{Bmatrix} = \begin{bmatrix} \cos\theta_a & -\sin\theta_a & 0 \\ \sin\theta_a & \cos\theta_a & 0 \\ 0 & 0 & 1 \end{bmatrix} \begin{Bmatrix} (\tilde{u}_r)_a \\ (\tilde{u}_\theta)_a \\ (\tilde{u}_z)_a \end{Bmatrix} = \boldsymbol{L}_a \tilde{\boldsymbol{u}}_{a'} \tag{2.111}$$

Applying this transformation to the residual yields

$$\boldsymbol{R}_{a'} = \boldsymbol{L}_a^{\mathrm{T}} \boldsymbol{R}_a \tag{2.112}$$

Applying this to mass and stiffness yields

$$\begin{aligned} \boldsymbol{M}_{a'b'} &= \boldsymbol{L}_a^{\mathrm{T}} \boldsymbol{M}_{ab} \boldsymbol{L}_b \\ \boldsymbol{M}_{a'c} &= \boldsymbol{L}_a^{\mathrm{T}} \boldsymbol{M}_{ac} \\ \boldsymbol{M}_{cb'} &= \boldsymbol{M}_{cb} \boldsymbol{L}_b \end{aligned} \tag{2.113}$$

Here, a and b are transformation nodes, while c is a node that retains its original orientation. It is usually convenient to perform these transformations on each individual unit; however, if necessary, they can also be applied to the assembled column structure.

Once the transformation has been performed, each individual displacement and traction boundary condition can be applied as described above.

2.6 Mixed Methods or Irreducible Forms

The method for computing the stiffness matrix given by Eq. (2.20) follows the so-called displacement formulation, which is widely used in many finite element solution methods. However, in certain cases, it is convenient to employ mixed finite element formulations, particularly when constraints such as (near) incompressibility arise, where these formulations become especially necessary. It is noted that constitutive models such as viscoelasticity and associative plasticity often appear in nearly incompressible forms. For such problems, it is necessary to re-derive the computational formulation. In this context, there are two options: one is the two-field formulation using $\mathbf{u}$ and $\boldsymbol{p}$ as variables (where $\boldsymbol{p}$ is the mean stress); the other is the three-field formulation using $\mathbf{u}, \boldsymbol{p}$ and $\boldsymbol{\vartheta}$ as variables (where $\boldsymbol{\vartheta}$ is the volumetric change). The specific choice depends on the constitutive model adopted. If volumetric change only affects the pressure, the two-field formulation can be used directly; if the response may involve coupling between the deviatoric and mean components of stress and strain, the three-field formulation yields a simpler form, which can be used to develop finite element models. To illustrate this, a general three-field mixed formulation is presented below, with a detailed explanation of how to easily incorporate such coupling effects without altering the previously introduced nonlinear solution procedure. This method also serves as a basis for further extension to finite deformation problems.

2.6.1 Deviatoric and Mean Components of Stress and Strain

By decomposing stress and strain into deviatoric (isochoric) components and mean components, the analysis of nearly incompressible materials is greatly simplified. Therefore, the mean stress (pressure) is defined as follows:

$$p = \frac{1}{3}[\sigma_{11} + \sigma_{22} + \sigma_{33}] = \frac{1}{3}\sigma_{ii} \tag{2.114}$$

The deviatoric stress is defined as

$$s_{ij} = \sigma_{ij} - \delta_{ij}p \tag{2.115}$$

Here, δ_{ij} is the Kronecker delta, that is,

$$\delta_{ij} = \begin{cases} 1, & \text{when } i = j \\ 0, & \text{when } i \neq j \end{cases}$$

Similarly, the average strain (volume change) is defined as

$$\vartheta = [\varepsilon_{11} + \varepsilon_{22} + \varepsilon_{33}] = \varepsilon_{ii} \tag{2.116}$$

The shear strain is defined as

$$e_{ij} = \varepsilon_{ij} - \frac{1}{3}\delta_{ij}\vartheta \tag{2.117}$$

Note: The position of the coefficient 1/3 differs in the two expressions.

2.6.2 Three-Variable Mixed Method for General Constitutive Models

For the mixed formulation of the coupled mean and deviatoric component constitutive model, the following matrix operators for the mean and deviatoric components are defined:

$$\boldsymbol{m} = [1\,1\,1\,0\,0\,0]^T \textit{and}\ \boldsymbol{I}_d = \boldsymbol{I} - \frac{1}{3}\boldsymbol{m}\boldsymbol{m}^T \tag{2.118}$$

Here, **I** denotes the identity matrix.

Thus, the strain can be expressed in a mixed form as follows:

$$\boldsymbol{\varepsilon} = \boldsymbol{I}_d(\mathbb{B}\boldsymbol{u}) + \frac{1}{3}\boldsymbol{m}\vartheta \tag{2.119}$$

Here, the first term represents the fluctuating component, while the second term represents the mean component. In the displacement calculation formula, ϑ can be expressed as

$$\vartheta = \boldsymbol{m}^{\mathrm{T}}(\mathbb{B}\boldsymbol{u})$$

Similarly, stress can be expressed in a mixed form as

$$\boldsymbol{\sigma} = \boldsymbol{I}_d\breve{\sigma} + \boldsymbol{m}p \tag{2.120}$$

Here, $\breve{\sigma}$ are directly derived from strain, incremental strain, or strain rate, depending on the specific constitutive model. Currently, this type of stress is expressed as

$$\breve{\sigma} = \boldsymbol{\sigma}(\boldsymbol{\varepsilon}) \tag{2.121}$$

Note: It is not necessary to separate it into mean and deviatoric components; the $\boldsymbol{\varepsilon}$ in Eq. (2.121) represents a mixed form of strain.

The weak form (variational Galerkin equation) incorporating transient behavior can be expressed as follows:

$$\begin{gathered}\int_\Omega \delta\boldsymbol{u}^{\mathrm{T}}\rho\ddot{\boldsymbol{u}}\mathrm{d}\Omega + \int_\Omega \delta(\mathbb{B}\boldsymbol{u})^{\mathrm{T}}\boldsymbol{\sigma}\mathrm{d}\Omega = \int_\Omega \delta\boldsymbol{u}^{\mathrm{T}}\boldsymbol{b}\mathrm{d}\Omega + \int_{\Gamma_t} \delta\boldsymbol{u}^{\mathrm{T}}\bar{\boldsymbol{t}}\mathrm{d}\Gamma \\ \int_\Omega \delta p\left[\boldsymbol{m}^{\mathrm{T}}(\mathbb{B}\boldsymbol{u}) - \vartheta\right]\mathrm{d}\Omega = 0 \\ \int_\Omega \delta\vartheta\left[\frac{1}{3}\boldsymbol{m}^{\mathrm{T}}\breve{\sigma} - p\right]\mathrm{d}\Omega = 0\end{gathered} \tag{2.122}$$

Introducing the finite element approximation to the above variables, i.e.,

$$\boldsymbol{u} \approx \hat{\boldsymbol{u}} = \boldsymbol{N}_u\tilde{\boldsymbol{u}}, \quad p \approx \hat{p} = \boldsymbol{N}_p\tilde{\boldsymbol{P}}, \quad \vartheta \approx \hat{\vartheta} = \boldsymbol{N}_\vartheta\tilde{\boldsymbol{\vartheta}}$$

The dummy variable can be similarly approximated as

$$\delta\boldsymbol{u} \approx \delta\hat{\boldsymbol{u}} = \boldsymbol{N}_u\delta\tilde{\boldsymbol{u}}, \quad \delta p \approx \delta\hat{p} = \boldsymbol{N}_p\delta\tilde{\boldsymbol{P}}, \quad \delta\vartheta \approx \delta\hat{\vartheta} = \boldsymbol{N}_\vartheta\delta\tilde{\boldsymbol{\vartheta}}$$

The strain of the element can be expressed as

$$\boldsymbol{\varepsilon} = \boldsymbol{I}_d \boldsymbol{B}\tilde{\boldsymbol{u}} + \frac{1}{3}\boldsymbol{m}\boldsymbol{N}_\vartheta \tilde{\boldsymbol{\vartheta}} \tag{2.123}$$

Here, **B** is the standard strain–displacement matrix given in Eq. (2.13). Similarly, the stress in each element can be calculated by the following formula:

$$\boldsymbol{\sigma} = \boldsymbol{I}_d \breve{\sigma} + \boldsymbol{m}\boldsymbol{N}_p \tilde{\boldsymbol{P}} \tag{2.124}$$

Here, $\breve{\sigma}$ is the stress calculated by Eq. (2.121), and the corresponding strain $\boldsymbol{\varepsilon}$ is calculated by Eq. (2.123).

Substituting the element strain and stress calculated from Eqs. (2.123) and (2.124) into (2.122), the following set of finite element equations can be obtained.

$$\begin{aligned}
\boldsymbol{\Psi}_u &= \boldsymbol{f} - \boldsymbol{M}\ddot{\tilde{\boldsymbol{u}}} - \boldsymbol{P} = \boldsymbol{0} \\
\boldsymbol{\Psi}_p &= \boldsymbol{K}_{p\vartheta}\tilde{\boldsymbol{\vartheta}} - \boldsymbol{K}_{pu}\tilde{\boldsymbol{u}} = \boldsymbol{0} \\
\boldsymbol{\Psi}_\vartheta &= \boldsymbol{K}_{\vartheta p}\tilde{\boldsymbol{P}} - \boldsymbol{P}_\vartheta = \boldsymbol{0}
\end{aligned} \tag{2.125}$$

where

$$\begin{aligned}
&\boldsymbol{P} = \int_\Omega \boldsymbol{B}^{\mathrm{T}}\boldsymbol{\sigma}\,\mathrm{d}\Omega, \quad \boldsymbol{P}_\vartheta = \frac{1}{3}\int_\Omega \boldsymbol{N}_\vartheta^{\mathrm{T}}\boldsymbol{m}^{\mathrm{T}}\breve{\sigma}\,\mathrm{d}\Omega \\
&\boldsymbol{K}_{\vartheta p} = \int_\Omega \boldsymbol{N}_\vartheta^{\mathrm{T}}\boldsymbol{N}_p\,\mathrm{d}\Omega = \boldsymbol{K}_{p\vartheta}^{\mathrm{T}}, \quad \boldsymbol{K}_{pu} = \int_\Omega \boldsymbol{N}_p^{\mathrm{T}}\boldsymbol{m}^{\mathrm{T}}\boldsymbol{B}\,\mathrm{d}\Omega = \boldsymbol{K}_{up}^{\mathrm{T}} \\
&\boldsymbol{f} = \int_\Omega \boldsymbol{N}_u^{\mathrm{T}}\boldsymbol{b}\,\mathrm{d}\Omega + \int_{\Gamma_t} \boldsymbol{N}_u^{\mathrm{T}}\bar{\boldsymbol{t}}\,\mathrm{d}\Gamma
\end{aligned} \tag{2.126}$$

2.6.3 *Local Approximation of P and Θ*

If approximate values of pressure and volumetric strain are used locally within each element, and $\mathbf{N}_\vartheta \equiv \mathbf{N}_p$, then the second and third equations in (2.125) can be solved independently within each element.

Note: $\mathbf{K}_{\vartheta p}$ is now symmetric and positive definite. The approximate values of pressure and volumetric strain can be expressed as

$$\begin{aligned}
\tilde{\boldsymbol{p}} &= \boldsymbol{K}_{\vartheta p}^{-1}\boldsymbol{P}_\vartheta \\
\tilde{\boldsymbol{\vartheta}} &= \boldsymbol{K}_{\vartheta p}^{-1}\boldsymbol{K}_{pu}\tilde{\boldsymbol{u}} = \boldsymbol{W}\tilde{\boldsymbol{u}}
\end{aligned} \tag{2.127}$$

The mixed strain of each element can thus be expressed as

$$\boldsymbol{\varepsilon} = \left[\boldsymbol{I}_d\boldsymbol{B} + \frac{1}{3}\boldsymbol{m}\boldsymbol{B}_\vartheta\right]\tilde{\boldsymbol{u}} = \left[\boldsymbol{I}_d \ \tfrac{1}{3}\boldsymbol{m}\right]\begin{bmatrix}\boldsymbol{B}\\ \boldsymbol{B}_\vartheta\end{bmatrix}\tilde{\boldsymbol{u}} \tag{2.128}$$

where

$$\boldsymbol{B}_\vartheta = \boldsymbol{N}_\vartheta \boldsymbol{W} \tag{2.129}$$

The above equation defines the mixed formulation of the volumetric strain–displacement equation.

Based on the results above, the vector **p** can be expressed in the following form:

$$\boldsymbol{p} = \int_\Omega \boldsymbol{B}^{\mathrm{T}}\boldsymbol{\sigma}\mathrm{d}\Omega = \int_\Omega \left[\boldsymbol{B}^{\mathrm{T}}\boldsymbol{I}_d + \frac{1}{3}\boldsymbol{B}_\vartheta^{\mathrm{T}}\boldsymbol{m}^{\mathrm{T}}\right]\breve{\sigma}\mathrm{d}\Omega = \int_\Omega \left[\boldsymbol{B}^{\mathrm{T}} \ \boldsymbol{B}_\vartheta^{\mathrm{T}}\right]\begin{bmatrix}\boldsymbol{I}_d\\ \frac{1}{3}\boldsymbol{m}^{\mathrm{T}}\end{bmatrix}\breve{\sigma}\mathrm{d}\Omega \tag{2.130}$$

From the above results, it can be seen that, apart from calculating the final result, there is no need to compute the mixed stresses. This is especially important when using inelastic and nonlinear material models.

The final step of this process is to compute the tangent vector of the equation. According to the form given in Eq. (2.121), it is straightforward to obtain

$$\mathrm{d}\breve{\sigma} = \breve{\boldsymbol{D}}_T\mathrm{d}\boldsymbol{\varepsilon}$$

According to the incremental mixed strain given by Eq. (2.128), we have

$$\mathrm{d}\breve{\sigma} = \breve{\boldsymbol{D}}_T\left[\boldsymbol{I}_d \ \tfrac{1}{3}\boldsymbol{m}\right]\begin{bmatrix}\boldsymbol{B}\\ \boldsymbol{B}_\vartheta\end{bmatrix}\mathrm{d}\boldsymbol{u} \tag{2.131}$$

The expression for the tangent stiffness is

$$\boldsymbol{K}_T = \int_\Omega \left[\boldsymbol{B}^{\mathrm{T}} \ \boldsymbol{B}_\vartheta^{\mathrm{T}}\right]\begin{bmatrix}\boldsymbol{I}_d\\ \frac{1}{3}\boldsymbol{m}^{\mathrm{T}}\end{bmatrix}\breve{\boldsymbol{D}}_T\left[\boldsymbol{I}_d \ \tfrac{1}{3}\boldsymbol{m}\right]\begin{bmatrix}\boldsymbol{B}\\ \boldsymbol{B}_\vartheta\end{bmatrix}\mathrm{d}\Omega \tag{2.132}$$

It should be noted that the corrected tangent modulus is given by the following equation:

$$\begin{aligned}\overline{\boldsymbol{D}}_T &= \begin{bmatrix}\boldsymbol{I}_d\\ \frac{1}{3}\boldsymbol{m}^{\mathrm{T}}\end{bmatrix}\breve{\boldsymbol{D}}_T\left[\boldsymbol{I}_d \ \tfrac{1}{3}\boldsymbol{m}\right]\\ &= \begin{bmatrix}\boldsymbol{I}_d\breve{\boldsymbol{D}}_T\boldsymbol{I}_d & \frac{1}{3}\boldsymbol{I}_d\breve{\boldsymbol{D}}_T\boldsymbol{m}\\ \frac{1}{3}\boldsymbol{m}^{\mathrm{T}}\breve{\boldsymbol{D}}_T\boldsymbol{I}_d & \frac{1}{9}\boldsymbol{m}^{\mathrm{T}}\breve{\boldsymbol{D}}_T\boldsymbol{m}\end{bmatrix} = \begin{bmatrix}\overline{\boldsymbol{D}}_{11} & \overline{\boldsymbol{D}}_{12}\\ \overline{\boldsymbol{D}}_{21} & \overline{\boldsymbol{D}}_{22}\end{bmatrix}\end{aligned} \tag{2.133}$$

Due to the sparsity and specific structure of $\mathbf{I}_d$ and $\mathbf{m}$, the above computation requires very few operations. Therefore, multiplying by the coefficients in this form is efficient. The computational efficiency of matrices $\mathbf{B}$ and $\mathbf{B}_\vartheta$ is much higher than that of constructing a $\overline{\boldsymbol{B}}$ matrix as shown below.

$$\overline{\boldsymbol{B}} = \boldsymbol{I}_d \boldsymbol{B} + \frac{1}{3}\boldsymbol{m}\boldsymbol{B}_\vartheta \tag{2.134}$$

The tangent stiffness matrix is then calculated using the following formula

$$\boldsymbol{K}_T = \int_\Omega \overline{\boldsymbol{B}}^{\mathrm{T}} \breve{\boldsymbol{D}}_T \overline{\boldsymbol{B}} \mathrm{d}\Omega \tag{2.135}$$

In this form, $\mathbf{B}$ has almost no zero entries.

Example 2.1 Linear Elastic Tangent Stiffness

Consider a linear elastic material, whose constitutive equation is expressed in matrix form as follows:

$$\sigma = \left[\left(K - \frac{2}{3}G\right)\boldsymbol{m}\boldsymbol{m}^{\mathrm{T}} + 2G\boldsymbol{I}_0\right]\varepsilon \tag{2.136}$$

where

$$\boldsymbol{I}_0 = \frac{1}{2}\begin{bmatrix} 2\,0\,0\,0\,0\,0 \\ 0\,2\,0\,0\,0\,0 \\ 0\,0\,2\,0\,0\,0 \\ 0\,0\,0\,1\,0\,0 \\ 0\,0\,0\,0\,1\,0 \\ 0\,0\,0\,0\,0\,1 \end{bmatrix} \tag{2.137}$$

Considering the transformation used to define strain. By replacing $\boldsymbol{\sigma}$, $\boldsymbol{\varepsilon}$ in Eq. (2.136) with $\mathrm{d}\boldsymbol{\sigma}$, $\mathrm{d}\boldsymbol{\varepsilon}$, its incremental form can be obtained. Note the following relationship.

$$\boldsymbol{I}_d\boldsymbol{I}_d = \boldsymbol{I}_d, \quad \boldsymbol{I}_d\boldsymbol{m} = \boldsymbol{0}, \quad \boldsymbol{I}_d\boldsymbol{I}_0 = \boldsymbol{I}_0 - \frac{1}{3}\boldsymbol{m}\boldsymbol{m}^{\mathrm{T}}, \quad \boldsymbol{m}^{\mathrm{T}}\boldsymbol{m} = 3$$

Thus, the corrected modulus can be expressed as

$$\overline{\boldsymbol{D}}_T = \begin{bmatrix} 2G\left(\boldsymbol{I}_0 - \frac{1}{3}\boldsymbol{m}\boldsymbol{m}^{\mathrm{T}}\right) & \boldsymbol{0} \\ \boldsymbol{0} & K \end{bmatrix} \tag{2.138}$$

Note: For isotropic linear elasticity, the deviatoric and volumetric components are decoupled. However, this is not the case when more general constitutive models are considered.

For quadrilateral and hexahedral elements, the aforementioned mixed elements are applicable to a wide range of linear and nonlinear constitutive models. When simulating the inelastic and nonlinear stress–strain behavior of materials using viscoelastic, classical plasticity, and generalized plasticity models, each of these models may encounter nearly incompressible conditions. Two finite element approximation methods are considered here: one uses 4-node quadrilateral or 8-node hexahedral isoparametric elements, with a unified constant interpolation for both N_ϑ and N_p within each element; the other uses 9-node quadrilateral or 27-node hexahedral isoparametric elements, with linear interpolation for both $\mathbf{N}_p$ and $\mathbf{N}_\vartheta$. For two-dimensional elements, use.

$$\boldsymbol{N}_p = \boldsymbol{N}_\vartheta = \left[\, 1\ \xi^1\ \xi^2 \,\right] \text{or} \left[\, 1\ x\ y \,\right].$$

For three-dimensional elements, use.

$$\boldsymbol{N}_p = \boldsymbol{N}_\vartheta = \left[\, 1\ \xi^1\ \xi^2\ \xi^3 \,\right] \text{or} \left[\, 1\ x\ y\ z \,\right].$$

The elements derived through the above process can be used to solve a wide range of solid mechanics problems, which will be discussed in detail in later chapters of this book.

2.6.4 Continuous u-p Approximation

The form of the above mixed problem is not applicable to triangular and tetrahedral elements. For these element shapes, interpolation is always exactly complete to the required order, with no additional terms (for example, 4-node quadrilaterals have extra terms ξ^1, ξ^2 in linear interpolation). By ensuring that the pressure p is continuous within the material (discontinuities are allowed at material interfaces) but interpolated at one order lower than the displacement, the mixed formulation remains valid. Therefore, this method can only be used with quadratic or higher-order interpolation for the displacement $\mathbf{u}$. The interpolation of ϑ is still performed independently within each element and is generally discontinuous between elements. This approach was originally introduced for higher-order elements.

To obtain a continuous solution for p, Eq. (2.125) can be linearized, resulting in the following semi-discrete system of equations:

$$\begin{bmatrix} \boldsymbol{M} & \mathbf{0} & \mathbf{0} \\ \mathbf{0} & \mathbf{0} & \mathbf{0} \\ \mathbf{0} & \mathbf{0} & \mathbf{0} \end{bmatrix} \begin{Bmatrix} \mathrm{d}\ddot{\tilde{\boldsymbol{u}}} \\ \mathrm{d}\ddot{\tilde{\boldsymbol{p}}} \\ \mathrm{d}\ddot{\tilde{\boldsymbol{\vartheta}}} \end{Bmatrix} + \begin{bmatrix} \boldsymbol{K}_{uu} & \boldsymbol{K}_{up} & \boldsymbol{K}_{u\vartheta} \\ \boldsymbol{K}_{pu} & \mathbf{0} & -\boldsymbol{K}_{p\vartheta} \\ \boldsymbol{K}_{\vartheta u} & -\boldsymbol{K}_{\vartheta p} & \boldsymbol{K}_{\vartheta\vartheta} \end{bmatrix} \begin{Bmatrix} \mathrm{d}\tilde{\boldsymbol{u}} \\ \mathrm{d}\tilde{\boldsymbol{p}} \\ \mathrm{d}\tilde{\boldsymbol{\vartheta}} \end{Bmatrix} = \begin{Bmatrix} \boldsymbol{\Psi}_u \\ \boldsymbol{\Psi}_p \\ \boldsymbol{\Psi}_\vartheta \end{Bmatrix} \tag{2.139}$$

Among them, in addition to the matrices defined by Eq. (2.126), $\mathbf{K}_{up} = \mathbf{K}^{\mathrm{T}}{}_{pu}$ and

$$\begin{aligned} \boldsymbol{K}_{uu} &= \int_{\Omega} \boldsymbol{B}^{\mathrm{T}} \overline{\boldsymbol{D}}_{11} \boldsymbol{B} \mathrm{d}\Omega, \quad \boldsymbol{K}_{\vartheta\vartheta} = \int_{\Omega} \boldsymbol{N}_{\vartheta}^{\mathrm{T}} \overline{\boldsymbol{D}}_{22} \boldsymbol{N}_{\vartheta} \mathrm{d}\Omega \\ \boldsymbol{K}_{u\vartheta} &= \int_{\Omega} \boldsymbol{B}^{\mathrm{T}} \overline{\boldsymbol{D}}_{12} \boldsymbol{N}_{\vartheta} \mathrm{d}\Omega, \quad \boldsymbol{K}_{\vartheta u} = \int_{\Omega} \boldsymbol{N}_{\vartheta}^{\mathrm{T}} \overline{\boldsymbol{D}}_{21} \boldsymbol{B} \mathrm{d}\Omega \end{aligned} \tag{2.140}$$

since each $\mathbf{K}_{\vartheta\vartheta}$ belongs to a single element, partial solutions can therefore be obtained using the following expression:

$$\mathrm{d}\tilde{\boldsymbol{\vartheta}} = \boldsymbol{K}_{\vartheta\vartheta}^{-1} \left[\boldsymbol{\Psi}_{\vartheta} - \boldsymbol{K}_{\vartheta u} \mathrm{d}\tilde{\boldsymbol{u}} + \boldsymbol{K}_{\vartheta p} \mathrm{d}\tilde{\boldsymbol{p}} \right] \tag{2.141}$$

Substituting Eqs. (2.141) into (2.139), we obtain

$$\begin{bmatrix} \boldsymbol{M} & \boldsymbol{0} \\ \boldsymbol{0} & \boldsymbol{0} \end{bmatrix} \begin{Bmatrix} \mathrm{d}\ddot{\tilde{\boldsymbol{u}}} \\ \mathrm{d}\ddot{\tilde{\boldsymbol{p}}} \end{Bmatrix} + \begin{bmatrix} \overline{\boldsymbol{K}}_{uu} & \overline{\boldsymbol{K}}_{up} \\ \overline{\boldsymbol{K}}_{pu} & \overline{\boldsymbol{K}}_{pp} \end{bmatrix} \begin{Bmatrix} \mathrm{d}\tilde{\boldsymbol{u}} \\ \mathrm{d}\tilde{\boldsymbol{p}} \end{Bmatrix} = \begin{Bmatrix} \overline{\boldsymbol{\Psi}}_{u} \\ \overline{\boldsymbol{\Psi}}_{p} \end{Bmatrix} \tag{2.142}$$

where

$$\begin{aligned} \overline{\boldsymbol{K}}_{uu} &= \boldsymbol{K}_{uu} - \boldsymbol{K}_{u\vartheta} \boldsymbol{K}_{\vartheta\vartheta}^{-1} \boldsymbol{K}_{\vartheta u}, \quad \overline{\boldsymbol{K}}_{pp} = -\boldsymbol{K}_{p\vartheta} \boldsymbol{K}_{\vartheta\vartheta}^{-1} \boldsymbol{K}_{\vartheta p} \\ \overline{\boldsymbol{K}}_{pu} &= \boldsymbol{K}_{pu} + \boldsymbol{K}_{p\vartheta} \boldsymbol{K}_{\vartheta\vartheta}^{-1} \boldsymbol{K}_{\vartheta u}, \quad \overline{\boldsymbol{K}}_{up} = \boldsymbol{K}_{up} + \boldsymbol{K}_{u\vartheta} \boldsymbol{K}_{\vartheta\vartheta}^{-1} \boldsymbol{K}_{\vartheta p} \\ \overline{\boldsymbol{\Psi}}_{u} &= \boldsymbol{\Psi}_{u} - \boldsymbol{K}_{u\vartheta} \boldsymbol{K}_{\vartheta\vartheta}^{-1} \boldsymbol{\Psi}_{\vartheta}, \quad \overline{\boldsymbol{\Psi}}_{p} = \boldsymbol{\Psi}_{p} + \boldsymbol{K}_{p\vartheta} \boldsymbol{K}_{\vartheta\vartheta}^{-1} \boldsymbol{\Psi}_{\vartheta} \end{aligned} \tag{2.143}$$

The parameters $\tilde{\boldsymbol{u}}, \tilde{\boldsymbol{p}}$ are solved using Eq. (2.142) in the standard manner, and then $\tilde{\boldsymbol{\vartheta}}$ is determined within each element.

For example, using the Newmark method, the following is required at time t_{n+1}:

$$\mathrm{d}\tilde{\boldsymbol{u}}_{n+1} = \beta \Delta t^2 \mathrm{d}\tilde{\boldsymbol{a}}_{n+1} \tag{2.144}$$

Thus, the following system of linear equations can be obtained:

$$\begin{bmatrix} \frac{1}{\beta \Delta t^2} \boldsymbol{M} + \overline{\boldsymbol{K}}_{uu} & \overline{\boldsymbol{K}}_{up} \\ \overline{\boldsymbol{K}}_{pu} & \overline{\boldsymbol{K}}_{pp} \end{bmatrix} \begin{Bmatrix} \mathrm{d}\tilde{\boldsymbol{u}} \\ \mathrm{d}\tilde{\boldsymbol{p}} \end{Bmatrix}_{n+1} = \begin{Bmatrix} \overline{\boldsymbol{\Psi}}_{u} \\ \overline{\boldsymbol{\Psi}}_{p} \end{Bmatrix}_{n+1} \tag{2.145}$$

Then, the solution is iteratively updated using the standard Newton method.

Example 2.2 Triangular Mixed u-p Element

Consider a 6-node triangular element, as shown in Fig. 2.13. All six nodes are used for interpolation of both the coordinates **x** and the displacement **u**, while the pressure is interpolated linearly using only the vertices, that is,

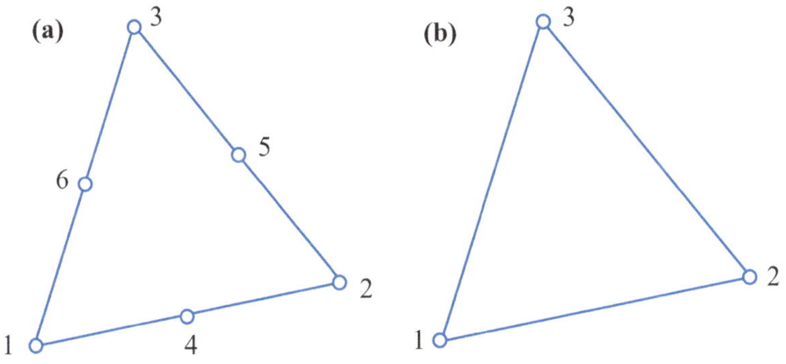

Fig. 2.13 Six-node triangular element for **u** and linear element for **p**, **a** Displacement nodes, **b** Pressure nodes

$$\begin{aligned} \boldsymbol{x} &= \boldsymbol{N}\tilde{\boldsymbol{x}} = \sum_{b=1}^{6} N_b(\boldsymbol{\xi})\tilde{\boldsymbol{x}}_b \\ \boldsymbol{u} &= \boldsymbol{N}\tilde{\boldsymbol{u}} = \sum_{b=1}^{6} N_b(\boldsymbol{\xi})\tilde{\boldsymbol{u}}_b \\ \boldsymbol{p} &= \boldsymbol{N}\tilde{\boldsymbol{p}} = \sum_{c=1}^{3} N_c(\boldsymbol{\xi})\tilde{p}_c \end{aligned} \tag{2.146}$$

Chapter 4 will present the shape function expressions in terms of area coordinates. Strain interpolation within each element is given by linear polynomials.

$$\boldsymbol{\vartheta} = \begin{bmatrix} 1 & (x - x_0) & (y - y_0) \end{bmatrix} \begin{Bmatrix} \tilde{\vartheta}_1 \\ \tilde{\vartheta}_2 \\ \tilde{\vartheta}_3 \end{Bmatrix}$$

Here, x_0 and y_0 are defined at the centroid of the triangle to improve numerical stability.

2.7 Nonlinear Quasi-Conforming Field Problems

The final two chapters of this book will explore nonlinear problems arising from inelastic constitutive behavior and finite deformations. Nonlinearity also appears in many other contexts, but the methods introduced in this chapter remain broadly applicable to such problems. Here, we present an example of a quasi-coordinated equation, which is commonly encountered in various engineering fields.

Consider a simple quasi-harmonic problem with the dependent variable ϕ (for example, heat conduction), whose governing equation is

$$\rho c\dot{\phi} + \nabla^{\mathrm{T}}\boldsymbol{q} - Q(\phi) = 0 \tag{2.147}$$

When solving it, it is also necessary to incorporate specific boundary conditions. This form can be used to determine the temperature response in solids and the flow in porous media, leakage, magnetic effects in solids, and issues such as vorticity flow. In Eq. (2.147), **q** is a flux, which can generally be expressed as

$$\boldsymbol{q} = \boldsymbol{q}(\phi, \nabla\phi) = -\boldsymbol{k}(\phi, \nabla\phi)\nabla\phi$$

Alternatively, after linearization, we obtain

$$\mathrm{d}\boldsymbol{q} = -\boldsymbol{k}^0\mathrm{d}\phi - \boldsymbol{k}^1\mathrm{d}(\nabla\phi)$$

where

$$k_i^0 = -\frac{\partial q_i}{\partial \phi}, k_{ij}^1 = -\frac{\partial q_i}{\partial \phi_{,j}}$$

In Eq. (2.147), $Q(\phi)$ can also be introduced as a nonlinear function.

Using the Galerkin method, after discretization and partial integration with respect to **q**, the problem can be expressed as

$$\mathrm{G} = \int_\Omega \delta\phi\rho c\dot{\phi}\mathrm{d}\Omega - \int_\Omega (\nabla\delta\phi)^{\mathrm{T}}\boldsymbol{q}\mathrm{d}\Omega - \int_\Omega \delta\phi Q(\phi)\mathrm{d}\Omega - \int_{\Gamma_q} \delta\phi\overline{q}_n\mathrm{d}\Gamma = 0 \tag{2.148}$$

If **q** and Q (which are actually boundary conditions) are functions of ϕ or its derivatives, the above equation still holds. Introducing interpolation

$$\phi = \boldsymbol{N}\tilde{\boldsymbol{\phi}}(t), \delta\phi = \boldsymbol{N}\delta\tilde{\boldsymbol{\phi}} \tag{2.149}$$

The discrete form can be obtained as

$$\boldsymbol{\Psi}_q = \boldsymbol{f}(\tilde{\boldsymbol{\phi}}) - \boldsymbol{C}\dot{\tilde{\boldsymbol{\phi}}} - \boldsymbol{P}_q(\tilde{\boldsymbol{\phi}}) = \boldsymbol{0} \tag{2.150}$$

where

$$\begin{aligned} \boldsymbol{C} &= \int_\Omega \boldsymbol{N}^{\mathrm{T}}\rho c\boldsymbol{N}\mathrm{d}\Omega, \quad \boldsymbol{P}_q = \int_\Omega (\nabla\boldsymbol{N})^{\mathrm{T}}\boldsymbol{q}\mathrm{d}\Omega, \\ \boldsymbol{f} &= \int_\Omega \boldsymbol{N}^{\mathrm{T}}Q(\phi)\mathrm{d}\Omega - \int_{\Gamma_q} \boldsymbol{N}^{\mathrm{T}}\overline{q}_n\mathrm{d}\Gamma \end{aligned} \tag{2.151}$$

Equation (2.150) can be solved using a procedure similar to the one described above. For example, as in the Newmark method, the generalized midpoint formula can be used, that is,

$$\boldsymbol{\phi}_{n+1} = \boldsymbol{\phi}_n + (1-\gamma)\Delta t\dot{\boldsymbol{\phi}}_n + \gamma\Delta t\dot{\boldsymbol{\phi}}_{n+1}, \quad 0 \leq \gamma \leq 1 \tag{2.152}$$

This is identical to the velocity expression in the Newmark method. In Eq. (2.152), $\Delta t = t_{n+1} - t_n$. Similarly, either ϕ_{n+1} and $\widetilde{\phi}_{n+1}$ can be chosen as the primary unknown. Therefore, the procedure for solving this transient problem is exactly the same as the one described in the previous section, and no further discussion is necessary here.

Note: Using ϕ_{n+1} as the primary variable allows this solution method to be applied to static (steady-state) problems. For static problems, the first term in Eq. (2.147) becomes zero.

The Newton–Raphson method is used below to solve Eq. (2.150) at each discrete time step t_{n+1}. In the discrete form of Eq. (2.150), $\tilde{\phi}_{n+1}$ can be used as the primary unknown, and the discrete result can be expressed as follows:

$$\boldsymbol{\Psi}_{n+1} = \boldsymbol{f} - \frac{1}{\gamma\Delta t}\boldsymbol{C}\left(\tilde{\boldsymbol{\phi}}_{n+1} - \hat{\tilde{\boldsymbol{\phi}}}_{n+1}\right) - \boldsymbol{P}_q = \boldsymbol{0} \tag{2.153}$$

Here, $\hat{\tilde{\boldsymbol{\phi}}}_{n+1} = \tilde{\boldsymbol{\phi}}_n + (1-\gamma)\Delta t\dot{\tilde{\boldsymbol{\phi}}}_n$ is linearized, so that each Newton iteration becomes

$$\left[\frac{1}{\gamma\Delta t}\boldsymbol{C} + \boldsymbol{H}_T\right]^k \mathrm{d}\tilde{\boldsymbol{\phi}}^k_{n+1} = \boldsymbol{\Psi}^k_{n+1} \tag{2.154}$$

and perform updates until convergence is achieved. The tangent term $\boldsymbol{H}_{\mathrm{T}}$ is given by the following expression

$$\boldsymbol{H}_T = \int_\Omega (\nabla N_a)^{\mathrm{T}} \boldsymbol{k}_T \nabla N_b \mathrm{d}\Omega \tag{2.155}$$

where $\boldsymbol{k}_T = -\frac{\partial \mathbf{q}}{\partial \nabla\phi}$.

2.8 Summary

This chapter presents the fundamental steps required to solve general small-strain solid mechanics problems as well as quasi-compatible field problems. The standard Newton method is used to solve the resulting nonlinear algebraic equations. For many problems involving nonlinear behaviour, additional solution strategies are necessary.

Although this chapter does not address large strain problems, the solution strategies described above remain applicable. The difference lies in the fact that the occurrence of finite deformation affects the calculation methods for stress, stress divergence, tangent modulus, and stiffness. For details, refer to Chap. 8 of this book.

Chapter 3
Finite Element Analysis of Trusses and Beams

Trusses and beams are two of the simplest and most widely used structural elements or components. Trusses are designed as straight members subjected only to axial forces, and therefore experience only axial deformation. Beams, on the other hand, are straight members with arbitrary cross-sections, but they deform only in directions perpendicular to their axes. The cross-section of a truss or beam can be arbitrary, but its dimensions should be much smaller than its axial length. The main difference between trusses and beams lies in the types of loads they carry. Beams are subjected to transverse loads, including transverse forces and moments that cause lateral deformation. This chapter introduces the finite element equations for truss and beam members, with the corresponding elements commonly referred to as truss (bar) elements and beam elements. Two other commonly used one-dimensional elements are the torsional shaft element and the deep beam bending element. Their finite element formulations are similar to those of truss elements, so only their weak forms are presented here.

In bar structures composed of truss elements, the truss members are connected by pins or hinges (rather than by welding), so only forces (and not moments) are transmitted between the bars. In beam structures, the beams are joined by welding (unlike truss elements, which are connected by pins or hinges), allowing both forces and moments to be transmitted between the beams. To clarify these concepts, this book assumes that the cross-sectional shape of truss and beam elements remains constant. Therefore, to address trusses or beams with varying cross-sections, it is necessary to develop variable cross-section truss or beam elements, which can be accomplished following the procedures for uniform cross-section members. It should be noted that, from a mechanical perspective, using variable cross-section bars is inappropriate, as the force distribution in bars is uniform. The beam elements developed in this chapter are based on the Euler–Bernoulli beam theory, which is applicable to slender beams.

B. Liu and C. Liu, *Introduction to Linear and Nonlinear Finite Element Analysis*,
https://doi.org/10.1007/978-981-95-9670-6_3

This chapter first introduces the energy functionals for truss (bar) elements, torsional shaft elements, beam bending elements, and deep (shear) beam bending elements, followed by a discussion of h- and p-type finite elements for trusses and beams.

3.1 Element Energy Functional

The equations of motion for any structural system can be derived from the system's energy functionals. These energy functionals include strain energy, kinetic energy, dissipation energy, and the virtual work done by external loads. This section will present the energy functionals for trusses and beams.

The following derivation is based on linear elasticity theory, where both the stress–strain and strain–displacement relationships are linear. The stress state in a three-dimensional elastic body is defined by the stress components (with reference to Cartesian coordinates x, y, z), namely,

$$[\sigma]=\begin{bmatrix} \sigma_x\, \tau_{xy}\, \tau_{xz} \\ \tau_{yx}\, \sigma_y\, \tau_{yz} \\ \tau_{zx}\, \tau_{zy}\, \sigma_z \end{bmatrix} \tag{3.1}$$

and

$$\tau_{yx}=\tau_{xy}, \quad \tau_{zx}=\tau_{xz}, \quad \tau_{zy}=\tau_{yz} \tag{3.2}$$

when relating stress and strain, it is important to consider whether the material is anisotropic, orthotropic, or isotropic. For one-dimensional elements (such as rods, shafts, and beams), it is generally sufficient to consider only the isotropic case.

The strain state in a three-dimensional elastic body is defined by its strain components, namely

$$[\boldsymbol{\varepsilon}]=\begin{bmatrix} \varepsilon_x\, \gamma_{xy}\, \gamma_{xz} \\ \gamma_{xy}\, \varepsilon_y\, \gamma_{yz} \\ \gamma_{xz}\, \gamma_{yz}\, \varepsilon_z \end{bmatrix} \tag{3.3}$$

If the displacement components in the coordinate axes directions are denoted by (u, v, w), then the strain–displacement relationship is

$$\begin{aligned} &\varepsilon_x=\frac{\partial u}{\partial x}, \quad \varepsilon_y=\frac{\partial v}{\partial y}, \quad \varepsilon_z=\frac{\partial w}{\partial z} \\ &\gamma_{xy}=\frac{\partial u}{\partial y}+\frac{\partial v}{\partial x}, \quad \gamma_{xz}=\frac{\partial u}{\partial z}+\frac{\partial w}{\partial x}, \quad \gamma_{yz}=\frac{\partial v}{\partial z}+\frac{\partial w}{\partial y} \end{aligned} \tag{3.4}$$

It should be noted that all displacement components are time-dependent. The dynamic equilibrium equation is

$$\begin{aligned}
&\frac{\partial \sigma_x}{\partial x}+\frac{\partial \tau_{yx}}{\partial y}+\frac{\partial \tau_{zx}}{\partial z}+f_x=\rho \ddot{u}\\
&\frac{\partial \tau_{xy}}{\partial x}+\frac{\partial \sigma_y}{\partial y}+\frac{\partial \tau_{zy}}{\partial z}+f_y=\rho \ddot{v}\\
&\frac{\partial \tau_{xz}}{\partial x}+\frac{\partial \tau_{yz}}{\partial y}+\frac{\partial \sigma_z}{\partial z}+f_z=\rho \ddot{w}
\end{aligned} \tag{3.5}$$

When considering the spatial variation of these quantities, the explicit representation of time is usually omitted.

3.1.1 Bar (Axial) Element

As shown in Fig. 3.1, consider a bar (axial) element with a cross-sectional area A and length $2a$. For elements whose length is more than ten times their cross-sectional dimension, it can be assumed that each cross-section remains planar during deformation. Furthermore, except for the axial component σ_x, all other stress components can be neglected, and σ_x is uniformly distributed across each cross-section.

In Fig. 3.1, the axial force on one face of the differential element $\mathrm{d}x$ is $\sigma_x A$, and the elongation of this element is $\varepsilon_x \mathrm{d}x$, where ε_x is the axial strain components. Therefore, the work done by the axial force on this infinitesimal element $\mathrm{d}W$ is

$$\mathrm{d}W=\frac{1}{2}\sigma_x A\cdot\varepsilon_x \mathrm{d}x=\frac{1}{2}\sigma_x\varepsilon_x A\mathrm{d}x \tag{3.6}$$

This work is stored as strain energy $\mathrm{d}U$, that is,

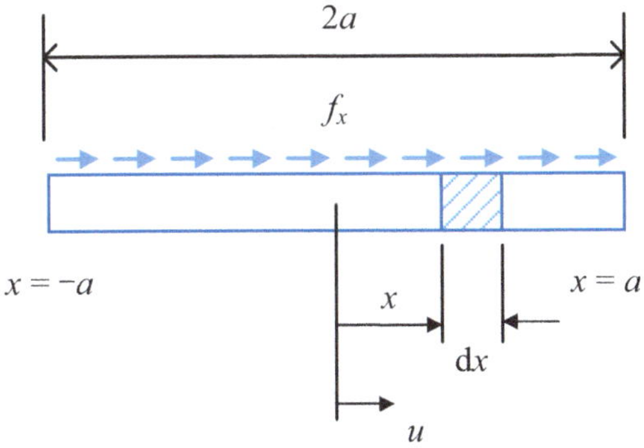

Fig. 3.1 Rod (axial) element in the local coordinate system

$$\mathrm{d}U = \frac{1}{2}\sigma_x \varepsilon_x A \mathrm{d}x \tag{3.7}$$

Therefore, the total strain energy in the entire unit is

$$U = \frac{1}{2}\int_{-a}^{+a} \sigma_x \varepsilon_x A \mathrm{d}x \tag{3.8}$$

Assuming a linear stress–strain relationship, the axial stress component σ_x can be directly expressed as follows:

$$\sigma_x = E\varepsilon_x \tag{3.9}$$

where E is the Young's modulus of the material. Substituting Eq. (3.9) into Eq. (3.8) yields

$$U = \frac{1}{2}\int_{-a}^{+a} EA\varepsilon_x^2 \mathrm{d}x \tag{3.10}$$

According to Eq. (3.4), the axial strain component can be expressed in terms of the axial displacement $u(x)$ as follows:

$$\varepsilon_x = \frac{\partial u}{\partial x} \tag{3.11}$$

Substituting Eq. (3.11) into Eq. (3.10), we obtain

$$U = \frac{1}{2}\int_{-a}^{+a} EA\left(\frac{\partial u}{\partial x}\right)^2 \mathrm{d}x \tag{3.12}$$

The kinetic energy of the infinitesimal element dx is $(1/2)\dot{u}^2\rho A\mathrm{d}x$, where ρ is the mass per unit volume of the material. Therefore, the total kinetic energy of the entire element is

$$T = \frac{1}{2}\int_{-a}^{+a} \rho A\dot{u}^2 \mathrm{d}x \tag{3.13}$$

If there is an external load of magnitude f_x per unit length, as shown in Fig. 3.1, then the external force acting on the infinitesimal element dx is $f_x\mathrm{d}x$, and the work done by this force during the virtual displacement δu is $\delta u \cdot f_x \mathrm{d}x$. Therefore, the total virtual work for the entire element is

$$\delta W = \int_{-a}^{+a} f_x \delta u \mathrm{d}x \tag{3.14}$$

Neglecting the y and z dimension terms in Eq. $(3.5)_1$, the one-dimensional dynamic equilibrium equation for a solid can be obtained as follows:

$$\frac{\partial \sigma_x}{\partial x} + f_x = \rho \ddot{u} \tag{3.15}$$

Substituting Eqs. (3.9) and (3.11) into Eq. (3.15), the governing equation for an isotropic elastic (E independent of x) truss can be obtained as follows:

$$E\frac{\partial^2 u}{\partial x^2} + f_x = \rho \ddot{u} \tag{3.16}$$

3.1.2 Torsional Shaft Element

As shown in Fig. 3.2, consider a torsion shaft element with a constant cross-sectional area A and length $2a$. Assume that the element undergoes torsional deformation only about the x-axis, and let the angle of twist at position x be denoted by θ_x.

If the length of a structural member is more than 10 times its cross-sectional dimensions, the Saint–Venant torsion theory is applicable. This theory assumes that the deformation of the torsional member consists of the rotation of the cross-section about the axis and the warping of the cross-section. Let the y-axis and z-axis be perpendicular to the x-axis; the displacements u, v and w in the x, y and z directions, respectively, can be expressed as follows:

$$\begin{aligned} u(x, y, z) &= \frac{\partial \theta_x(x)}{\partial x}\psi(y, z) \\ v(x, y, z) &= -\theta_x(x)z \\ w(x, y, z) &= \theta_x(x)y \end{aligned} \tag{3.17}$$

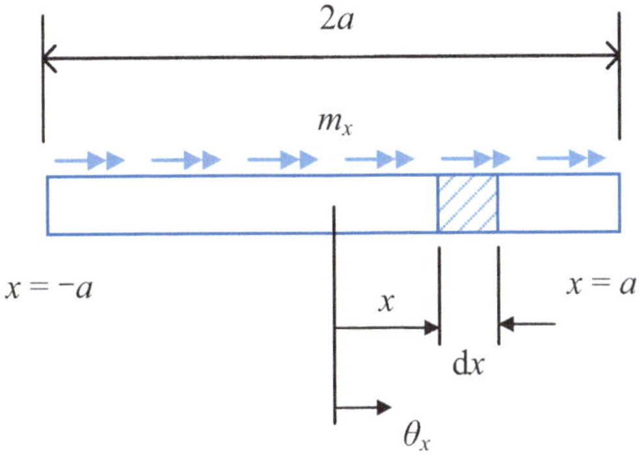

Fig. 3.2 Torsional shaft element in the local coordinate system

Here, $\psi(y, z)$ is a function representing the degree of cross-sectional warping. Therefore, the strain components can be expressed as

$$\begin{aligned}
\varepsilon_y &= \varepsilon_z = \gamma_{yz} = 0 \\
\varepsilon_x &= \frac{\partial^2 \theta}{\partial x^2}\psi \\
\gamma_{xy} &= \frac{\partial \theta_x}{\partial x}\left(\frac{\partial \psi}{\partial y} - z\right) \\
\gamma_{xz} &= \frac{\partial \theta_x}{\partial x}\left(\frac{\partial \psi}{\partial z} + y\right)
\end{aligned} \tag{3.18}$$

In most cases, the axial strain component ε_x can be neglected (an exception is thin-walled open-section members). Therefore, the only strain components that need to be considered are the shear strains γ_{xy} and γ_{xz}, and their corresponding stress components can be expressed as follows:

$$\tau_{xy} = G\gamma_{xy}, \quad \tau_{xz} = G\gamma_{xz} \tag{3.19}$$

Here, G denotes the shear modulus of the material.

As shown in Fig. 3.3, on an element with thickness dz, the shear stress τ_{xy} generates a pair of torques with magnitude τ_{xy}dzdx, resulting in a rotation of $\partial v/\partial x$. Similarly, the shear stress τ_{yx} produces a pair of torques with magnitude τ_{yx}dxdzdy, leading to a rotation of $\partial u/\partial y$. Since $\tau_{yx} = \tau_{xy}$, the work done by the shear stress on the element is

$$\mathrm{d}W = \frac{1}{2}\tau_{xy}\left(\frac{\partial v}{\partial x} + \frac{\partial u}{\partial y}\right)\mathrm{d}x\mathrm{d}y\mathrm{d}z = \frac{1}{2}\tau_{xy}\gamma_{xy}\mathrm{d}A\mathrm{d}x \tag{3.20}$$

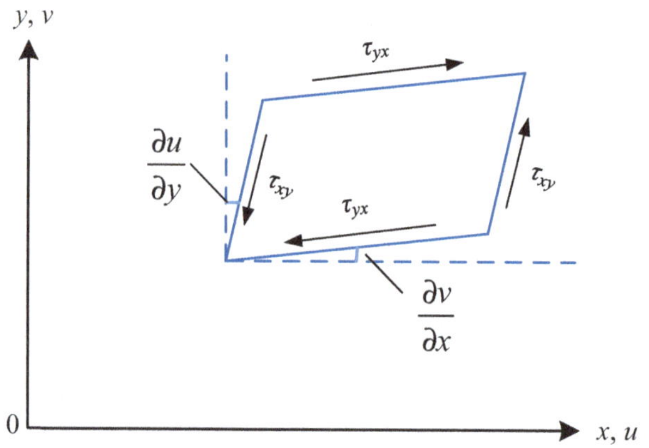

Fig. 3.3 Stress and deformation in the xy-plane coordinates

Here, dA represents the area of an infinitesimal cross-sectional element.

Therefore, for $\tau_{xz} = \tau_{zx}$ the work done by the shear stress τ_{xz} is

$$\mathrm{d}W = \frac{1}{2}\tau_{xz}\gamma_{xz}\mathrm{d}A\mathrm{d}x \tag{3.21}$$

Since the strain energy is equal to the total work done, we have

$$U = \frac{1}{2}\int_{-a}^{+a}\int_{A}\left(\tau_{xy}\gamma_{xy} + \tau_{xz}\gamma_{xz}\right)\mathrm{d}A\mathrm{d}x \tag{3.22}$$

Substituting Eqs. (3.18) and (3.19) into Eq. (3.22) yields

$$U = \frac{1}{2}\int_{-a}^{+a} GJ\left(\frac{\partial\theta_x}{\partial x}\right)^2\mathrm{d}x \tag{3.23}$$

where

$$J = \int_{A}\left[\left(\frac{\partial\psi}{\partial y} - z\right)^2 + \left(\frac{\partial\psi}{\partial z} + y\right)^2\right]\mathrm{d}A \tag{3.24}$$

is the torsion constant of the cross-section. When $\psi = 0$, J reduces to the polar moment of area I_x; for other solid cross-sections, J can be approximately given as

$$J \simeq 0.025A^4/I_x \tag{3.25}$$

Note: This formula is only applicable to slender structures.

When deriving the expression for the kinetic energy of an element, the longitudinal displacement caused by cross-sectional warping can be neglected. Thus, the kinetic energy of the infinitesimal element dx is $(1/2)\dot{\theta}^2\rho I_x\mathrm{d}x$, where I_x is the moment of inertia of the cross-section about the x-axis. Therefore, the total kinetic energy of the entire element is

$$T = \frac{1}{2}\int_{-a}^{+a}\rho I_x\dot{\theta}_x^2\mathrm{d}x \tag{3.26}$$

If a distributed torsional moment of magnitude m_x acts per unit length, as shown in Fig. 3.2, then the moment acting on the infinitesimal element dx is $m_x\mathrm{d}x$, and the virtual work done during the virtual displacement $\delta\theta_x$ is $\delta\theta_x m_x\mathrm{d}x$. Therefore, the total virtual work for the entire element is

$$\delta W = \int_{-a}^{+a} m_x\delta\theta_x^2\mathrm{d}x \tag{3.27}$$

3.1.3 Beam Bending Element

In deriving the energy functional for a beam bending element, it is assumed that deformation occurs within one of the principal planes of the beam. As shown in Fig. 3.4, a straight beam element with length $2a$ and constant cross-sectional area A is considered, where the xy-plane is the principal plane of deformation and the x-axis coincides with the centroidal axis.

If the length of a structural member is more than 10 times its cross-sectional dimension, the fundamental assumptions of beam bending theory apply. This theory assumes that the stress components σ_y, σ_z, τ_{yz} and τ_{xz} are zero, and that plane cross-sections perpendicular to the undeformed centroidal axis remain plane and perpendicular to the deformed axis after bending. Under these assumptions, the axial displacement u at a distance y from the centroidal axis is given by

$$u(x, y) = -y\frac{\partial v}{\partial x} \tag{3.28}$$

Here, $v = v(x)$ is the displacement function of the centroidal axis in the y-direction at position x. Therefore, the strain components ε_x and γ_{xy} can be expressed as follows:

$$\begin{aligned} \varepsilon_x &= \frac{\partial u}{\partial x} = -y\frac{\partial^2 u}{\partial x^2} \\ \gamma_{xy} &= \frac{\partial u}{\partial y} + \frac{\partial v}{\partial x} = 0 \end{aligned} \tag{3.29}$$

The normal stress can be expressed as

$$\sigma_x = E\varepsilon_x \tag{3.30}$$

The strain energy stored in the element is

$$U = \frac{1}{2}\int_V \sigma_x \varepsilon_x \mathrm{d}V \tag{3.31}$$

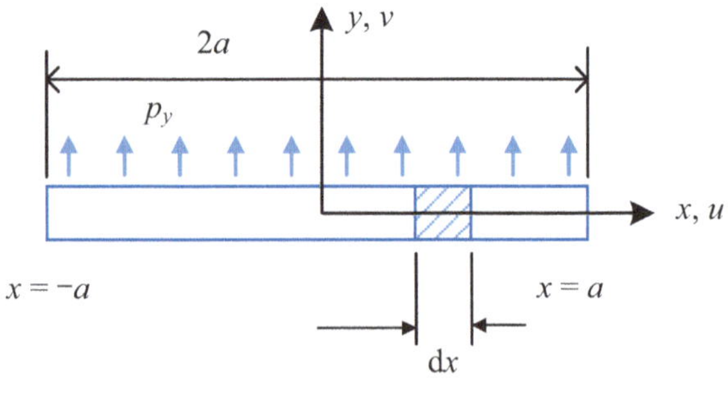

Fig. 3.4 Straight beam element in the local coordinate system

Substituting Eqs. (3.29) and (3.30) into Eq. (3.31), and combining with $\mathrm{d}V = \mathrm{d}A \cdot \mathrm{d}x$, we obtain

$$U = \frac{1}{2}\int_{-a}^{+a} EI_z\left(\frac{\partial^2 v}{\partial x^2}\right)^2 \mathrm{d}x \tag{3.32}$$

where

$$I_z = \int_A y^2 \mathrm{d}A \tag{3.33}$$

Here, I_z is the cross-sectional moment of inertia about the z-axis.

The stress–strain relationship is

$$\tau_{xy} = G\gamma_{xy} \tag{3.34}$$

Substituting Eq. (3.29) into Eq. (3.34) yields τ_{xy} is zero. In fact, this stress component is nonzero and can be proved using the equilibrium equation.

The kinetic energy of the infinitesimal element $\mathrm{d}x$ is $(1/2)\dot{v}^2\rho A\mathrm{d}x$. Therefore, the total kinetic energy of the entire element is

$$T = \frac{1}{2}\int_{-a}^{+a} \rho A\dot{v}^2 \mathrm{d}x \tag{3.35}$$

If there is a distributed load of magnitude p_y per unit length, as shown in Fig. 3.4, then the force on the infinitesimal element $\mathrm{d}x$ is $p_y\mathrm{d}x$, and the work done during the virtual displacement δv is $\delta v p_y \mathrm{d}x$. Therefore, the total virtual work for the entire element is

$$\delta W = \int_{-a}^{+a} p_y \delta v \, \mathrm{d}x \tag{3.36}$$

Since the load on the beam is transverse, there will be a moment and corresponding shear force on the cross-section of the beam. In addition, applying a pure moment instead of a transverse load can also cause the beam to bend. As shown in Fig. 3.5, a beam element of length $\mathrm{d}x$ is illustrated. It is subjected to the combined effects of external force F_z, moment M, shear force Q, and inertial force $\rho A\,\ddot{w}$, where ρ is the material density and A is the cross-sectional area.

The moment generated by the normal stress on the cross-section can be calculated by integrating over the cross-sectional area, that is,

$$M = \int_A \sigma_{xx} z \mathrm{d}A = -E\left(\int_A z^2 \mathrm{d}A\right)\frac{\partial^2 w}{\partial x^2} = -EI_y\frac{\partial^2 w}{\partial x^2} \tag{3.37}$$

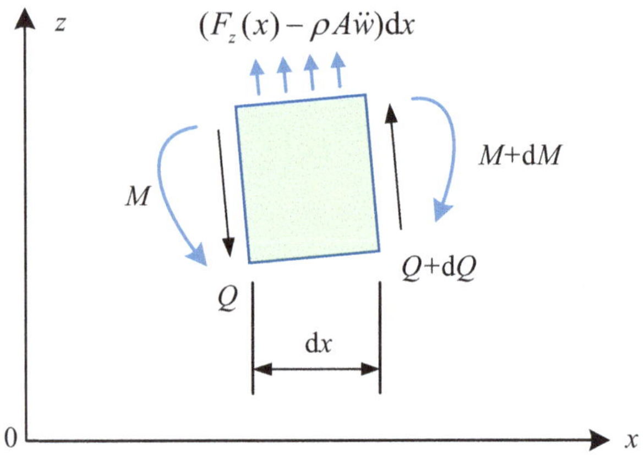

Fig. 3.5 Beam element of length dx

Here, I_y is the moment of inertia of the cross-section with respect to the y-axis, which can be calculated based on the given cross-sectional shape, namely

$$I_y = \int_A z^2 \mathrm{d}A \tag{3.38}$$

The force equilibrium of the beam element in the z-direction can be expressed as

$$\mathrm{d}Q + (F_z(x) - \rho A\ddot{w})\mathrm{d}x = 0 \tag{3.39}$$

or

$$\frac{\mathrm{d}Q}{\mathrm{d}x} = -F_z(x) + \rho A\ddot{w} \tag{3.40}$$

The moment equilibrium of the beam element with respect to any point on the right surface of the differential element can be expressed as

$$\mathrm{d}M - Q\mathrm{d}x + \frac{1}{2}(F_z(x) - \rho A\ddot{w})(\mathrm{d}x)^2 = 0 \tag{3.41}$$

By neglecting the second-order infinitesimal term containing $(\mathrm{d}x)^2$ in Eq. (3.41), we obtain

$$\frac{\mathrm{d}M}{\mathrm{d}x} = Q \tag{3.42}$$

Substituting Eq. (3.37) into Eq. (3.42), we obtain

$$Q = -EI_y \frac{\partial^3 w}{\partial x^3} \tag{3.43}$$

Equations (3.42) and (3.43) describe the relationship between the bending moment and shear force of a beam and the deflection of an Euler–Bernoulli beam. Substituting Eq. (3.43) into Eq. (3.40) yields the dynamic equilibrium equation for the beam.

$$EI_y \frac{\partial^4 w}{\partial x^4} + \rho A \ddot{w} = F_z \tag{3.44}$$

3.1.4 Shear Bending Beam Element

The velocity of flexural waves is much lower than that of longitudinal or torsional waves. At low frequencies, the wavelength of flexural waves in a beam may be less than ten times the size of the beam's cross-sectional dimension. This situation can arise when analyzing the low-frequency vibration of deep beams or the high-frequency vibration of slender beams. Therefore, in such cases, the deformation caused by transverse shear and the kinetic energy due to cross-sectional rotation become non-negligible.

In deriving the energy expressions that account for shear deformation and rotary inertia, it is still assumed that cross-sections perpendicular to the undeformed axis remain planar after bending. However, it is no longer assumed that these sections remain perpendicular to the deformed axis. Therefore, the axial displacement u at a distance y from the centroidal axis can be expressed as

$$u(x, y) = -y\theta_z(x) \tag{3.45}$$

Here, $\theta_z(x)$ denotes the rotation angle of the cross-section at position x.

The strain components ε_x and γ_{xy} can be expressed as follows:

$$\begin{aligned} \varepsilon_x &= \frac{\partial u}{\partial x} = -y\frac{\partial \theta_z}{\partial x} \\ \gamma_{xy} &= \frac{\partial u}{\partial y} + \frac{\partial v}{\partial x} = -\theta_z + \frac{\partial v}{\partial x} \end{aligned} \tag{3.46}$$

Therefore, the strain energy stored in the beam is the sum of the energies generated by bending and shear deformation, namely,

$$U = \frac{1}{2}\int_V \sigma_x \varepsilon_x \mathrm{d}V + \frac{1}{2}\int_V \tau_{xy}\gamma_{xy}\mathrm{d}V \tag{3.47}$$

The normal stress can be expressed as

$$\sigma_x = E\varepsilon_x \tag{3.48}$$

The shear stress τ_{xy} corresponding to a given shear force varies across the cross-section. Consequently, the associated shear strain also varies across the cross-section. When assuming that the cross-section remains planar, the variation of strain across the section is neglected. To account for the resulting error, a correction factor κ, which depends on the shape of the cross-section, can be introduced. Thus, we have

$$\tau_{xy} = \kappa G\gamma_{xy} \tag{3.49}$$

Here, τ_{xy} denotes the average shear stress on the cross-section; the value of the cross-sectional shape correction factor κ varies in the literature, for example, 5/6, $\pi^2/12$ etc. Substituting Eqs. (3.48) and (3.49) into Eq. (3.47), we obtain

$$U = \frac{1}{2}\int_V E\varepsilon_x^2 \mathrm{d}V + \frac{1}{2}\int_V \kappa G\gamma_{xy}^2 \mathrm{d}V \tag{3.50}$$

Substituting Eq. (3.46) into Eq. (3.50), and using $\mathrm{d}V = \mathrm{d}A\mathrm{d}x$, we obtain

$$U = \frac{1}{2}\int_{-a}^{+a} EI_z\left(\frac{\partial\theta_z}{\partial x}\right)^2 \mathrm{d}x + \frac{1}{2}\int_{-a}^{+a} \kappa AG\left(\frac{\partial v}{\partial x} - \theta_z\right)^2 \mathrm{d}x \tag{3.51}$$

The total kinetic energy of the beam consists of translational kinetic energy and rotational kinetic energy, and can be expressed as

$$T = \frac{1}{2}\int_{-a}^{+a} \rho A\dot{v}^2 \mathrm{d}x + \frac{1}{2}\int_{-a}^{+a} \rho I_z\dot{\theta}_z^2 \mathrm{d}x \tag{3.52}$$

The total virtual work of the distributed load is given by Eq. (3.36).

3.2 Truss Finite Elements

3.2.1 *Construction of Shape Functions*

Consider a structure composed of numerous trusses or bars. Each element is defined by two nodes ($n_d = 2$) and is of uniform cross-section. As shown in Fig. 3.6, each element has two end nodes (node 1 and 2), with the element length denoted as l_e. The local coordinate x is oriented along the axis of the element, with the origin located at local node 1. In the local coordinate system, each node of the element has only one degree of freedom, namely the axial displacement. Therefore, the element has a total of two degrees of freedom. That is, $n_e = 2$. According to the finite element method discussed in Chap. 2, the displacement of the element can be expressed as

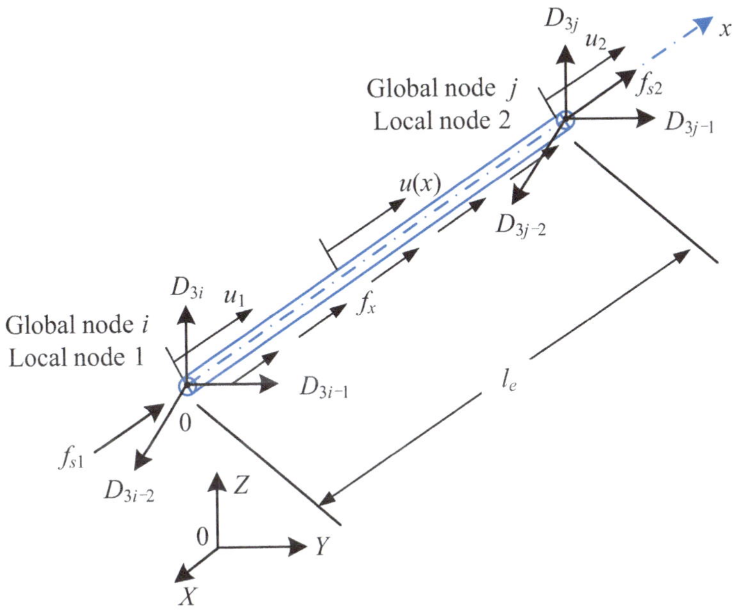

Fig. 3.6 Truss/bar element and coordinate system

$$u^h(x) = \mathbf{N}(x)\mathbf{d}_e \tag{3.53}$$

Here, u^h represents the approximate value of the axial displacement within the element, $\mathbf{N}$ denotes the shape function with the properties described in Chap. 1, and $\mathbf{d}_e$ is the displacement vector at the two nodes of the element.

$$\mathbf{d}_e = \begin{bmatrix} u_1 \\ u_2 \end{bmatrix} \tag{3.54}$$

In the following, shape functions are constructed according to the standard procedure described in Sect. 1.4.3, assuming that the axial displacement of the truss element can be expressed in the following general form

$$u^h(x) = \alpha_0 + \alpha_1 x = \underbrace{[1 \quad x]}_{\mathbf{p}^{\mathrm{T}}} \underbrace{\begin{bmatrix} u_1 \\ u_2 \end{bmatrix}}_{\alpha} = \mathbf{p}^{\mathrm{T}}\boldsymbol{\alpha} \tag{3.55}$$

Here, u^h denotes the approximate value of the displacement, $\boldsymbol{\alpha}$ is a vector consisting of two unknown constants α_0 and α_1, and $\boldsymbol{p}$ represents the polynomial (or monomial) basis function vector. For this specific problem, a first-order polynomial basis function vector is used. Depending on the particular problem, higher-order polynomial basis function vectors may also be employed. The polynomial basis function vector of order n is given by

$$\mathbf{p}^{\mathrm{T}}=\begin{bmatrix}1 & x & \cdots & x^{n}\end{bmatrix} \tag{3.56}$$

The number of polynomial (or monomial) basis functions per element depends on the number of element nodes and the degrees of freedom within the element. Since there are only two nodes here, there are a total of two degrees of freedom within the element. Therefore, two basis functions are selected, which results in Eq. (3.55).

Note: Typically, using a complete set of polynomial basis functions means that no lower-order terms are omitted when constructing Eq. (3.55). This ensures that the resulting shape functions can reproduce polynomials up to degree n. If a k-th order polynomial basis function is omitted, the constructed shape functions will only be complete up to order (k-1), regardless of how many higher-order monomials are included in the bases. This is determined by the consistency property of shape functions discussed in Sect. 1.4.4. According to properties 3, 4, and 5 of shape functions discussed in Sect. 1.4.4, it can be anticipated that the complete linear basis functions used in Eq. (3.55) ensure that the constructed shape functions satisfy the sufficiency requirements for finite element shape functions: the delta function property, partition of unity, and linear field reproduction.

During the derivation of the shape functions, the conditions $u(x=0)=u_1$ at $x=0$ and $u(x=l_e)=u_2$ at $x=l_e$ are substituted into Eq. (3.55), yielding:

$$\begin{bmatrix}u_1\\u_2\end{bmatrix}=\begin{bmatrix}1 & 0\\1 & l_e\end{bmatrix}\begin{bmatrix}\alpha_0\\\alpha_1\end{bmatrix} \tag{3.57}$$

Solving for $\boldsymbol{\alpha}$ in Eq. (3.57) yields

$$\begin{bmatrix}\alpha_0\\\alpha_1\end{bmatrix}=\begin{bmatrix}1 & 0\\-1/l_e & 1/l_e\end{bmatrix}\begin{bmatrix}u_1\\u_2\end{bmatrix} \tag{3.58}$$

Substituting Eq. (3.58) into Eq. (3.55), we obtain

$$u(x)=\mathbf{p}^{\mathrm{T}}\boldsymbol{\alpha}=\begin{bmatrix}1 & x\end{bmatrix}\begin{bmatrix}1 & 0\\-\dfrac{1}{l_e} & \dfrac{1}{l_e}\end{bmatrix}\begin{bmatrix}u_1\\u_2\end{bmatrix}=\underbrace{\begin{bmatrix}\underbrace{1-\dfrac{x}{l_e}}_{N_1(x)} & \underbrace{\dfrac{x}{l_e}}_{N_2(x)}\end{bmatrix}}_{\mathbf{N}(x)}\underbrace{\begin{bmatrix}u_1\\u_2\end{bmatrix}}_{\mathbf{d}_e}=\mathbf{N}(x)\mathbf{d}_e \tag{3.59}$$

This is exactly the desired Eq. (3.53). The shape function matrix is then obtained as:

$$\mathbf{N}(x)=[N_1(x)\quad N_2(x)] \tag{3.60}$$

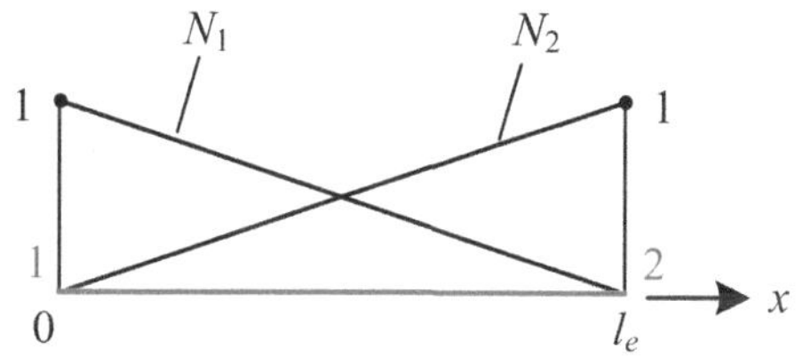

Fig. 3.7 Linear shape function

The shape function of the truss element is

$$N_1(x) = 1 - \frac{x}{l_e}, \quad N_2(x) = \frac{x}{l_e} \tag{3.61}$$

Since a truss element has two degrees of freedom, it therefore contains two shape functions.

It is easy to verify that these two shape functions satisfy the δ function property and the partition of unity characteristic required of finite element shape functions. We leave this verification as a simple exercise for the reader. The graphical representation of the linear shape functions is shown in Fig. 3.7. The figure clearly illustrates that N_i represents the contribution of the displacement at node i, which is why these functions are called shape functions. In this example, the shape functions vary linearly within the element, and are therefore referred to as linear shape functions. Substituting Eqs. (3.60) and (3.54) into Eq. (3.53) yields

$$u(x) = N_1(x)u_1 + N_2(x)u_2 = u_1 + \frac{u_2 - u_1}{l_e}x \tag{3.62}$$

This clearly indicates that the displacement within the element varies linearly. Therefore, this element is referred to as a linear element.

3.2.2 Strain Matrix

As described in Sect. 3.1, there is only one stress component σ_x in the truss. According to Eq. (3.62), the corresponding strain is

$$\varepsilon_x = \frac{\partial u}{\partial x} = \frac{u_2 - u_1}{l_e} \tag{3.63}$$

Note: The strain in Eq. (3.63) is constant within the element.

As mentioned in Sect. 2.1, once the strain matrix **B** is obtained, the stiffness and mass matrices can be determined. Equation (3.63) can be rewritten in matrix form as follows:

$$\varepsilon_x = \frac{\partial u}{\partial x} = L\mathbf{N}\mathbf{d}_e = \mathbf{B}\mathbf{d}_e \tag{3.64}$$

where $L = \partial/\partial x$,, and the strain matrix $\boldsymbol{B}$ of the truss element is

$$\mathbf{B} = L\mathbf{N} = \frac{\partial}{\partial x}\begin{bmatrix} 1 - \frac{x}{l_e} & \frac{x}{l_e} \end{bmatrix} = \begin{bmatrix} -\frac{1}{l_e} & \frac{1}{l_e} \end{bmatrix} \tag{3.65}$$

3.2.3 Element Matrices in the Local Coordinate System

Once the strain matrix $\boldsymbol{B}$ is obtained, the stiffness matrix of the truss element can be determined using Eq. (1.125) from Sect. 1.4.5.

$$\mathbf{k}_e = \int_{V_e} \mathbf{B}^{\mathrm{T}}\mathbf{C}\mathbf{B}\,\mathrm{d}V = A\int_0^{l_e} \begin{bmatrix} -\frac{1}{l_e} \\ \frac{1}{l_e} \end{bmatrix} E \begin{bmatrix} -\frac{1}{l_e} & \frac{1}{l_e} \end{bmatrix} \mathrm{d}x = \frac{AE}{l_e}\begin{bmatrix} 1 & -1 \\ -1 & 1 \end{bmatrix} \tag{3.66}$$

Here, A denotes the cross-sectional area of the truss element. It should be noted that for one-dimensional truss elements, the material constant matrix $\mathbf{c}$ degenerates to the modulus E (see Eq. (3.9)). It should be noted that the element stiffness matrix, as shown in Eq. (3.66), is symmetric. By taking advantage of the symmetry of the stiffness matrix, only half of the matrix needs to be computed and stored during calculations.

The mass matrix of the truss element can be obtained using Eq. (1.129) in Sect. 1.4.5.

$$\mathbf{m}_e = \int_{V_e} \rho\mathbf{N}^{\mathrm{T}}\mathbf{N}\,\mathrm{d}V = A\rho\int_0^{l_e} \begin{bmatrix} N_1N_1 & N_1N_2 \\ N_2N_1 & N_2N_2 \end{bmatrix} \mathrm{d}x = \frac{A\rho l_e}{6}\begin{bmatrix} 2 & 1 \\ 1 & 2 \end{bmatrix} \tag{3.67}$$

Similarly, the element mass matrix is also symmetric.

The nodal force vector of a truss element can be obtained using Eqs. (1.132) and (1.133) from Sect. 1.4.5. Suppose the element is subjected to a uniformly distributed force f_x along the x-axis, and two concentrated forces f_{s1} and f_{s2} are applied at local nodes 1 and 2, respectively, as shown in Fig. 3.6. The total nodal force vector is then given by

$$\mathbf{f}_e = \int_{V_e} \mathbf{N}^{\mathrm{T}} f_b\,\mathrm{d}V + \int_{S_e} \mathbf{N}^{\mathrm{T}} f_s\,\mathrm{d}S = f_x\int_0^{l_e} \begin{bmatrix} N_1 \\ N_2 \end{bmatrix} \mathrm{d}x + \begin{bmatrix} f_{s1} \\ f_{s2} \end{bmatrix} = \begin{bmatrix} \frac{f_x l_e}{2} + f_{s1} \\ \frac{f_x l_e}{2} + f_{s2} \end{bmatrix} \tag{3.68}$$

3.2.4 Element Matrix in the Global Coordinate System

As shown in Fig. 3.6, the element matrices given in Eqs. (3.66), (3.67), and (3.68) are all based on the local coordinate system, where the x-axis coincides with the centroidal axis of node 1–2 of the bar. In actual truss structures, there are many members oriented in different directions and positions. To assemble all element matrices into the global system matrix, it is necessary to perform a coordinate transformation for each element, thereby obtaining the element matrices in the global coordinate system of the entire truss structure. The coordinate transformation methods for both spatial and planar truss structures are presented in the following.

1. **Spatial Truss**

Assume that local nodes 1 and 2 of the element correspond to global nodes i and j, respectively, as shown in Fig. 3.6. At each global node in space, the displacement has three components in the X, Y, and Z directions, numbered sequentially. For example, the three components at node i are denoted as D_{3i-2}, D_{3i-1} and D_{3i}. The coordinate transformation provides the relationship between the displacement vector $\mathbf{d}_e$ in the local coordinate system and the displacement vector $\mathbf{D}_e$ of the same element in the global XYZ coordinate system as follows:

$$\mathbf{d}_e = \mathbf{T}\mathbf{D}_e \tag{3.69}$$

where

$$\mathbf{D}_e = \begin{bmatrix} D_{3i-2} \\ D_{3i-1} \\ D_{3i} \\ D_{3j-2} \\ D_{3j-1} \\ D_{3j} \end{bmatrix} \tag{3.70}$$

and $\mathbf{T}$ is the transformation matrix for the truss element that can be expressed as

$$\mathbf{T} = \begin{bmatrix} l_{ij} & m_{ij} & n_{ij} & 0 & 0 & 0 \\ 0 & 0 & 0 & l_{ij} & m_{ij} & n_{ij} \end{bmatrix}_e \tag{3.71}$$

where

$$
\begin{aligned}
l_{ij} &= \cos(x, X) = \frac{X_j - X_i}{l_e} \\
m_{ij} &= \cos(x, Y) = \frac{Y_j - Y_i}{l_e} \\
n_{ij} &= \cos(x, Z) = \frac{Z_j - Z_i}{l_e}
\end{aligned}
\tag{3.72}
$$

are the direction cosines of the unit axis. Since

$$
\mathbf{T}\mathbf{T}^{\mathrm{T}} = \mathbf{I} \tag{3.73}
$$

where, $\mathbf{I}$ is the 2×2 identity matrix, the matrix $\mathbf{T}$ is an orthogonal matrix. The element length l_e can be expressed in terms of the global coordinates of the two nodes of the element as follows:

$$
l_e = \sqrt{(X_j - X_i)^2 + (Y_j - Y_i)^2 + (Z_j - Z_i)^2} \tag{3.74}
$$

Equation (3.69) is easily verified, as it simply defines d_1 at node i as the sum of all projections of $D_{3i\text{-}2}$, $D_{3i\text{-}1}$ and D_{3i} on the local x-axis; the same applies to node j. The transformation matrix $\mathbf{T}$ of the truss element converts a 6×1 vector in the global coordinate system to a 2×1 vector in the local coordinate system.

The transformation matrix also applies to force vectors between local and global coordinate systems, that is,

$$
\mathbf{f}_e = \mathbf{T}\mathbf{F}_e \tag{3.75}
$$

where

$$
\mathbf{F}_e = \begin{bmatrix} F_{3i-2} \\ F_{3i-1} \\ F_{3i} \\ F_{3j-2} \\ F_{3j-1} \\ F_{3j} \end{bmatrix} \tag{3.76}
$$

Here, $F_{3i\text{-}2}$, $F_{3i\text{-}1}$ and F_{3i} represent the three components of the force vector at node i in the global coordinate system.

Substituting Eq. (3.69) into Eq. (1.136), the element equation based on the global coordinate system can be obtained as follows:

$$
\mathbf{k}_e\mathbf{T}\mathbf{D}_e + \mathbf{m}_e\mathbf{T}\ddot{\mathbf{D}}_e = \mathbf{f}_e \tag{3.77}
$$

Multiplying both sides of the above equation on the left by $\mathbf{T}^{\mathrm{T}}$ yields

$$\left(\mathbf{T}^{\mathrm{T}}\mathbf{k}_e\mathbf{T}\right)\mathbf{D}_e + \left(\mathbf{T}^{\mathrm{T}}\mathbf{m}_e\mathbf{T}\right)\ddot{\mathbf{D}}_e = \mathbf{T}^{\mathrm{T}}\mathbf{f}_e \tag{3.78}$$

or

$$\mathbf{K}_e\mathbf{D}_e + \mathbf{M}_e\ddot{\mathbf{D}}_e = \mathbf{F}_e \tag{3.79}$$

where

$$\mathbf{K}_e = \mathbf{T}^{\mathrm{T}}\mathbf{k}_e\mathbf{T}$$
$$= \frac{AE}{l_e}\begin{bmatrix} l_{ij}^2 & l_{ij}m_{ij} & l_{ij}n_{ij} & -l_{ij}^2 & -l_{ij}m_{ij} & -l_{ij}n_{ij} \\ l_{ij}m_{ij} & m_{ij}^2 & m_{ij}n_{ij} & -l_{ij}m_{ij} & -m_{ij}^2 & -m_{ij}n_{ij} \\ l_{ij}n_{ij} & m_{ij}n_{ij} & n_{ij}^2 & -l_{ij}n_{ij} & -m_{ij}n_{ij} & -n_{ij}^2 \\ -l_{ij}^2 & -l_{ij}m_{ij} & -l_{ij}n_{ij} & l_{ij}^2 & l_{ij}m_{ij} & l_{ij}n_{ij} \\ -l_{ij}m_{ij} & -m_{ij}^2 & -m_{ij}n_{ij} & l_{ij}m_{ij} & m_{ij}^2 & m_{ij}n_{ij} \\ -l_{ij}n_{ij} & -m_{ij}n_{ij} & -n_{ij}^2 & l_{ij}n_{ij} & m_{ij}n_{ij} & n_{ij}^2 \end{bmatrix} \tag{3.80}$$

and

$$\mathbf{M}_e = \mathbf{T}^{\mathrm{T}}\mathbf{m}_e\mathbf{T}$$
$$= \frac{A\rho l_e}{6}\begin{bmatrix} 2l_{ij}^2 & 2l_{ij}m_{ij} & 2l_{ij}n_{ij} & l_{ij}^2 & l_{ij}m_{ij} & l_{ij}n_{ij} \\ 2l_{ij}m_{ij} & 2m_{ij}^2 & 2m_{ij}n_{ij} & l_{ij}m_{ij} & m_{ij}^2 & m_{ij}n_{ij} \\ 2l_{ij}n_{ij} & 2m_{ij}n_{ij} & 2n_{ij}^2 & l_{ij}n_{ij} & m_{ij}n_{ij} & n_{ij}^2 \\ l_{ij}^2 & l_{ij}m_{ij} & l_{ij}n_{ij} & 2l_{ij}^2 & 2l_{ij}m_{ij} & 2l_{ij}n_{ij} \\ l_{ij}m_{ij} & m_{ij}^2 & m_{ij}n_{ij} & 2l_{ij}m_{ij} & 2m_{ij}^2 & 2m_{ij}n_{ij} \\ l_{ij}n_{ij} & m_{ij}n_{ij} & n_{ij}^2 & 2l_{ij}n_{ij} & 2m_{ij}n_{ij} & 2n_{ij}^2 \end{bmatrix} \tag{3.81}$$

Note: Coordinate transformation preserves the symmetry properties of the stiffness and mass matrices.

For the load vector given in Eq. (3.68), we have

$$\mathbf{F}_e = \mathbf{T}^{\mathrm{T}}\mathbf{f}_e$$
$$= \begin{bmatrix} (f_x l_e/2 + f_{s1})l_{ij} \\ (f_x l_e/2 + f_{s1})m_{ij} \\ (f_x l_e/2 + f_{s1})n_{ij} \\ \left(f_y l_e/2 + f_{s2}\right)l_{ij} \\ \left(f_y l_e/2 + f_{s2}\right)m_{ij} \\ \left(f_y l_e/2 + f_{s2}\right)n_{ij} \end{bmatrix} \tag{3.82}$$

Note: The stiffness matrix $\mathbf{K}_e$ and the mass matrix $\mathbf{M}_e$ have dimensions of 6×6 in the three-dimensional global coordinate system, while the displacement vector $\mathbf{D}_e$ and the force vector $\mathbf{F}_e$ have dimensions of 6×1.

2. **Planar Truss**

For planar trusses, the global XY coordinate system can be used to represent the plane of the truss. All coordinate transformation formulas can be derived from the corresponding formulas for spatial trusses by simply removing the rows and/or columns associated with the z- (or Z-) axis. The displacement at global node i should have only two components in the X and Y directions: $D_{2i\text{-}1}$ and D_{2i}. The coordinate transformation provides the relationship between the local displacement vector $\mathbf{d}_e$ and the global displacement vector $\mathbf{D}_e$, which takes the same form as Eq. (3.69), where

$$\mathbf{D}_e = \begin{bmatrix} D_{2i-1} \\ D_{2i} \\ D_{2j-1} \\ D_{2j} \end{bmatrix} \tag{3.83}$$

The transformation matrix $\mathbf{T}$ can be expressed as

$$\mathbf{T} = \begin{bmatrix} l_{ij} & m_{ij} & 0 & 0 \\ 0 & 0 & l_{ij} & m_{ij} \end{bmatrix} \tag{3.84}$$

The force vector in the global coordinate system can be expressed as

$$\mathbf{F}_e = \begin{bmatrix} F_{2i-1} \\ F_{2i} \\ F_{2j-1} \\ F_{2j} \end{bmatrix} \tag{3.85}$$

All other equations for planar trusses have the same form as the corresponding equations for spatial trusses. The stiffness matrix $\mathbf{K}_e$ and mass matrix $\mathbf{M}_e$ of a planar truss are 4×4 matrices in the global coordinate system, as shown below:

$$\mathbf{K}_e = \mathbf{T}^{\mathrm{T}}\mathbf{k}_e\mathbf{T} = \begin{bmatrix} K_{11}^e & K_{12}^e & K_{13}^e & K_{14}^e \\ K_{12}^e & K_{22}^e & K_{23}^e & K_{24}^e \\ K_{13}^e & K_{23}^e & K_{33}^e & K_{34}^e \\ K_{14}^e & K_{24}^e & K_{34}^e & K_{44}^e \end{bmatrix} = \frac{AE}{l_e}\begin{bmatrix} l_{ij}^2 & l_{ij}m_{ij} & -l_{ij}^2 & -l_{ij}m_{ij} \\ l_{ij}m_{ij} & m_{ij}^2 & -l_{ij}m_{ij} & -m_{ij}^2 \\ -l_{ij}^2 & -l_{ij}m_{ij} & l_{ij}^2 & l_{ij}m_{ij} \\ -l_{ij}m_{ij} & -m_{ij}^2 & l_{ij}m_{ij} & m_{ij}^2 \end{bmatrix} \tag{3.86}$$

$$\mathbf{M}_e = \mathbf{T}^{\mathrm{T}}\mathbf{m}_e\mathbf{T} = \begin{bmatrix} M_{11}^e & M_{12}^e & M_{13}^e & M_{14}^e \\ M_{12}^e & M_{22}^e & M_{23}^e & M_{24}^e \\ M_{13}^e & M_{23}^e & M_{33}^e & M_{34}^e \\ M_{14}^e & M_{24}^e & M_{34}^e & M_{44}^e \end{bmatrix} = \frac{A\rho l_e}{6}\begin{bmatrix} 2l_{ij}^2 & 2l_{ij}m_{ij} & l_{ij}^2 & l_{ij}m_{ij} \\ 2l_{ij}m_{ij} & 2m_{ij}^2 & l_{ij}m_{ij} & m_{ij}^2 \\ l_{ij}^2 & l_{ij}m_{ij} & 2l_{ij}^2 & 2l_{ij}m_{ij} \\ l_{ij}m_{ij} & m_{ij}^2 & 2l_{ij}m_{ij} & 2m_{ij}^2 \end{bmatrix} \tag{3.87}$$

3.2.5 Boundary Conditions

The stiffness matrix $\mathbf{K}_e$ in Eq. (3.79) is typically singular because the entire structure is capable of rigid body motion. A planar truss has two degrees of freedom for rigid body motion, while a spatial truss has three. These rigid body motions are restricted by supports or displacement constraints. In engineering practice, truss structures are anchored to the ground or the main structure at a certain number of nodes. When a node is fixed, the displacement at that node must be zero. This fixed displacement boundary condition can be applied to Eq. (3.79). The constraints result in the removal of the corresponding rows and columns from the stiffness matrix. If sufficient displacement constraints are imposed, the reduced stiffness matrix becomes symmetric and positive definite.

3.2.6 Stress and Strain Calculations

After applying the boundary conditions, the displacements at all nodes can be obtained by solving Eq. (3.79). The displacement at any location other than the nodes can also be interpolated using the shape functions. The stress in the truss element can be expressed as follows:

$$\sigma_x = E\mathbf{B}\mathbf{d}_e = E\mathbf{B}\mathbf{T}\mathbf{D}_e \tag{3.88}$$

In deriving the above equation, Hooke's law in the form $\sigma = E\varepsilon$, as well as Eqs. (3.64) and (3.69), were used.

3.2.7 Calculation Example

3.2.7.1 Example 3.1: Uniform Straight Bar Subjected to Axial Load

As shown in Fig. 3.8, a bar with a uniform cross-sectional area is depicted. One end of the bar is fixed, while the free end is subjected to a horizontal load of magnitude P. The bar is made of an isotropic material with Young's modulus E and has a length l.

1. **Exact Solution**

According to the strong form of the truss governing equation given in Eq. (3.16), we have

$$\frac{\partial^2 u}{\partial x^2} = 0 \tag{3.89}$$

It should be noted that, for the current example, the rod is weightless, so $f_x = 0$. The general form of the solution to Eq. (3.89) is

$$u(x) = c_0 + c_1 x \tag{3.90}$$

Here, c_0 and c_1 are unknown constants that must be determined by the boundary conditions. The displacement boundary condition for this example can be expressed as $u = 0$ at $x = 0$. Therefore, with $c_0 = 0$, Eq. (3.90) becomes:

$$u(x) = c_1 x \tag{3.91}$$

Combining Eqs. (3.9), (3.11), and (3.91), we obtain

$$\sigma_x = E\frac{\partial u}{\partial x} = c_1 E \tag{3.92}$$

The force boundary condition for the bar can be expressed as $\sigma_x = \frac{P}{A}$ at $x = l$. Substituting this boundary condition into Eq. (3.92) yields:

$$c_1 = \frac{P}{EA} \tag{3.93}$$

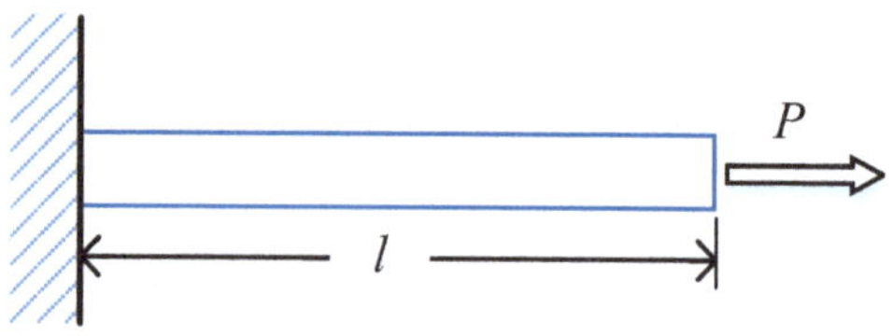

Fig. 3.8 Cantilever rod under static load

Fig. 3.9 A cantilever rod under static load

The stress in the rod can be obtained by substituting Eq. (3.93) into Eq. (3.92), yielding

$$\sigma_x = \frac{P}{A} \tag{3.94}$$

Substituting Eq. (3.93) into Eq. (3.92) yields the displacement solution of the rod.

$$u(x) = \frac{P}{EA}x \tag{3.95}$$

At $x = l$, we have

$$u(x = l) = \frac{Pl}{EA} \tag{3.96}$$

2. **Finite Element Solution**

As shown in Fig. 3.9, a truss element is used to model a cantilevered bar under static loading. According to Eq. (3.66), the stiffness matrix of the bar is given by

$$\mathbf{K} = \mathbf{k}_e = \frac{AE}{l}\begin{bmatrix} 1 & -1 \\ -1 & 1 \end{bmatrix} \tag{3.97}$$

In this case, no coordinate transformation is required, as the local and global coordinate systems are identical; assembly is also unnecessary since there is only one element. Thus, the finite element equation can be expressed as

$$\frac{AE}{l}\begin{bmatrix} 1 & -1 \\ -1 & 1 \end{bmatrix}\begin{bmatrix} u_1 \\ u_2 \end{bmatrix} = \begin{bmatrix} F_1 = ? \\ F_2 = P \end{bmatrix} \tag{3.98}$$

where F_1 is the reaction force applied at node 1, which is currently unknown. Conversely, the displacement boundary condition equation at node 1 is known, i.e., $u_1 = 0$. Thus, substituting into Eq. (3.98) yields:

$$\frac{AE}{l}\begin{bmatrix} 1 & -1 \\ -1 & 1 \end{bmatrix}\begin{bmatrix} 0 \\ u_2 \end{bmatrix} = \begin{bmatrix} F_1 = ? \\ F_2 = P \end{bmatrix}$$

we obtain

$$u_2 = \frac{Pl}{AE} \tag{3.99}$$

This is the finite element solution of the bar, which is exactly the same as the exact solution obtained in Eq. (3.96). By substituting the known boundary condition $u_1 = 0$ and Eq. (3.99) into Eq. (3.53), the displacement distribution of the bar can be obtained as

$$u(x) = \mathbf{N}(x)\mathbf{d}_e = \begin{bmatrix} 1 - \frac{x}{l} & \frac{x}{l} \end{bmatrix} \begin{bmatrix} u_1 \\ u_2 \end{bmatrix} = \begin{bmatrix} 1 - \frac{x}{l} & \frac{x}{l} \end{bmatrix} \begin{bmatrix} 0 \\ \frac{Pl}{EA} \end{bmatrix} = \frac{P}{EA}x \tag{3.100}$$

This is also exactly the same as the exact solution obtained in Eq. (3.95).

Combining Eqs. (3.88) and (3.65), the stress distribution in the rod can be obtained as follows:

$$\sigma_x = E\mathbf{B}\mathbf{d}_e = E\begin{bmatrix} -\frac{1}{l} & \frac{1}{l} \end{bmatrix} \begin{bmatrix} 0 \\ u_2 \end{bmatrix} = \frac{P}{A} \tag{3.101}$$

Similarly, this is exactly the same as the exact solution given in Eq. (3.94).

The following is a discussion of the finite element properties demonstrated in Example 3.1.

1. **Reproducibility of Finite Elements**

The finite element method typically yields only an approximate solution. However, in Example 3.1, we obtained the exact solution because the exact solution for beam deformation is a first-order polynomial (see Eq. (3.95)). The shape functions used in finite element analysis are also first-order polynomials, constructed using a complete set of first-order monomials. Therefore, the exact solution in this example is included within the assumed displacement set represented by the finite element shape functions. It is well known that the finite element method based on the principle of minimum potential energy guarantees that the chosen shape functions produce the optimal solution. In Example 3.1, due to the representational property of the shape functions, the best possible solution that the shape functions can produce is the exact solution, and the finite element method indeed reproduces it accurately. Thus, the representational property of the finite element method is verified: if the exact solution can be formed by the basis functions used to construct the finite element shape functions, then the finite element method will always yield the exact solution, provided that there are no numerical errors in the finite element computation process.

By leveraging this property, it is possible, when feasible, to incorporate basis functions that form exact or partially exact solutions, thereby achieving higher accuracy in the finite element method.

2. **Convergence of the Finite Element Method**

For complex problems, the solution cannot be expressed as a combination of monomials. Therefore, finite elements using polynomial shape functions will not yield exact solutions for such problems. How, then, can we ensure that the finite element method provides a good approximation to the solution of complex problems? This is guaranteed by the convergence property of the finite element method: as long as the exact solution is continuous, when the element size becomes sufficiently small, and as long as a complete set of linear polynomial basis functions is included in the basis functions forming the finite element shape functions, the finite element solution will always converge to the exact solution. The theoretical foundation for this convergence property lies in the fact that any continuous function can always be approximated by a first-order polynomial with a second-order refinement error. This theory can be demonstrated using a local Taylor expansion, based on which a continuous (displacement) function $u(x)$ can always be approximated as follows:

$$u = u_i + \left.\frac{\partial u}{\partial x}\right|_i (x - x_i) + O\left(h^2\right) \tag{3.102}$$

Here, h represents the characteristic length (or the size of the element) associated with $(x - x_i)$.

Some may argue that, according to Eq. (3.102), it is not necessary to use $O(h^1)$; a constant is sufficient to reconstruct the function u. However, unless the entire displacement field is constant (rigid body motion), the constant displacement generated by an element may be discontinuous between elements. Therefore, to ensure the convergence of a continuous solution, it is necessary to use at least a first-order complete polynomial.

3. **Convergence Rate of Finite Element Results**

The Taylor expansion of order p can be expressed as

$$u = u_i + \left.\frac{\partial u}{\partial x}\right|_i (x - x_i) + \left.\frac{1}{2!}\frac{\partial^2 u}{\partial x^2}\right|_i (x - x_i)^2 + \cdots + \left.\frac{1}{p!}\frac{\partial^p u}{\partial x^p}\right|_i (x - x_i)^p + O\left(h^{p+1}\right) \tag{3.103}$$

If a complete polynomial of degree up to p is used to construct the shape functions, then the first $(p + 1)$ terms in Eq. (3.103) can be reproduced by the finite element shape functions with an error of order $O\left(h^{p+1}\right)$. Therefore, the convergence rate of the displacement is $O\left(h^{p+1}\right)$. For a linear element, $p = 1$, so the convergence order of the displacement is $O\left(h^2\right)$. This means that if the element size is halved, the error in the displacement result reduces 1/4.

The reproducibility and convergence of the finite element method are crucial for providing reliable numerical results in mechanics problems, as they allow for the prediction of expected outcomes. For simple problems where the exact solution is a polynomial, the finite element method can reproduce the exact solution using the minimum number of elements, provided that a complete set of basis functions of appropriate order—including the order of the exact solution—is used. In Example 3.1, a single first-order element is sufficient. For more complex problems, where the exact solution is a high-order polynomial or, more commonly, a non-polynomial function, analysts must select an appropriate mesh density to achieve the desired accuracy in the finite element results, with the displacement convergence rate being $O(h^{p+1})$.

In recent years, the concepts of so-called *h*-adaptivity and *p*-adaptivity have been widely used in finite element analysis. Typically, *h* denotes the characteristic size of an element, while *p* represents the order of the polynomial basis functions. *h*-adaptive analysis achieves the desired finite element accuracy by employing finer meshes (smaller *h*), whereas *p*-adaptive analysis utilizes higher-order shape functions (larger *p*) to reach the required precision.

3.2.7.2 Example 3.2: Triangular Truss Structure Subjected to Vertical Loads

As shown in Fig. 3.10, a planar truss structure is composed of three planar truss members, with a vertical downward load of 1,000 N applied at node 2. In the figure, element numbers are indicated by squares, and node numbers are indicated by circles.

The numbering of local coordinates and global degrees of freedom (D_1, D_2, …, D_6, corresponding to the three nodes in the structure) for the three truss elements is shown in Fig. 3.11. It should be noted that there are a total of six global degrees of freedom, with each node having two degrees of freedom in the *X* and *Y* directions. However, for each element, each node in the local coordinate system actually has only one degree of freedom. As can be clearly seen from the figure, the degrees of

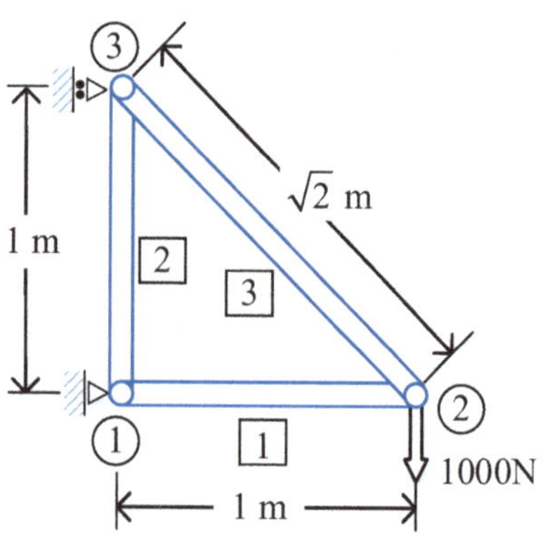

Fig. 3.10 Planar truss structure of the member

freedom at each node are shared by multiple elements. For example, at node 1, global degrees of freedom D_1, and D_2 are contributed by both elements 1 and 2. This plays an important role in the assembly of the final finite element matrix. The dimensions and material properties of the truss members in the structure are listed in Table 3.1. The calculation process is as follows.

Step 1: Obtain the direction cosines of the element

After obtaining the coordinates of the nodes in the global coordinate system, it is necessary to consider the orientation of the element relative to the global coordinate system. This can be achieved by calculating the direction cosines according to Eq. (3.72). Since this example is a planar problem, there is no need to calculate n_{ij}. The global coordinates of all nodes, as well as the direction cosines l_{ij} and m_{ij}, are listed in Table 3.2.

Fig. 3.11 Local coordinates and degrees of freedom for three truss elements

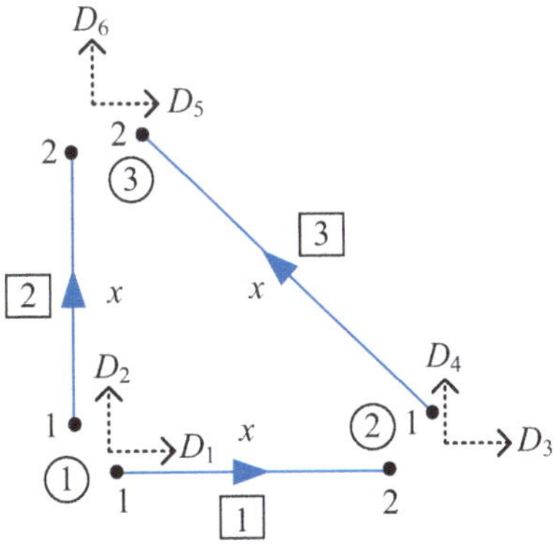

Table 3.1 Dimensions and performance of truss members

Element number	Cross-sectional area A_e (m^2)	Length l_e (m)	Young's modulus E (N/m^2)
1	0.1	1	70×10^9
2	0.1	1	70×10^9
3	0.1	$\sqrt{2}$	70×10^9

Table 3.2 Global coordinates of nodes and direction cosines of elements

Element ID	Local nodes corresponding to global nodes		Global coordinates		Direction cosines	
	Node 1 (i)	Node 2 (j)	X_i, Y_i	X_j, Y_j	l_{ij}	m_{ij}
1	1	2	0,0	1,0	1	0
2	1	3	0,0	0,1	0	1
3	2	3	1,0	0,1	−1/2	1/2

Step 2: Calculate the element matrix in the global coordinate system

After obtaining the direction cosines, the element matrix in the global coordinate system can be determined. It is important to note that this example concerns a static problem, so there is no need to compute the element mass matrix; only the element stiffness matrix needs to be calculated. Since each element has a total of 2 degrees of freedom, the element stiffness matrix in the local coordinate system is a 2×2 matrix. However, after transforming from the local to the global coordinate system, each element has 4 degrees of freedom, so the element stiffness matrix in the global coordinate system becomes a 4×4 matrix. The stiffness matrix is calculated according to Eq. (3.86) as follows:

$$\mathbf{K}^{e1} = \frac{(0.1)\left(70 \times 10^9\right)}{1.0}\begin{bmatrix} 1 & 0 & -1 & 0 \\ 0 & 0 & 0 & 0 \\ -1 & 0 & 1 & 0 \\ 0 & 0 & 0 & 0 \end{bmatrix} = \begin{bmatrix} 7 & 0 & -7 & 0 \\ 0 & 0 & 0 & 0 \\ -7 & 0 & 7 & 0 \\ 0 & 0 & 0 & 0 \end{bmatrix} \times 10^9\ \mathrm{N \cdot m^{-1}} \tag{3.104}$$

$$\mathbf{K}_{e2} = \frac{(0.1)\left(70 \times 10^9\right)}{1.0}\begin{bmatrix} 0 & 0 & 0 & 0 \\ 0 & 1 & 0 & -1 \\ 0 & 0 & 0 & 0 \\ 0 & -1 & 0 & 1 \end{bmatrix} = \begin{bmatrix} 0 & 0 & 0 & 0 \\ 0 & 7 & 0 & -7 \\ 0 & 0 & 0 & 0 \\ 0 & -7 & 0 & 7 \end{bmatrix} \times 10^9\ \mathrm{N \cdot m^{-1}} \tag{3.105}$$

$$\mathbf{K}_{e3} = \frac{(0.1)\left(70 \times 10^9\right)}{\sqrt{2}}\begin{bmatrix} \frac{1}{2} & -\frac{1}{2} & -\frac{1}{2} & \frac{1}{2} \\ -\frac{1}{2} & \frac{1}{2} & \frac{1}{2} & -\frac{1}{2} \\ -\frac{1}{2} & \frac{1}{2} & \frac{1}{2} & -\frac{1}{2} \\ \frac{1}{2} & -\frac{1}{2} & -\frac{1}{2} & \frac{1}{2} \end{bmatrix} = \begin{bmatrix} \frac{7}{2\sqrt{2}} & -\frac{7}{2\sqrt{2}} & -\frac{7}{2\sqrt{2}} & \frac{7}{2\sqrt{2}} \\ -\frac{7}{2\sqrt{2}} & \frac{7}{2\sqrt{2}} & \frac{7}{2\sqrt{2}} & -\frac{7}{2\sqrt{2}} \\ -\frac{7}{2\sqrt{2}} & \frac{7}{2\sqrt{2}} & \frac{7}{2\sqrt{2}} & -\frac{7}{2\sqrt{2}} \\ \frac{7}{2\sqrt{2}} & -\frac{7}{2\sqrt{2}} & -\frac{7}{2\sqrt{2}} & \frac{7}{2\sqrt{2}} \end{bmatrix} \times 10^9\ \mathrm{N \cdot m^{-1}} \tag{3.106}$$

Step 3: Assembly of the Global Finite Element Matrix

Once the element matrices have been obtained, they can be assembled into a global finite element matrix. Since the total number of global degrees of freedom in the structure is six, the global stiffness matrix is a 6×6 matrix. Assembly is accomplished by summing the contributions of each element to the shared nodes. For example, as shown in Fig. 3.11, element 1 contributes to the degrees of freedom D_1 and D_2 at node O, as well as D_3 and D_4 at node O. Element 2 contributes to the degrees of freedom D_1 and D_2 at node O, and D_5 and D_6 at node O. By adding the contributions of each element matrix to the corresponding positions in the global matrix according to their contributions to the global degrees of freedom, the global matrix can be obtained. This assembly process is referred to as direct assembly.

Before assembly begins, the entire global stiffness matrix is first initialized to zero. The matrix of element 1 is then added to the global matrix, yielding

$$
\begin{array}{c}
\begin{array}{cccc} D_1 & D_2 & D_3 & D_4 \\ \uparrow & \uparrow & \uparrow & \uparrow \end{array} \\
\mathbf{K}_{e2} = 10^9 \times \begin{bmatrix}
\boxed{7} & \boxed{0} & \boxed{-7} & \boxed{0} & 0 & 0 \\
\boxed{0} & \boxed{1} & \boxed{0} & \boxed{0} & 0 & 0 \\
\boxed{-7} & \boxed{0} & \boxed{7} & \boxed{0} & 0 & 0 \\
\boxed{0} & \boxed{0} & \boxed{0} & \boxed{0} & 0 & 0 \\
0 & 0 & 0 & 0 & 0 & 0 \\
0 & 0 & 0 & 0 & 0 & 0
\end{bmatrix}
\begin{array}{l} \rightarrow D_1 \\ \rightarrow D_2 \\ \rightarrow D_3 \\ \rightarrow D_4 \\ \\ \\ \end{array}
\end{array}
\tag{3.107}
$$

Note that element 1 also contributes to the degrees of freedom D_1 through D_4. Then, the element matrix of element 2 is added to the new global matrix, yielding

$$
\begin{array}{c}
\begin{array}{cccccc} D_1 & D_2 & & & D_5 & D_6 \\ \uparrow & \uparrow & & & \uparrow & \uparrow \end{array} \\
\mathbf{K} = 10^9 \times \begin{bmatrix}
7+\boxed{0} & 0+\boxed{0} & -7 & 0 & \boxed{0} & \boxed{0} \\
0+\boxed{0} & 0+\boxed{7} & 0 & 0 & \boxed{0} & \boxed{-7} \\
-7 & 0 & 7 & 0 & 0 & 0 \\
0 & 0 & 0 & 0 & 0 & 0 \\
\boxed{0} & \boxed{0} & 0 & 0 & \boxed{0} & \boxed{0} \\
\boxed{0} & \boxed{-7} & 0 & 0 & \boxed{0} & \boxed{7}
\end{bmatrix}
\begin{array}{l} \rightarrow D_1 \\ \rightarrow D_2 \\ \\ \\ \rightarrow D_5 \\ \rightarrow D_6 \end{array}
\end{array}
\tag{3.108}
$$

It should be noted that Element 2 also contributes to the degrees of freedom D_1, D_2, D_5 and D_6. Finally, by adding the element matrix of element 3 to the current global matrix, we obtain

$$
\begin{array}{c}
\begin{array}{cccccc} & & D_3 & D_4 & D_5 & D_6 \\ & & \uparrow & \uparrow & \uparrow & \uparrow \end{array} \\
\mathbf{K} = 10^9 \times \begin{bmatrix}
7 & 0 & -7 & 0 & 0 & 0 \\
0 & 7 & 0 & 0 & 0 & -7 \\
-7 & 0 & 7+\boxed{7/2\sqrt{2}} & \boxed{-7/2\sqrt{2}} & \boxed{-7/2\sqrt{2}} & \boxed{7/2\sqrt{2}} \\
0 & 0 & \boxed{-7/2\sqrt{2}} & \boxed{7/2\sqrt{2}} & \boxed{7/2\sqrt{2}} & \boxed{-7/2\sqrt{2}} \\
0 & 0 & \boxed{-7/2\sqrt{2}} & \boxed{7/2\sqrt{2}} & \boxed{7/2\sqrt{2}} & \boxed{-7/2\sqrt{2}} \\
0 & -7 & \boxed{7/2\sqrt{2}} & \boxed{-7/2\sqrt{2}} & \boxed{-7/2\sqrt{2}} & 7+\boxed{7/2\sqrt{2}}
\end{bmatrix}
\begin{array}{l} \\ \\ \rightarrow D_3 \\ \rightarrow D_4 \\ \rightarrow D_5 \\ \rightarrow D_6 \end{array}
\end{array}
\tag{3.109}
$$

Note that element 3 also contributes to the degrees of freedom D_3 through D_6. The resulting global stiffness matrix is

$$\mathbf{K} = 10^9 \times \begin{array}{c} \begin{array}{cccccc} D_1 & D_2 & D_3 & D_4 & D_5 & D_6 \\ \uparrow & \uparrow & \uparrow & \uparrow & \uparrow & \uparrow \end{array} \\ \begin{bmatrix} 7 & 0 & -7 & 0 & 0 & 0 \\ 0 & 7 & 0 & 0 & 0 & -7 \\ -7 & 0 & 7+\frac{7}{2\sqrt{2}} & -\frac{7}{2\sqrt{2}} & -\frac{7}{2\sqrt{2}} & \frac{7}{2\sqrt{2}} \\ 0 & 0 & -\frac{7}{2\sqrt{2}} & \frac{7}{2\sqrt{2}} & \frac{7}{2\sqrt{2}} & -\frac{7}{2\sqrt{2}} \\ 0 & 0 & -\frac{7}{2\sqrt{2}} & \frac{7}{2\sqrt{2}} & \frac{7}{2\sqrt{2}} & -\frac{7}{2\sqrt{2}} \\ 0 & -7 & \frac{7}{2\sqrt{2}} & -\frac{7}{2\sqrt{2}} & -\frac{7}{2\sqrt{2}} & 7+\frac{7}{2\sqrt{2}} \end{bmatrix} \end{array} \begin{array}{l} \\ \\ \to D_1 \\ \to D_2 \\ \to D_3 \\ \to D_4 \\ \to D_5 \\ \to D_6 \end{array} \tag{3.110}$$

The direct assembly process described above is straightforward and can be easily implemented in computer programs. By utilizing the index mapping between local element node numbers and global node numbers, the entries of the element matrix can be added to the corresponding entries in the global matrix.

How can we prove that simply adding the unit matrices to the global matrix in this way indeed yields the global stiffness matrix? This can be readily demonstrated by considering the global equilibrium conditions for all nodes within the entire problem domain. A straightforward proof is provided below.

Here, we choose to demonstrate the assembly result of the third row of the global stiffness matrix given in Eq. (3.110). For the third row of Eq. (3.110), we consider the force equilibrium in the x-direction at node O of elements 1 and 3, which corresponds to the third global degree of freedom of the truss structure connecting elements 1 and 3. For the static problem, the finite element equation for element 1 can be expressed in the following general form (in the global coordinate system):

$$\begin{bmatrix} K_{11}^{e1} & K_{12}^{e1} & K_{13}^{e1} & K_{14}^{e1} \\ K_{12}^{e1} & K_{22}^{e1} & K_{23}^{e1} & K_{24}^{e1} \\ K_{13}^{e1} & K_{23}^{e1} & K_{33}^{e1} & K_{34}^{e1} \\ K_{14}^{e1} & K_{24}^{e1} & K_{34}^{e1} & K_{44}^{e1} \end{bmatrix} \begin{bmatrix} D_1 \\ D_2 \\ D_3 \\ D_4 \end{bmatrix} = \begin{bmatrix} F_1^{e1} \\ F_2^{e1} \\ F_3^{e1} \\ F_4^{e1} \end{bmatrix} \tag{3.111}$$

The third equation in (3.111) corresponds to the third global degree of freedom, which is,

$$K_{13}^{e1} D_1 + K_{23}^{e1} D_2 + K_{33}^{e1} D_3 + K_{34}^{e1} D_4 = F_3^{e1} \tag{3.112}$$

The finite element equation for element 3 can be expressed in the following general form:

$$\begin{bmatrix} K_{11}^{e3} & K_{12}^{e3} & K_{13}^{e3} & K_{14}^{e3} \\ K_{12}^{e3} & K_{22}^{e3} & K_{23}^{e3} & K_{24}^{e3} \\ K_{13}^{e3} & K_{23}^{e3} & K_{33}^{e3} & K_{34}^{e3} \\ K_{14}^{e3} & K_{24}^{e3} & K_{34}^{e3} & K_{44}^{e3} \end{bmatrix} \begin{bmatrix} D_3 \\ D_4 \\ D_5 \\ D_6 \end{bmatrix} = \begin{bmatrix} F_3^{e3} \\ F_4^{e3} \\ F_5^{e3} \\ F_6^{e3} \end{bmatrix} \tag{3.113}$$

The first row in the above Eq. (3.113) corresponds to the third global degree of freedom, that is,

$$K_{13}^{e3}D_1 + K_{23}^{e3}D_2 + K_{33}^{e3}D_3 + K_{34}^{e3}D_4 = F_3^{e3} \tag{3.114}$$

The loads on node 2 in the x-direction include the element force ${F_3}^{e3}$ from element 3 and the possible external force F_3. All these forces must satisfy the following equilibrium equation:

$$F_3^{e1} + F_3^{e3} = F_3 \tag{3.115}$$

Substituting Eqs. (3.112) and (3.114) into Eq. (3.115), we obtain

$$K_{13}^{e1}D_1 + K_{23}^{e1}D_2 + \left(K_{33}^{e1} + K_{11}^{e3}\right)D_3 + \left(K_{34}^{e1} + K_{12}^{e3}\right)D_4 + K_{13}^{e3}D_5 + K_{14}^{e3}D_6 = F_3 \tag{3.116}$$

The coefficient on the left-hand side of Eq. (3.116) corresponds to the third row of the global stiffness matrix given in Eq. (3.110). Similarly, the above proof process is also applicable to all other rows in the global stiffness matrix.

Step 4: Apply Boundary Conditions

After applying the boundary conditions, the size of the global matrix is typically reduced. In this example, D_1, D_2, D_5 are constrained, that is,

$$D_1 = D_2 = D_5 = 0\ \text{m} \tag{3.117}$$

This means that rows and columns 1, 2, and 5 actually have no effect on solving the matrix equation. Therefore, the corresponding rows and columns can simply be removed, that is,

$$\begin{array}{c} \\ \\ \mathbf{K} = 10^9 \times \end{array} \begin{array}{cccccc} D_1 & D_2 & D_3 & D_4 & D_5 & D_6 \\ \uparrow & \uparrow & \uparrow & \uparrow & \uparrow & \uparrow \end{array}$$

$$\mathbf{K} = 10^9 \times \begin{bmatrix} 7 & 0 & -7 & 0 & 0 & 0 \\ 0 & 7 & 0 & 0 & 0 & -7 \\ -7 & 0 & 7+7/2\sqrt{2} & -7/2\sqrt{2} & -7/2\sqrt{2} & 7/2\sqrt{2} \\ 0 & 0 & -7/2\sqrt{2} & 7/2\sqrt{2} & 7/2\sqrt{2} & -7/2\sqrt{2} \\ 0 & 0 & -7/2\sqrt{2} & 7/2\sqrt{2} & 7/2\sqrt{2} & -7/2\sqrt{2} \\ 0 & -7 & 7/2\sqrt{2} & -7/2\sqrt{2} & -7/2\sqrt{2} & 7+7/2\sqrt{2} \end{bmatrix} \begin{array}{c} \rightarrow D_1 \\ \rightarrow D_2 \\ \rightarrow D_3 \\ \rightarrow D_4 \\ \rightarrow D_5 \\ \rightarrow D_6 \end{array} \tag{3.118}$$

The condensed global stiffness matrix is a 3×3 matrix, that is,

$$\mathbf{K} = 10^9 \times \begin{bmatrix} 7+7/2\sqrt{2} & -7/2\sqrt{2} & 7/2\sqrt{2} \\ -7/2\sqrt{2} & 7/2\sqrt{2} & -7/2\sqrt{2} \\ 7/2\sqrt{2} & -7/2\sqrt{2} & 7+7/2\sqrt{2} \end{bmatrix} \mathrm{N \cdot m^{-1}} \tag{3.119}$$

It can be readily verified that this compressed global stiffness matrix is symmetric and positive definite.

The constrained global finite element equation is

$$\mathbf{KD} = \mathbf{F} \tag{3.120}$$

where

$$\mathbf{D}^{\mathrm{T}} = [D_3\, D_4\, D_6] \tag{3.121}$$

and the load **F** vector

$$\mathbf{F} = \begin{bmatrix} 0 \\ -1000 \\ 0 \end{bmatrix} \mathrm{N} \tag{3.122}$$

It should be noted that the only applied nodal force is at node O, acting in the downward direction of D_4. Equation (3.120) is actually equivalent to a system of three simultaneous equations for the unknowns D_3, D_4 and D_6, namely,

$$\begin{aligned} &\left[\left(7+7/2\sqrt{2}\right)D_3 - \left(7/2\sqrt{2}\right)D_4 + \left(7/2\sqrt{2}\right)D_6\right] \times 10^9 = 0 \\ &\left[\left(-7/2\sqrt{2}\right)D_3 - \left(7/2\sqrt{2}\right)D_4 - \left(7/2\sqrt{2}\right)D_6\right] \times 10^9 = -1000 \\ &\left[\left(7/2\sqrt{2}\right)D_3 - \left(7/2\sqrt{2}\right)D_4 + \left(7+7/2\sqrt{2}\right)D_6\right] \times 10^9 = 0 \end{aligned} \tag{3.123}$$

Step 5: Solving the finite element equations

Solving the finite element Eq. (3.120) or the system of Eqs. (3.123) yields the solutions for D_3, D_4 and D_6. It is possible to solve this system manually, as system (3.123) contains only three equations with three unknowns. The solutions are as follows:

$$D_3 = -1.429 \times 10^{-7}\mathrm{m}$$
$$D_4 = -6.898 \times 10^{-7}\mathrm{m}$$
$$D_6 = -1.429 \times 10^{-7}\mathrm{m}$$

To obtain the stresses within the element, Eq. (3.88) is used, yielding the following results:

$$\sigma_x^1 = E\mathbf{BTD}_e = 70 \times 10^9 \begin{bmatrix} -1 & 1 \end{bmatrix} \begin{bmatrix} 1 & 0 & 0 & 0 \\ 0 & 0 & 1 & 0 \end{bmatrix} \begin{bmatrix} 0 \\ 0 \\ -1.429 \times 10^{-7} \\ -6.898 \times 10^{-7} \end{bmatrix} = -10003 \text{ Pa}$$

$$\sigma_x^2 = E\mathbf{BTD}_e = 70 \times 10^9 \begin{bmatrix} -1 & 1 \end{bmatrix} \begin{bmatrix} 1 & 0 & 0 & 0 \\ 0 & 0 & 1 & 0 \end{bmatrix} \begin{bmatrix} 0 \\ 0 \\ 0 \\ -1.429 \times 10^{-7} \end{bmatrix} = -10003 \text{ Pa}$$

$$\sigma_x^3 = E\mathbf{BTD}_e = 70 \times 10^9 \begin{bmatrix} -1/\sqrt{2} & 1/\sqrt{2} \end{bmatrix} \begin{bmatrix} -1/\sqrt{2} & 1/\sqrt{2} & 0 & 0 \\ 0 & 0 & -1/\sqrt{2} & 1/\sqrt{2} \end{bmatrix} \begin{bmatrix} -1.429 \times 10^{-7} \\ -6.898 \times 10^{-7} \\ 0 \\ -1.429 \times 10^{-7} \end{bmatrix} = 14140 \text{ Pa}$$

In engineering practice, problems can be much larger in scale, with a significantly greater number of unknowns or degrees of freedom. Typical real-world engineering problems may involve hundreds of thousands or even millions of degrees of freedom. Therefore, numerical methods or solvers must be used to solve the finite element equations. Most mathematical or algorithm libraries in computer systems include various such solvers.

3.2.8 *Higher-Order One-Dimensional Elements*

In the previous section, it was demonstrated that refining the mesh can achieve higher accuracy, which increases the number of elements while reducing their size h. Therefore, this approach is commonly referred to as the h-type finite element method.

Another approach to improving accuracy is to keep the finite element mesh unchanged while increasing the number of displacement functions within each element. When polynomials are used, this method involves increasing their order p, and is therefore referred to as the p-version finite element method. This section will present some examples of p-version finite elements.

3.2.8.1 Higher-Order Lagrange Elements

For truss members not subjected to body forces, higher-order elements are unnecessary, as linear elements can already provide exact solutions, as demonstrated in Example 3.1 of Sect. 3.2.7.1. However, for truss members subjected to body forces that are arbitrarily distributed along the axial direction within the truss element, higher-order elements are required for more accurate analysis. The derivation process for such higher-order one-dimensional elements is similar to that of linear elements, with the only difference being in the derivation of the shape functions.

When deriving higher-order shape functions, the natural coordinate ξ is typically used instead of the physical coordinate x. The natural coordinate ξ is defined as follows:

$$\xi = 2\frac{x - x_c}{l_e} \tag{3.124}$$

Here, x_c represents the physical coordinates of the nodes in a one-dimensional element. In the natural coordinate system, the element is defined over $-1 \leq \xi \leq 1$. As shown in Fig. 3.12, for an n-th order one-dimensional element with $(n+1)$ nodes, the shape functions of the element can be expressed using Lagrange interpolation function as follows:

$$N_k(\xi) = l_k^n(\xi) \tag{3.125}$$

Here, $l_k^n(x)$ is the Lagrange interpolation function, defined as

$$l_k^n(\xi) = \frac{(\xi - \xi_0)(\xi - \xi_1)\cdots(\xi - \xi_{k-1})(\xi - \xi_{k+1})\cdots(\xi - \xi_n)}{(\xi_k - \xi_0)(\xi_k - \xi_1)\cdots(\xi_k - \xi_{k-1})(\xi_k - \xi_{k+1})\cdots(\xi_k - \xi_n)} \tag{3.126}$$

It can be seen from the above Eq. (3.125) that

$$\mathrm{N_k}(\xi) = \begin{cases} 1 \text{ at node k where } \xi = \xi_\mathrm{k} \\ 0 \text{ at other nodes} \end{cases} \tag{3.127}$$

Therefore, the higher-order shape functions defined by Eq. (3.125) possess the δ function property.

According to Eq. (3.125), the shape functions for the 3-node one-dimensional quadratic element shown in Fig. 3.13a can be expressed as follows:

$$N_1(\xi) = -\frac{1}{2}\xi(1 - \xi)$$

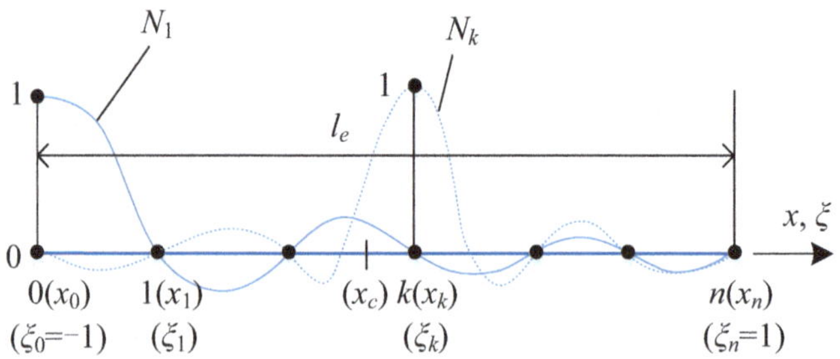

Fig. 3.12 One-dimensional element of order n with $(n + 1)$ nodes

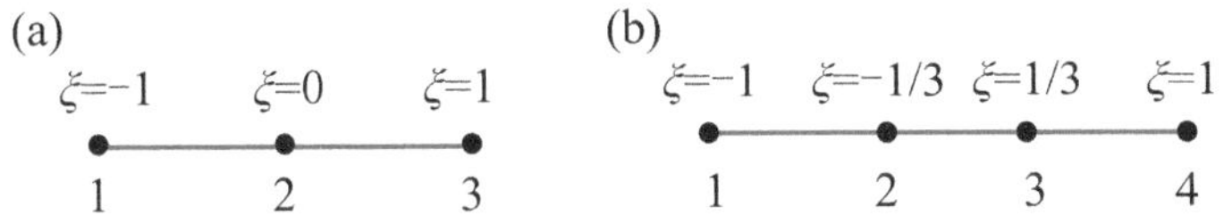

Fig. 3.13 One-dimensional element with uniformly distributed nodes

$$N_2(\xi) = \frac{1}{2}\xi(1+\xi)$$
$$N_3(\xi) = (1+\xi)(1-\xi) \tag{3.128}$$

As shown in Fig. 3.13b, the shape functions of the 4-node one-dimensional cubic element can be expressed as

$$\begin{aligned} N_1(\xi) &= -\frac{1}{16}(1-\xi)\left(1-9\xi^2\right) \\ N_2(\xi) &= -\frac{1}{16}(1+\xi)\left(1-9\xi^2\right) \\ N_3(\xi) &= \frac{9}{16}(1-3\xi)\left(1-\xi^2\right) \\ N_4(\xi) &= \frac{9}{16}(1+3\xi)\left(1-\xi^2\right) \end{aligned} \tag{3.129}$$

3.2.8.2 Weak Form Quadrature Element Method

In Lagrange interpolation with more than three nodes, attention must be paid to the distribution of the nodes. It is well known that Gauss–Lobatto quadrature nodes provide optimal interpolation accuracy. In the Quadrature Element Method (QEM), Gauss–Lobatto points are used as element nodes, and the element stiffness and mass matrices are obtained using Gauss–Lobatto quadrature. As a result, the number of element nodes is not restricted, and the Differential Quadrature Method (DQM) formulas can be used to explicitly compute derivatives at the nodes. This reduces the difficulty and rounding errors associated with calculating derivatives for higher-order elements, as well as the complexity of programming. However, to use the Differential Quadrature Method, the element nodes must coincide with the quadrature points, so only Gauss–Lobatto points can be used as element nodes. The Gauss–Lobatto quadrature is defined on the interval $[-1, 1]$ and can be expressed as

$$\int_{-1}^{1} f(x)\mathrm{d}x = \sum_{j=1}^{n} C_j f\left(x_j\right) \tag{3.130}$$

Here, the integration weighting coefficients C_j are given by

$$C_1 = C_n = \frac{2}{n(n-1)},\ C_j = \frac{2}{n(n-1)[P_{n-1}(x_j)]^2}\ (j \neq 1, n) \tag{3.131}$$

Here, x_j denotes the (j-1)th zero of $P'_{n-1}(x)$, where $P_n(x)$ is the Legendre orthogonal polynomial of degree n.

The commonly used term "quadrature" primarily refers to the numerical approximation of integrals, as in Eq. (3.130). The same approximation methods can be extended to more general linear functionals and used to approximate derivatives, that is,

$$f'(x_i) \cong \sum_{j=1}^{N} A_{ij}^{(1)} f(x_j),\quad i = 1,\ 2,\ \cdots,\ N \tag{3.132}$$

In this equation, the weighting coefficient matrix $\{A_{ij}^{(1)}\}$ can be computed using various methods. Equation (3.132) is the well-known Differential Quadrature Method (DQM) formula. It should be noted that the Differential Quadrature Method is based on Lagrange interpolation, that is,

$$f(x) = \sum_{j=1}^{N} l_j(x) f_j \tag{3.133}$$

Therefore, the weighting coefficient matrix of the first-order derivatives can be expressed as

$$A_{ij}^{(1)} = l'_j(x_i) = \begin{cases} \prod\limits_{k=1,\,k\neq i,j}^{N} (x_i - x_k) \Big/ \prod\limits_{k=1,\,k\neq j}^{N} (x_j - x_k),\ (i \neq j) \\ \sum\limits_{k=1,\,k\neq i}^{N} \dfrac{1}{(x_i - x_k)},\ (i = j) \end{cases} \tag{3.134}$$

For the weighting coefficient matrix of higher-order derivatives, the off-diagonal elements can be obtained through a recurrence relation, namely

$$A_{ij}^{(n)} = n\left(A_{ii}^{(n-1)} A_{ij}^{(1)} - \frac{A_{ij}^{(n-1)}}{x_i - x_j}\right) \tag{3.135}$$

where $i, j = 1, 2, \ldots, N$ and $i \neq j, 2 \leq n \leq N-1$. The diagonal elements of the weight coefficient matrix can be expressed as

$$A_{ii}^{(n)} = -\sum_{j=1, j\neq i}^{N} A_{ij}^{(n)} \tag{3.136}$$

where $i = 1, 2, \cdots, N$ and $1 \leq n \leq N - 1$.

Note: The weighting coefficient matrix for higher-order derivatives can be derived from lower-order derivatives as $\mathbf{A}^{(r)} = \mathbf{A}^{(1)}\mathbf{A}^{(r-1)}$. This formula is computationally less efficient but very simple, thus frequently used.

Applying the differential quadrature rule along with the Gauss–Lobatto quadrature rule, the strain energy of the rod and the potential of external forces can be discretized as:

$$U = \frac{1}{2}\int_0^L ES\left(\frac{\partial u}{\partial x}\right)^2 dx = \frac{ES}{2}\mathbf{u}^T\mathbf{A}^{(1)T}\mathbf{C}\mathbf{A}^{(1)}\mathbf{u} \tag{3.137}$$

$$W = -\int_0^L u\left(-\rho S\frac{\partial^2 u}{\partial t^2}\right)dx - \int_0^L qudx = \mathbf{u}^T(\rho SC)\ddot{\mathbf{u}} - \mathbf{u}^T(\mathbf{Cq}) \tag{3.138}$$

where E is the Young's modulus; $\mathbf{u}^{\mathrm{T}} = \begin{bmatrix} u_1 \ u_2 \ \ldots \ u_N \end{bmatrix}$ is the nodal displacement vector; $\mathbf{q}^{\mathrm{T}} = \left[q(x_1)q(x_2)\ldots q(x_N)\right]$ is the nodal load vector; $\mathbf{A}^{(1)\mathrm{T}} = \left(\mathbf{A}^{(1)}\right)^{\mathrm{T}}$ is the matrix of differential quadrature weighting coefficients for the first derivative;

$$\mathbf{C} = \mathrm{diag}\left(C_1 \ C_2 \ \cdots \ C_N\right) \tag{3.139}$$

Here, C_j denotes the Gauss–Lobatto integration weighting coefficient. Therefore, the stiffness matrix **K**, mass matrix **M**, and load vector **R** can be expressed as follows:

$$\mathbf{K} = ES\mathbf{A}^{(1)T}\mathbf{C}\mathbf{A}^{(1)}, \ \mathbf{M} = \rho S\mathbf{C}, \ \mathbf{R} = \mathbf{Cq} \tag{3.140}$$

Note: The finite element matrix of the weak form quadrature element can be obtained through simple algebraic operations involving the differential quadrature method and the Gauss–Lobatto integration weighting coefficient matrix, and the mass matrix **M** is diagonal.

To compare the standard finite element method with the weak form quadrature element method, the standard finite element matrices for a three-degree-of-freedom bar element subjected to a uniformly distributed load q_0 are given below.

$$\mathbf{K} = \frac{ES}{3L}\begin{bmatrix} 7 & -8 & 1 \\ -8 & 16 & -8 \\ 1 & -8 & 7 \end{bmatrix}, \ \mathbf{M} = \frac{\rho SL}{30}\begin{bmatrix} 4 & 2 & -1 \\ 2 & 16 & 2 \\ -1 & 2 & 4 \end{bmatrix}, \ \mathbf{R} = \frac{q_0 L}{6}\begin{bmatrix} 1 \\ 4 \\ 1 \end{bmatrix} \tag{3.141}$$

The corresponding matrices for the weak-form quadrature element are as follows:

$$\mathbf{K} = \frac{ES}{3L}\begin{bmatrix} 7 & -8 & 1 \\ -8 & 16 & -8 \\ 1 & -8 & 7 \end{bmatrix}, \ \mathbf{M} = \frac{\rho SL}{6}\begin{bmatrix} 1 & 0 & 0 \\ 0 & 4 & 0 \\ 0 & 0 & 1 \end{bmatrix}, \ \mathbf{R} = \frac{q_0 L}{6}\begin{bmatrix} 1 \\ 4 \\ 1 \end{bmatrix} \tag{3.142}$$

Note: The stiffness matrix and load vector obtained by both methods are identical when the nodes coincide. The mass matrix of the weak form quadrature element for C^0 elements is diagonal, which can reduce the computational cost in dynamic problems.

3.2.8.3 Hierarchical Finite Element Method

If the function set corresponding to the p-th order approximation forms a subset of the function set corresponding to the $(p + 1)$-th order approximation, then the p-version finite element method is referred to as the hierarchical finite element method (HFEM).

Compared with h-version finite elements, the advantages of the hierarchical finite element method are:

- o There is no need to refine the mesh to improve the accuracy of the solution.
- o The stiffness and mass matrices exhibit a nested property; that is, the associated matrix of a displacement function of order p_1 is always a submatrix of the associated matrix of a displacement function of order p_2, where $p_2 > p_1$.
- o The nesting property can be used to prove that the eigenvalues corresponding to $(p_1 + 1)$ displacement functions include those corresponding to p_1 functions; this is known as the inclusion principle.

④ Simple structures can be modeled using a single element, thereby eliminating the need for element assembly.

⑤ Since connecting elements with polynomials of different orders is no longer difficult, local p-refinement can be implemented as needed.

⑥ More accurate results can be obtained with fewer degrees of freedom.

The most commonly used hierarchical shape functions employ the Rodrigues form of Legendre polynomials, namely,

$$P_m(\xi) = \frac{1}{m!(-2)^m} \frac{d^m}{d\xi^m}\left\{\left(1 - \xi^2\right)^m\right\} \quad -1 \le \xi \le +1 \tag{3.143}$$

The above expression can be expanded as follows:

$$P_m(\xi) = \sum_{n=0}^{[m/2]} \frac{(-1)^n}{2^n n!} \frac{(2m - 2n - 1)!!}{(m - 2n)!} \xi^{m-2n} \tag{3.144}$$

where

$$m!! = m(m - 2)\ \ldots\ (2\, or\, 1)$$
$$0!! = 1, \quad (-1)!! = 1$$

and $[m/2]$ denotes the integer part of the number inside the brackets.

Let $P_m^s(\xi)$ denote the s-fold integral of $P_{m+s-1}(\xi)$, that is,

$$P_m^s(\xi) = \int_{-1}^{\xi} \cdots \int_{-1}^{\xi} P_{m+s-1}(\xi)\mathrm{d}\xi \cdots \mathrm{d}\xi \tag{3.145}$$

Therefore

$$P_m^1(\xi) = \int_{-1}^{\xi} P_m(\xi)d\xi = \frac{(\xi^2-1)}{m(m+1)}\frac{dP_m(\xi)}{d\xi},\ \xi \in [-1,\ 1]$$
$$P_m^2(\xi) = \int_{-1}^{\xi}\int_{-1}^{\xi} P_{m+1}(\xi)d\xi = \frac{(\xi^2-1)^2}{m(m+1)(m+2)(m+3)}\frac{d^2P_{m+1}(\xi)}{d\xi^2},\ \xi \in [-1,\ 1]$$

Substituting Eq. (3.144) into Eq. (3.145), we obtain

$$P_m^s(\xi) = \sum_{n=0}^{[m/2]} \frac{(-1)^n}{2^n n!}\frac{(2m-2n-2s-1)!!}{(m-2n)!}\xi^{m-2n} \tag{3.146}$$

All derivatives of these functions of order less than s vanish at $\xi = \pm 1$. They are not only orthogonal, but can also be shown to be k-orthogonal.

The truncation error of the Taylor series expression (3.144) is relatively large, and severe numerical issues may arise at higher orders. Therefore, the following recursive expression should be used for higher-order terms.

$$P_{m+1}(\xi) = \frac{2m+1}{m+1}\xi P_m(\xi) - \frac{m}{m+1}P_{m-1}(\xi) \tag{3.147}$$

The formulas are $P_0(\xi) = 1$, $P_1(\xi) = \xi$. Taking the n-th derivative yields the recurrence relation for the n-th derivative of the Legendre polynomial.

$$P_{m+1}^{(n)}(\xi) = \xi P_m^{(n)}(\xi) + (n+m)P_m^{(n-1)}(\xi) \tag{3.148}$$

Computation shows that using the recurrence formula above, no numerical issues arise even when the order m reaches several thousand.

In the higher-order Hierarchical Finite Element Method, the axial displacement u of the rod can be expressed as

$$u = \sum_{r=1}^{p} g_r(\xi)q_r \tag{3.149}$$

Here, p denotes the number of basis functions, and $-1 \le \xi \le 1$. The first two basis functions are given by the following expressions:

$$g_1(\xi) = \frac{1}{2}(1-\xi),\ g_2(\xi) = \frac{1}{2}(1+\xi) \tag{3.150}$$

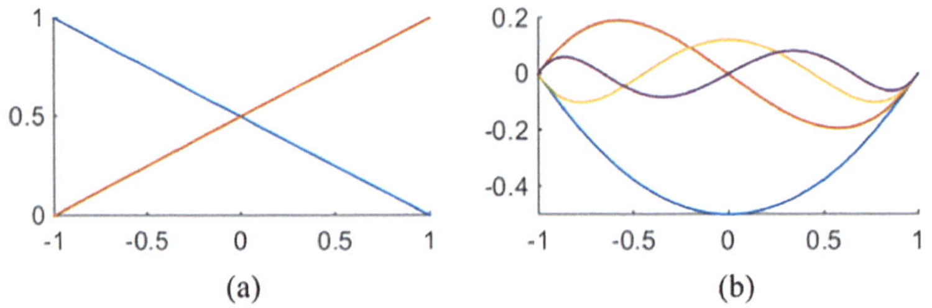

Fig. 3.14 One-dimensional basis functions for the hierarchical finite element method: **a** interpolation functions at two ends, **b** hierarchical basis functions within the domain

By substituting $s = 1$ and $m = (r - 1)$ into Eq. (3.146), the hierarchical functions $g_r(\xi)$ for $r > 2$ can be obtained as

$$g_r(\xi) = \sum_{n=0}^{[(r-1)/2]} \frac{(-1)^n}{2^n n!} \frac{(2r-2n-5)!!}{(r-2n-1)!} \xi^{r-2n-1} \tag{3.151}$$

The first four of these functions can be expressed as

$$\begin{aligned} g_3(\xi) &= -(1/2) + (1/2)\xi^2 \\ g_4(\xi) &= -(1/2)\xi + (1/2)\xi^3 \\ g_5(\xi) &= (1/8) - (3/4)\xi^2 + (5/8)\xi^4 \\ g_6(\xi) &= (3/8)\xi - (5/4)\xi^3 + (7/8)\xi^5 \end{aligned} \tag{3.152}$$

Note: All of these functions are zero at $\xi = \pm 1$. Therefore, the number of nodal degrees of freedom does not increase, i.e., $q_1 = u_1$ and $q_2 = u_2$. The first six functions are shown in Fig. 3.14.

Equation (3.149) can be expressed in matrix form as

$$u = [\mathbf{N}(\xi)]\{\mathbf{q}\} \tag{3.153}$$

Here, the row matrix $\mathbf{N}(\xi)$ contains the functions $g_r(\xi)$, while the column vector $\{\mathbf{q}\}$ consists of the degrees of freedom q_r.

Similarly, the element stiffness matrix can be expressed as

$$[\mathbf{k}]_e = \frac{2EA}{a} \int_{-1}^{+1} [\mathbf{N}'(\xi)]^{\mathrm{T}} [\mathbf{N}'(\xi)] d\xi \tag{3.154}$$

where a denotes the length of the unit. Substituting Eqs. (3.150) and (3.152) into Eq. (3.154) yields

$$[\mathbf{k}]_e = \frac{2EA}{a}\begin{bmatrix} 1/2 & -1/2 & 0 & 0 \\ -1/2 & 1/2 & 0 & 0 \\ 0 & 0 & 2/3 & 0 \\ 0 & 0 & 0 & 2/5 \end{bmatrix} \tag{3.155}$$

where only two hierarchical functions are included. This indicates that the hierarchical functions used are k-orthogonal. The preceding 2×2 submatrix is the sub-stiffness matrix without hierarchical functions, as derived in Sect. 3.2.3, while the remaining diagonal terms correspond to the hierarchical functions. It should be noted that there is no coupling either among these functions or between them and the nodal degrees of freedom. This is because

$$\int_{-1}^{1} P_n(\xi)P_m(\xi)\mathrm{d}\xi = 0, \quad \text{where } n \neq m \tag{3.156}$$

and

$$\int_{-1}^{1} \xi^n P_m(\xi) = 0, \quad \text{where } 0 \leq n \leq (m-1) \tag{3.157}$$

The above also demonstrates the nesting property: including a new higher-order hierarchical function introduces an additional row and column, while the previous rows and columns remain unchanged.

Similarly, the element mass matrix can be expressed as

$$[\mathbf{m}]_e = \frac{\rho A a}{2}\int_{-1}^{+1} \lfloor \mathbf{N}(\xi) \rfloor^{\mathrm{T}} \lfloor \mathbf{N}(\xi) \rfloor \mathrm{d}\xi \tag{3.158}$$

Substituting Eqs. (3.150) and (3.152) into Eq. (3.158), we obtain

$$[\mathbf{m}]_e = \frac{\rho A a}{2}\begin{bmatrix} 2/3 & 1/3 & -1/3 & 1/15 \\ 1/3 & 2/3 & -1/3 & -1/15 \\ -1/3 & -1/3 & 4/15 & 0 \\ 1/15 & -1/15 & 0 & 4/105 \end{bmatrix} \tag{3.159}$$

Note that the functions $g_r(\xi)$ are not m-orthogonal.

Table 3.3 presents the first two dimensionless natural frequencies of the fixed rod shown in Fig. 3.8.

The following example considers a simply supported beam with both ends fixed. The results of the hierarchical finite element method are compared with those obtained from isogeometric analysis (IGA) based on non-uniform rational B-splines (NURBS). The relative errors across the entire frequency spectrum are shown in Fig. 3.15. It can be seen that, for the first 60% of the natural frequencies of the fixed

Table 3.3 Dimensionless natural frequencies $\omega\left(\rho L^2/E\right)^{1/2}$ of a cantilever beam

Modes	The results of hierarchical finite elements			
	1 element	1 element + 1 function	1 element + 2 function	Exact solution
1	1.732	1.577	1.571	1.571
2	–	5.673	4.837	4.712

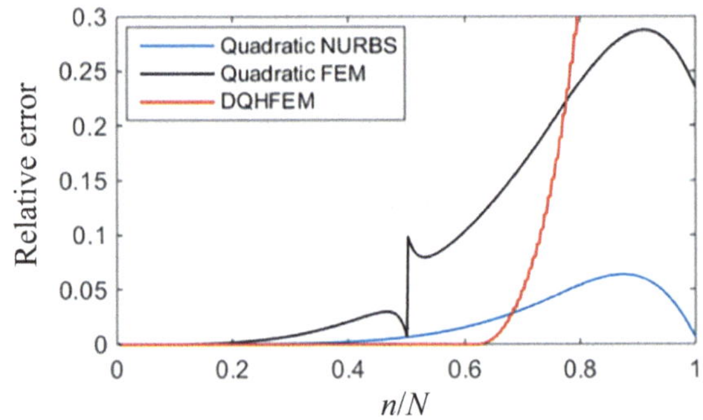

Fig. 3.15 Discrete frequency spectrum of a clamped beam: comparison of relative errors among Finite Element Method (FEM), Isogeometric Analysis (NURBS), and Hierarchical Finite Element Method (HFEM)

beam, the results from the hierarchical finite element method are highly accurate compared to those from isogeometric analysis.

3.3 Beam Finite Elements

3.3.1 Construction of Shape Functions

In the local coordinate system of a planar beam element, each node has two degrees of freedom: the deflection v in the y-direction and the rotation θ_z about the z-axis in the x–y plane (see Sect. 3.1.3). Therefore, each beam element has a total of four degrees of freedom.

As shown in Fig. 3.16, a beam element of length $l = 2a$ has nodes 1 and 2 at its ends, with the x-axis of the local coordinate system taken as the direction along the element axis, and its origin located at the middle position of the beam. The finite element equation must be derived from the interpolation shape functions of the nodal variables. Since a beam element has four degrees of freedom, there should be four shape functions accordingly. It is often more convenient to derive shape functions starting from a set of special local coordinates, commonly referred to as natural coordinate systems. The origin of this natural coordinate system is located at the center of the element, with a range from -1 to 1 (see Fig. 3.16).

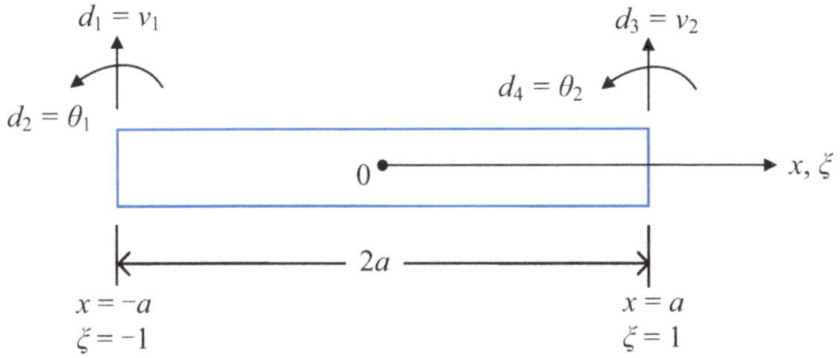

Fig. 3.16 Beam element and its local coordinate systems (physical coordinate x and natural coordinate ξ)

The relationship between the natural coordinate system and the local coordinate system can be simply expressed as follows:

$$\xi = \frac{x}{a} \tag{3.160}$$

To derive the four shape functions in the natural coordinate system, the displacement within the element is first assumed to be a cubic polynomial in ξ, that is,

$$v(\xi) = \alpha_0 + \alpha_1 \xi + \alpha_2 \xi^2 + \alpha_3 \xi^3 \tag{3.161}$$

Here, α_0 to α_3 are four unknown constants. A third-degree polynomial is chosen because it contains four unknowns, which can correspond to the four nodal degrees of freedom in a beam element. The matrix form of Eq. (3.161) is

$$v(\xi) = \begin{bmatrix} 1 & \xi & \xi^2 & \xi^3 \end{bmatrix} \begin{bmatrix} \alpha_0 \\ \alpha_1 \\ \alpha_2 \\ \alpha_3 \end{bmatrix} \tag{3.162}$$

or

$$v(\xi) = \mathbf{p}^{\mathrm{T}}(\xi)\boldsymbol{\alpha} \tag{3.163}$$

Here, $\mathbf{p}$ is the basis function vector, and $\boldsymbol{\alpha}$ is the coefficient vector. Combining with Eq. (3.160), the rotation angle θ can be obtained by differentiating Eq. (1.161), that is,

$$\theta = \frac{\partial v}{\partial x} = \frac{\partial v}{\partial \xi}\frac{\partial \xi}{\partial x} = \frac{1}{a}\frac{\partial v}{\partial \xi} = \frac{1}{a}\left(\alpha_1 + 2\alpha_2 \xi + 3\alpha_3 \xi^2\right) \tag{3.164}$$

Four unknown constants α_0 to α_3 can be determined using the following conditions:

At $x = -a$ or $\xi = -1$:

$$\begin{aligned} &(1)\ v(-1) = v_1 \\ &(2)\ \left.\frac{\mathrm{d}v}{\mathrm{d}x}\right|_{\xi=-1} = \theta_1 \end{aligned} \tag{3.165}$$

At $x = +a$ or $\xi = +1$:

$$\begin{aligned} &(3)\ v(1) = v_2 \\ &(4)\ \left.\frac{\mathrm{d}v}{\mathrm{d}x}\right|_{\xi=1} = \theta_2 \end{aligned} \tag{3.166}$$

The application of the above four conditions gives

$$\begin{bmatrix} v_1 \\ \theta_1 \\ v_2 \\ \theta_2 \end{bmatrix} = \begin{bmatrix} 1 & -1 & 1 & -1 \\ 0 & 1/a & -2/a & 3/a \\ 1 & 1 & 1 & 1 \\ 0 & 1/a & 2/a & 3/a \end{bmatrix} \begin{bmatrix} \alpha_0 \\ \alpha_1 \\ \alpha_2 \\ \alpha_3 \end{bmatrix} \tag{3.167}$$

or

$$\mathbf{d}_e = \mathbf{A}_e \boldsymbol{\alpha} \tag{3.168}$$

Solving Eq. (3.168) yields

$$\boldsymbol{\alpha} = \mathbf{A}_e^{-1} \mathbf{d}_e \tag{3.169}$$

where

$$\mathbf{A}_e^{-1} = \frac{1}{4} \begin{bmatrix} 2 & a & 2 & -a \\ -3 & -a & 3 & -a \\ 0 & -a & 0 & a \\ 1 & a & -1 & a \end{bmatrix} \tag{3.170}$$

Therefore, substituting Eq. (3.169) into Eq. (3.163) yields

$$v = \mathbf{N}(\xi) \mathbf{d}_e \tag{3.171}$$

where $\mathbf{N}$ is the shape function matrix

$$\mathbf{N}(\xi) = \mathbf{p} \mathbf{A}_e^{-1} = [N_1(\xi)\ N_2(\xi)\ N_3(\xi)\ N_4(\xi)] \tag{3.172}$$

where the shape functions are expressed as

$$
\begin{aligned}
N_1(\xi) &= \frac{1}{4}\left(2 - 3\xi + \xi^3\right)\\
N_2(\xi) &= \frac{a}{4}\left(1 - \xi - \xi^2 + \xi^3\right)\\
N_3(\xi) &= \frac{1}{4}\left(2 + 3\xi - \xi^3\right)\\
N_4(\xi) &= \frac{1}{4}a\left(-1 - \xi + \xi^2 + \xi^3\right)
\end{aligned}
\tag{3.173}
$$

It is easy to verify that the two translational shape functions, N_1 and N_3, satisfy the delta function property of finite element shape functions (Eq. (1.99)) and the partition of unity condition (Eq. (1.100)). However, the two rotational shape functions, N_2 and N_4, do not satisfy the delta function property (Eq. (1.99)) or the partition of unity condition (Eq. (1.100)). This is because these two rotational shape functions are associated with the rotational degrees of freedom derived from the deflection function. The fact that N_1 and N_3 satisfy the requirements of Eq. (1.99) ensures the correct representation of the rigid body motion of the beam element.

3.3.2 Strain Matrix

Substituting Eq. (3.171) into Eq. (3.29) yields the relationship between strain and deflection, namely

$$
\boldsymbol{\varepsilon}_{xx} = \mathbf{B}\mathbf{d}_e \tag{3.174}
$$

where, combined with Eq. (3.160), the strain matrix can be obtained as:

$$
\mathbf{B} = -yL\mathbf{N} = -y\frac{\partial^2}{\partial x^2}\mathbf{N} = -\frac{y}{a^2}\frac{\partial^2}{\partial \xi^2}\mathbf{N} = -\frac{y}{a^2}\mathbf{N}'' \tag{3.175}
$$

From Eq. (3.173), it can be obtained that:

$$
\mathbf{N}'' = \left[N_1''\ \ N_2''\ \ N''\ \ N_4''\right] \tag{3.176}
$$

where

$$
\begin{aligned}
N_1'' &= \frac{3}{2}\xi,\ N_2'' = \frac{a}{2}(-1 + 3\xi)\\
N_3'' &= -\frac{3}{2}\xi,\ N_4'' = = \frac{a}{2}(1 + 3\xi)
\end{aligned}
\tag{3.177}
$$

3.3.3 Element Matrix in the Local Coordinate System

After obtaining the strain matrix, the element stiffness and mass matrices can be determined. Substituting Eq. (3.175) into Eq. (1.125), the stiffness matrix is given by

$$
\begin{aligned}
\mathbf{k}_e &= \int_V \mathbf{B}^T \mathbf{c}\mathbf{B}\mathrm{d}V = E\int_A y^2\mathrm{d}A \int_{-a}^{a}\left[\frac{\partial^2}{\partial x^2}\mathbf{N}\right]^{\mathrm{T}}\left[\frac{\partial^2}{\partial x^2}\mathbf{N}\right]\mathrm{d}x \\
&= EI_z\int_{-1}^{1}\frac{1}{a^4}\left[\frac{\partial^2}{\partial \xi^2}\mathbf{N}\right]^{\mathrm{T}}\left[\frac{\partial^2}{\partial \xi^2}\mathbf{N}\right]a\mathrm{d}\xi = \frac{EI_z}{a^3}\int_{-1}^{1}\mathbf{N}''^{\mathrm{T}}\mathbf{N}''\mathrm{d}\xi
\end{aligned}
\tag{3.178}
$$

Here, $I_z = \int_A y^2 \mathrm{d}A$ is the moment of inertia of the beam's cross-section with respect to the z-axis. Substituting Eq. (3.176) into Eq. (3.178) yields

$$
\mathbf{k}_e = \frac{EI}{a^3}\int_{-1}^{1}\begin{bmatrix} N_1''N_1'' & N_1''N_2'' & N_1''N_3'' & N_1''N_4'' \\ N_2''N_1'' & N_2''N_2'' & N_2''N_3'' & N_2''N_4'' \\ N_3''N_1'' & N_3''N_2'' & N_3''N_3'' & N_3''N_4'' \\ N_4''N_1'' & N_4''N_2'' & N_4''N_3'' & N_4''N_4'' \end{bmatrix}\mathrm{d}x \tag{3.179}
$$

Evaluating the integral of the above equation yields,

$$
\mathbf{k}_e = \frac{EI_z}{2a^3}\begin{bmatrix} 3 & 3a & -3 & 3a \\ & 4a^2 & -3a & 2a^2 \\ & & 3 & -3a \\ sy. & & & 4a^2 \end{bmatrix} \tag{3.180}
$$

To obtain the mass matrix, substitute Eq. (3.172) into Eq. (1.129) yields

$$
\begin{aligned}
\mathbf{m}_e &= \int_V \rho\mathbf{N}^{\mathrm{T}}\mathbf{N}\mathrm{d}V = \rho\int_A \mathrm{d}A\int_{-a}^{a}\mathbf{N}^{\mathrm{T}}\mathbf{N}\mathrm{d}x = \rho A\int_A \mathbf{N}^{\mathrm{T}}\mathbf{N}a\mathrm{d}\xi \\
&= \rho A a\int_{-1}^{1}\begin{bmatrix} N_1N_1 & N_1N_2 & N_1N_3 & N_1N_4 \\ N_2N_1 & N_2N_2 & N_2N_3 & N_2N_4 \\ N_3N_1 & N_3N_2 & N_3N_3 & N_3N_4 \\ N_4N_1 & N_4N_2 & N_4N_3 & N_4N_4 \end{bmatrix}\mathrm{d}x
\end{aligned}
\tag{3.181}
$$

where A is the cross-sectional area of the beam. Evaluating the integral above yields

$$
\mathbf{m}_e = \frac{\rho A a}{105}\begin{bmatrix} 78 & 22a & 27 & -13a \\ & 8a^2 & 13a & -6a^2 \\ & & 78 & -22a \\ sy. & & & 8a^2 \end{bmatrix} \tag{3.182}
$$

The element load vector can be obtained from Eq. (1.132) and (1.133s) in Sect. 1.4.5. Assuming the element is subjected to a distributed external force f_y along the x-axis, two concentrated forces f_{s1} and f_{s2}, and concentrated moments m_{s1} and m_{s2} at nodes 1 and 2, respectively, the total nodal load vector becomes:

$$\mathbf{f}_e = f_y a \begin{bmatrix} N_1 \\ N_2 \\ N_3 \\ N_4 \end{bmatrix} + \begin{bmatrix} f_{s1} \\ m_{s1} \\ f_{s2} \\ m_{s1} \end{bmatrix} = \begin{bmatrix} f_y a + f_{s1} \\ f_y a^2/3 + m_{s1} \\ f_y a + f_{s2} \\ -f_y a^2/3 + m_{s1} \end{bmatrix} \tag{3.183}$$

3.3.4 Coordinate Transformation of Beam Element Matrices

Theoretically, coordinate transformation can also be used to convert the beam element matrix from the local coordinate system to the global coordinate system. However, this transformation is only necessary when the beam structure contains multiple beam elements, and at least two of these elements are oriented in different directions. A beam structure with at least two beam elements in different directions is typically referred to as a frame or frame structure. In order to analyze frames, it is essential to use frame elements that can carry both axial and bending forces, and coordinate transformation is generally required.

3.3.5 Calculation Example

3.3.5.1 Example 3.3: Uniform Cantilever Beam Subjected to a Downward Load

As shown in Fig. 3.17, a cantilever beam with a uniform cross-section is fixed at one end, while the free end is subjected to a downward static load of $P = 1,000\text{N}$ to bend downwardly. The beam is made of aluminum, and its material properties are listed in Table 3.4.

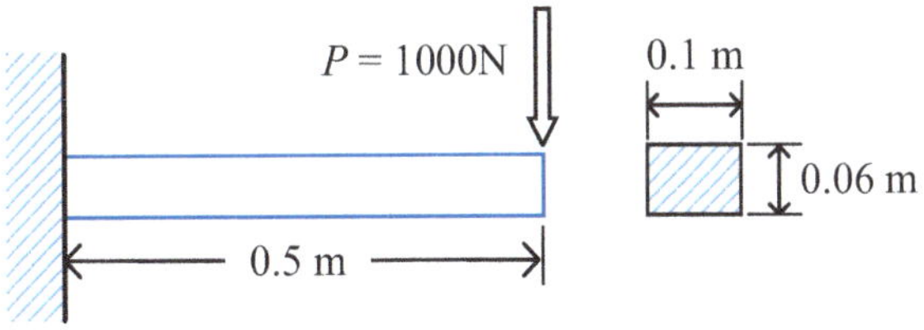

Fig. 3.17 Cantilever beam under static load

Table 3.4 Material properties of aluminum

Young's modulus E/GPa	Poisson's ratio
69. 0	0. 33

To clarify the steps required to solve this example, we will first use a single beam element to calculate the deflection. The beam element will have the following properties: Degrees of freedom as shown in Fig. 3.16.

Step 1: Obtain the element matrix

In this example, since the element matrix is the only matrix used, it effectively serves as the global finite element matrix, eliminating the need for assembly. The shape functions for the four degrees of freedom are given by Eq. (3.173). The element stiffness matrix can be obtained using Eq. (3.1801). It should be noted that, as this is a static problem, there is no need to compute the mass matrix. The second moment of area (moment of inertia) of the beam element's cross-section about the z-axis is

$$I_z = \frac{1}{12}bh^3 = \frac{1}{12}(0.1)(0.06)^3 = 1.8 \times 10^{-6}\ \mathrm{m}^4 \tag{3.184}$$

Since only a single element is used, the global stiffness matrix of the beam is identical to the element stiffness matrix, that is,

$$\begin{aligned}\mathbf{K} = \mathbf{k}_e &= \frac{(69 \times 10^9)(1.8 \times 10^{-6})}{2 \times 0.25^3}\begin{bmatrix} 3 & 0.75 & -3 & 0.75 \\ 0.75 & 0.25 & -0.75 & 0.125 \\ -3 & -0.75 & 3 & -0.75 \\ 0.75 & 0.125 & -0.75 & 0.25 \end{bmatrix} \\ &= 3,974 \times 10^6 \begin{bmatrix} 3 & 0.75 & -3 & 0.75 \\ 0.75 & 0.25 & -0.75 & 0.125 \\ -3 & -0.75 & 3 & -0.75 \\ 0.75 & 0.125 & -0.75 & 0.25 \end{bmatrix} \mathrm{Nm}^{-2}\end{aligned} \tag{3.185}$$

The finite element equation for the beam can thus be expressed as

$$\underbrace{3,974 \times 10^6 \begin{bmatrix} 3 & 0.75 & -3 & 0.75 \\ 0.75 & 0.25 & -0.75 & 0.125 \\ -3 & -0.75 & 3 & -0.75 \\ 0.75 & 0.125 & -0.75 & 0.25 \end{bmatrix}}_{\mathbf{K}} \underbrace{\begin{Bmatrix} v_1 \\ \theta_1 \\ v_2 \\ \theta_2 \end{Bmatrix}}_{\mathbf{D}} = \underbrace{\begin{Bmatrix} Q_1 = ? \\ M_1 = ? \\ Q_2 = P \\ M_2 = 0 \end{Bmatrix}}_{\mathbf{F}} \begin{matrix} \rightarrow \text{Unknown shear force} \\ \rightarrow \text{Unknown torque} \\ \\ \\ \end{matrix} \tag{3.186}$$

Note: The beam is fixed at node 1, so the shear force and moment at this node should be considered as reaction shear force and moment, which are unknown prior to solving for displacements in the finite element equations.

Step 2: Apply Boundary Conditions

One end of the beam is fixed or clamped, meaning that both the deflection v_1 and the rotation θ_1 at that end are zero. To apply this displacement boundary condition, the first and second rows and columns of the stiffness matrix must be removed, i.e.,

$$3,974 \times 10^6 \begin{bmatrix} \cancel{3} & \cancel{0.75} & \cancel{-3} & \cancel{0.75} \\ \cancel{0.75} & \cancel{0.25} & \cancel{-0.75} & \cancel{0.125} \\ \cancel{-3} & \cancel{-0.75} & 3 & -0.75 \\ \cancel{0.75} & \cancel{0.125} & -0.75 & 0.25 \end{bmatrix} \begin{Bmatrix} \cancel{v_1 = 0} \\ \cancel{\theta_1 = 0} \\ v_2 \\ \theta_2 \end{Bmatrix} = \begin{Bmatrix} \cancel{Q_1} \\ \cancel{M_1} \\ Q_2 = P \\ M_2 = 0 \end{Bmatrix} \tag{3.187}$$

The degenerated stiffness matrix thus becomes a 2 × 2 matrix, that is,

$$\mathbf{K} = 3,974 \times 10^6 \begin{bmatrix} 3 & -0.75 \\ -0.75 & 0.25 \end{bmatrix} \text{Nm}^{-2} \tag{3.188}$$

Therefore, after applying the displacement conditions, the finite element equation can be expressed as

$$\mathbf{Kd} = \mathbf{F} \tag{3.189}$$

where

$$\mathbf{d}^{\mathrm{T}} = \begin{bmatrix} v_2 \ \theta_2 \end{bmatrix}$$

and the load vector **F** is

$$\mathbf{F} = \begin{bmatrix} -1000 \\ 0 \end{bmatrix} \text{N}$$

Note: Although the reaction shear force Q_1 and moment M_1 are unknowns, this does not affect the solution of the finite element equations, because the known values of v_1 and θ_1 can be used to eliminate the unknowns Q_1 and M_1 from the original finite element equations. After solving the finite element equations for all displacements (deflections and rotations), Q_1 and M_1 can then be calculated in turn.

Step 3: Solve the Finite Element Equations

In the above case, Eq. (3.189) is actually a system of two simultaneous equations with two unknowns, which can be easily solved by hand. Of course, when there are more unknowns or degrees of freedom, it may be necessary to solve the matrix equation numerically. The solution of Eq. (3.189) is

$$\begin{aligned} v_2 &= -3.355 \times 10^{-4} \text{ m} \\ \theta_2 &= -1.007 \times 10^{-3} \text{ rad} \end{aligned} \tag{3.190}$$

Substituting v_2 and θ_2 into the first two equations of (3.186) yields the reaction shear force at node 1,

$$\begin{aligned} Q_1 &= 3.974 \times 10^6(-3v_2 + 0.75\theta_2) \\ &= 3.974 \times 10^6\left[-3 \times \left(-3.355 \times 10^{-4}\right) + 0.75 \times \left(-1.007 \times 10^{-3}\right)\right] \\ &= 998.47 \text{ N} \end{aligned}$$

and moment at node 1

$$\begin{aligned} M_1 &= 3.974 \times 10^6(-0.75v_2 + 0.125\theta_2) \\ &= 3.974 \times 10^6\left[-0.75 \times \left(-3.355 \times 10^{-4}\right) + 0.125 \times \left(-1.007 \times 10^{-3}\right)\right] \\ &= 499.73 \text{ Nm} \end{aligned}$$

This completes the entire solution process for this example.

Note: This finite element solution is entirely consistent with the analytical solution. The solving process in this example once again demonstrates the reproducibility characteristic of the finite element method.

In this case, this occurs because the exact solution for the deflection of the slender cantilever beam is a third-order polynomial, which can be obtained by solving the strong form of the beam system equation given by Eq. (3.44) when $F_z = 0$. On the other hand, the shape functions used in our finite element analysis are also third-order polynomials (refer to Eqs. (3.173) or (3.161)). Therefore, the exact solution to this problem is included in the set of assumed deflection basis functions. The finite element method, based on the principle of minimum potential energy, indeed reproduces the exact solution. Of course, if the deflection at any location other than the nodes is to be computed, it will also be exact. For example, to compute the deflection at the center of the beam, Eq. (3.171) can be used with $x = 0.25$, or it can be computed at the node in the natural coordinate system by substituting $\xi = 0$):

$$v_{\xi=0} = \mathbf{N}_{\xi=0}\mathbf{d}_e = \begin{bmatrix} \frac{1}{2} & \frac{1}{16} & \frac{1}{2} & -\frac{1}{16} \end{bmatrix} \begin{bmatrix} 0 \\ 0 \\ -3.355 \times 10^{-4} \\ -1.007 \times 10^{-3} \end{bmatrix}$$

$$= -1.048 \times 10^{-4} \text{ m}$$

To calculate the rotation angle at the center of the beam, first determine the derivative of the displacement function, and then perform the calculation as follows:

$$\theta_{\xi=0} = \left(\frac{\mathrm{d}v}{\mathrm{d}x}\right)_{\xi=0} = \left(\frac{\mathrm{d}\mathbf{N}}{\mathrm{d}x}\right)_{\xi=0} \mathbf{d}_e = \begin{bmatrix} -3 & -\frac{1}{4} & 3 & -\frac{1}{4} \end{bmatrix} \begin{bmatrix} 0 \\ 0 \\ -3.355 \times 10^{-4} \\ -1.007 \times 10^{-3} \end{bmatrix}$$

$$= -7.548 \times 10^{-4} \text{ rad}$$

Note: When obtaining d**N**/dx above, the chain rule of differentiation must be applied together with the relationship between x and ξ as given in Eq. (3.160).

3.3.6 *Higher-Order Hierarchical Finite Element Method*

In the higher-order hierarchical finite element method, the transverse displacement v of the beam can be expressed as

$$v = \sum_{r=1}^{p} f_r(\xi) q_r \tag{3.191}$$

Here, p is the number of basis functions, and $-1 \leq \xi \leq +1$. The first four interpolation functions are given by Eqs. (3.172) and (3.173), and a is replaced by $a/2$, that is,

$$\begin{aligned} f_1(\xi) &= \frac{1}{4}(2 - 3\xi + \xi^3) \\ f_2(\xi) &= \frac{1}{8}(1 - \xi - \xi^2 + \xi^3) \\ f_3(\xi) &= \frac{1}{4}(2 + 3\xi - \xi^3) \\ f_4(\xi) &= \frac{1}{8}(-1 - \xi + \xi^2 + \xi^3) \end{aligned} \tag{3.192}$$

Substituting $s = 2$ and $m = (r - 1)$ into Eq. (3.146) yields the higher-order hierarchical functions $f_r(\xi)$ for $r > 4$ as

$$f_r(\xi) = \sum_{n=0}^{[(r-1)/2]} \frac{(-1)^n}{2^n n!} \frac{(2r-2n-7)!!}{(r-2n-1)!} \xi^{(r-2n-1)} \tag{3.193}$$

The first four higher-order hierarchical basis functions are

$$\begin{aligned} f_5(\xi) &= \frac{1}{8}\left(1 - 2\xi^2 + \xi^4\right) \\ f_6(\xi) &= \frac{1}{8}\left(\xi - 2\xi^3 + \xi^5\right) \\ f_7(\xi) &= \frac{1}{48}\left(-1 + 9\xi^2 - 15\xi^4 + 7\xi^6\right) \\ f_8(\xi) &= \frac{1}{6}\left(-\xi - 5\xi^3 + 7\xi^5 + 3\xi^7\right) \end{aligned} \tag{3.194}$$

Note that these functions and their first derivatives vanish at $\xi = \pm 1$. Therefore, the first four degrees of freedom are:

$$q_1 = v_1,\ q_2 = \theta_{z1},\ q_3 = v_2,\ q_4 = \theta_{z2} \tag{3.195}$$

The first 8 hierarchical polynomial functions of the beam are shown in Fig. 3.18. Equation (3.191) can be more concisely expressed in matrix form as

$$v = [\mathbf{N}(\xi)]^{\mathrm{T}}\{\mathbf{q}\} \tag{3.196}$$

Here, the vector $[\mathbf{N}(\xi)]$ contains the function $f_r(\xi)$, and the vector $\{\boldsymbol{q}\}$ contains the degrees of freedom q_r.

As shown in Sect. 3.3.3, the element stiffness matrix can be expressed as

$$[\mathbf{k}]_e = \frac{9EI_z}{a^3} \int_{-1}^{1} \left[\mathbf{N}''(\xi)\right]^{\mathrm{T}} \left[\mathbf{N}''(\xi)\right] \mathrm{d}\xi \tag{3.197}$$

Here, a denotes the length of the unit cell.

Substituting Eqs. (3.192) and (3.194) into Eq. (3.197), we obtain, for the case of six basis functions,

$$[\mathbf{k}]_e = \frac{EI_z}{a^3} \begin{bmatrix} 12 & 6a & -12 & 6a & 0 & 0 \\ 6a & 4a^2 & -6a & 2a^2 & 0 & 0 \\ -12 & -6a & 12 & -6a & 0 & 0 \\ 6a & 2a^2 & -6a & 4a^2 & 0 & 0 \\ 0 & 0 & 0 & 0 & 16/5 & 0 \\ 0 & 0 & 0 & 0 & 0 & 16/7 \end{bmatrix} \tag{3.198}$$

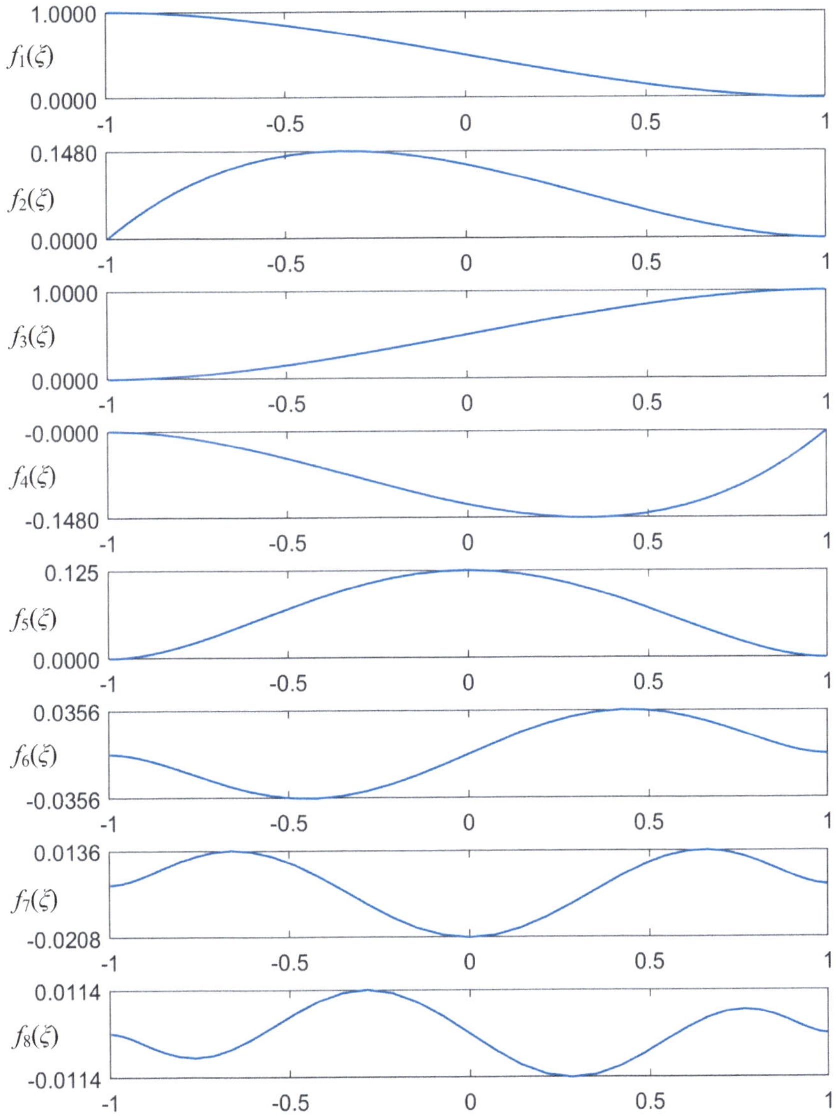

Fig. 3.18 The first 8 ascending order hierarchical polynomial functions of the beam

From the above equation, it can be seen that the higher-order hierarchical functions are k-orthogonal, and the preceding 4×4 submatrix is the stiffness matrix without higher-order hierarchical functions.

According to Sect. 3.3.3, the element mass matrix can be expressed as

$$[\mathbf{m}]_e = \frac{\rho Aa}{2}\int_{-1}^{1}[\mathbf{N}(\xi)]^{\mathrm{T}}[\mathbf{N}(\xi)]\mathrm{d}\xi \tag{3.199}$$

Substituting Eqs. (3.192) and (3.194) into Eq. (3.199), and using the six basis functions, we have

$$[\mathbf{m}]_e = \frac{\rho Aa}{210}\begin{bmatrix} 78 & 11a & 27 & -(13/2)a & 7 & -4/3 \\ 11a & 2a^2 & (13/2)a & -(3/2)a^2 & 15a & -a/6 \\ 27 & (13/2)a & 78 & -11a & 7 & 4/3 \\ -(13/2)a & -(3/2)a^2 & -11a & 2a^2 & -15a & -a/6 \\ 7 & 15a & 7 & -15a & 4/3 & 0 \\ -4/3 & -a/6 & 4/3 & -a/6 & 0 & 4/33 \end{bmatrix} \tag{3.200}$$

In high-frequency analysis, the number of nodes required may be very large. In such cases, the recursive formulas given by Eqs. (3.147) and (3.148) can be used to compute high-order polynomials and their derivatives. Table 3.5 presents the results for a simply supported Euler–Bernoulli beam, comparing the hierarchical finite element method (HFEM) results with the discrete singular convolution (DSC) results from the literature. The formula for the dimensionless frequency parameter Ω is provided in the table caption. The exact value of the frequency parameter is $(n/100)^2$, where n is the mode number. It is evident that HFEM is more accurate than DSC for the first 60% of the frequencies. The accuracy of HFEM frequencies decreases beyond 60% because the number of basis functions is insufficient to describe the mode shapes. The HFEM results for this problem are also compared with those of the finite element method (FEM) using cubic Euler–Bernoulli beam elements. A total of 1000(N) frequencies are computed. The relative errors of both methods with respect to the frequency parameter n are plotted in Fig. 3.19. As can be seen from Fig. 3.19, the relative error of FEM begins to increase after the first 20% and rises sharply by the first 50%. The relative error of HFEM only starts to increase beyond the first 60%. The frequencies within the first 60% are highly accurate. DSC can provide results with higher accuracy for frequencies beyond 60%. This may be attributed to the use of singular basis functions. However, DSC currently appears to be primarily applicable to regular domains and, like DQM or the weak-form QEM, does not possess the hierarchical property.

Table 3.5 Dimensionless frequencies of simply supported Euler–Bernoulli beams

Mode number	Grid size							
	1001		2001		3001		4001	
	HFEM	DSC	HFEM	DSC	HFEM	DSC	HFEM	DSC
500	25.0000	25.0002	25.0000	25.0000	25.0000	25.0000	25.0000	25.0000
1000	–	–	100.000	100.001	100.000	100.000	100.000	100.000
2000	–	–	–	–	410.976	401.206	400.000	400.004

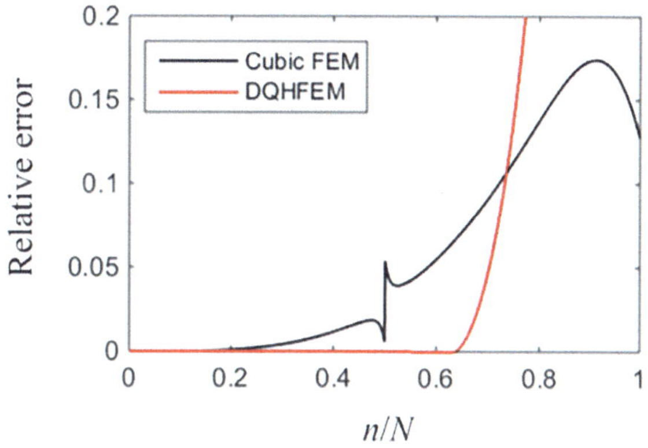

Fig. 3.19 Simply supported beam: normalized discrete frequency spectrum using Hierarchical Finite Element Method (HFEM) and cubic Finite Element Method (FEM)

3.4 Summary

This chapter introduces the finite element methods for trusses and beams, which are two typical one-dimensional finite elements. The construction of torsional shaft elements and deep beam bending elements is similar to that of truss elements. Deep beam elements are also C^0 elements, but when used for very slender beams, shear locking may occur. In practice, *h*-version finite elements are widely used in engineering due to their simplicity. However, for high-precision analysis, *p*-version finite elements are more effective. There are many approaches to *p*-version finite elements; the hierarchical finite element method (HFEM) was developed around the same time as the finite element method (FEM) and has become nearly mature. Therefore, this chapter provides a more detailed introduction to HFEM.

Chapter 4
Finite Element Analysis of Two- and Three-Dimensional Solids

Although true two-dimensional structures do not exist in real life, experienced analysts often idealize many practical problems as two-dimensional cases for analysis using 2D models, achieving satisfactory results. Compared to three-dimensional modeling, this approach is significantly more efficient and cost-effective. In-plane deformation includes both plane stress and plane strain; accordingly, plane stress and plane strain elements are required for analysis. For example, as shown in Fig. 4.1, when a load is applied within the plane of a plate structure, two-dimensional plane stress elements (membrane elements) should be used. When simulating the effect of water pressure on a dam (see Fig. 4.2), two-dimensional plane strain elements are necessary.

Three-dimensional solid elements can be regarded as the most general form of all solid finite elements, since all field variables depend on x, y, and z. A 3D solid can have arbitrary shapes, material properties, and boundary conditions in space. Therefore, six possible stress components must be considered: three normal components and three shear components. Typically, three-dimensional solid elements may take the form of tetrahedra, pyramids, triangular prisms, or hexahedra, with either planar or curved faces. Each node of the element possesses three translational degrees of freedom. As a result, the element can deform in all three spatial directions. The formulation for three-dimensional solid elements is essentially an extension of that for two-dimensional solid elements, and all techniques used in two-dimensional elements can be applied, except that all variables must be extended to functions of x, y, and z.

B. Liu and C. Liu, *Introduction to Linear and Nonlinear Finite Element Analysis*,
https://doi.org/10.1007/978-981-95-9670-6_4

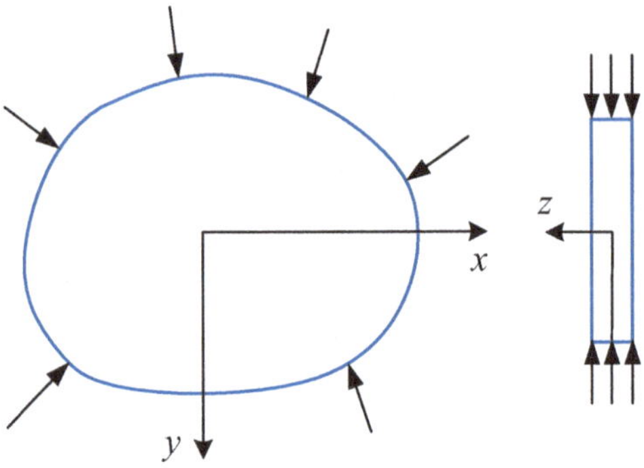

Fig. 4.1 Typical two-dimensional plane stress problem

Fig. 4.2 Typical two-dimensional plane strain problem

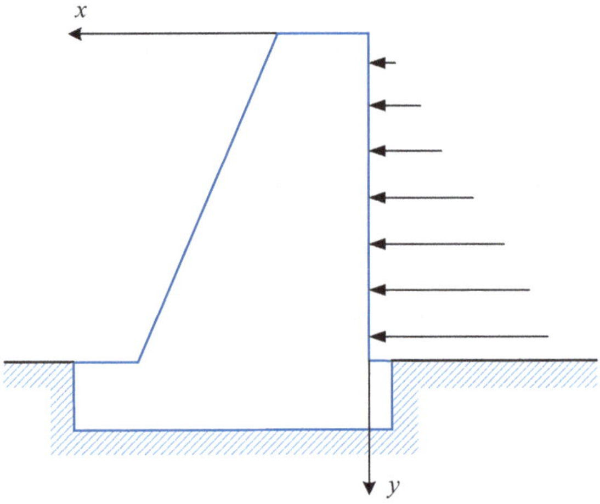

4.1 Element Energy Functional

4.1.1 Plane Stress/Strain Element

As shown in Fig. 4.3, a thin plate of constant thickness h is subjected to distributed boundary loads. These loads are applied in the plane parallel to the mid-surface of the plate and are uniformly distributed throughout the thickness. Since there are no forces acting on the surfaces $z = \pm h/2$, the stress components σz, τzx and τzy are zero on these surfaces. Under these conditions, it is reasonable to assume that the stresses at various points within the plate are sufficiently small to be neglected. Thus, the stress state is defined by the components σx, σy and τxy, which are assumed to be independent of z. This state is referred to as plane stress, and the corresponding element is called a membrane element. A plane strain body refers to a solid whose

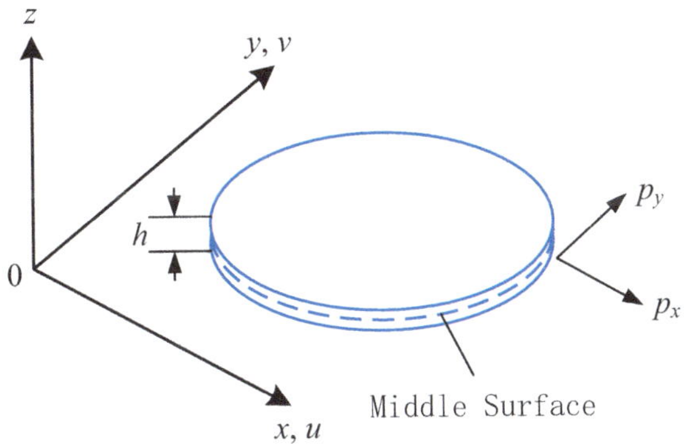

Fig. 4.3 Membrane element located in the x–y plane

thickness in the z-direction is much greater than its dimensions in the x and y directions. External forces are uniformly applied along the z-axis, and the motion at any point in the z-direction is constrained. Therefore, the strain components (εzz, εx, εyz) in the z-direction are all zero, as shown in Fig. 4.2.

The strain energy stored in the membrane element can be expressed as

$$U = \frac{1}{2}\int_V \left(\sigma_x\varepsilon_x + \sigma_y\varepsilon_y + \tau_{xy}\gamma_{xy}\right) \mathrm{d}V \tag{4.1}$$

It can also be expressed in matrix form as

$$U = \frac{1}{2}\int_V \boldsymbol{\sigma}^{\mathrm{T}}\, \boldsymbol{\varepsilon} \mathrm{d}V \tag{4.2}$$

where

$$\begin{aligned} \boldsymbol{\sigma} &= \left[\sigma_x\ \sigma_y\ \tau_{xy}\right]^{\mathrm{T}} \\ \boldsymbol{\varepsilon} &= \left[\varepsilon_x\ \varepsilon_y\ \gamma_{xy}\right]^{\mathrm{T}} \end{aligned} \tag{4.3}$$

The stress–strain relationship can be expressed as

$$\boldsymbol{\sigma} = \mathbf{D}\boldsymbol{\varepsilon} \tag{4.4}$$

For anisotropic materials, which exhibit different material properties in each direction,

$$\mathbf{D} = \begin{bmatrix} d_{11} & d_{12} & d_{13} \\ & d_{22} & d_{23} \\ \text{Sym} & & d_{33} \end{bmatrix} \tag{4.5}$$

The coefficients $d_{ij}(i, j = 1, 2, 3)$ are material constants.

If the material is orthotropic, it will have two axes of symmetry. Taking these two axes of symmetry as the $\overline{x}$ and $\overline{y}$ axes, the material constant matrix can be expressed as follows:

$$\mathbf{D}^* = \begin{bmatrix} E'_{\overline{x}} & E'_{\overline{x}}\upsilon_{\overline{xy}} & 0 \\ & E'_{\overline{y}} & 0 \\ \text{Sym} & & G_{\overline{xy}} \end{bmatrix} \tag{4.6}$$

where

$$E'_{\overline{x}} = \frac{E_{\overline{x}}}{1 - \upsilon_{\overline{xy}}\upsilon_{\overline{yx}}}, \quad E'_{\overline{y}} = \frac{E_{\overline{y}}}{1 - \upsilon_{\overline{xy}}\upsilon_{\overline{yx}}} \tag{4.7}$$

in which

$E_{\overline{x}}$ = Elastic modulus in the x-direction.

$E_{\overline{y}}$ = Elastic modulus in the y-direction.

$\upsilon_{\overline{xy}}$ = Strain in the y-direction due to a unit strain in the x-direction.

$\upsilon_{\overline{yx}}$ = Strain in the x-direction due to a unit strain in the y-direction.

$G_{\overline{xy}}$ = Shear modulus corresponding to the x and y directions.

The relationships among these constants are as follows:

$$E_{\overline{x}}\upsilon_{\overline{xy}} = E_{\overline{y}}\upsilon_{\overline{yx}}, \quad E'_{\overline{x}}\upsilon_{\overline{xy}} = E'_{\overline{y}}\upsilon_{\overline{yx}} \tag{4.8}$$

In general, there is an inclination angle of magnitude β between the material coordinates and the geometric coordinates. By considering the relationship between the strains associated with the material and geometric coordinates, it can be shown that the material constant matrix in the geometric coordinate system can be expressed as follows:

$$\mathbf{D} = \mathbf{R}^{*\mathrm{T}}\mathbf{D}\mathbf{R}^* \tag{4.9}$$

The transformation matrix $\mathbf{R}^*$ is given by the following expression:

$$\mathbf{R}^* = \begin{bmatrix} \cos^2\beta & \sin^2\beta & \frac{1}{2}\sin 2\beta \\ \sin^2\beta & \cos^2\beta & -\frac{1}{2}\sin 2\beta \\ -\sin 2\beta & \sin 2\beta & \cos 2\beta \end{bmatrix} \tag{4.10}$$

For isotropic materials, the elastic properties are identical in all directions. Therefore, the material constant matrix is simplified to

$$\mathbf{D} = \begin{bmatrix} E' & E'\upsilon & 0 \\ & E' & 0 \\ \text{Sym} & & G \end{bmatrix} \tag{4.11}$$

where

$$E' = \frac{E}{1-\upsilon^2}, \quad G = \frac{E}{2(1+\upsilon)} \tag{4.12}$$

in which E is the Young's modulus, and υ is the Poisson's ratio.

To obtain the material constant matrix **D** under the above plane stress conditions for isotropic materials, it is necessary to refer to the generalized Hooke's law and apply the conditions $\sigma_{zz} = \sigma_{xz} = \sigma_{yz} = 0$. For plane strain problems, the conditions $\varepsilon_{zz} = \varepsilon_{xz} = \varepsilon_{yz} = 0$ must be imposed. Alternatively, replace E and υ in Eq. (4.12) with $E/(1-\upsilon^2)$ and $\upsilon/(1-\upsilon)$, respectively.

Substituting Eqs. (4.4) into (4.2), we obtain

$$U = \frac{1}{2}\int_V \boldsymbol{\varepsilon}^{\mathrm{T}}\mathbf{D}\boldsymbol{\varepsilon}\mathrm{d}V \tag{4.13}$$

Since the assumed stress σ is independent of z, the strain **ε** and displacement components u and v are also independent of z. Therefore, integrating Eq. (4.13) with respect to z yields

$$U = \frac{1}{2}\int_A h\boldsymbol{\varepsilon}^{\mathrm{T}}\mathbf{D}\boldsymbol{\varepsilon}\mathrm{d}A \tag{4.14}$$

where A is the area of the midplane. In Eq. (4.14), the strain can be expressed in terms of displacement as

$$\boldsymbol{\varepsilon} = \begin{bmatrix} \frac{\partial u}{\partial x} \\ \frac{\partial v}{\partial x} \\ \frac{\partial u}{\partial y} + \frac{\partial v}{\partial x} \end{bmatrix} \tag{4.15}$$

The kinetic energy of the membrane element can be expressed as

$$T = \frac{1}{2}\int_A \rho h\left(\dot{u}^2 + \dot{v}^2\right)\mathrm{d}A \tag{4.16}$$

If p_x and p_y are the components of the boundary force applied per unit arc length on the boundary, then the virtual work can be expressed as

$$\delta W = \int_s \left(p_x\delta u + p_y\delta v\right)\mathrm{d}s \tag{4.17}$$

Here, s denotes the boundary of the element.

By eliminating the terms related to the z-coordinate in Eq. (3.5), the dynamic equilibrium equations for the two-dimensional problem can be obtained.

$$\begin{aligned} \frac{\partial \sigma_x}{\partial x} + \frac{\partial \tau_{yx}}{\partial y} + p_x = \rho \ddot{u} \\ \frac{\partial \tau_{xy}}{\partial x} + \frac{\partial \sigma_y}{\partial y} + p_y = \rho \ddot{v} \end{aligned} \tag{4.18}$$

4.1.2 Three-Dimensional Solid Elements

As shown in Fig. 4.4, the three-dimensional solid V is enclosed by the surface S. The stress and strain state at any point is defined by the six independent components given in Eqs. (3.1) and (3.3). Therefore, the strain energy stored in the three-dimensional solid element can be expressed as.

$$U = \frac{1}{2}\int_V \{\sigma_x \varepsilon_x + \sigma_y \varepsilon_y + \sigma_z \varepsilon_z + \tau_{xy}\gamma_{xy} + \tau_{xz}\gamma_{xz} + \tau_{yz}\gamma_{yz}\} \mathrm{d}V \tag{4.19}$$

It can also be expressed in matrix form as

$$U = \frac{1}{2}\int_V \boldsymbol{\sigma}^{\mathrm{T}} \boldsymbol{\varepsilon} \mathrm{d}V \tag{4.20}$$

where

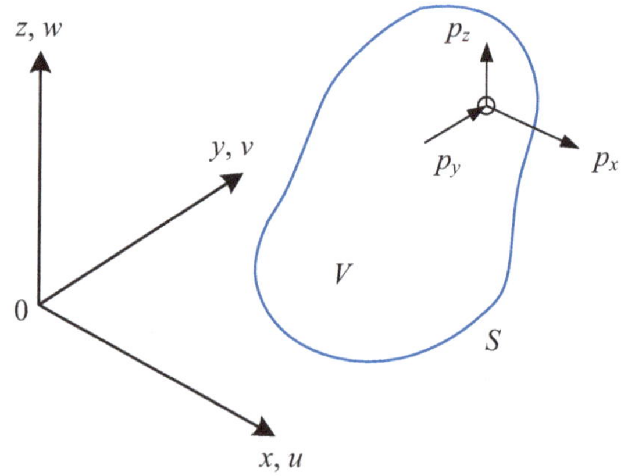

Fig. 4.4 Three-dimensional solid

$$\begin{aligned}\boldsymbol{\sigma} &= \begin{bmatrix}\sigma_x & \sigma_y & \sigma_z & \tau_{xy} & \tau_{xz} & \tau_{yz}\end{bmatrix}^{\mathrm{T}} \\ \boldsymbol{\varepsilon} &= \begin{bmatrix}\varepsilon_x & \varepsilon_y & \varepsilon_z & \gamma_{xy} & \gamma_{xz} & \gamma_{yz}\end{bmatrix}^{\mathrm{T}}\end{aligned} \tag{4.21}$$

The stress–strain relationship can be expressed as

$$\boldsymbol{\sigma} = \mathbf{D}\boldsymbol{\varepsilon} \tag{4.22}$$

Here, **D** is the symmetric matrix of material constants. For anisotropic materials, it contains 21 independent material constants; for isotropic materials, it can be expressed as

$$\mathbf{D} = \frac{E}{(1+\upsilon)(1-2\upsilon)}\begin{bmatrix} 1-\upsilon & \upsilon & \upsilon & 0 & 0 & 0 \\ & 1-\upsilon & \upsilon & 0 & 0 & 0 \\ & & 1-\upsilon & 0 & 0 & 0 \\ & & & \frac{1}{2}(1-2\upsilon) & 0 & 0 \\ & & & & \frac{1}{2}(1-2\upsilon) & 0 \\ \textit{symmetric} & & & & & \frac{1}{2}(1-2\upsilon) \end{bmatrix} \tag{4.23}$$

where E is Young's modulus and υ is Poisson's ratio.

Substituting Eqs. (4.22) into (4.20) yields

$$U = \frac{1}{2}\int_V \boldsymbol{\varepsilon}^{\mathrm{T}}\mathbf{D}\boldsymbol{\varepsilon}\mathrm{d}V \tag{4.24}$$

In Eq. (4.24), strain can be expressed in terms of displacement as follows:

$$\{\boldsymbol{\varepsilon}\} = \begin{bmatrix} \frac{\partial u}{\partial x} \\ \frac{\partial v}{\partial y} \\ \frac{\partial w}{\partial z} \\ \frac{\partial u}{\partial y} + \frac{\partial v}{\partial x} \\ \frac{\partial u}{\partial z} + \frac{\partial w}{\partial x} \\ \frac{\partial v}{\partial z} + \frac{\partial w}{\partial y} \end{bmatrix} \tag{4.25}$$

The kinetic energy of a three-dimensional solid element can be expressed as

$$T = \frac{1}{2}\int_V \rho\left(\dot{u}^2 + \dot{v}^2 + \dot{w}^2\right)\mathrm{d}V \tag{4.26}$$

If px, py and pz are the components of the surface force per unit area, then the virtual work can be expressed as

$$\delta W = \int_S (p_x \delta u + p_y \delta v + p_z \delta w) \mathrm{d}S \tag{4.27}$$

4.2 Two-Dimensional Solid Finite Element Method

Two-dimensional solid elements, whether plane strain or plane stress elements, can take the form of triangles, rectangles, or quadrilaterals with either straight or curved edges. In engineering practice, linear elements are the most commonly used. Quadratic elements are generally employed when higher accuracy is required, but are not frequently used in practical applications. Higher-order elements have also been developed, but are typically not used except for certain specific problems. The order of a two-dimensional solid element is determined by the order of the shape functions employed. Linear elements use linear shape functions, resulting in straight element edges. Quadratic elements use quadratic shape functions, allowing the element edges to be curved. Cubic or higher-order elements can be defined analogously.

In two-dimensional models, elements can only deform within the plane in which the model is defined, which is most often the x–y plane. At any given point, variables such as displacement have components in both the x and y directions, as do external forces. For plane strain problems, the actual thickness of the structure is not significant and is typically treated as a unit value in two-dimensional models. However, for plane stress problems, thickness is an important parameter for calculating the stiffness matrix and stresses. In this chapter, it is assumed that the elements have a uniform thickness h. If the structure with varying thickness is to be analyzed, it should be divided into small elements, each of which can be assigned a uniform thickness. Moreover, the formulation for two-dimensional elements with variable thickness can be readily derived, following a process similar to that for uniform elements.

Compared to one-dimensional elements, two-dimensional elements are more complex due to their higher dimensionality. However, the process of deriving these equations is very similar to that for one-dimensional truss elements, as detailed in Sect. 3.2. The derivation steps can be summarized as follows:

- Construct the shape function matrix $\mathbf{N}$ that satisfies Eqs. (1.99) and (1.100).
- Construct the strain matrix $\mathbf{B}$.
- Use $\mathbf{N}$ and $\mathbf{B}$ together with Eq. (1.19) to calculate $\mathbf{k}_\mathrm{e}$, $\mathbf{m}_\mathrm{e}$ and $\mathbf{f}_\mathrm{e}$.

In the following, three simple yet highly important element types: linear triangles, linear rectangles, and linear quadrilaterals, will be introduced. Once these three element types are understood, the same procedures can be directly applied to develop other types of higher-order elements (such as higher-order triangles and higher-order quadrilaterals).

4.2.1 Linear Triangular Element

The linear triangular element was the first type of element developed for two-dimensional solids and is the simplest among all 2D solid elements. In practice, it has been shown that linear triangular elements are less accurate than linear quadrilateral elements. For this reason, quadrilateral elements are generally considered preferable; however, in reality, triangular elements remain highly useful due to their ability to conform to complex geometries. When meshing two-dimensional models with intricate shapes and sharp angles, triangular elements are often employed. Additionally, triangles possess the simplest topological characteristics, making mesh generation software easier to develop. Today, analysts expect mesh generation to be fully automated, as complex analyses frequently require repeated or even adaptive remeshing. Most automatic mesh generation software can only produce triangular meshes. While some automatic mesh generators can create quadrilateral meshes, they still require triangular elements to handle special cases, ultimately resulting in hybrid meshes. Therefore, for many practical engineering problems, the use of triangular elements remains unavoidable.

As shown in Fig. 4.5, a two-dimensional model is presented in the x–y plane. This two-dimensional domain is appropriately partitioned into multiple triangular elements. In a mesh of linear triangular elements, each triangle has three nodes and three straight edges.

4.2.1.1 Field Variable Interpolation

As shown in Fig. 4.6, the triangular element has a thickness of h, with its nodes numbered 1, 2, and 3 in a counterclockwise direction. For two-dimensional solid elements, the field variable is displacement, which has two components (u and v). Therefore, each node has two degrees of freedom (DOF). Since a linear triangular element has three nodes, the total number of degrees of freedom for a linear triangular element is six. For linear triangular elements, the local coordinates of each element

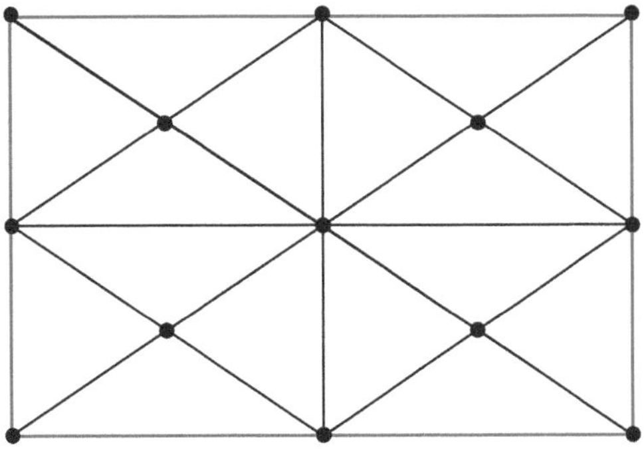

Fig. 4.5 Rectangular domain partitioned with a triangular mesh

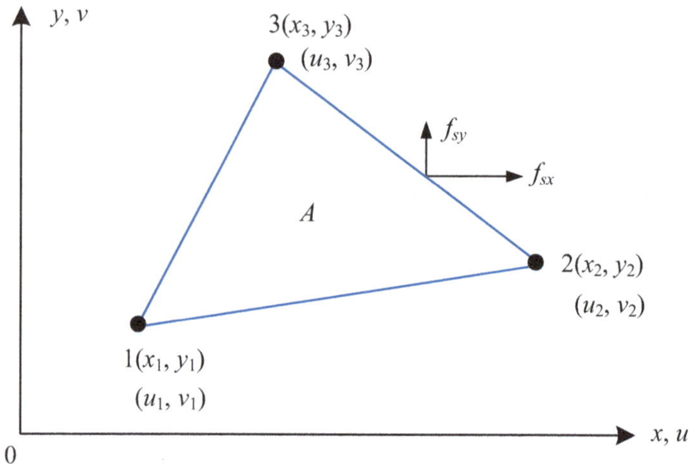

Fig. 4.6 Linear triangular element

can be considered identical to the global coordinates, as assigning a separate local coordinate system to each element offers no advantage.

When constructing triangular elements, the displacement **U** is typically a function of the coordinates x and y. Therefore, the displacement at any point within the element is expressed using the nodal displacements and shape functions. Thus, we assume

$$\mathbf{U}^h(x, y) = \mathbf{N}(x, y)\mathbf{d}_e \tag{4.28}$$

Here, the superscript h indicates that the displacement is an approximate value; $\mathbf{d}_e$ is the nodal displacement vector, arranged in the following order:

$$\mathbf{d}_e = \begin{bmatrix} u_1 \\ v_1 \\ u_2 \\ v_2 \\ u_3 \\ v_3 \end{bmatrix} \begin{matrix} \left.\vphantom{\begin{matrix} u_1 \\ v_1 \end{matrix}}\right\} \text{Displacement of Node 1} \\ \left.\vphantom{\begin{matrix} u_2 \\ v_2 \end{matrix}}\right\} \text{Displacement of Node 2} \\ \left.\vphantom{\begin{matrix} u_3 \\ v_3 \end{matrix}}\right\} \text{Displacement of Node 3} \end{matrix} \tag{4.29}$$

The shape function matrix **N** is arranged as follows

$$\mathbf{N} = \begin{bmatrix} N_1 & 0 & N_2 & 0 & N_3 & 0 \\ 0 & N_1 & 0 & N_2 & 0 & N_3 \end{bmatrix} \tag{4.30}$$

with braces beneath labelled *Node*1, *Node*2, *Node*3.

Here, $N_i (i = 1, 2, 3)$ are the three nodal shape functions corresponding to the triangular element. Equation (4.28) can be explicitly expressed as

$$
\begin{aligned}
u^h(x, y) =& N_1(x, y)u_1 + N_2(x, y)u_2 + N_3(x, y)u_3 \\
v^h(x, y) =& N_1(x, y)v_1 + N_2(x, y)v_2 + N_3(x, y)v_3
\end{aligned}
\tag{4.31}
$$

Since these two displacement components are essentially independent of each other, this means that each displacement component at any point within the element can be approximated by interpolation using the nodal displacements and shape functions. The following section introduces the construction of shape functions for triangular elements that satisfy the properties of the delta function, partition of unity, and linear field reproduction.

4.2.1.2 Shape Function Construction

Constructing shape functions is typically the first and most crucial step in deriving finite element equations. When determining the shape functions $N_i (i = 1, 2, 3)$ for triangular elements, we can fully follow the standard procedure: first, assume the displacement using polynomial basis functions with unknown constants, and then determine these unknown constants through nodal displacements at the element nodes. This standard procedure is, in principle, applicable to constructing any type of element, but it may not always be the most convenient approach. Here, an alternative, slightly different method for constructing shape functions is presented. Start by directly assuming the shape functions as polynomial basis functions with unknown coefficients and derive from there. Then, determine these unknown constants by leveraging the properties of the shape functions. The only distinction here is that we directly assume the shape functions rather than the displacement. For linear triangular elements, assume the shape functions are linear functions of x and y. Thus, they should take the following form:

$$
\begin{aligned}
N_1 &= a_1 + b_1 x + c_1 y \\
N_2 &= a_2 + b_2 x + c_2 y \\
N_3 &= a_3 + b_3 x + c_3 y
\end{aligned}
\tag{4.32}
$$

where a_i, b_i and c_i $(i = 1, 2, 3)$ are constants to be determined. Equation (4.32) can be written in a concise form as:

$$
N_i = a_i + b_i x + c_i y, \quad i = 1,\ 2,\ 3 \tag{4.33}
$$

Expressed in matrix form as

$$N_i = \underbrace{\{1\ x\ y\}}_{\mathbf{p}^{\mathrm{T}}} \underbrace{\begin{Bmatrix} a_i \\ b_i \\ c_i \end{Bmatrix}}_{\boldsymbol{\alpha}} = \mathbf{p}^{\mathrm{T}}\boldsymbol{\alpha} \tag{4.34}$$

Here, $\boldsymbol{\alpha}$ is a vector of three unknown constants, and $\mathbf{p}$ is a vector of polynomial basis functions (or monomials). According to Eq. (1.89), the matrix $\mathbf{P}$ corresponding to $\mathbf{p}$ can be expressed as

$$\mathbf{P} = \begin{bmatrix} 1\ x_1\ y_1 \\ 1\ x_2\ y_2 \\ 1\ x_3\ y_3 \end{bmatrix} \tag{4.35}$$

Note: The above equation is given for the shape function, not for the displacement.

For this particular problem, we use at most first-order polynomial basis functions. Depending on the specific problem, higher-order polynomial basis functions may be employed. In two-dimensional space, the complete order of polynomial basis functions can be represented by Pascal's triangle, as shown in Fig. 1.3. The number of parameters used in **p** depends on the number of nodes in the two-dimensional element. Typically, the lowest order is chosen to ensure the basis functions are as complete as possible. Specific higher-order terms can also be selected for different types of elements. For this triangular element, there are only three nodes, so the lowest-order complete linear polynomial is used, as shown in Eq. (4.34). Equation (4.32) indicates that displacement varies linearly within the element. In Eq. (4.33), there are a total of nine undetermined constants, and the following section explains how these constants are determined.

If the constructed shape functions possess the delta function property and the moment matrix given in Eq. (4.35) is of full rank, then the resulting shape functions will exhibit both the partition of unity and linear field reproduction properties. Therefore, it can be anticipated that the complete set of linear basis functions used in Eq. (4.34) ensures that the constructed shape functions satisfy all the requirements of finite element shape functions. The remaining task is simply to impose the delta function property on the assumed shape functions in order to determine the unknown constants a_i, b_i, and c_i.

According to the properties of the δ function, the shape function must be equal to 1 at its associated node and 0 at all other nodes. For two-dimensional problems, this can be expressed as

$$N_i(x_j, y_j) = \begin{cases} 1, & i = j \\ 0, & i \neq j \end{cases} \tag{4.36}$$

Therefore, in a triangular element, for the shape function N_1, we have

$$N_1(x_1, y_1) = 1$$

$$
\begin{aligned}
N_1(x_2, y_2) &= 0 \\
N_1(x_3, y_3) &= 0
\end{aligned} \tag{4.37}
$$

This is because the node at (x_1, y_1) is the primary node of N_1, while the node at (x_2, y_2) and the node at (x_3, y_3) are the edge nodes of N_1. By combining Eqs. (4.32) and (4.37), we obtain

$$
\begin{aligned}
N_1(x_1, y_1) &= a_1 + b_1 x_1 + c_1 y_1 = 1 \\
N_1(x_2, y_2) &= a_1 + b_1 x_2 + c_1 y_2 = 0 \\
N_1(x_3, y_3) &= a_1 + b_1 x_3 + c_1 y_3 = 0
\end{aligned} \tag{4.38}
$$

The coefficients a_1, b_1 and c_1 are obtained as follows:

$$
a_1 = \frac{x_2 y_3 - x_3 y_2}{2A_e}, \quad b_1 = \frac{y_2 - y_3}{2A_e}, \quad c_1 = \frac{x_3 - x_2}{2A_e} \tag{4.39}
$$

Here, A_e can be calculated using the determinant of a matrix and represents the area of the triangular element, that is,

$$
\begin{aligned}
A_e &= \frac{1}{2}|\mathbf{P}| = \frac{1}{2}\begin{vmatrix} 1 & x_1 & y_1 \\ 1 & x_2 & y_2 \\ 1 & x_3 & y_3 \end{vmatrix} \\
&= \frac{1}{2}\left[(x_2 y_3 - x_3 y_2) + (y_2 - y_3)x_1 + (x_3 - x_2)y_1\right]
\end{aligned} \tag{4.40}
$$

Note: As long as the area of the triangular element is not zero, or as long as the three nodes are not collinear, the matrix **P** will always be of full rank.

Substituting Eqs. (4.39) into $(4.32)_1$ yields

$$
N_1 = \frac{1}{2A_e}\left[(x_2 y_3 - x_3 y_2) + (y_2 - y_3)x + (x_3 - x_2)y\right] \tag{4.41}
$$

which can also be rewritten as

$$
N_1 = \frac{1}{2A_e}\left[(y_2 - y_3)(x - x_2) + (x_3 - x_2)(y - y_2)\right] \tag{4.42}
$$

Equation (4.42) clearly shows that N_1 is a plane in the (x, y, N) space. This plane passes through line 2–3 and is zero at node 2, located at (x_2, y_2), and at node 3, located at (x_3, y_3). The plane also passes through the point $(x_1, y_1, 1)$ in space to ensure that the shape function takes a unit value at the primary node. Since the shape function varies linearly within the element, N_1 can be easily plotted, as illustrated in Fig. 4.7a. Based on these properties of the shape function N_1, the other two shape functions for nodes 2 and 3 can be derived. For the shape function $N2$, we have

$$
\begin{aligned}
N_2(x_1, y_1) &= 0 \\
N_2(x_2, y_2) &= 1 \\
N_2(x_3, y_3) &= 0
\end{aligned} \tag{4.43}
$$

Use the condition that the shape function N_2 should pass through line 3–1, one obtains

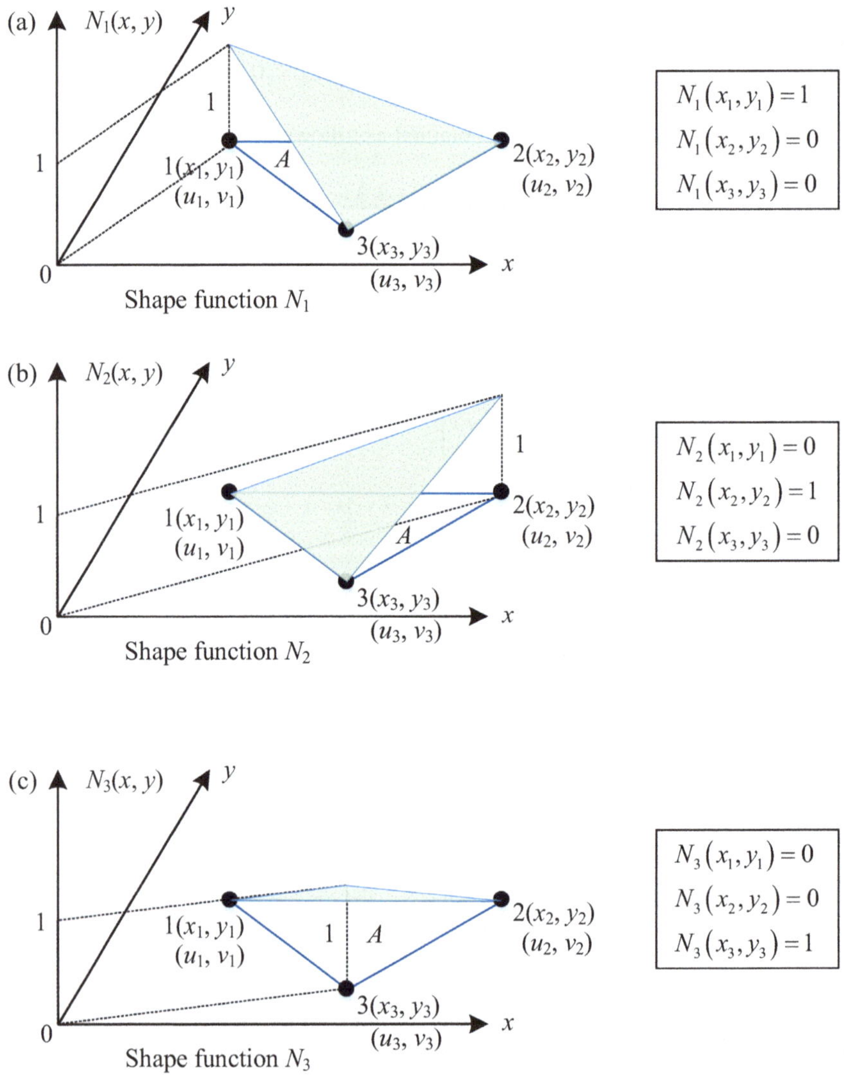

Fig. 4.7 Linear triangular element and its shape functions

$$
\begin{aligned}
N_2 &= \frac{1}{2A_e}\left[(x_3y_1 - x_1y_3) + (y_3 - y_1)x + (x_1 - x_3)y\right] \\
&= \frac{1}{2A_e}\left[(y_3 - y_1)(x - x_3) + (x_1 - x_3)(y - y_3)\right]
\end{aligned} \tag{4.44}
$$

as shown in Fig. 4.7b.

For the shape function $N3$, we have

$$
\begin{aligned}
N_3(x_1, y_1) &= 0 \\
N_3(x_2, y_2) &= 0 \\
N_3(x_3, y_3) &= 1
\end{aligned} \tag{4.45}
$$

Use the condition that the shape function N_3 should pass through line 1–2, one obtains

$$
\begin{aligned}
N_3 &= \frac{1}{2A_e}\left[(x_1y_2 - x_2y_1) + (y_1 - y_2)x + (x_2 - x_1)y\right] \\
&= \frac{1}{2A_e}\left[(y_1 - y_2)(x - x_1) + (x_2 - x_1)(y - y_1)\right]
\end{aligned} \tag{4.46}
$$

as shown in Fig. 4.7c.

The process of determining the above constants is essentially some straightforward algebra computation. Several shape functions can be uniformly expressed in the following concise form:

$$
\begin{aligned}
N_i &= a_i + b_i x + c_i y \quad i = 1,\ 2,\ 3 \\
a_i &= \frac{1}{2A_e}(x_j y_k - x_k y_j) \\
b_i &= \frac{1}{2A_e}(y_j - y_k) \\
c_i &= \frac{1}{2A_e}(x_k - x_j)
\end{aligned} \tag{4.47}
$$

Here, the subscript i ranges from 1 to 3, while the values of j and k are determined by the cyclic permutation of the sequence i, j, k. For example, if $i = 1$, then $j = 2$ and $k = 3$. When $i = 2$, then $j = 3$ and $k = 1$.

4.2.1.3 Area Coordinates

Another equivalent and effective method for constructing shape functions of triangular elements is based on the area coordinates L_1, L_2 and L_3. Using area coordinates, the shape functions of triangular elements can be directly obtained. The definition of area coordinates is given below.

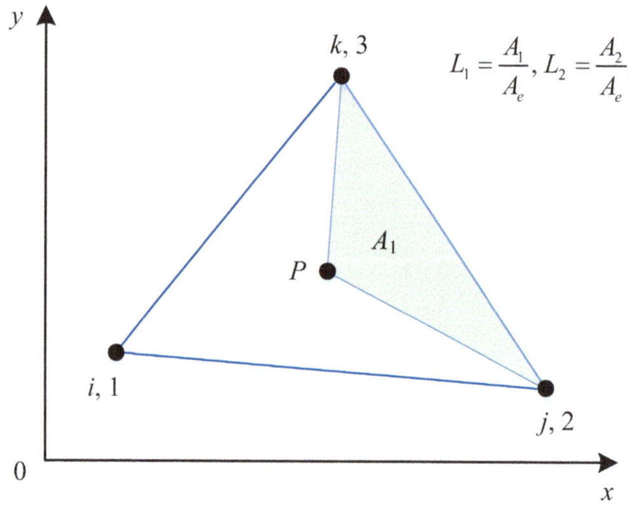

Fig. 4.8 Definition of area coordinates

When defining L_1, select a point **P** at (x, y) inside the triangle, as shown in Fig. 4.8, forming the subtriangle 2–3-P. Denote the area of this subtriangle as A_1, then

$$\begin{aligned} A_1 &= \frac{1}{2}\begin{vmatrix} 1 & x & y \\ 1 & x_2 & y_2 \\ 1 & x_3 & y_3 \end{vmatrix} \\ &= \frac{1}{2}\left[(x_2y_3 - x_3y_2) + (y_2 - y_3)x + (x_3 - x_2)y\right] \end{aligned} \tag{4.48}$$

Thus, the area coordinate L_1 is defined as

$$L_1 = \frac{A_1}{A_e} \tag{4.49}$$

Similarly, for L_2 forming the subtriangle 3–1-P with area A_2, we have

$$\begin{aligned} A_2 &= \frac{1}{2}\begin{vmatrix} 1 & x & y \\ 1 & x_3 & y_3 \\ 1 & x_1 & y_1 \end{vmatrix} \\ &= \frac{1}{2}\left[(x_3y_1 - x_1y_3) + (y_3 - y_1)x + (x_1 - x_3)y\right] \end{aligned} \tag{4.50}$$

The area coordinate L_2 can be defined as

$$L_2 = \frac{A_2}{A_e} \tag{4.51}$$

Similarly, the area coordinate L_3 can be defined as

$$L_3 = \frac{A_3}{A_e} \tag{4.52}$$

Here, A_3 is the area of subtriangle 1–2-P, that is,

$$\begin{aligned} A_3 &= \frac{1}{2}\begin{vmatrix} 1 & x & y \\ 1 & x_1 & y_1 \\ 1 & x_2 & y_2 \end{vmatrix} \\ &= \frac{1}{2}\left[(x_1 y_2 - x_2 y_1) + (y_1 - y_2)x + (x_2 - x_1)y\right] \end{aligned} \tag{4.53}$$

In the following we will verify that the area coordinates L_1, L_2 and L_3 possess the property of partition of unity and the delta function property. First, we verify that the area coordinates satisfy the partition of unity, that is,

$$L_1 + L_2 + L_3 = 1 \tag{4.54}$$

This can be proven using the definition of area coordinates, namely

$$L_1 + L_2 + L_3 = \frac{A_1}{A_e} + \frac{A_2}{A_e} + \frac{A_3}{A_e} = \frac{A_1 + A_2 + A_3}{A_e} = 1 \tag{4.55}$$

Next, we verify that the area coordinates possess the properties of the δ function. For example, if point P is located on edge nodes 2 and 3, then L_1 is necessarily zero; if P is at its primary node 1, then L_1 takes the unit value. The same conclusion holds for L_2 and L_3.

These two properties are precisely those required to define shape functions. Therefore, we have

$$\begin{aligned} N_1 &= L_1 \\ N_2 &= L_2 \\ N_3 &= L_3 \end{aligned} \tag{4.56}$$

The above equation can also be verified by comparing Eqs. (4.41) with (4.49), (4.44) with (4.51), and Eqs. (4.46) with (4.52). Area coordinates are also very convenient for constructing higher-order triangular elements.

Once the shape function matrix is obtained, the displacement at any point within the element can be expressed in terms of the nodal displacements as shown in Eq. (4.28). Based on the nodal displacements, the strain and stress at any point within the element can be calculated, thereby allowing the determination of the strain matrix and the element matrix.

4.2.1.4 Strain Matrix

According to the discussion in Sect. 4.1.1, for two-dimensional problems, there are only three principal stress components, $\boldsymbol{\sigma}^{\mathrm{T}} = \{\sigma_{xx}\ \sigma_{yy}\ \sigma_{xy}\}$ and the corresponding strain $\boldsymbol{\varepsilon}^{\mathrm{T}} = \{\varepsilon_{xx}\varepsilon_{yy}\varepsilon_{xy}\}$, both of which can be expressed as

$$\begin{aligned} \varepsilon_{xx} &= \frac{\partial u}{\partial x} \\ \varepsilon_{yy} &= \frac{\partial v}{\partial y} \\ \varepsilon_{xy} &= \frac{\partial u}{\partial y} + \frac{\partial v}{\partial x} \end{aligned} \tag{4.57}$$

Alternatively, it can be expressed in matrix form as

$$\boldsymbol{\varepsilon} = \mathbf{LU} \tag{4.58}$$

Here, **L** is referred to as the differential operator matrix, which can be directly obtained by observing Eq. (4.57), that is,

$$\mathbf{L} = \begin{bmatrix} \frac{\partial}{\partial x} & 0 \\ 0 & \frac{\partial}{\partial y} \\ \frac{\partial}{\partial y} & \frac{\partial}{\partial x} \end{bmatrix} \tag{4.59}$$

Substituting Eqs. (4.28) into (4.58), we obtain

$$\boldsymbol{\varepsilon} = \mathbf{LU} = \mathbf{LNd}_e = \mathbf{Bd}_e \tag{4.60}$$

Here, **B** is referred to as the strain matrix. Once the shape functions are obtained,

$$\mathbf{B} = \mathbf{LN} = \begin{bmatrix} \frac{\partial}{\partial x} & 0 \\ 0 & \frac{\partial}{\partial y} \\ \frac{\partial}{\partial y} & \frac{\partial}{\partial x} \end{bmatrix} \mathbf{N} \tag{4.61}$$

Equation (4.60) indicates that we can already express the strain of the element in terms of the nodal displacements of the element through the strain matrix. Equations (4.60) and (4.61) are applicable to various types of two-dimensional elements.

Using Eqs. (4.30), (4.47), and (4.61), it is straightforward to obtain the strain matrix **B** for the linear triangular element, which can be expressed in the following simple form:

$$\mathbf{B} = \begin{bmatrix} a_1 & 0 & a_2 & 0 & a_3 & 0 \\ 0 & b_1 & 0 & b_2 & 0 & b_3 \\ b_1 & a_1 & b_2 & a_2 & b_3 & a_3 \end{bmatrix} \tag{4.62}$$

4.2.1.5 Element Matrix

Once the shape functions and strain matrix are obtained, displacement, strain, and stress can all be expressed in terms of the nodal displacements of the element. Element matrices, such as the stiffness matrix **k***e*, mass matrix **m***e*, and nodal load vector **f***e*, can be calculated using the equations provided in Chap. 1.

The stiffness matrix **k***e* of a two-dimensional solid element can be expressed as

$$\begin{aligned} \mathbf{k}_e &= \int_{V_e} \mathbf{B}^{\mathrm{T}}\mathbf{c}\mathbf{B}\mathrm{d}V = \int_{A_e} \left(\int_0^h \mathrm{d}z \right) \mathbf{B}^{\mathrm{T}}\mathbf{c}\mathbf{B}\mathrm{d}A \\ &= \int_{A_e} h\mathbf{B}^{\mathrm{T}}\mathbf{c}\mathbf{B}\mathrm{d}A \end{aligned} \tag{4.63}$$

Note: The material constant matrix **c** has been given in Eq. (4.11).

Since the strain matrix **B** is a constant matrix and the thickness of the element is assumed to be uniform, the integral in Eq. (4.63) can be easily evaluated, yielding

$$\mathbf{k}_e = hA_e\mathbf{B}^{\mathrm{T}}\mathbf{c}\mathbf{B} \tag{4.64}$$

The mass matrix $\mathbf{m}_e$ for a two-dimensional solid element can be expressed as

$$\begin{aligned} \mathbf{m}_e &= \int_{V_e} \rho\mathbf{N}^{\mathrm{T}}\mathbf{N}\mathrm{d}V = \int_{A_e} \int_0^h \mathrm{d}x\rho\mathbf{N}^{\mathrm{T}}\mathbf{N}\mathrm{d}A \\ &= \int_{A_e} h\rho\mathbf{N}^{\mathrm{T}}\mathbf{N}\mathrm{d}A \end{aligned} \tag{4.65}$$

For elements with uniform thickness and density, Eq. (4.65) can also be expressed as

$$\mathbf{m}_e = h\rho \int_{A_e} \begin{bmatrix} N_1N_1 & 0 & N_1N_2 & 0 & N_1N_3 & 0 \\ 0 & N_1N_1 & 0 & N_1N_2 & 0 & N_1N_3 \\ N_2N_1 & 0 & N_2N_2 & 0 & N_2N_3 & 0 \\ 0 & N_2N_1 & 0 & N_2N_2 & 0 & N_2N_3 \\ N_3N_1 & 0 & N_3N_2 & 0 & N_3N_3 & 0 \\ 0 & N_3N_1 & 0 & N_3N_2 & 0 & N_3N_3 \end{bmatrix} \mathrm{d}A \tag{4.66}$$

All terms in the mass matrix can be easily integrated using the following formula.

$$\int_A L_1^m L_2^n L_3^p \mathrm{d}A = \frac{m!n!p!}{(m+n+p+2)!} 2A \tag{4.67}$$

Here, the area coordinates of the triangular element, $L_i = N_i$, are equivalent to the shape functions, as we saw in Sect. 4.2.1.2. The mass matrix **m***e* can be expressed as:

$$\mathbf{m}_e = \frac{\rho h A}{12} \begin{bmatrix} 2 & 0\,1\,0\,1\,0 \\ & 2\,0\,1\,0\,1 \\ & 2\,0\,1\,0 \\ & 2\,0\,1 \\ & 2\,0 \\ \text{Sym.} & 2 \end{bmatrix} \tag{4.68}$$

Assuming the element is subjected to a distributed force $\mathbf{f}_s$ along edge 2–3, as shown in Fig. 4.6, the nodal load vector is:

$$\mathbf{f}_e = \int_l [\mathbf{N}]^{\mathrm{T}} \Big|_{2-3} \begin{Bmatrix} f_{sx} \\ f_{sy} \end{Bmatrix} \mathrm{d}l \tag{4.69}$$

If the load is uniformly distributed, then fsx and f_{sy} are constant within the element. Therefore, Eq. (4.69) can be expressed as

$$\mathbf{f}_e = \frac{1}{2} l_{2-3} \begin{Bmatrix} 0 \\ 0 \\ f_x \\ f_y \\ f_x \\ f_y \end{Bmatrix} \tag{4.70}$$

Here, $l_{2\text{-}3}$ is the length of side 2–3 of the element.

Once the element stiffness matrix **k***e*, mass matrix **m***e*, and nodal load vector **f***e* have been determined, the element matrices can be assembled by summing the contributions of all adjacent elements at shared nodes, thereby obtaining the global finite element equations.

4.2.2 Linear Rectangular Element

Today, many analysts generally prefer not to use triangular elements unless it is necessary to mesh complex geometries or when remeshing is particularly challenging. The main reason is that triangular elements are typically less accurate than rectangular or quadrilateral elements. As discussed in the previous section, the inaccuracy arises because the strain matrix of linear triangular elements is constant. With advances in mesh generation algorithms, many complex geometries with sharp angles or curved edges can now also be meshed using quadrilateral grids. Since the strain matrix of rectangular elements is not constant, this allows for a more realistic representation of strain, resulting in a more accurate stress distribution throughout the structure. Compared to triangular elements, the equations for rectangular elements are simpler, as the regular shape of rectangular elements makes it easier to derive the shape functions.

4.2.2.1 Shape Function Construction

As shown in Fig. 4.9, the two-dimensional domain is discretized into multiple rectangular elements, each with four nodes and four straight edges. Similarly, the nodes 1, 2, 3, and 4 of each element are numbered in a counterclockwise direction, as illustrated in Fig. 4.10.

Note: Since each node has two degrees of freedom, the total number of degrees of freedom for a linear rectangular element is 8.

The dimensions of the rectangular element are $2a \times 2b \times h$. As shown in Fig. 4.10b, a local natural coordinate system (ξ, η) is defined with its origin at the center of the rectangular element. The relationship between the physical coordinate system (x, y) and the local natural coordinate system (ξ, η) can be expressed as follows:

$$\xi = \frac{x}{a}, \quad \eta = \frac{y}{b} \tag{4.71}$$

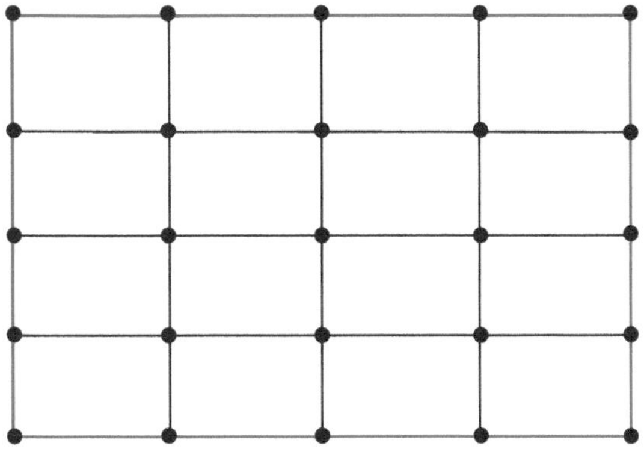

Fig. 4.9 Rectangular domain partitioned into rectangular elements

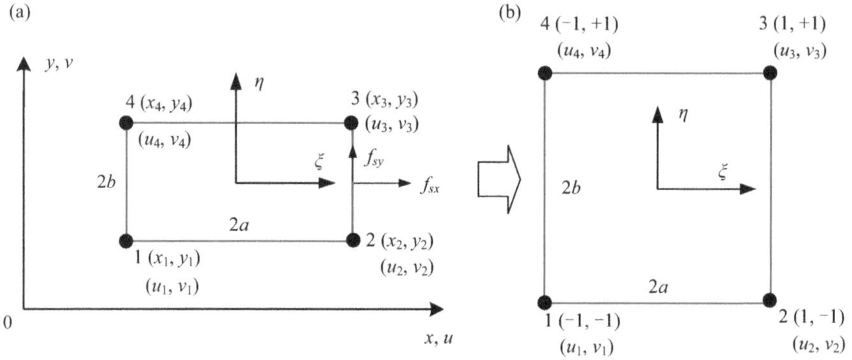

Fig. 4.10 Rectangular element and coordinate system

As shown in Fig. 4.10, there is a very simple coordinate mapping relationship between the physical coordinate system and the natural coordinate system for a rectangular element. Therefore, the formulas can be transformed into the natural coordinate system for computation. Using natural coordinates simplifies the construction of shape functions and the integration of element matrices. This coordinate mapping technique is one of the most commonly used methods in finite element analysis. It is particularly useful when developing elements with complex geometries.

The field variables are interpolated below, expressing the displacement within the element by interpolating the nodal displacements using shape functions. Assume that the displacement vector **U** takes the following form:

$$\mathbf{U}^h(x, y) = \mathbf{N}(x, y)\mathbf{d}_e \tag{4.72}$$

where the nodal displacement vector **de** is given by

$$\mathbf{d}_e = \begin{bmatrix} u_1 \\ v_1 \\ u_2 \\ v_2 \\ u_3 \\ v_3 \\ u_4 \\ v_4 \end{bmatrix} \begin{matrix} \left.\begin{matrix} \\ \\ \end{matrix}\right\} \text{Displacement of Node 1} \\ \left.\begin{matrix} \\ \\ \end{matrix}\right\} \text{Displacement of Node 2} \\ \left.\begin{matrix} \\ \\ \end{matrix}\right\} \text{Displacement of Node 3} \\ \left.\begin{matrix} \\ \\ \end{matrix}\right\} \text{Displacement of Node 4} \end{matrix} \tag{4.73}$$

The shape function matrix takes of the form

$$\mathbf{N} = \begin{bmatrix} \underbrace{\begin{matrix} N_1 & 0 \\ 0 & N_1 \end{matrix}}_{Node\,1} & \underbrace{\begin{matrix} N_2 & 0 \\ 0 & N_2 \end{matrix}}_{Node\,2} & \underbrace{\begin{matrix} N_3 & 0 \\ 0 & N_3 \end{matrix}}_{Node\,3} & \underbrace{\begin{matrix} N_4 & 0 \\ 0 & N_4 \end{matrix}}_{Node\,4} \end{bmatrix} \tag{4.74}$$

where the shape functions $N_i (i = 1, 2, 3, 4)$ correspond to the four nodes of the rectangular element.

When determining these shape functions $N_i (i = 1, 2, 3, 4)$, one can fully follow the procedures used in Sects. 3.2.1 or 4.2.1.2. That is, first assume the displacement function or shape functions using polynomial basis functions with unknown constant coefficients, and then determine these unknown constants by leveraging the properties of displacements or shape functions at the element nodes. The only difference here is that for rectangular elements, a basis of four monomial terms is required. As seen in Sect. 4.2.1.2, this process can be quite cumbersome and lengthy. In many cases, people often resort to certain "shortcut" methods to construct shape functions. One such method involves intuitive observation and utilization of the properties of shape functions.

Since the square element is highly regular in natural coordinates, the shape functions in Eq. (4.74) can be directly expressed as follows: The formula does not require the detailed process of constructing triangular elements as described in the previous section.

$$\begin{aligned} N_1 &= \frac{1}{4}(1-\xi)(1-\eta) \\ N_2 &= \frac{1}{4}(1+\xi)(1-\eta) \\ N_3 &= \frac{1}{4}(1+\xi)(1+\eta) \\ N_4 &= \frac{1}{4}(1-\xi)(1+\eta) \end{aligned} \tag{4.75}$$

It is easy to verify that all shape functions given in Eq. (4.75) satisfy the delta function property. For example, for the shape function $N3$, we have

$$\begin{aligned} N3|_{At\;node\;1} &= \frac{1}{4}(1+\xi)(1+\eta)\Big|_{\substack{\xi=-1\\ \eta=-1}} = 0 \\ N3|_{At\;node\;2} &= \frac{1}{4}(1+\xi)(1+\eta)\Big|_{\substack{\xi=1\\ \eta=-1}} = 0 \\ N3|_{At\;node\;3} &= \frac{1}{4}(1+\xi)(1+\eta)\Big|_{\substack{\xi=1\\ \eta=1}} = 1 \\ N3|_{At\;node\;4} &= \frac{1}{4}(1+\xi)(1+\eta)\Big|_{\substack{\xi=-1\\ \eta=1}} = 0 \end{aligned} \tag{4.76}$$

Similarly, the shape functions $N1$, $N2$ and $N4$ possess the same properties.

It is also easy to verify that all shape functions given by Eq. (4.75) satisfy the partition of unity property, that is,

$$\begin{aligned}\sum_{i=1}^{4} N_i &= N_1 + N_2 + N_3 + N_4 \\ &= \frac{1}{4}[(1-\xi)(1-\eta) + (1+\xi)(1-\eta) + (1+\xi)(1+\eta) + (1-\xi)(1+\eta)] \\ &= \frac{1}{4}[2(1-\xi) + 2(1+\xi)] = 1\end{aligned}$$

By Lemma 1 in Sect. 1.4.4, the partition of unity property is satisfied.

Strictly speaking, Eq. (4.75) should be referred to as a bilinear shape function, as it varies linearly along the ξ and η directions, but exhibits quadratic variation in any direction other than ξ and η. Let (ξ_j, η_j) denote the natural coordinates of node j, the bilinear shape function N_j can then be expressed in the following concise form:

$$N_j = \frac{1}{4}\left(1 + \xi_j\xi\right)\left(1 + \eta_j\eta\right) \tag{4.77}$$

4.2.2.2 Strain Matrix

By applying the same procedure for deriving the strain matrix as used for triangular elements, the strain matrix **B** for rectangular elements can be obtained in a form similar to Eq. (4.61), namely,

$$\mathbf{B} = \mathbf{LN} = \begin{bmatrix} -\frac{1-\eta}{a} & 0 & \frac{1-\eta}{a} & 0 & \frac{1+\eta}{a} & 0 & -\frac{1+\eta}{a} & 0 \\ 0 & -\frac{1-\xi}{b} & 0 & -\frac{1+\xi}{b} & 0 & \frac{1+\xi}{b} & 0 & \frac{1-\xi}{b} \\ -\frac{1-\xi}{b} & -\frac{1-\eta}{a} & -\frac{1+\xi}{b} & \frac{1-\eta}{a} & \frac{1+\xi}{b} & \frac{1+\eta}{a} & \frac{1-\xi}{b} & -\frac{1+\eta}{a} \end{bmatrix} \tag{4.78}$$

From the above equation, it can be seen that the strain matrix of a bilinear rectangular element is not a constant matrix. This implies that the strain and stress within a linear rectangular element are not constant values.

4.2.2.3 Element Matrix

After obtaining the shape functions and strain matrix, the element stiffness matrix **k***e*, mass matrix **m***e*, and nodal force vector **f***e* can be determined using the methods described in Sects. 2.1 and 4.1.1. First, according to the relationship given in Eq. (4.71), we have

$$dxdy = ab d\xi d\eta \tag{4.79}$$

Next, substituting Eqs. (4.79) into (4.63), we obtain

$$\mathbf{k}_e = \int_{A_e} h\mathbf{B}^{\mathrm{T}}\mathbf{c}\mathbf{B}dA = \int_{-1}^{1}\int_{-1}^{1} abh\mathbf{B}^{\mathrm{T}}\mathbf{c}\mathbf{B}d\xi d\eta \tag{4.80}$$

Here, the material constant matrix **c** is given by Eq. (4.11). Since the strain matrix **B** is a function of ξ and η, the integration in Eq. (4.80) is not straightforward. However, a closed-form expression for the stiffness matrix can still be obtained analytically. In practice, numerical integration is often used to evaluate the integral, with Gaussian quadrature being the most commonly employed method. Gaussian quadrature is a simple and efficient numerical integration technique, which has been briefly introduced in Sect. 2.2.

By employing the Gaussian quadrature method described in Sect. 2.2, Eq. (2.23), the element stiffness matrix **k***e* can be obtained by numerically evaluating the integral in Eq. (4.80). Since the strain–displacement matrix **B** is a linear function of either ξ or η, the use of a 2 × 2 Gaussian point scheme, as shown in Fig. 4.11b, is sufficient to yield the exact solution for the stiffness matrix given by Eq. (4.80). The integrand in Eq. (4.80) consists of $\mathbf{B}^{\mathrm{T}}\mathbf{c}\mathbf{B}$, which is the product of two linear functions and thus forms a quadratic function. Selecting two Gaussian points in each direction, as listed in Table 2.1, is sufficient to achieve exact integration for polynomial functions up to third order in that direction. Figure 4.11a, c illustrate several other possible distributions of integration points within the square domain.

To obtain the element mass matrix **m***e*, substitute Eqs. (4.79) into (4.65) to obtain

$$\begin{aligned}\mathbf{m}_e &= \int_{V_e} \rho\mathbf{N}^{\mathrm{T}}\mathbf{N}dV = \int_{A_e}\int_0^h dx\rho\mathbf{N}^{\mathrm{T}}\mathbf{N}dA = \int_{A_e} h\rho\mathbf{N}^{\mathrm{T}}\mathbf{N}dA \\ &= \int_{-1}^{1}\int_{-1}^{1} abh\rho\mathbf{N}^{\mathrm{T}}\mathbf{N}d\xi d\eta\end{aligned} \tag{4.81}$$

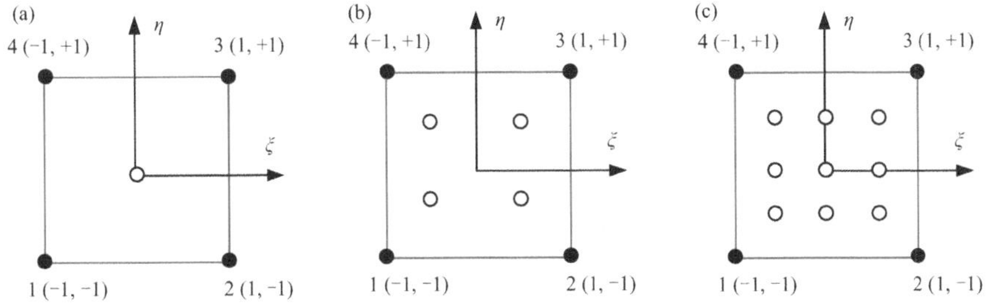

Fig. 4.11 Integration points for $n_x = n_y = 1, 2,$ and 3 in a square domain

After evaluating the integral, substituting Eqs. (4.74) into (4.81) yields the element mass matrix **me** as follows:

$$\mathbf{m}_e = \frac{\rho hab}{9}\begin{bmatrix} 4 & 0 & 2 & 0 & 1 & 0 & 2 & 0 \\ & 4 & 0 & 2 & 0 & 1 & 0 & 2 \\ & & 4 & 0 & 2 & 0 & 1 & 0 \\ & & & 4 & 0 & 2 & 0 & 1 \\ & & & & 4 & 0 & 2 & 0 \\ & & & & & 4 & 0 & 2 \\ & & & & & & 4 & 0 \\ \text{Sym.} & & & & & & & 4 \end{bmatrix} \tag{4.82}$$

To obtain m_{ij} in the element mass matrix, the following integration process must be repeatedly performed

$$\begin{aligned} m_{ij} &= \rho hab \int_{-1}^{+1}\int_{-1}^{+1} N_i N_j \mathrm{d}\xi \mathrm{d}\eta \\ &= \frac{\rho hab}{16}\int_{-1}^{+1}(1+\xi_i\xi)(1+\xi_j\xi)\mathrm{d}\xi \int_{-1}^{+1}(1+\eta_i\eta)(1+\eta_j\eta)\mathrm{d}\eta \\ &= \frac{\rho hab}{4}\left(1+\frac{1}{3}\xi_i\xi_j\right)\left(1+\frac{1}{3}\eta_i\eta_j\right) \end{aligned} \tag{4.83}$$

For example, when calculating m_{33}, using Eq. (4.83), we obtain

$$m_{33} = \frac{\rho hab}{4}\left(1+\frac{1}{3}\times 1\times 1\right)\left(1+\frac{1}{3}\times 1\times 1\right) = 4\times\frac{\rho hab}{9} \tag{4.84}$$

In practice, the integral in Eq. (4.81) is typically evaluated numerically using the Gaussian quadrature method.

As shown in Fig. 4.10, suppose the element is subjected to a distributed force **f***s* along edge 2–3. The nodal load vector for the rectangular element can then be expressed as

$$\mathbf{f}_e = \int_l [\mathbf{N}]^{\mathrm{T}}\Big|_{2-3}\begin{bmatrix} f_{sx} \\ f_{sy} \end{bmatrix}\mathrm{d}l \tag{4.85}$$

If the load is uniformly distributed within the element and f_{sx} and f_{sy} are constants, then Eq. (4.85) becomes

$$\mathbf{f}_e = b \begin{bmatrix} 0 \\ 0 \\ f_x \\ f_y \\ f_x \\ f_y \\ 0 \\ 0 \end{bmatrix} \tag{4.86}$$

Here, b is half the length of side 2–3. Equation (4.86) indicates that the uniformly distributed load is equally shared between nodes 2 and 3.

The element stiffness matrix $\mathbf{k}e$, mass matrix $\mathbf{m}e$, and nodal load vector $\mathbf{f}_e$ can be directly assembled to obtain the global finite element equations. If the direction of the local natural coordinates does not align with that of the global coordinate system, a coordinate transformation is required. In such cases, quadrilateral elements are typically used, which will be discussed in the next section.

4.2.3 Linear Quadrilateral Element

Although rectangular elements are highly useful and often more accurate than triangular elements, they are difficult to apply to problems involving arbitrary geometries and are generally limited to rectangular domains. As a result, their practical applications are quite restricted. In practice, quadrilateral elements, whose opposite sides need not be parallel, are more versatile. However, because the integration domain of a quadrilateral element is irregular, the computation of its mass and stiffness matrices can present challenges. Gaussian integration schemes cannot be directly applied to quadrilateral elements. Therefore, it is necessary to map the quadrilateral element to a natural coordinate system, transforming it into a square element so that the shape functions and integration methods for rectangular elements can be used. Thus, the key to constructing quadrilateral elements lies in coordinate mapping. Once the mapping is established, the remaining procedures are identical to those used for constructing rectangular elements as described in the previous section.

4.2.3.1 Coordinate Mapping

As shown in Fig. 4.12, the domain has the shape of an aircraft wing. It is impossible to partition such a domain into rectangular elements with parallel opposite sides, but it can be completely divided into quadrilateral elements with four non-parallel straight edges. When developing quadrilateral elements, the same coordinate mapping as used for rectangular elements in the previous section is employed. Due

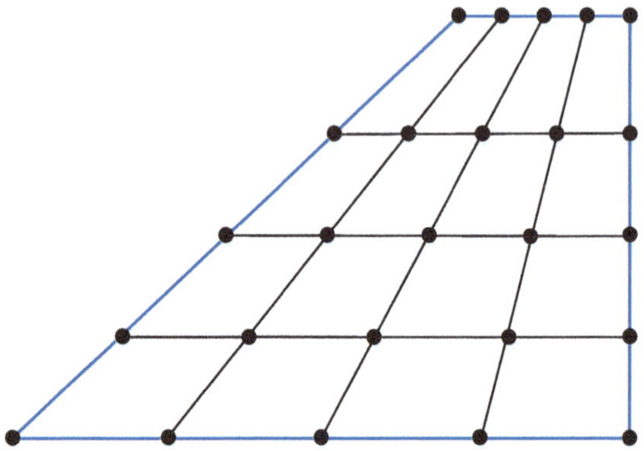

Fig. 4.12 Two-dimensional domain discretized with quadrilateral elements

to the increased complexity of the element shapes, the mapping becomes somewhat more complicated, but the overall process remains essentially the same.

As shown in Fig. 4.13, a quadrilateral element has four nodes numbered 1, 2, 3, and 4 in a counterclockwise direction. The physical coordinates of the four nodes are illustrated in Fig. 4.13a. The physical coordinate system may coincide with the global coordinate system of the entire structure. Since each node has two degrees of freedom, a linear quadrilateral element has a total of eight degrees of freedom, the same as a rectangular element. As shown in Fig. 4.13b, the origin of the local natural coordinate system (ξ, η) is located at the center of the unit rectangular domain. The local natural coordinate system is used to construct the shape functions. The displacement interpolation within the element can be expressed as

$$\mathbf{U}^h(\xi, \eta) = \mathbf{N}(\xi, \eta)\mathbf{d}_e \tag{4.87}$$

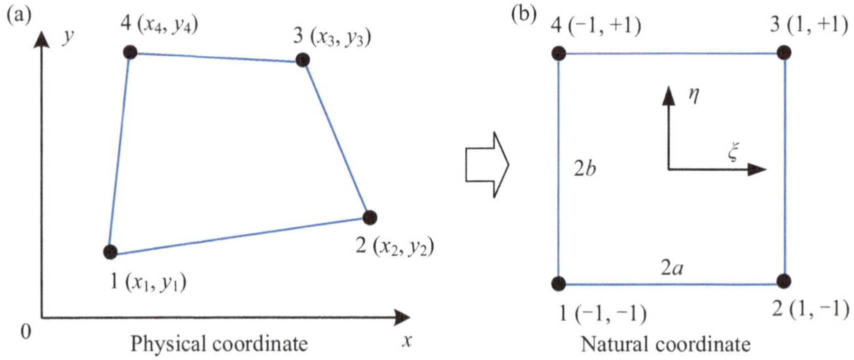

Fig. 4.13 Coordinate mapping between coordinate systems

Equation (4.87) represents the interpolation of field variables using nodal displacements. Based on a similar concept, we can also interpolate the coordinates x and y. In other words, the coordinates x and y can be expressed as interpolations of the nodal coordinates using shape functions in the natural coordinate system. Mathematically, this coordinate interpolation is expressed as

$$\mathbf{X}(\xi, \eta) = \mathbf{N}(\xi, \eta)\mathbf{x}_e \tag{4.88}$$

Here, **X** is the physical coordinate vector, that is,

$$\mathbf{X} = \begin{bmatrix} x \\ y \end{bmatrix} \tag{4.89}$$

and **N** is the shape function matrix given by Eq. (4.74); **x**e represents the physical coordinates at the element nodes, given by the following expression:

$$\mathbf{x}_e = \begin{bmatrix} x_1 \\ y_1 \\ x_2 \\ y_2 \\ x_3 \\ y_3 \\ x_4 \\ y_4 \end{bmatrix} \begin{matrix} \left.\vphantom{\begin{matrix} x \\ y \end{matrix}}\right\} \text{displacement of node 1} \\ \left.\vphantom{\begin{matrix} x \\ y \end{matrix}}\right\} \text{displacement of node 2} \\ \left.\vphantom{\begin{matrix} x \\ y \end{matrix}}\right\} \text{displacement of node 3} \\ \left.\vphantom{\begin{matrix} x \\ y \end{matrix}}\right\} \text{displacement of node 4} \end{matrix} \tag{4.90}$$

Equation (4.88) can also be expressed as

$$x = \sum_{i=1}^{4} N_i(\xi, \eta)x_i$$
$$y = \sum_{i=1}^{4} N_i(\xi, \eta)y_i \tag{4.91}$$

where N_i are the shape functions for the rectangular element given in Eq. (4.77). Note that due to the unique properties of the shape functions, interpolation at these nodes is exact. For example, substituting $\xi = 1$ and $\eta = -1$ into Eq. (4.91) yields $x = x_2$ and $y = y_2$, as shown in Fig. 4.13. Physically, this means that point 2 in the natural coordinate system maps to point 2 in the physical coordinate system, and vice versa. The same conclusion can be easily observed for points 1, 3, and 4.

Now, let us analyze this mapping more carefully. Substituting $\xi = 1$ into Eq. (4.91), we obtain:

$$
\begin{aligned}
x &= \frac{1}{2}(1-\eta)x_2 + \frac{1}{2}(1+\eta)x_3 \\
y &= \frac{1}{2}(1-\eta)y_2 + \frac{1}{2}(1+\eta)y_3
\end{aligned}
\tag{4.92}
$$

or

$$
\begin{aligned}
x &= \frac{1}{2}(x_2+x_3) + \frac{1}{2}\eta(x_3-x_2) \\
y &= \frac{1}{2}(y_2+y_3) + \frac{1}{2}\eta(y_3-y_2)
\end{aligned}
\tag{4.93}
$$

Eliminating η from Eq. (4.92) or (4.93) yields

$$
y = \frac{x_3 - x_2}{y_3 - y_2}\left\{x - \frac{1}{2}(x_2+x_3)\right\} + \frac{1}{2}(y_2+y_3) \tag{4.94}
$$

This represents a straight line connecting the points (x_2, y_2) and (x_3, y_3), meaning that edge 2–3 in the physical coordinate system is mapped to edge 2–3 in the natural coordinate system. Analogous mappings exist for the other three edges. Therefore, the four straight edges of the quadrilateral in the physical coordinate system correspond to the four straight edges of the square in the natural coordinate system, that is, the entire domain of the quadrilateral element is mapped onto a square.

4.2.3.2 Strain Matrix

After completing the coordinate mapping, the strain matrix **B** can be computed. Since the relationship between the x and y coordinates and the natural coordinates is no longer as straightforward as in rectangular elements, it is necessary to express the differentials in terms of natural coordinates. By applying the chain rule for partial derivatives, we have

$$
\begin{aligned}
\frac{\partial N_i}{\partial \xi} &= \frac{\partial N_i}{\partial x}\frac{\partial x}{\partial \xi} + \frac{\partial N_i}{\partial y}\frac{\partial y}{\partial \xi} \\
\frac{\partial N_i}{\partial \eta} &= \frac{\partial N_i}{\partial x}\frac{\partial x}{\partial \eta} + \frac{\partial N_i}{\partial y}\frac{\partial y}{\partial \eta}
\end{aligned}
\tag{4.95}
$$

It can also be expressed in matrix form, that is,

$$
\begin{bmatrix} \frac{\partial N_i}{\partial \xi} \\ \frac{\partial N_i}{\partial \eta} \end{bmatrix} = \mathbf{J} \begin{bmatrix} \frac{\partial N_i}{\partial x} \\ \frac{\partial N_i}{\partial y} \end{bmatrix} \tag{4.96}
$$

where **J** is the Jacobian matrix defined by the following equation

$$\mathbf{J} = \begin{bmatrix} \frac{\partial x}{\partial \xi} & \frac{\partial y}{\partial \xi} \\ \frac{\partial x}{\partial \eta} & \frac{\partial y}{\partial \eta} \end{bmatrix} \tag{4.97}$$

Substituting the coordinate interpolation defined by Eqs. (4.91) into (4.97) yields:

$$\mathbf{J} = \begin{bmatrix} \frac{\partial N_1}{\partial \xi} & \frac{\partial N_2}{\partial \xi} & \frac{\partial N_3}{\partial \xi} & \frac{\partial N_4}{\partial \xi} \\ \frac{\partial N_1}{\partial \eta} & \frac{\partial N_2}{\partial \eta} & \frac{\partial N_3}{\partial \eta} & \frac{\partial N_4}{\partial \eta} \end{bmatrix} \begin{bmatrix} x_1 & y_1 \\ x_2 & y_2 \\ x_3 & y_3 \\ x_4 & y_4 \end{bmatrix} \tag{4.98}$$

Rewriting Eq. (4.96), we obtain

$$\begin{bmatrix} \frac{\partial N_i}{\partial x} \\ \frac{\partial N_i}{\partial y} \end{bmatrix} = \mathbf{J}^{-1} \begin{bmatrix} \frac{\partial N_i}{\partial \xi} \\ \frac{\partial N_i}{\partial \eta} \end{bmatrix} \tag{4.99}$$

The above equation provides the relationship between the derivatives of the shape functions with respect to x and y and those with respect to ξ and η. Next, we can use the equation $\mathbf{B} = \mathbf{LN}$ to calculate the strain matrix $\mathbf{B}$. According to Eq. (4.99), all the differentials of the shape functions with respect to x and y should be replaced with differentials with respect to ξ and η. This process must be performed numerically by a computer.

4.2.3.3 Element Matrix

After obtaining the strain matrix $\mathbf{B}$, the element matrix can be computed. The element stiffness matrix can be calculated using Eq. (4.63). To perform the integration (see Sect. 2.2.1), it is necessary to use

$$dA = \det|\mathbf{J}| d\xi d\eta \tag{4.100}$$

Here, $\det|\mathbf{J}|$ is the determinant of the Jacobian matrix. Therefore, the element stiffness matrix can be expressed as

$$\mathbf{k}_e = \int_{-1}^{1} \int_{-1}^{1} h\mathbf{B}^{\mathrm{T}}\mathbf{cB} \det|\mathbf{J}| d\xi d\eta \tag{4.101}$$

The Gaussian integral introduced in the previous section can be used to evaluate the above integral.

Note: The use of coordinate mapping allows Gaussian integration to be performed over a simple rectangular region.

Since the shape functions defined by Eq. (4.75) are bilinear functions of ξ and η, and the components of the strain matrix $\mathbf{B}$ are obtained by differentiating these bilinear functions with respect to ξ and η and dividing by the Jacobian matrix, the resulting expressions are also bilinear functions. Therefore, the components of $\mathbf{B}^{\mathrm{T}}\mathbf{cB}$ det|$\mathbf{J}$| are rational functions, which generally cannot be represented by polynomials. This implies that the stiffness matrix may not be exactly evaluated using Gaussian quadrature, which is different from the case of rectangular elements.

The element mass matrix $\mathbf{m}e$ can also be calculated in a manner similar to that used for the rectangular element given in Eq. (4.81), that is,

$$\begin{aligned}\mathbf{m}_e &= \int_{V_e} \rho \mathbf{N}^{\mathrm{T}}\mathbf{N}\mathrm{d}V = \int_{A_e}\int_0^h \mathrm{d}x\rho\mathbf{N}^{\mathrm{T}}\mathbf{N}\mathrm{d}A = \int_{A_e} h\rho\mathbf{N}^{\mathrm{T}}\mathbf{N}\mathrm{d}A \\ &= \int_{-1}^{1}\int_{-1}^{1} h\rho\mathbf{N}^{\mathrm{T}}\mathbf{N}\,\mathrm{det}|\mathbf{J}|\mathrm{d}\xi\mathrm{d}\eta \qquad (4.102)\end{aligned}$$

Since the integration of the load vector is a one-dimensional line integral, the calculation method for the element load vector is the same as that for rectangular elements. After obtaining the element matrix, the standard element matrix assembly method is used to construct the global finite element matrix.

4.2.3.4 Relevant Notes

In Eq. (4.91), the shape functions used for interpolating coordinates are the same as those used for interpolating displacements, and such elements are referred to as isoparametric elements. However, the shape functions required for coordinate and displacement interpolation can also be different. When different shape functions are used for interpolating coordinates and displacements, the resulting elements are called subparametric or superparametric elements. These elements have been studied academically, but are rarely used in practical applications. Such cases may arise in higher-order elements discussed below.

4.2.4 Higher-Order Triangular Elements

Traditional high-order elements use equally spaced nodes, which can lead to numerical instability when the order of the polynomial basis functions exceeds three. In the Quadrature Element Method (QEM) based on the weak form, non-uniformly distributed nodes such as the roots of orthogonal polynomials are used, allowing for very high polynomial orders. However, the tensor product nature of the elements causes mesh refinement to propagate across elements. The Hierarchical Finite Element Method (HFEM) can easily connect elements of different polynomial orders and exhibits nested properties as the order increases. Nevertheless, the non-interpolatory nature of boundary degrees of freedom complicates element assembly and the imposition of boundary conditions. Therefore, this section introduces the

Hierarchical Quadrature Element Method (HQEM), which combines the advantages of the weak-form QEM and HFEM. Traditional high-order elements, the weak-form QEM, and HFEM can all be regarded as special cases of the HQEM.

4.2.4.1 Shape Function Construction

The hierarchical quadrature element method transforms the boundary basis functions of the hierarchical finite element method into interpolation basis functions using non-uniformly distributed nodes. Using area coordinates, the displacement field u of a hierarchical element on a triangle can be expressed as

$$u\left[x(\xi,\eta),y(\xi,\eta)\right]=\sum_{k=1}^{3}V^{k}(\xi,\eta)u_{k}$$
$$+\sum_{e=1}^{3}\sum_{i=0}^{n_e-1}E_{i}^{e}(\xi,\ \eta)a_{i}^{e}+\sum_{i=0}^{n}\sum_{j=0}^{n-i}F_{ij}(\xi,\eta)c_{ij} \tag{4.103}$$

Here, $k=1,2,3$ denotes the vertices, and $e=1,2,3$ denotes the edges (see Fig. 4.14); n_e $(e=1,2,3)$ represents the number of elements on each edge, n denotes the number of surface (basis) functions in one direction on the triangular face (see Fig. 4.15); u_k represents the vertex displacement; ${a^e}_i$ and c_{ij} are the coefficients of the higher-order hierarchical shape functions on the edges and faces of the triangular element, respectively (the coefficients of the higher-order hierarchical shape functions do not have clear physical meaning).

The vertex function V^k defined at the three vertices of a unit triangle can be expressed as

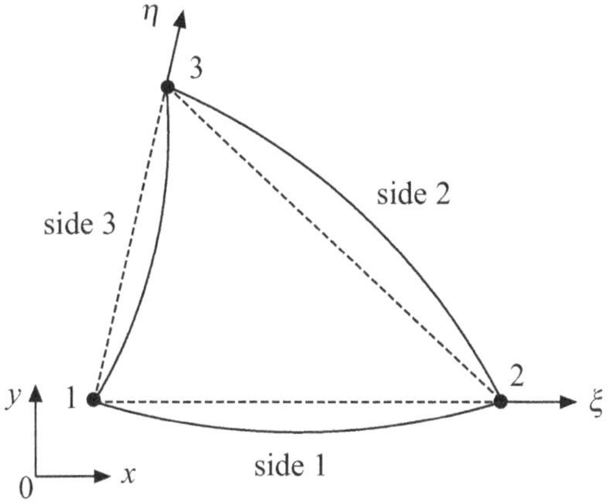

Fig. 4.14 Curved triangular element

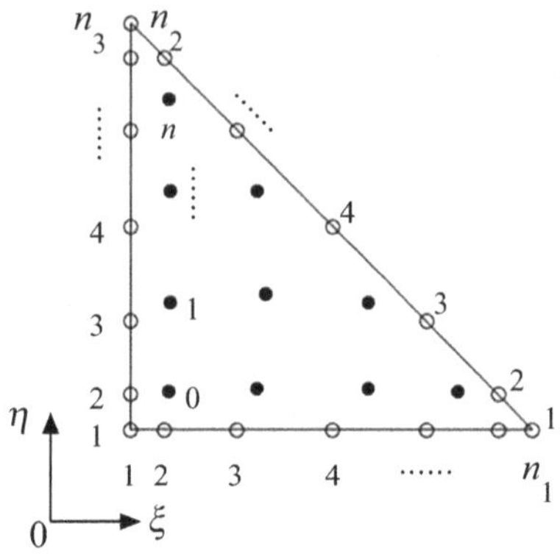

Fig. 4.15 Basis functions (or nodes) on the unit triangular domain

$$V^1 = 1 - \xi - \eta, \quad V^2 = \xi, \quad V^3 = \eta \tag{4.104}$$

It can be seen that they are similar to the linear triangular elements defined by Eq. (4.56) and (4.54) in Sect. 4.2.1.

The basis functions E_i^e and F_{ij} are defined on the three edges and the face of the unit triangle, respectively. They should be orthogonal over the reference triangular domain, which can improve the numerical properties of the basis functions. Using the Duffy transformation, the unit square domain $[-1, 1] \times [-1, 1]$ can be mapped to the unit triangular domain ($\xi^3 0$, $\eta \geq 0$, $\xi + \eta \leq 1$), that is,

$$\xi = \frac{(1+s)(1-t)}{4}, \quad \eta = \frac{(1+s)(1+t)}{4} \tag{4.105}$$

The mapping relationship of this unit domain is shown in Fig. 4.16. Through the inverse transformation, the triangular domain can be mapped to a rectangular domain, that is,

$$s = 2(\xi + \eta) - 1, \quad t = \frac{\eta - \xi}{\eta + \xi} \tag{4.106}$$

The n-th order Jacobi polynomial with the following orthogonality relation on the interval $[-1, 1]$ is denoted as $P_n^{(\alpha,\beta)}(x)$,

$$\int_{-1}^{1} (1-x)^\alpha (1+x)^\beta P_m^{(\alpha,\beta)}(x) P_n^{(\alpha,\beta)}(x) \mathrm{d}x = \delta_n^m C_n^{(\alpha,\beta)} \tag{4.107}$$

Here, δ_n^m is the Kronecker symbol, and

$$C_n^{(\alpha,\beta)} = \frac{2^{\alpha+\beta+1}}{2n+\alpha+\beta+1} \frac{\Gamma(n+\alpha+1)\Gamma(n+\beta+1)}{n!\Gamma(n+\alpha+\beta+1)} \tag{4.108}$$

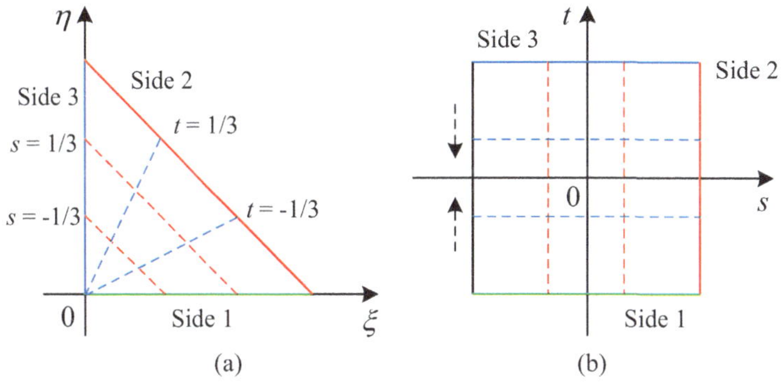

Fig. 4.16 Mapping relationship of the unit domain

The Dubiner orthogonal basis functions satisfy the following orthogonality condition:

$$\int_S g_{ij} g_{pq} \mathrm{d}S = \int_0^1 \int_0^{1-\xi} g_{ij} g_{pq} \mathrm{d}\xi \mathrm{d}\eta = \delta_i^p \delta_j^q C_{ij} \tag{4.109}$$

Here, C_{ij} is a constant. The Jacobian determinant of the coordinates (ξ, η) with respect to (s, t) is

$$|\mathbf{J}| = \left| \frac{\partial(\xi,\ \eta)}{\partial(s,\ t)} \right| = \begin{vmatrix} \frac{1-t}{4} & -\frac{1+s}{4} \\ \frac{1+t}{4} & \frac{1+s}{4} \end{vmatrix} = \frac{1+s}{2^3} \tag{4.110}$$

Consequently, substituting Eqs. (4.105) into (4.109) yields

$$\int_S g_{ij} g_{pq} \mathrm{d}S = \int_{-1}^1 \int_{-1}^1 \frac{1+s}{2^3} g_{ij}(s, t) g_{pq}(s, t) \mathrm{d}s\mathrm{d}t \tag{4.111}$$

Using the orthogonality property of Jacobi polynomials as given in Eq. (4.108), the basis functions $g_{ij}(s, t)$ can be determined.

$$g_{ij}(s, t) = \left[(1-s)^a (1+s)^b P_i^{(2a,\ 2b+1)}(s)\right]\left[(1-t)^c (1+t)^d P_j^{(2c,\ 2d)}(t)\right] \tag{4.112}$$

Here, a, b, c and d should be determined based on convenience and the requirements of the Jacobi polynomials. In order to eliminate the denominator $(\xi + \eta)$ of t in Eq. (4.113), b should be set to 1; for the C^0 problem considered here, $a = c = d = 0$ can be chosen, thus yielding

$$g_{ij}(s, t) = (1+s)^j P_i^{(0,\ 2j+1)}(s) P_j^{(0,\ 0)}(t) \tag{4.113}$$

or

$$g_{ij}(\xi,\eta) = 2^j P_i^{(0,\,2j+1)}[2(\xi+\eta)-1]\left[(\xi+\eta)^j P_j^{(0,\,0)}\left(\frac{\eta-\xi}{\eta+\xi}\right)\right] \tag{4.114}$$

These are the Dubiner orthogonal basis functions. The constant coefficient 2^j in Eq. (4.115) is often omitted; this is referred to as the twisted product, to distinguish it from the standard tensor product associated with quadrilateral domains.

In Eq. (4.103), the edge function E_i^e and the face function F_{ij} are typically represented using the modified Dubiner triangular basis functions. The modified edge function is

$$\begin{aligned} E_i^1(\xi,\eta) &= (1-\xi-\eta)\xi P_i^{(2,\,3)}[2(\xi+\eta)-1] \\ E_j^2(\xi,\eta) &= \xi\eta(\eta+\xi)^j P_j^{(2,\,2)}\left(\frac{\eta-\xi}{\eta+\xi}\right) \\ E_i^3(\xi,\eta) &= (1-\xi-\eta)\eta P_i^{(2,\,3)}[2(\xi+\eta)-1] \end{aligned} \tag{4.115}$$

The first few terms of the edge function are

$$\begin{aligned} &P_0^{(2,\,3)}(s) = 1, \quad P_1^{(2,\,3)}(s) = 7(\xi+\eta)-4 \\ &P_2^{(2,\,3)}(s) = 36(\xi+\eta)^2 - 40(\xi+\eta)+10 \\ &E_0^2(\xi,\eta) = \xi\eta, \quad E_1^2(\xi,\eta) = \xi\eta(\eta-\xi) \\ &E_2^2(\xi,\eta) = \xi\eta\left[6(\eta-\xi)^2 - 4\xi\eta\right] \end{aligned} \tag{4.116}$$

Figure 4.17 shows plots of several edge functions for the triangular element, calculated using Eq. (4.115).

The modified surface function is

$$F_{ij}(\xi,\eta) = (1-\xi-\eta)\xi\eta H_{ij}(\xi,\eta) \tag{4.117}$$

Here

$$H_{ij}(\xi,\eta) = P_i^{(2,\,2j+5)}[2(\eta+\xi)-1]\left[(\eta+\xi)^j P_j^{(2,\,2)}\left(\frac{\eta-\xi}{\eta+\xi}\right)\right] \tag{4.118}$$

The first few terms are

$$\begin{aligned} H_{00}(\xi,\eta) &= 1 \\ H_{01}(\xi,\eta) &= 3(\eta-\xi) \\ H_{10}(\xi,\eta) &= 9(\xi+\eta)-6 \\ H_{02}(\xi,\eta) &= 6(\eta-\xi)^2 - 4\xi\eta \\ H_{11}(\xi,\eta) &= (\eta-\xi)[33(\eta+\xi)-24] \\ H_{20}(\xi,\eta) &= 55(\xi+\eta)^2 - 70(\xi+\eta)+21 \end{aligned} \tag{4.119}$$

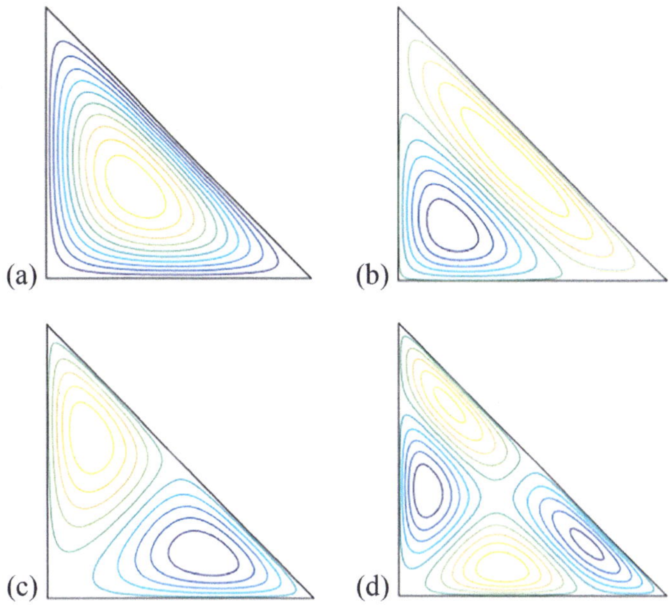

Fig. 4.17 Edge functions on a triangular element: **a** F_{00}, **b** F_{01}, **c** F_{10}, **d** F_{11}

Figure 4.18 shows the plots of several face functions of the triangular element calculated using Eq. (4.117).

Based on the above derivation, the general form of the distortion basis function, as well as the edge and face functions given above, can be uniformly expressed as

$$N_{ij}(\xi, \eta) = (1 - \xi - \eta)^h \xi^l \eta^m H_{ij}(\xi, \eta) \tag{4.120}$$

These functions should satisfy the following orthogonality condition

$$\int_0^1 \int_0^{1-\xi} N_{ij} N_{pq} \mathrm{d}\xi \mathrm{d}\eta = \int_{-1}^1 \int_{-1}^1 \frac{(1-s)^{2h}(1+s)^{2l+2m+1}(1-t)^{2l}(1+t)^{2m}}{2^{3+2h+4l+4m}} H_{ij}(s,t) H_{pq}(s,t) \mathrm{d}s\mathrm{d}t = \delta_i^p \delta_j^q C_{ij} \tag{4.121}$$

Among them, H_{ij} can be determined in a manner similar to Eqs. (4.105) through (4.114) as follows:

$$H_{ij}(s, t) = (1 + s)^j J_i^{(2h,\, 2l+2m+1+2j)}(s) J_j^{(2l,\, 2m)}(t) \tag{4.122}$$

or

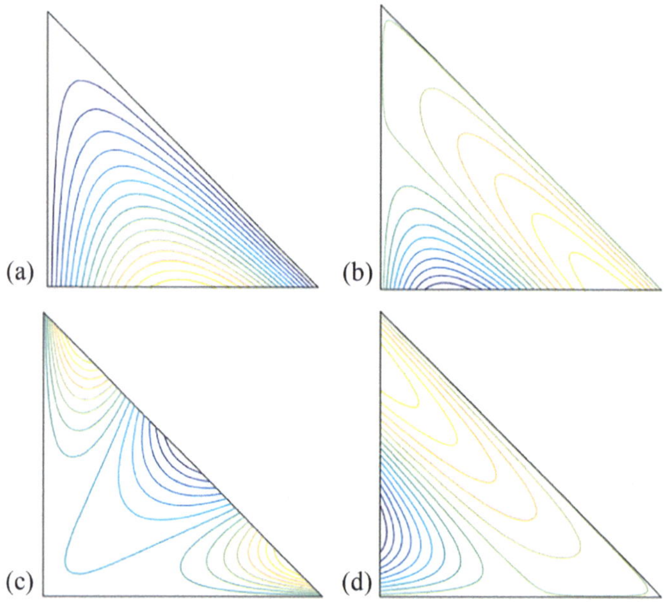

Fig. 4.18 Surface functions on a triangular element: **a** $E^1{}_0$, **b** $E^1{}_1$, **c** $E^2{}_2$, **d** $E^3{}_1$

$$H_{ij}(\xi, \eta) = 2^j (\xi + \eta)^j P_i^{(2h,\ 2j+2l+2m+1)}[2(\xi + \eta) - 1] P_j^{(2l,\ 2m)}\left(\frac{\eta - \xi}{\eta + \xi}\right) \tag{4.123}$$

where h, l and m are equal to 0 or 1. The constant coefficient 2^j in Eq. (4.123) is usually omitted. The constant in Eq. (4.121) is

$$C_{ij} = \frac{C_i^{(2h,\ 2j+2l+2m+1)} C_j^{(2l,\ 2m)}}{2^{3+2h+4l+4m}}$$

When $h = l = m = 1$, it can be observed that Eq. (4.120) reduces to the face functions given by Eq. (4.117). The edge functions above can be obtained in a similar manner. The Dubiner orthogonal basis functions given by Eq. (4.113) or (4.114) can be obtained by setting $h = l = m = 0$ in Eqs. (4.122) or (4.123).

Note: The orthogonal basis described above appears to be rational in form; however, as seen from Eqs. (4.116) and (4.119), they are not inherently so.

To avoid computational issues with the denominator approaching $\xi + \eta = 0$, the recurrence formula for the Jacobi polynomials can be reformulated. For convenience in derivation, the three-term recurrence relation for Jacobi polynomials is first presented as follows:

$$a_{1n} P_{n+1}^{(\alpha,\beta)}(\xi) = (a_{2n} + a_{3n}\xi) P_n^{(\alpha,\beta)}(\xi) - a_{4n} P_{n-1}^{(\alpha,\beta)}(\xi) \tag{4.124}$$

Here, n *is* a non-negative integer, $\alpha > -1, \beta > -1$, and

$$\begin{aligned} P_0^{(\alpha,\beta)}(\xi) &= 1 \\ P_1^{(\alpha,\beta)}(\xi) &= \frac{1}{2}[\alpha - \beta + (\alpha + \beta + 2)\xi] \end{aligned} \tag{4.125}$$

$$\begin{aligned} a_{1n} &= 2(n+1)(n+\alpha+\beta+1)(2n+\alpha+\beta) \\ a_{2n} &= (2n+\alpha+\beta+1)(\alpha^2-\beta^2) \\ a_{3n} &= (2n+\alpha+\beta)(2n+\alpha+\beta+1)(2n+\alpha+\beta+2) \\ a_{4n} &= 2(n+\alpha)(n+\beta)(2n+\alpha+\beta+2) \end{aligned} \tag{4.126}$$

To eliminate $\xi + \eta = 0$ in the denominator, we perform the following variable substitution, that is, replace ξ in the expression with $\tilde{s}/\tilde{t}$, and denote

$$\tilde{P}_n^{(\alpha,\beta)}(\tilde{s},\tilde{t}) = \tilde{t}^n P_n^{(\alpha,\beta)}(\tilde{s}/\tilde{t}), \quad \tilde{s} = \eta - \xi, \quad \tilde{t} = \eta + \xi \tag{4.127}$$

Substituting Eqs. (4.127) into (4.124) yields

$$a_{1n}\tilde{P}_{n+1}^{(\alpha,\beta)}(\tilde{s},\tilde{t}) = (a_{2n}\tilde{t} + a_{3n}\tilde{s})\tilde{P}_n^{(\alpha,\beta)}(\tilde{s},\tilde{t}) - a_{4n}\tilde{t}^2\tilde{P}_{n-1}^{(\alpha,\beta)}(\tilde{s},\tilde{t}) \tag{4.128}$$

where

$$\tilde{P}_0^{(\alpha,\beta)}(\tilde{s},\tilde{t}) = 1, \quad \tilde{P}_1^{(\alpha,\beta)}(\tilde{s},\tilde{t}) = \frac{1}{2}\big[(\alpha-\beta)\tilde{t} + (\alpha+\beta+2)\tilde{s}\big] \tag{4.129}$$

This technique can also be applied to tetrahedral and pyramidal elements, which will be discussed in detail later.

As shown in Fig. 4.14, the displacement field Eq. (4.103) for the curved triangular element can be expressed as

$$u[x(\xi,\eta), y(\xi,\eta)] = \left[\tilde{\mathbf{S}}_1^{\mathrm{T}}(\xi,\eta)\ \tilde{\mathbf{S}}_2^{\mathrm{T}}(\xi,\eta)\right]\begin{Bmatrix} \tilde{\mathbf{u}}_1 \\ \tilde{\mathbf{u}}_2 \end{Bmatrix} = \tilde{\mathbf{S}}^{\mathrm{T}}(\xi,\eta)\tilde{\mathbf{u}} \tag{4.130}$$

where

$$\begin{aligned} \tilde{\mathbf{S}}_1^{\mathrm{T}}(\xi,\eta) &= \{1-\xi-\eta\ \xi\ \eta\ E_0^1(\xi,\ \eta) \cdots E_{n_1-1}^1(\xi,\ \eta) \\ &\qquad E_0^2(\xi,\ \eta) \cdots E_{n_2-1}^2(\xi,\ \eta)\ E_0^3(\xi,\ \eta) \cdots E_{n_3-1}^2(\xi,\ \eta)\} \\ \tilde{\mathbf{S}}_2^{\mathrm{T}}(\xi,\eta) &= \left\{ F_{0,0}(\xi,\eta) \cdots F_{0,n}(\xi,\eta)\ F_{1,0}(\xi,\eta) \cdots F_{n,0}(\xi,\eta) \right\} \\ \tilde{\mathbf{u}}_1^{\mathrm{T}} &= \left\{ u_1\ u_2\ u_3\ a_0^1 \cdots a_{n_1-1}^1\ a_0^2 \cdots a_{n_2-1}^2\ a_0^3 \cdots a_{n_3-1}^1 \right\} \\ \tilde{\mathbf{u}}_2^{\mathrm{T}} &= \left\{ c_{0,0} \cdots c_{0,n}\ c_{1,0} \cdots c_{n,0} \right\} \end{aligned} \tag{4.131}$$

Here, $\tilde{\mathbf{S}}_1$ and $\tilde{\mathbf{u}}_1$ are the edge function vector and its coefficient vector, respectively; $\tilde{\mathbf{S}}_2$ and $\tilde{\mathbf{u}}_2$ are the face function vector and its coefficient vector, respectively.

To convert basis functions into interpolation functions, non-uniformly distributed nodes on the domain are required. Several analytical formulas have been derived to map one-dimensional Gauss–Lobatto nodes onto the unit triangle. One such method is presented below. Let the transformation from uniform nodes $(\xi, \eta, 1-\xi-\eta)$ to non-uniform nodes $(\lambda_1, \lambda_2, \lambda_3)$ be given by the following formula:

$$\lambda_1 = f(\xi), \quad \lambda_2 = f(\eta), \quad \lambda_3 = f(1-\xi-\eta) \tag{4.132}$$

Note that λ_3 is not equal to $1-\lambda_1-\lambda_2$. For one-dimensional Gauss–Lobatto Chebyshev nodes defined on [0, 1], Eq. (4.132) can be specifically expressed as follows:

$$\lambda_1 = \sin(\pi\xi/2), \lambda_2 = \sin(\pi\eta/2) \text{ and } \lambda_3 = \sin[\pi(1-\xi-\eta)/2]$$

If $(\lambda_1, \lambda_2, \lambda_3)$ are also area coordinates, then the variables λ_i $(i = 1, 2, 3)$ are symmetric with respect to each other, allowing:

$$\begin{aligned} s(\lambda_1, \lambda_2, \lambda_3) &= \frac{\lambda_1 + \alpha\lambda_1\lambda_2\lambda_3}{\lambda_1+\lambda_2+\lambda_3+3\alpha\lambda_1\lambda_2\lambda_3}, \\ t(\lambda_1, \lambda_2, \lambda_3) &= \frac{\lambda_2 + \alpha\lambda_1\lambda_2\lambda_3}{\lambda_1+\lambda_2+\lambda_3+3\alpha\lambda_1\lambda_2\lambda_3} \end{aligned} \tag{4.133}$$

where α is an arbitrary constant. Equation (4.133) represents non-uniformly distributed nodes on the unit triangle. If one-dimensional Gauss–Lobatto-Legendre nodes are used, for example, $\boldsymbol{\alpha} = [\alpha_1, \alpha_2, \cdots, \alpha_n]$ are the one-dimensional Gauss–Lobatto-Legendre nodes defined on the interval [0, 1]. Let $\lambda_1 = \alpha_i$ and $\lambda_2 = \alpha_j$, then we have $\lambda_3 = \alpha_{n+2-i-j}$. Empirically, it has been found that $\alpha = 3$ is a good choice. It is easy to verify that s and t can be used as coordinates for the unit triangle (Fig. 4.16a), similar to ξ and η. In particular,

$$s(\lambda_1, 0, \lambda_3) = \lambda_1, \quad t(0, \lambda_2, \lambda_3) = \lambda_2 \tag{4.134}$$

This means that if non-uniformly distributed nodes in the unit triangle are generated using Gauss–Lobatto nodes, the nodes generated along the edges of the triangle will also be Gauss–Lobatto nodes. This property is important because the Fekete points on the edges of the triangle have been shown to be Gauss–Lobatto points.

Using the indices of the one-dimensional Gauss–Lobatto nodes and uniform nodes shown in Fig. 4.19a, non-uniform nodes can be generated via Eq. (4.133), as illustrated in Fig. 4.19b. The nodes on the equilateral triangle shown in Fig. 4.19c are derived from those in Fig. 4.19b. The mapping relationship is $\tilde{\eta} = \sqrt{3}\eta$, $\tilde{\xi} = 2\xi + \eta - 1$.

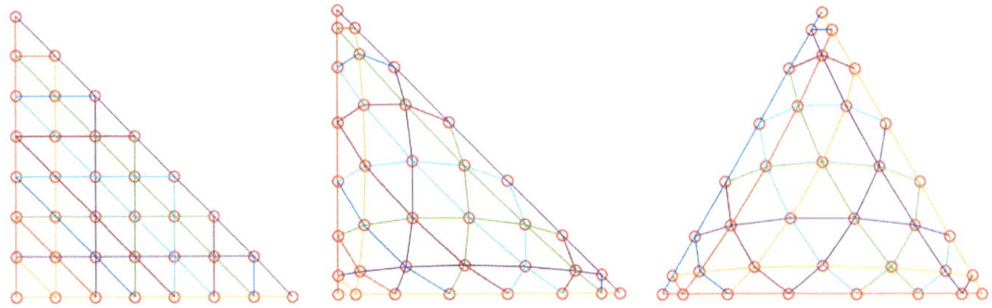

Fig. 4.19 Node distribution of the unit triangle

Let the number of non-uniform nodes on the edges and surface of the triangle be equal to the number of corresponding basis functions, and denote the resulting node vectors as $(\boldsymbol{\xi}, \boldsymbol{\eta})$. A generalized Vandermonde matrix can then be constructed.

$$\mathbf{V} = \tilde{\mathbf{S}}^{\mathrm{T}}(\boldsymbol{\xi}, \boldsymbol{\eta}) \tag{4.135}$$

Therefore, we have

$$\begin{aligned} u\big[x(\xi, \eta), y(\xi, \eta)\big] &= \tilde{\mathbf{S}}^{\mathrm{T}}(\xi, \eta)\mathbf{V}^{-1}u(\boldsymbol{\xi}, \boldsymbol{\eta}) \\ &= \sum_{i=1}^{N_t} L_i(\xi, \eta)u(\xi_i, \eta_i) = \mathbf{S}^{\mathrm{T}}(\xi, \eta)\mathbf{u} \end{aligned} \tag{4.136}$$

Here, N_t denotes the total number of basis functions or nodes, and $L_i(\xi, \eta)$ is defined as the Lagrange function corresponding to the point (ξ_i, η_i) on the triangle. The condition number of the generalized Vandermonde matrix given in Eq. (4.135) can be approximately expressed as $e^{0.45\,m}$, where $(m + 1)$ is the order of the element. The Fekete points $(\boldsymbol{\xi}, \boldsymbol{\eta})$ should satisfy the following condition:

$$|L_i(\xi, \eta)| \leq 1 \text{ and } L_i(\xi_i, \eta_i) = 1 \tag{4.137}$$

This condition can usually be replaced by

$$\frac{\partial L_i(\xi, \eta)}{\partial \xi_i} = 0, \quad \frac{\partial L_i(\xi, \eta)}{\partial \eta_i} = 0 \tag{4.138}$$

The above formula can be used to optimize the distribution of nodes. Since the nodes on the edges of the triangle have already been selected as Gauss–Lobatto nodes, and the Fekete points on the edges of the triangle have been proven to be Gauss–Lobatto points, only the interior nodes of the triangle need to be optimized. A comparison between the computed Fekete points and the nodes explicitly generated by Eq. (4.133) is shown in Fig. 4.20, indicating that they are very close to each other.

Figure 4.21 shows four Lagrange interpolation basis functions obtained using Fekete points and orthogonal polynomials.

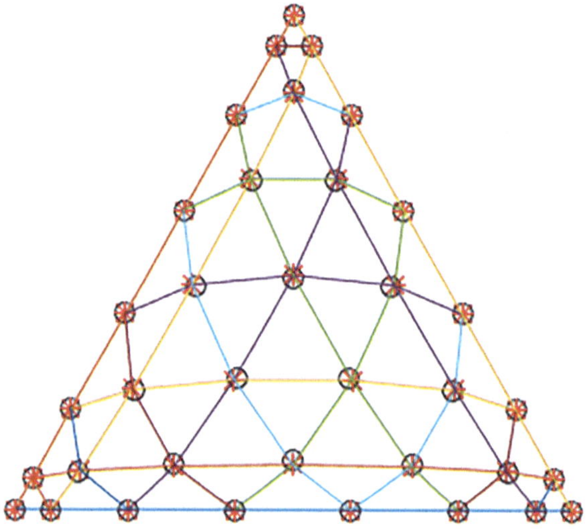

Fig. 4.20 Comparison of Fekete points (*) obtained by computation and explicitly generated nodes (○)

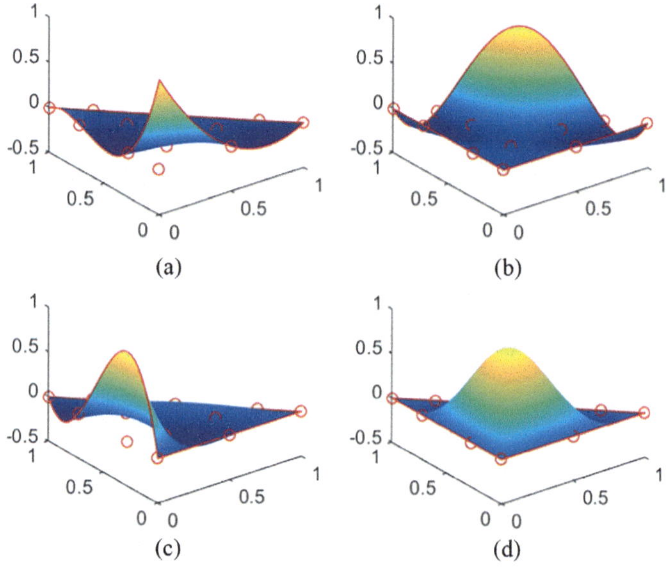

Fig. 4.21 The four Lagrange interpolation basis functions on the unit triangle

4.2.4.2 Strain Matrix

Using the Fekete points described above, it is possible to construct triangular elements with the nodal distribution shown in Fig. 4.15, where the number of nodes on each edge of the triangle can differ and is independent of the number of interior nodes (referring here only to the node count). In general, it is only necessary to convert the edge functions into interpolation functions. The interpolation basis functions shown in Fig. 4.21 utilize the nodal distribution depicted in Fig. 4.15. Taking the derivative of Eq. (4.136) with respect to the parametric coordinates and substituting the Fekete points yields the differential quadrature (DQ) rule for non-uniformly distributed nodes on the unit triangle, namely:

$$\frac{\partial u\left(\overline{\xi},\overline{\eta}\right)}{\partial \xi}=\mathbf{A}^{(1)}u(\boldsymbol{\xi},\boldsymbol{\eta}),\quad \frac{\partial u\left(\overline{\xi},\overline{\eta}\right)}{\partial \eta}=\mathbf{B}^{(1)}u(\boldsymbol{\xi},\boldsymbol{\eta}) \tag{4.139}$$

Higher-order derivatives can be expressed as

$$\begin{aligned}&\frac{\partial^{r} u\left(\overline{\xi},\overline{\eta}\right)}{\partial \xi^{r}}=\mathbf{A}^{(r)}u(\boldsymbol{\xi},\boldsymbol{\eta}),\quad \frac{\partial^{s} u\left(\overline{\xi},\overline{\eta}\right)}{\partial \eta^{s}}=\mathbf{B}^{(s)}u(\boldsymbol{\xi},\boldsymbol{\eta}),\\ &\frac{\partial^{r+s} u\left(\overline{\xi},\overline{\eta}\right)}{\partial \xi^{r}\partial \eta^{s}}=\mathbf{F}^{(r+s)}u(\boldsymbol{\xi},\boldsymbol{\eta})\end{aligned} \tag{4.140}$$

Note: The calculation points $(\overline{\xi},\overline{\eta})$ and the nodes (ξ, η) may be different.
In this case, the following interpolation relations may be required.

$$u\left(\overline{\xi},\overline{\eta}\right)=Gu(\xi,\eta)\text{ or }\overline{u}=Gu \tag{4.141}$$

If the calculation points coincide with the nodes, then

$$\mathbf{A}^{(r)}=\mathbf{A}^{(1)}\mathbf{A}^{(r-1)},\quad \mathbf{B}^{(s)}=\mathbf{B}^{(1)}\mathbf{B}^{(s-1)},\quad \mathbf{F}^{(r+s)}=\mathbf{A}^{(r)}\mathbf{B}^{(s)} \tag{4.142}$$

Direct application of Eq. (4.136) is generally more efficient and accurate than Eq. (4.142). However, it is clear that Eq. (4.142) is simpler than Eq. (4.136).

The first-order derivatives of $u=u[x(\xi,\eta),y(\xi,\eta)]$ with respect to the Cartesian coordinates x and y can be obtained by using the chain rule of differentiation.

$$\frac{\partial u}{\partial x}=\frac{1}{|\mathbf{J}|}\left(\frac{\partial y}{\partial \eta}\frac{\partial u}{\partial \xi}-\frac{\partial y}{\partial \xi}\frac{\partial u}{\partial \eta}\right),\quad \frac{\partial u}{\partial y}=\frac{1}{|\mathbf{J}|}\left(\frac{\partial x}{\partial \xi}\frac{\partial u}{\partial \eta}-\frac{\partial x}{\partial \eta}\frac{\partial u}{\partial \xi}\right) \tag{4.143}$$

Among them, the determinant $|\mathbf{J}|$ of the Jacobian matrix $\mathbf{J}=\partial(x,y)/\partial(\xi,\eta)$ is

$$|\mathbf{J}| = \frac{\partial x}{\partial \xi}\frac{\partial y}{\partial \eta} - \frac{\partial y}{\partial \xi}\frac{\partial x}{\partial \eta} \tag{4.144}$$

Therefore, the first-order derivatives of $u(x, y)$ with respect to x and y at the discrete points $x_i = x(\xi_i, \eta_i), y_i = y(\xi_i, \eta_i)(i = 1, 2, \ldots, N_t)$ are

$$\begin{aligned} \left(\frac{\partial u}{\partial x}\right)_i &= \frac{1}{|\mathbf{J}|_i}\left[\left(\frac{\partial y}{\partial \eta}\right)_i\left(\sum_{n=1}^{N_t} A_{in}^{(1)} u_n\right) - \left(\frac{\partial y}{\partial \xi}\right)_i\left(\sum_{n=1}^{N_t} B_{in}^{(1)} u_n\right)\right] \\ \left(\frac{\partial u}{\partial y}\right)_i &= \frac{1}{|\mathbf{J}|_i}\left[\left(\frac{\partial x}{\partial \xi}\right)_i\left(\sum_{n=1}^{N_t} B_{in}^{(1)} u_n\right) - \left(\frac{\partial x}{\partial \eta}\right)_i\left(\sum_{n=1}^{N_t} A_{in}^{(1)} u_n\right)\right] \end{aligned} \tag{4.145}$$

where $A_{in}{}^{(1)}$ and $B_{in}{}^{(1)}$ are the weighting coefficients associated with the first-order derivatives with respect to ξ and η, respectively, as given in Eq. (4.139). This can be further expressed in a single-index notation form Eq. (4.145) as

$$\begin{aligned} \frac{\partial u\left[x\left(\overline{\xi}, \overline{\eta}\right), y\left(\overline{\xi}, \overline{\eta}\right)\right]}{\partial x} &= \mathbf{G}_x u[x(\boldsymbol{\xi}, \boldsymbol{\eta}), y(\boldsymbol{\xi}, \boldsymbol{\eta})] \\ \frac{\partial u\left[x\left(\overline{\xi}, \overline{\eta}\right), y\left(\overline{\xi}, \overline{\eta}\right)\right]}{\partial y} &= \mathbf{G}_y u[x(\boldsymbol{\xi}, \boldsymbol{\eta}), y(\boldsymbol{\xi}, \boldsymbol{\eta})] \end{aligned} \tag{4.146}$$

or

$$\overline{\mathbf{u}}_x = \mathbf{G}_x\mathbf{u}, \quad \overline{\mathbf{u}}_y = \mathbf{G}_y\mathbf{u} \tag{4.147}$$

Equation (4.147) presents the generalized differential quadrature (DQ) rule for triangular elements. Since the number of nodes on the element boundary and within the element can be specified independently, this method is referred to as the hierarchical differential quadrature (HDQ) rule.

The strain–displacement relationship given by Eq. (4.15) can be discretized using the generalized differential quadrature rule in Eq. (4.147) as follows:

$$\{\boldsymbol{\varepsilon}\} = \begin{bmatrix} \mathbf{G}_x & \mathbf{0} \\ \mathbf{0} & \mathbf{G}_y \\ \mathbf{G}_y & \mathbf{G}_x \end{bmatrix}\begin{Bmatrix} \mathbf{u} \\ \mathbf{v} \end{Bmatrix} = \mathbf{B}\mathbf{d}_e \tag{4.148}$$

It can be observed that the above equation is similar to the lower-order case given in Eq. (4.60), with the only difference being that the partial derivatives are computed using the weighting coefficient matrices.

4.2.4.3 Element Matrices

The application of the hierarchical differential quadrature (HDQ) rule in the weak form of structural mechanics requires integration rules for irregular domains. Let the one-dimensional integration rule be

$$\int_0^1 f(\xi)\mathrm{d}\xi = \sum_{j=1}^{n} C_j f(\xi_j) \tag{4.149}$$

Here, ξ_j and $C_j (j = 1, 2, \ldots, n)$ represent the integration points and weighting coefficients, respectively. Therefore, the integration of a two-dimensional function defined on the unit triangle is

$$\int_0^1 \int_0^{1-\eta} u(\xi, \eta)\mathrm{d}\xi\mathrm{d}\eta = \sum_{i=1}^{N_\xi} \sum_{j=1}^{N_\eta} C_i^\xi C_j^{\eta *} u(\xi_i^*, \eta_j) \tag{4.150}$$

where

$$\xi_i^* = (1 - \eta_j)\xi_i, \quad C_j^{\eta *} = (1 - \eta_j)C_j^\eta \tag{4.151}$$

Here, ξ_i, C^ξ_i and η_j, C^η_j, represent the integration points and weight coefficients in the ξ and η directions, respectively. On an irregular domain, Eq. (4.150) can be expressed as

$$\int_\Omega u[x(\xi, \eta), y(\xi, \eta)]\mathrm{d}x\mathrm{d}y = \sum_{i=1}^{N_\xi} \sum_{j=1}^{N_\eta} C_i^\xi C_j^{\eta *} |\mathbf{J}|_{ij} u(\xi_i^*, \eta_j) = \overline{\mathbf{C}}^\mathrm{T}\overline{\mathbf{u}} \tag{4.152}$$

where

$$\overline{\mathbf{C}}^\mathrm{T} = \left[C_1^\xi C_1^{\eta *} |\mathbf{J}|_{1,1} \; C_2^\xi C_1^{\eta *} |\mathbf{J}|_{2,1} \; \cdots \; C_{N_\xi}^\xi C_{N_\eta}^{\eta *} |\mathbf{J}|_{N_\xi, N_\eta} \right] \tag{4.153}$$

Note: The above integration rules are based on the tensor product form of univariate Gaussian quadrature rules. In a certain sense, product formulas are inefficient because they use far more points and weights than are actually required to exactly integrate a polynomial of degree *d*. When applied to elements such as triangles and tetrahedra, these rules are asymmetric and result in a clustering of integration points near the vertices, a phenomenon that becomes more pronounced as the degree *d* increases.

The strain energy of a plane stress problem can be discretized as

$$U = \frac{1}{2} h \mathbf{d}_e^\mathrm{T} \mathbf{B}^\mathrm{T} \overline{\mathbf{D}} \mathbf{B} \mathbf{d}_e = \frac{1}{2} \mathbf{d}_e^\mathrm{T} \mathbf{k}_e \mathbf{d}_e \tag{4.154}$$

where

$$\overline{\mathbf{D}} = \begin{bmatrix} E' & E'\upsilon & 0 \\ & E' & 0 \\ \text{Sym} & & G \end{bmatrix} \otimes \mathbf{C} \tag{4.155}$$

and

$$E' = \frac{E}{1-\upsilon^2}, \quad G = \frac{E}{2(1+\upsilon)}, \quad \mathbf{C} = \text{diag}(\overline{\mathbf{C}}) \tag{4.156}$$

The element stiffness matrix can thus be expressed as

$$\mathbf{k}_e = h\mathbf{B}^{\mathrm{T}}\overline{\mathbf{D}}\mathbf{B} = \frac{Eh}{1-\upsilon^2}\begin{bmatrix} \mathbf{G}_x^{\mathrm{T}}\mathbf{C}\mathbf{G}_x + \upsilon_1\mathbf{G}_y^{\mathrm{T}}\mathbf{C}\mathbf{G}_y & \upsilon\mathbf{G}_x^{\mathrm{T}}\mathbf{C}\mathbf{G}_y + \upsilon_1\mathbf{G}_y^{\mathrm{T}}\mathbf{C}\mathbf{G}_x \\ \upsilon\mathbf{G}_y^{\mathrm{T}}\mathbf{C}\mathbf{G}_x + \upsilon_1\mathbf{G}_x^{\mathrm{T}}\mathbf{C}\mathbf{G}_y & \mathbf{G}_y^{\mathrm{T}}\mathbf{C}\mathbf{G}_y + \upsilon_1\mathbf{G}_x^{\mathrm{T}}\mathbf{C}\mathbf{G}_x \end{bmatrix} \tag{4.157}$$

Here, E denotes the Young's modulus, υ is the Poisson's ratio, and $\upsilon_1 = (1-\upsilon)/2$.

Similarly, the element mass matrix and load vector can be expressed as

$$\mathbf{m}_e = \rho h\begin{bmatrix} \mathbf{G}^{\mathrm{T}}\mathbf{C}\mathbf{G} & \mathbf{0} \\ \mathbf{0} & \mathbf{G}^{\mathrm{T}}\mathbf{C}\mathbf{G} \end{bmatrix} \tag{4.158}$$

$$\mathbf{f}_e = h\begin{bmatrix} \mathbf{G}^{\mathrm{T}}\mathbf{C}\mathbf{q}_u \\ \mathbf{G}^{\mathrm{T}}\mathbf{C}\mathbf{q}_v \end{bmatrix} \tag{4.159}$$

Here, ρ denotes the density, while $\boldsymbol{q}_u$ and $\boldsymbol{q}_v$ represent the distributed nodal load vectors in the x and y directions, respectively.

4.2.4.4 In-Plane Vibration Analysis

Unlike out-of-plane vibration, in-plane vibration has two distinct types of simply supported boundary conditions: one with the normal stress and tangential displacement being zero (S1), and the other with both the normal displacement and tangential stress being zero (S2). First, consider the sector plate shown in Fig. 4.22, with a radius $R = 1$ m. The boundary conditions on its two straight edges are S2, and the circumferential edge of the sector is free. The computational results are presented in Table 4.1 and compared with exact results from the literature and the p-version finite element method (p-FEM). The number of internal hierarchical basis functions is denoted by P. The number of nodes on each of the three edges is assumed to be the same and denoted as N_e, i.e., $N_1 = N_2 = N_3 = N_e$. First, $N_e = P + 3$ is chosen. As can be seen from Table 4.1, the results from the Hierarchical Quadrature Element Method (HQEM) converge monotonically and very rapidly with an increasing number of

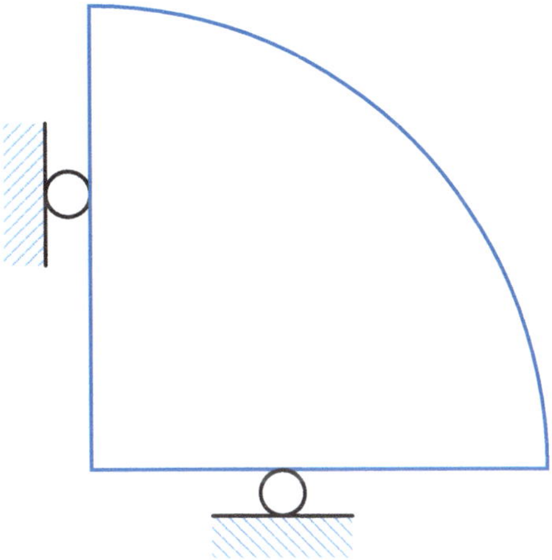

Fig. 4.22 A sector plate with two straight edges under S2 boundary conditions

nodes. When $(P, N_e) = (10, 13)$, the first four digits of the first ten frequencies are fully converged. Moreover, it is evident that HQEM remains highly stable even with a large number of nodes. Next, consider $(P, N_e) = (6, 13)$, as shown in the bold italicized row in Table 4.1. It can be observed that HQEM maintains good accuracy even when there are relatively few nodes inside the element but a sufficient number of nodes on its boundaries. This property is similar to the fixed-interface modal synthesis method, but the HQEM does not require modal analysis and achieves very high accuracy.

The sector plate analyzed in Table 4.1 is further examined in Fig. 4.23, where the effect of geometric error on simulation accuracy is investigated through the fundamental frequency. The mesh lengths are taken as $h_0 = 0.1, 0.2, 0.3, 0.4, 0.5, 0.7$. It is clear that using higher-order basis functions to represent the geometry significantly improves simulation accuracy, while this does not lead to a substantial increase in the number of degrees of freedom.

4.2.5 Higher-Order Quadrilateral Elements

4.2.5.1 Shape Function Construction

The shape functions of the quadrilateral hierarchical quadrature element can be constructed in the same manner as those for triangular elements. For quadrilateral elements, the displacement field u in the hierarchical finite element method (HFEM) is given by

$$u\big[x(\xi, \eta), y(\xi, \eta)\big] = \sum_{k=1}^{4} V^k(\xi, \eta) u_k$$

Table 4.1 The first ten frequency parameters of a sector plate (Fig. 4.22) with S2-S2-F boundary conditions, where $\upsilon = 0.3$, $R = 1$ m, $\Omega = \omega\sqrt{\rho/G}$

(P, N_e)	Degrees of freedom	Modal order									
		1	2	3	4	5	6	7	8	9	10
(2, 5)	20	2.350	3.464	4.277	4.905	–	–	–	–	–	–
(3, 6)	30	2.347	3.464	4.249	4.742	7.310	7.570	7.683	8.899	–	–
(4, 7)	42	2.346	3.463	4.246	4.702	6.905	7.468	7.652	8.814	–	–
(5, 8)	56	2.346	3.463	4.245	4.691	6.760	7.446	7.645	8.797	9.110	9.147
(6, 9)	72	2.346	3.463	4.245	4.689	6.707	0.440	7.643	8.795	8.804	9.110
(7, 10)	90	2.346	3.463	4.245	4.689	6.698	7.439	7.642	8.666	8.795	9.110
(8, 11)	110	2.346	3.463	4.245	4.689	6.696	7.439	7.642	8.634	8.795	9.110
(9, 12)	132	2.346	3.463	4.245	4.689	6.696	7.439	7.642	8.622	8.795	9.110
(10, 13)	156	2.346	3.463	4.245	4.689	6.695	7.439	7.642	8.620	8.795	9.110
(15, 18)	306	2.346	3.463	4.245	4.689	6.695	7.439	7.642	8.620	8.795	9.110
(25, 28)	756	2.346	3.463	4.245	4.689	6.695	7.439	7.642	8.620	8.795	9.110
(6, 13)	88	2.346	3.463	4.245	4.689	6.704	7.439	7.642	8.721	8.795	9.110
Exact solution	–	2.346	3.463	4.245	4.689	6.695	7.439	7.642	8.620	8.795	9.110
p-FEM	–	2.346	3.463	4.245	4.689	6.695	7.439	7.642	8.620	8.795	9.110

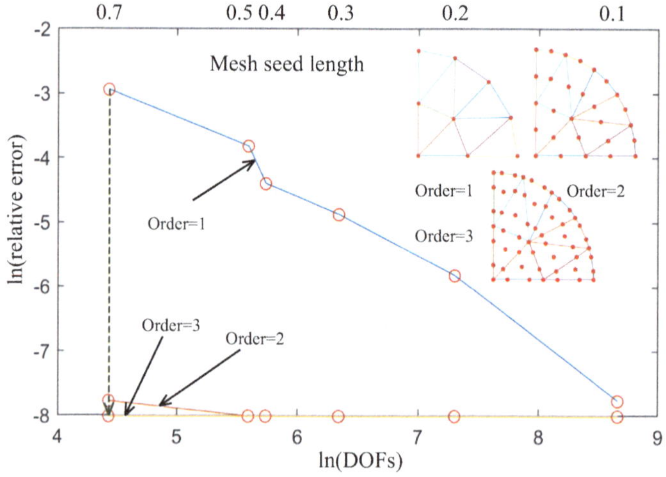

Fig. 4.23 Effect of geometric error (mesh order) on simulation accuracy analyzed via fundamental frequency

$$+\sum_{e=1}^{4}\sum_{i=0}^{n_e-1} E_i^e(\xi,\ \eta)a_i^e + \sum_{i=0}^{m}\sum_{j=0}^{n} F_{ij}(\xi,\eta)c_{ij} \tag{4.160}$$

where $k = 1, 2, 3, 4$ denotes the vertices, $e = 1, 2, 3, 4$ denotes the edges (Fig. 4.24), n_e $(e = 1, 2, 3, 4)$ represents the number of edge functions on each edge, and m and n indicate the orders of the face functions in the η and ξ directions, respectively, for the quadrilateral face. In Eq. (4.160), u_k is the displacement at the vertices, while $a^e{}_i$ and c_{ij} are the coefficients of the edge functions and face functions of the quadrilateral element, respectively. The coefficients of the hierarchical functions have no explicit physical meaning.

The shape functions V^k at the four vertices of an element can be expressed as

$$\begin{aligned} V^1 &= \frac{1}{4}(1-\xi)(1-\eta) \\ V^2 &= \frac{1}{4}(1+\xi)(1-\eta) \\ V^3 &= \frac{1}{4}(1+\xi)(1+\eta) \\ V^4 &= \frac{1}{4}(1-\xi)(1+\eta) \end{aligned} \tag{4.161}$$

For quadrilateral elements, the edge functions and face functions can be directly given as follows:

$$E_i^1(\xi,\eta) = \frac{1}{2}(1-\xi)\varphi_i(\eta)$$

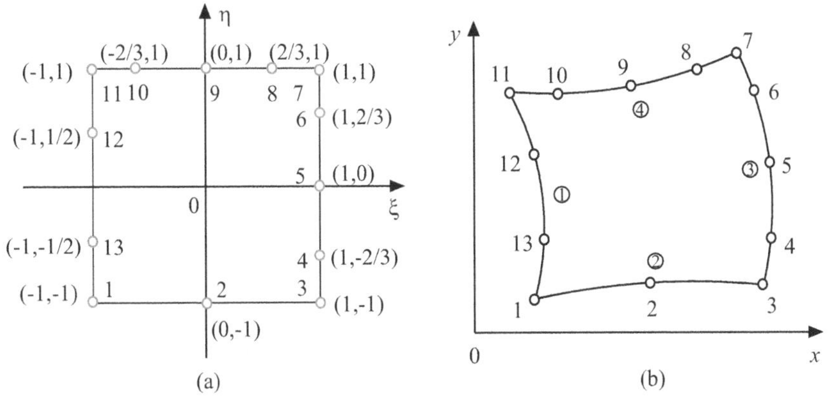

Fig. 4.24 Geometric mapping: **a** the parametric domain in the natural ξ-η coordinate plane, **b** the curved quadrilateral region in the Cartesian x–y coordinate plane

$$
\begin{aligned}
E_i^2(\xi,\eta) &= \frac{1}{2}(1-\eta)\varphi_i(\xi) \\
E_i^3(\xi,\eta) &= \frac{1}{2}(1+\xi)\varphi_i(\eta) \\
E_i^4(\xi,\eta) &= \frac{1}{2}(1+\eta)\varphi_i(\xi)
\end{aligned} \tag{4.162}
$$

$$
F_{ij}(\xi,\eta) = \varphi_i(\xi)\varphi_j(\eta) \tag{4.163}
$$

where

$$
\varphi_n(\xi) = \left(1-\xi^2\right)P_n^{(2,\,2)}(\xi),\ \xi \in [-1,\ 1] \tag{4.164}
$$

It should be noted that the truss hierarchical element in Sect. 3.2.8.3 can also use Eq. (4.164). $\varphi_n(\xi)$ can also be the integrated Legendre polynomial given by Eq. (3.145), that is,

$$
\varphi_n(\xi) = \int_{-1}^{\xi} P_n(\xi)\mathrm{d}\xi = \frac{(\xi^2-1)}{n(n+1)}\frac{\mathrm{d}P_n(\xi)}{\mathrm{d}\xi},\ \xi \in [-1,\ 1] \tag{4.165}
$$

The nodes on the quadrilateral domain are given by the tensor product of Gauss–Lobatto nodes. The following procedure is similar to that for triangular elements.

The basis functions on the unit quadrilateral element can also be constructed using the mixed function method and Lagrange functions. For any quadrilateral domain, as shown in Fig. 4.24, the geometric mapping is given by

$$
\begin{aligned}
V(\xi,\eta) = &\frac{1-\xi}{2}M(-1,\eta) + \frac{1+\xi}{2}M(1,\eta) + \frac{1-\eta}{2}M(\xi,-1) + \frac{1+\eta}{2}M(\xi,1) \\
&- \frac{(1-\xi)(1-\eta)}{4}M(-1,-1) - \frac{(1-\xi)(1+\eta)}{4}M(-1,1) \\
&- \frac{(1+\xi)(1-\eta)}{4}M(1,-1) - \frac{(1+\xi)(1+\eta)}{4}M(1,1)
\end{aligned} \tag{4.166}
$$

where $-1 \le \xi, \eta \le 1$; the functions $M(\xi_i, \eta)$ and $M\left(\xi, \eta_j\right)$ represent the four parametric curves that form the boundaries of the quadrilateral domain; the function $M\left(\xi, \eta_j\right)$ denotes the coordinate points in the xy-plane corresponding to the parametric coordinates (ξ_i, η). The blending function method provides an accurate geometric mapping. Using the blending function given by Eq. (4.166), the vertex basis functions can be constructed as follows:

$$
\begin{aligned}
V^1 &= \frac{1-\eta}{2}L_1^M(\xi) + \frac{1-\xi}{2}L_1^N(\eta) - \frac{1}{4}(1-\xi)(1-\eta) \\
V^2 &= \frac{1-\eta}{2}L_M^M(\xi) + \frac{1+\xi}{2}L_1^N(\eta) - \frac{1}{4}(1+\xi)(1-\eta)
\end{aligned}
$$

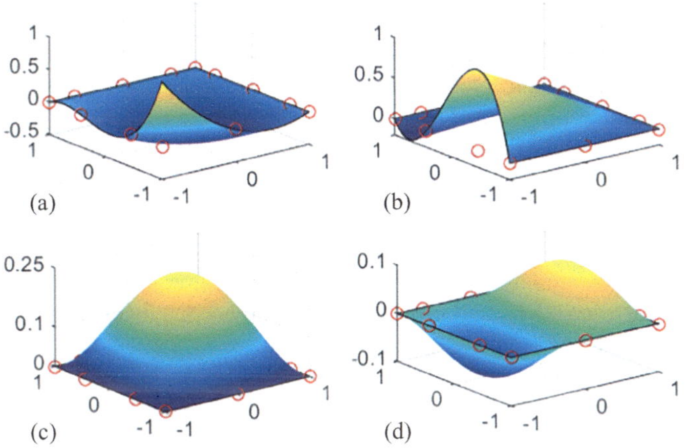

Fig. 4.25 The four basis functions of a unit quadrilateral HQEM element

$$
\begin{aligned}
V^3 &= \frac{1+\eta}{2}L_M^M(\xi) + \frac{1+\xi}{2}L_N^N(\eta) - \frac{1}{4}(1+\xi)(1+\eta) \\
V^4 &= \frac{1+\eta}{2}L_1^M(\xi) + \frac{1-\xi}{2}L_N^N(\eta) - \frac{1}{4}(1-\xi)(1+\eta)
\end{aligned} \tag{4.167}
$$

The boundary function can be constructed as follows:

$$
\begin{aligned}
E_i^1(\xi,\eta) &= \frac{1-\xi}{2}L_i^N(\eta) \\
E_i^2(\xi,\eta) &= \frac{1-\eta}{2}L_i^M(\xi) \\
E_i^3(\xi,\eta) &= \frac{1+\xi}{2}L_i^N(\eta) \\
E_i^4(\xi,\eta) &= \frac{1+\eta}{2}L_i^M(\xi)
\end{aligned} \tag{4.168}
$$

The face shape functions are the same as those in Eq. (4.163). It should be noted that although the basis functions constructed by this method are less computationally efficient, they are more intuitive. Figure 4.25 shows the four basis functions of a unit quadrilateral element.

4.2.5.2 In-Plane Vibration Analysis

As an example, consider the application of the hierarchical quadrature element method (HQEM) to the analysis of in-plane vibrations of isotropic plates in curved regions, to illustrate the characteristics of quadrilateral elements. The geometric mapping from the square parametric domain to the computational domain is achieved using either the blending function method or the basis functions of the hierarchical quadrature element method. For all of the significant digits used in comparison, there is no difference between the two approaches. For numerical convenience, the same number of sampling points, denoted as N_e, is taken along the ξ and η directions on the edges; the number of hierarchical shape functions in both in-plane directions is also set to be equal, denoted as N_h. To avoid the influence of rounding errors, the natural frequencies are expressed using the same frequency parameters as in the literature. Unlike out-of-plane vibrations, in-plane vibrations have two distinct simply supported boundary conditions: one where the normal stress and tangential displacement are zero (S1), and the other where the normal displacement and shear stress are zero (S2).

Table 4.2 analyzes the in-plane vibration of a square plate under two combinations of the following boundary conditions: simply supported (S), clamped (C), and free (F). The boundary condition S1-C-S2-F indicates that the plate has S1-S2 boundary conditions along the x-direction and CF boundary conditions along the y-direction. The natural frequencies are expressed in terms of the same frequency parameter in Table 4.2. As shown in the table, for the S1-S1-S1-S1 plate, the results from the Hierarchical Quadrature Element Method (HQEM) with sampling points $(N_e, N_h) = (13, 10)$ are very close to but not exactly the same as the exact solutions in the literature. This discrepancy arises from the treatment of corner boundary conditions. In analytical methods, corners are typically considered part of the edges, whereas in numerical methods, corners can be handled separately. When the number of sampling points is sufficiently large, this effect becomes negligible, as demonstrated in Table 4.2. For an S1S1S1S1 plate with fixed corners, the HQEM results with $(N_e, N_h) = (10, 8)$ match the exact comparative results to all significant digits, as shown in Table 4.2. For the S1-C-S2-F plate, where the influence of corner boundary conditions is considered, the HQEM results converge very rapidly. It can be observed that HQEM is similar to the fixed-interface modal synthesis method, as it allows structural analysis using a few nodes on each element boundary and a limited number of fixed-interface modes within the element. However, HQEM does not require modal analysis and can achieve very high computational accuracy.

To demonstrate the convergence across the entire spectrum and investigate the influence of corner boundary conditions, the S1S1S1S1 plate is further examined. The results from the Hierarchical Quadrature Element Method (where DQHFEM is equivalent to HQEM) with a nodal distribution of $(N_e, N_h) = (30, 28)$ are compared with those obtained using the Finite Element Method (FEM) from the commercial software ANSYS. The plate has a Young's modulus of 71 GPa, a density of 2700 kg/m^3, and a Poisson's ratio of 0.3. The element used in ANSYS is the SHELL 181 quadrilateral element. The element length in the FEM is $l = 1/30$ m, which is the same as in HQEM. The relative errors between FEM and HQEM are shown in Fig. 4.26a. It can be observed that HQEM yields better results than FEM for up to 80% of the frequencies, particularly for the first 20 natural frequencies, where HQEM achieves high accuracy. In Fig. 4.26a, DQHFEM* denotes the results without considering the effects of corner boundary conditions. This curve nearly coincides with the curve

Table 4.2 Comparison of dimensionless frequency parameters for square plates ($\gamma = (\omega a/\pi)\sqrt{\rho/G}$)

Boundary conditions	(N_e, N_h)	Modal number							
		1	2	3	4	5	6	7	8
S1-S1-S1-S1	(10, 8)	0.9995	0.9995	1.4142	1.9980	2.0000	2.2353	2.2353	2.3897
	(13, 10)	0.9999	0.9999	1.4142	1.9996	2.0000	2.2358	2.2360	2.3902
	(10, 8)*	1.0000	1.0000	1.4142	2.0000	2.0000	2.2361	2. 2361	2.3905
	(25, 20)	1.0000	1.0000	1.4142	2.0000	2.0000	2.2361	2. 2361	2.3905
	Exact solution	1	1	1.4142	2	2	2.2361	2. 2361	2.3905
S1-C-S2-F	(5, 3)	0.8600	0.9468	1.4309	1.7482	2.2306	2.4229	2.4897	2.7977
	(8, 3)	0.8600	0.9468	1.4276	1.7421	2.2273	2.3529	2.3627	2.7935
	(10, 3)	0.8600	0.9468	1.4276	1.7421	2.2273	2.3529	2.3626	2.7935
	(10, 5)	0.8600	0.9468	1.4258	1.7420	2.2193	2.2978	2.3610	2.7525
	(10, 8)	0.8600	0.9468	1.4258	1.7420	2.2193	2.2967	2.3610	2.7517
	(25, 20)	0.8600	0.9468	1.4258	1.7420	2.2193	2.2967	2.3610	2.7517
	Exact solution	0.8600	0.9468	1.4258	1.7420	2.2193	2.2967	2.3610	2.7517

Note The influence of corner boundary conditions has been taken into account

that accounts for corner boundary conditions, indicating that the influence of corner boundary conditions on the overall hierarchical accuracy is minimal.

To compare HQEM results with those from Isogeometric Analysis (IGA) using NURBS basis functions, the previously analyzed S1S1S1S1 square plate is evaluated again, and the relative errors across the entire spectrum are plotted in Fig. 4.26. As shown in Fig. 4.26b, HQEM results are highly accurate compared to IGA results for the first 20% of the natural frequencies of the S1S1S1S1 square plate. Two important considerations should be noted when performing calculations with NURBS elements: (1) integration should be performed over each knot interval of the NURBS, rather than across the entire domain as in HQEM; and (2) NURBS knots should be uniformly distributed, unlike the non-uniform nodal distribution used in HQEM. Otherwise, NURBS-based IGA cannot provide satisfactory results for higher-order frequencies, which is relevant for explicit transient analysis. As shown in Fig. 4.26c, the relative errors of higher-order frequencies in HQEM can also be reduced through mesh refinement, with a nodal distribution of $(N_e, N_h) = (6, 4)$ per element. Unlike HQEM or methods such as the Differential Quadrature Method (DQM) and its weak forms, NURBS-based IGA does not possess hierarchical convergence properties.

To verify the applicability of the Hierarchical Quadrature Element Method (HQEM) to complex planar geometries, the HQEM results for in-plane vibration of a square plate with a central circular opening (see Fig. 4.27) (All four edges of the plate are free) are shown in Tables 4.3 and 4.4. Due to the symmetry of the geometry and boundary conditions, it is sufficient to consider one quarter of the

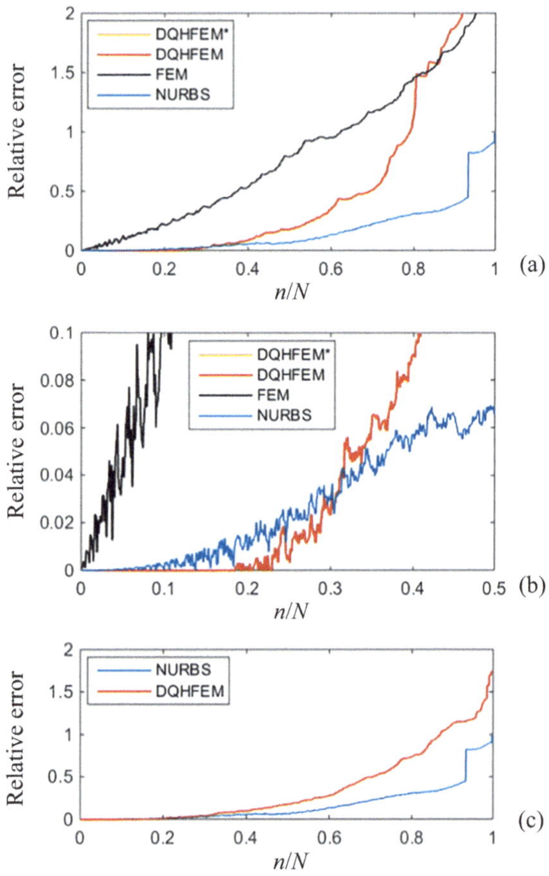

Fig. 4.26 Simply supported (S1) plate: **a** relative errors of the discrete spectrum using Finite Element Method (FEM), Isogeometric Analysis (NURBS), and the Hierarchical Quadrature Element Method (DQHFEM); **b** local enlargement of Figure (**a**); **c** HQEM results using 66 elements

plate 1/4 (see Fig. 4.28). Tables 4.3 and 4.4 present the first ten frequency parameters corresponding to the symmetric-symmetric and symmetric-antisymmetric modes, respectively, and compare them with *p*-FEM results. Figure 4.29 shows the first four mode shapes of the symmetric-symmetric modes. For symmetric-symmetric modes, the displacements u and v are even functions of the x and y coordinates, respectively. For symmetric-antisymmetric modes, the displacements u and v are even functions of the x coordinate and odd functions of the y coordinate, respectively. As shown in Fig. 4.29a, the first symmetric mode shape is a rigid body mode, but it is still regarded as a normal mode in the literature.

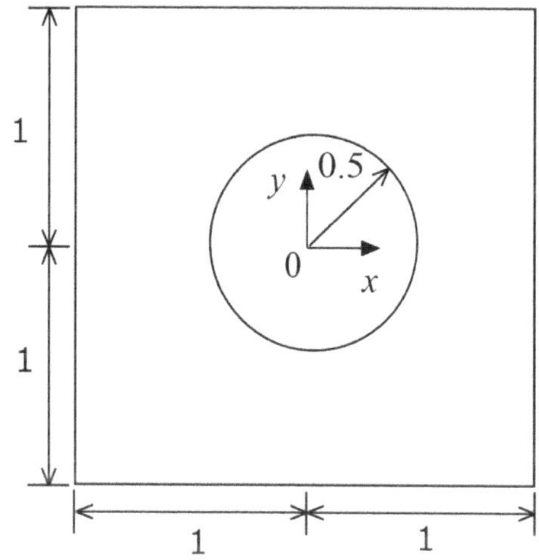

Fig. 4.27 Square plate with a central circular opening

Table 4.3 Comparison of the first 10 frequency parameters of the present method for symmetry-symmetry modes of a spare plate with a central circular opening with *p*-FEM results ($\gamma = \omega\sqrt{\rho/G}$, $\upsilon = 0.3$)

(N_e, N_h)	Modal number									
	1	2	3	4	5	6	7	8	9	10
(5,6)	0.0000	0.9220	3.2337	4.4344	4.6215	5.6997	6.2583	7.0118	7.0829	8.5128
(8,6)	0.0000	0.9215	3.2312	4.4262	4.6201	5.6806	6.1842	6.9940	7.0093	8.4452
(10,8)	0.0000	0.9215	3.2311	4.4261	4.6201	5.6805	6.1841	6.9935	7.0093	8.4450
(12,10)	0.0000	0.9215	3.2311	4.4261	4.6201	5.6805	6.1841	6.9934	7.0093	8.4450
(18,16)	0.0000	0.9215	3.2311	4.4261	4.6201	5.6805	6.1841	6.9934	7.0093	8.4450
p-FEM	0.003	0.922	3.233	4.427	4.620	5.682	6.187	7.000	7.010	8.448

Table 4.4 Comparison of the first 10 frequency parameters of the present method for symmetry-antisymmetric modes of a spare plate with a central circular opening with *p*-FEM results ($\gamma = \omega\sqrt{\rho/G}$, υ= 0.3)

(N_e, N_h)	Modal number									
	1	2	3	4	5	6	7	8	9	10
(5,6)	2.0483	2.2723	4.2939	5.0010	5.4476	6.5981	7.2555	7.8318	8.3722	8.9076
(8,6)	2.0476	2.2711	4.2898	4.9964	5.4196	6.5160	7.2213	7.8087	8.1745	8.7333
(10,8)	2.0476	2.2711	4.2897	4.9963	5.4196	6.5158	7.2211	7.8085	8.1729	8.7317
(18,16)	2.0476	2.2711	4.2897	4.9963	5.4196	6.5158	7.2211	7.8085	8.1729	8.7317
p-FEM	2.049	2.272	4.292	4.998	5.421	6.521	7.223	7.810	8.180	8.735
(5,6)	2.0483	2.2723	4.2939	5.0010	5.4476	6.5981	7.2555	7.8318	8.3722	8.9076

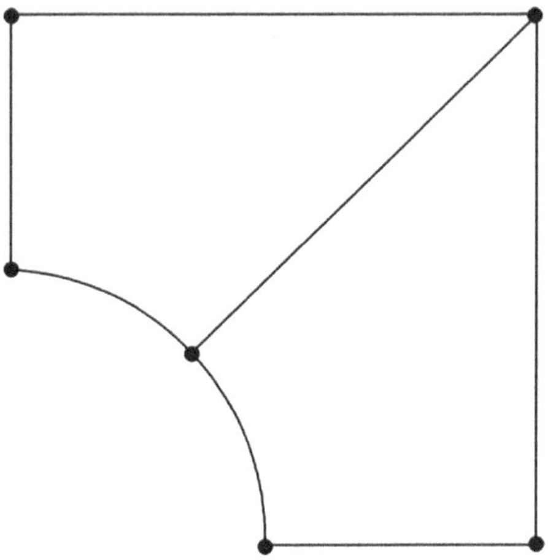

Fig. 4.28 HQEM mesh of 1/4 of the plate

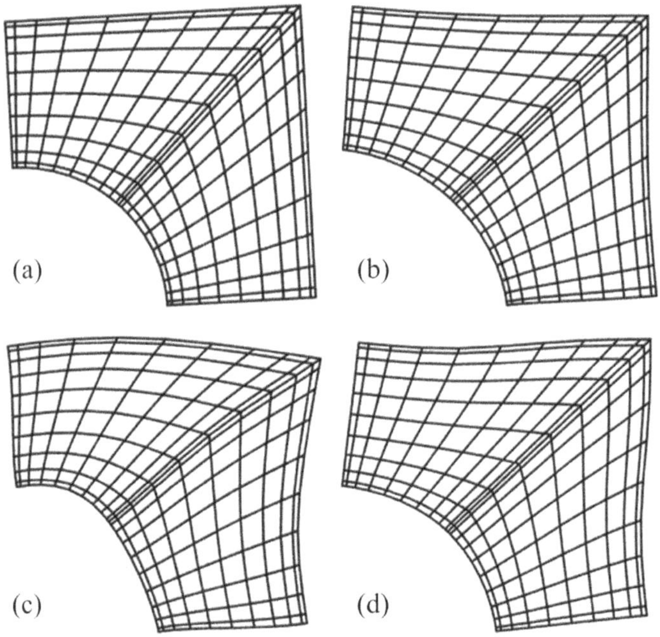

Fig. 4.29 The four modes for the symmetry-symmetry case in Table 4.3: **a** Mode 1, **b** Mode 2, **c** Mode 3, **d** Mode 4

4.3 Three-Dimensional Solid Finite Element Method

4.3.1 Construction of Shape Functions

Through two-dimensional solid finite element analysis, it can be observed that the most critical process in finite element analysis is the construction of shape functions. The remaining construction procedures for various element types are generally similar. Therefore, this section will focus on the methods for constructing shape functions for different three-dimensional elements. Since the hierarchical quadrature element method (HQEM) is a general approach, other methods can be considered as special cases of HQEM. Thus, this section will primarily introduce the three-dimensional high-order quadrature element method. The element types discussed include tetrahedral, triangular prism, hexahedral, and pyramidal elements. The hexahedral element, analogous to the quadrilateral element in two dimensions, is the simplest and most accurate three-dimensional element. However, there are currently no tools capable of automatically generating hexahedral meshes for complex engineering problems, and most commercial codes only support automatic generation of tetrahedral meshes. Therefore, triangular prism and pyramidal elements serve as transitional elements and are essential for mesh generation schemes that predominantly use hexahedral elements.

The displacement fields of tetrahedral, triangular prism, hexahedral, and pyramid elements can be expressed in the following general form

$$\begin{aligned} & u\left[x(\xi,\ \eta,\ \zeta), y(\xi,\ \eta,\ \zeta), z(\xi,\ \eta,\ \zeta)\right] \\ & = \sum_{i=1}^{N_{\mathrm{v}}} S_i^{\mathrm{v}}(\xi,\ \eta,\ \zeta) u_i + \sum_{i=1}^{L_k} \sum_{k=1}^{N_{\mathrm{e}}} S_i^{\mathrm{e},\ k}(\xi,\ \eta,\ \zeta) a_i^{\mathrm{e},\ k} \\ & + \sum_{i=1}^{M_k} \sum_{j=1}^{N_k} \sum_{k=1}^{N_{\mathrm{f}}} S_{ij}^{\mathrm{f},\ k}(\xi,\ \eta,\ \zeta) a_{ij}^{\mathrm{f},\ k} + \sum_{i=1}^{L} \sum_{j=1}^{M} \sum_{k=1}^{N} S_{ijk}^{\mathrm{b}}(\xi,\ \eta,\ \zeta) a_{ijk}^{\mathrm{b}} \end{aligned} \tag{4.169}$$

Here, v, e, f, and b represent vertex, edge, face, and body, respectively; N_{v}, N_{e} and N_{f} denote the number of vertices, edges, and faces, respectively; i, j, and k are integer indices; $S_i^{\mathrm{v}}(\xi, \eta, \zeta)$ is the vertex function; $S_i^{\mathrm{e},k}(\xi, \eta, \zeta)$ is the edge function; $S_{ij}^{\mathrm{f},k}(\xi, \eta, \zeta)$ denotes the surface function; $S_{ijk}^{\mathrm{b}}(\xi, \eta, \zeta)$ denotes the volume function.

4.3.1.1 Tetrahedral Element

A tetrahedral element has 4 vertices, 6 edges, and 4 faces (see Fig. 4.30). The vertex functions of the tetrahedral element are

$$S_1^{\mathrm{v}}(\xi,\ \eta,\ \zeta) = 1 - \xi - \eta - \zeta$$

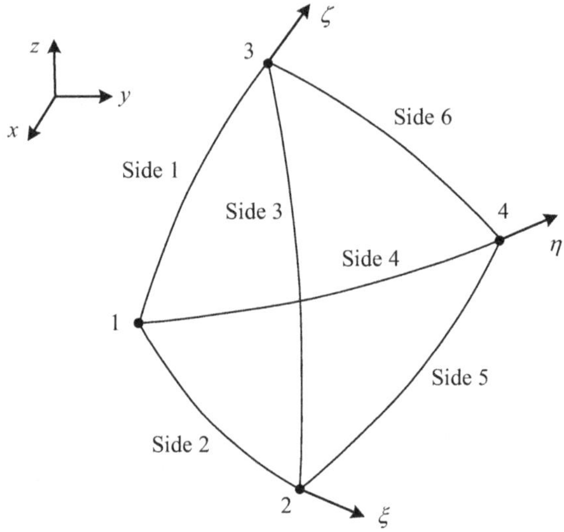

Fig. 4.30 Curved tetrahedral element

$$
\begin{aligned}
S_2^{\mathrm{v}}(\xi,\ \eta,\ \zeta) &= \xi \\
S_3^{\mathrm{v}}(\xi,\ \eta,\ \zeta) &= \zeta \\
S_4^{\mathrm{v}}(\xi,\ \eta,\ \zeta) &= \eta
\end{aligned} \tag{4.170}
$$

Based on the derivation process of the hierarchical quadrature element for triangles, the general form of the modified Dubiner tetrahedral basis function can be expressed as

$$N_{ijk}(\xi,\ \eta,\ \zeta) = (1-\xi-\eta-\zeta)^h \xi^l \eta^m \zeta^n H_{ijk}(\xi,\ \eta,\ \zeta) \tag{4.171}$$

where h, l, m, and n are non-negative integers. In order to apply Jacobi orthogonal polynomials to tetrahedral elements, a coordinate transformation is required. As shown in Fig. 4.31, the Duffy transformation from the unit hexahedral domain to the unit tetrahedral domain can be expressed as follows:

$$\xi = \frac{(1+r)(1+t)}{4},\quad \eta = \frac{(1-r)(1+s)(1+t)}{8},\quad \zeta = \frac{(1-r)(1-s)(1+t)}{8} \tag{4.172}$$

The corresponding inverse transformation is

$$r = 1 - \frac{2(\eta+\zeta)}{(\xi+\eta+\zeta)},\quad s = \frac{\eta-\zeta}{\eta+\zeta},\quad t = 2(\xi+\eta+\zeta) - 1 \tag{4.173}$$

Based on the transformation given by Eq. (4.172), it can be seen that the face $r = -1$ of the unit hexahedron corresponds to the face $\xi = 0$ of the unit tetrahedron,

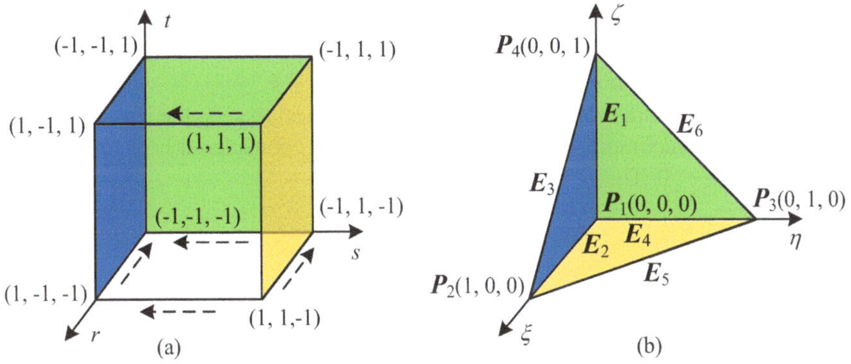

Fig. 4.31 A unit hexahedral domain (**a**) and a tetrahedral parametric domain (**b**)

$s = -1$ corresponds to $\eta = 0$, $s = 1$ corresponds to $\zeta = 0$, $t = 1$ corresponds to $\xi + \eta + \zeta = 1$. Here, the face $t = -1$ degenerates into a point, while the face $r = 1$ degenerates into a line. According to the orthogonality requirement, the shape functions on the unit tetrahedron should satisfy the following formulas:

$$\int_V N_{ijk} N_{pqw} \mathrm{d}V = \int_0^1 \int_0^{1-\xi} \int_0^{1-\xi-\eta} N_{ijk} N_{pqw} \mathrm{d}\zeta \, \mathrm{d}\eta \mathrm{d}\xi = \delta_i^p \delta_j^q \delta_k^w C_{ijk} \tag{4.174}$$

Substituting Eqs. (4.171) and (4.172) into (4.174), we obtain

$$\int_0^1 \int_0^{1-\xi} \int_0^{1-\xi-\eta} N_{ijk} N_{pqw} \mathrm{d}\zeta \, \mathrm{d}\eta \mathrm{d}\xi = \int_{-1}^1 \int_{-1}^1 \int_{-1}^1 \tilde{J} H_{ijk} H_{pqw} \mathrm{d}r \mathrm{d}s \mathrm{d}t \tag{4.175}$$

where

$$\tilde{J} = \frac{(1+r)^{2l}(1-r)^{2m+2n+1}(1+s)^{2m}(1-s)^{2n}(1+t)^{2l+2m+2n+2}(1-t)^{2h}}{2^{2h+4l+6m+6n+6}} \tag{4.176}$$

According to the orthogonality property of Jacobi polynomials, $H_{ijk}(r, s, t)$ can be determined as follows:

$$\begin{aligned} H_{ijk}(r, s, t) = & \left[(1-r)^a (1+r)^b P_i^{(2m+2n+1+2a,\ 2l+2b)}(r)\right] \\ & \left[(1-s)^c (1+s)^d P_j^{(2n+2c,\ 2m+2d)}(s)\right] \\ & \times \left[(1-t)^e (1+t)^f P_k^{(2h+2e,\ 2l+2m+2n+2+2f)}(t)\right] \end{aligned} \tag{4.177}$$

where a, b, c, d, e and f should be determined based on the convenience of application and the requirements of Jacobi polynomials. To eliminate the term $(\eta + \zeta)$ in the denominator of s in Eq. (4.177), it is necessary to set $a = j$. Similarly, to eliminate $(\xi + \eta + \zeta)$ in the denominators of r and $(1-r)$ in Eq. (4.177), $f = i + j$ must be set. The indices i, j, k should satisfy the relation $i + j + k \leq \gamma - 4$, where $\gamma \geq 4$ is the intended maximum order of the basis functions, used to keep the basis functions below a certain maximum order. For the C^0 problem considered here, we can set $b = c = d = e = 0$. Thus,

$$H_{ijk}(r, s, t) = \left[(1-r)^j P_i^{(2m+2n+1+2j,\ 2l)}(r)\right] P_j^{(2n,\ 2m)}(s)\left[(1+t)^{i+j} P_k^{(2h,\ 2(l+m+n+1)+2(i+j))}(t)\right] \tag{4.178}$$

Substituting Eqs. (4.173) into (4.178), we obtain

$$\begin{aligned} H_{ijk}(\xi,\ \eta,\ \zeta) &= \left[2^{i+2j}(\xi+\eta+\zeta)^i P_i^{(2m+2n+1+2j,\ 2l)}\left(\frac{\xi-\eta-\zeta}{\xi+\eta+\zeta}\right)\right]\times \\ &\left[(\eta+\zeta)^j P_j^{(2n,\ 2m)}\left(\frac{\eta-\zeta}{\eta+\zeta}\right)\right] P_k^{(2h,\ 2(l+m+n+1)+2(i+j))}[2(\xi+\eta+\zeta)-1] \end{aligned} \tag{4.179}$$

The constant 2^{i+2j} in the above equation is generally disregarded in calculations. The computation of the rational terms in the denominator of Eq. (4.179) can be avoided by following the procedure shown in Eqs. (4.124) to (4.129). The constant in Eq. (4.174)

$$C_{ijk} = \frac{C_i^{(2m+2n+1+2j,\ 2l)} C_j^{(2n,\ 2m)} C_k^{(2h,\ 2(l+m+n+1)+2(i+j))}}{2^{2h+4l+6m+6n+6}} \tag{4.180}$$

The warping orthogonal basis functions of Dubiner can be obtained by setting $h = l = m = n = 0$. The modified Dubiner basis functions can be obtained by setting at least one of h, l, m or n to a nonzero value. As shown in Fig. 4.32, several basis functions on the unit tetrahedral domain are illustrated.

4.3.1.2 Triangular Prism Element

The triangular prism (wedge) element (see Fig. 4.33) is the product of a triangular element and a one-dimensional element. The vertex functions of the triangular prism element are

$$\begin{aligned} S_1^{\mathrm{v}}(\xi,\eta,\zeta) &= (1-\xi-\eta)(1-\zeta), \\ S_2^{\mathrm{v}}(\xi,\eta,\zeta) &= \xi(1-\zeta), \\ S_3^{\mathrm{v}}(\xi,\eta,\zeta) &= \eta(1-\zeta), \\ S_4^{\mathrm{v}}(\xi,\eta,\zeta) &= (1-\xi-\eta)\zeta, \\ S_5^{\mathrm{v}}(\xi,\eta,\zeta) &= \xi\zeta, \\ S_6^{\mathrm{v}}(\xi,\eta,\zeta) &= \eta\zeta \end{aligned} \tag{4.181}$$

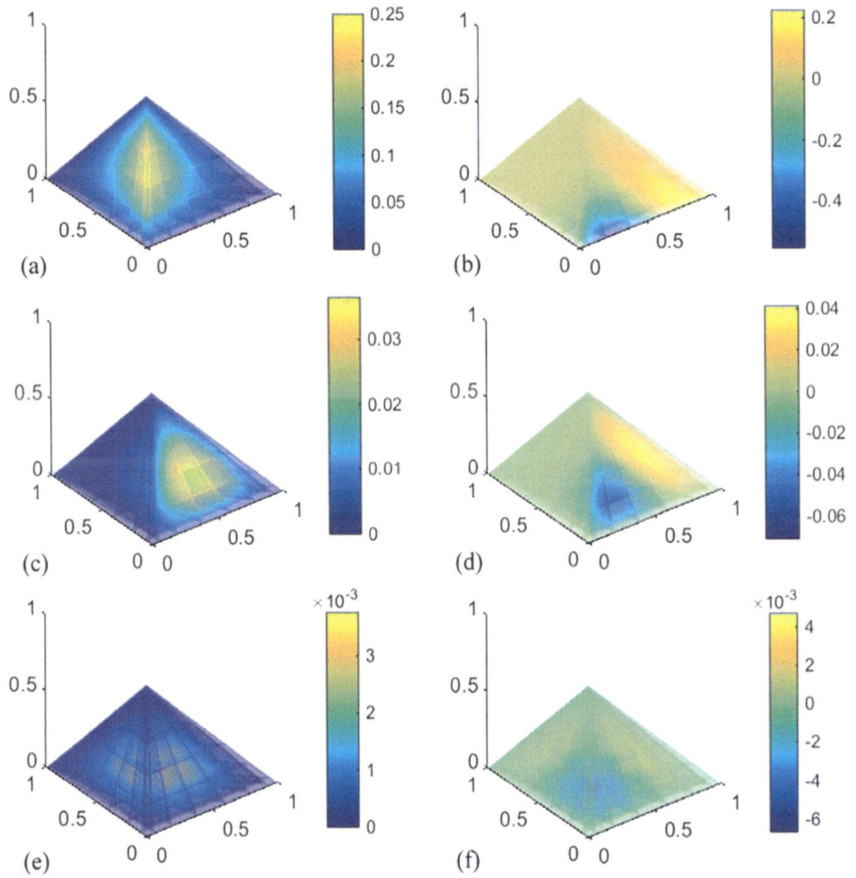

Fig. 4.32 Basis functions on the unit tetrahedral domain

The orthogonal polynomials in the thickness direction of the triangular prism element can be expressed as

$$H_k(\zeta) = \zeta(1-\zeta)P_k^{(2,\,2)}(2\zeta - 1) \tag{4.182}$$

Therefore, the orthogonal basis functions $S_i^{\mathrm{e},k}(\xi, \eta, \zeta)$ on the 9 edges of the triangular prism element can be expressed as

$$\begin{aligned}
S_i^{\mathrm{e},1}(\xi, \eta, \zeta) &= (1-\zeta)E_i^1(\xi, \eta) \\
S_i^{\mathrm{e},2}(\xi, \eta, \zeta) &= (1-\zeta)E_i^2(\xi, \eta) \\
S_i^{\mathrm{e},3}(\xi, \eta, \zeta) &= (1-\zeta)E_i^3(\xi, \eta) \\
S_i^{\mathrm{e},4}(\xi, \eta, \zeta) &= \zeta E_i^1(\xi, \eta)
\end{aligned}$$

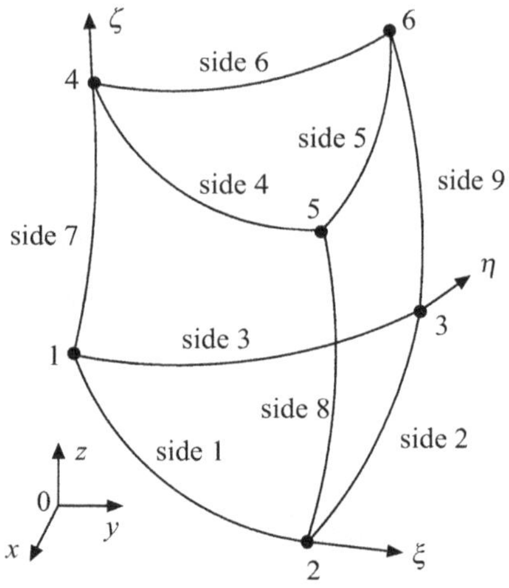

Fig. 4.33 Curved triangular prism element

$$S_i^{e,5}(\xi,\eta,\zeta) = \zeta E_i^2(\xi,\eta)$$
$$S_i^{e,6}(\xi,\eta,\zeta) = \zeta E_i^3(\xi,\eta)$$
$$S_i^{e,7}(\xi,\eta,\zeta) = (1-\xi-\eta)H_i(\zeta)$$
$$S_i^{e,8}(\xi,\eta,\zeta) = \xi H_i(\zeta)$$
$$S_i^{e,9}(\xi,\eta,\zeta) = \eta H_i(\zeta) \tag{4.183}$$

The orthogonal basis functions $S_{ij}^{f,k}(\xi,\eta,\zeta)$ on the five faces can be expressed as

$$S_{ij}^{f,1}(\xi,\eta,\zeta) = (1-\zeta)F_{ij}(\xi,\eta)$$
$$S_{ij}^{f,2}(\xi,\eta,\zeta) = \zeta F_{ij}(\xi,\eta)$$
$$S_{ij}^{f,3}(\xi,\eta,\zeta) = E_i^1(\xi,\eta)H_j(\zeta)$$
$$S_{ij}^{f,4}(\xi,\eta,\zeta) = E_i^2(\xi,\eta)H_j(\zeta)$$
$$S_{ij}^{f,5}(\xi,\eta,\zeta) = E_i^3(\xi,\eta)H_j(\zeta) \tag{4.184}$$

The orthogonal basis $S_{ijk}^{b}(\xi,\eta,\zeta)$ within the element can be expressed as

$$S_{ijk}^{b}(\xi,\eta,\zeta) = F_{ij}(\xi,\eta)H_k(\zeta) \tag{4.185}$$

The edge, face, and volume functions of the triangular prism element described above utilize the edge and face functions from Sect. 4.2.4.1. As shown in Fig. 4.34, two orthogonal basis functions and two Lagrange interpolation basis functions on the triangular prism element are illustrated.

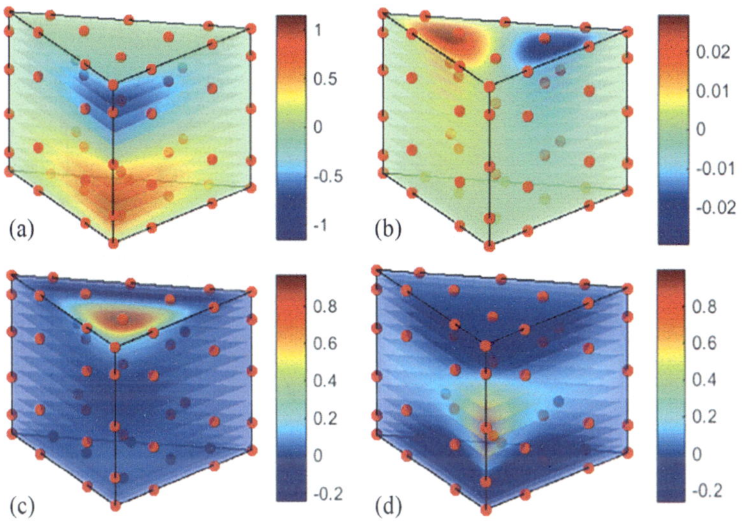

Fig. 4.34 Shape function diagrams on a unit wedge element: (**a**, **b**) orthogonal basis functions, (**c**, **d**) interpolation basis functions

4.3.1.3 Hexahedral Element

The derivation process for the basis functions of the hexahedral element (Fig. 4.35) is similar to that of the triangular prism element described above. When constructing the edge, face, and volume functions of the hexahedral element, only the one-dimensional orthogonal basis given by Eq. (4.182) is required. The vertex basis functions $S_i^{\mathrm{v}}(\xi, \eta, \zeta)$ of the element are given by:

$$
\begin{aligned}
S_1^{\mathrm{v}}(\xi, \eta, \zeta) &= (1-\xi)(1-\eta)(1-\zeta) \\
S_2^{\mathrm{v}}(\xi, \eta, \zeta) &= \xi(1-\eta)(1-\zeta) \\
S_3^{\mathrm{v}}(\xi, \eta, \zeta) &= \xi\eta(1-\zeta) \\
S_4^{\mathrm{v}}(\xi, \eta, \zeta) &= (1-\xi)\eta(1-\zeta) \\
S_5^{\mathrm{v}}(\xi, \eta, \zeta) &= (1-\xi)(1-\eta)\zeta \\
S_6^{\mathrm{v}}(\xi, \eta, \zeta) &= \xi(1-\eta)\zeta \\
S_7^{\mathrm{v}}(\xi, \eta, \zeta) &= \xi\eta\zeta \\
S_8^{\mathrm{v}}(\xi, \eta, \zeta) &= (1-\xi)\eta\zeta
\end{aligned}
\tag{4.186}
$$

The orthogonal basis function $S_{ijk}^{\mathrm{b}}(\xi, \eta, \zeta)$ within the element can be expressed as

$$
S_{ijk}^{\mathrm{b}}(\xi, \eta, \zeta) = \xi(1-\xi)\eta(1-\eta)\zeta(1-\zeta)P_i^{(2,\,2)}(2\xi-1)P_j^{(2,\,2)}(2\eta-1)P_k^{(2,\,2)}(2\zeta-1)
\tag{4.187}
$$

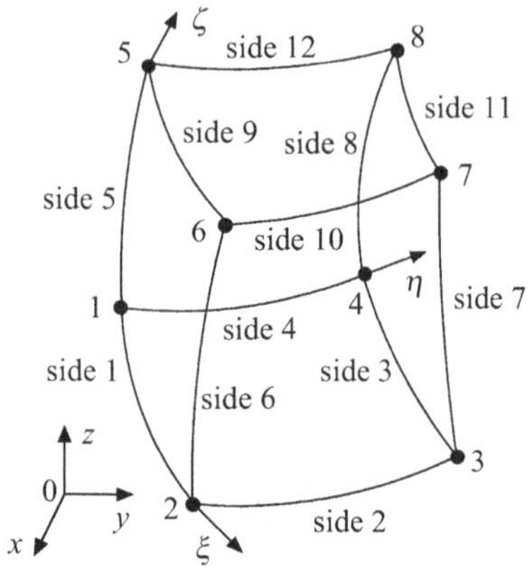

Fig. 4.35 A curved hexahedral element

The edge and face functions of the hexahedral element can be obtained by omitting certain terms from Eq. (4.187). This process is left to the reader.

4.3.1.4 Pyramid Element

The selection of degrees of freedom and basis function types for pyramid elements is constrained by the requirement for compatibility with adjacent hexahedral and tetrahedral elements. In fact, the need for compatibility between pyramid elements and hexahedral or tetrahedral elements determines the choice of surface degrees of freedom for the pyramid element. When attempting to select a basis function for the pyramid element, these issues must be addressed. In 1971, Zienkiewicz was the first to propose merging all the nodes on one face of a hexahedral element into a single node to form a pyramid element (see Fig. 4.36). The shape functions for the non-degenerate nodes remain the same as before, while the shape function for the degenerate node is the sum of the original shape functions on that face. Thus, the vertex functions of the pyramid element are

$$
\begin{aligned}
S_1^{\mathrm{v}} &= \frac{1}{4}(1-r)(1-s)(1-t)\\
S_2^{\mathrm{v}} &= \frac{1}{4}(1+r)(1-s)(1-t)\\
S_3^{\mathrm{v}} &= \frac{1}{4}(1+r)(1+s)(1-t)\\
S_4^{\mathrm{v}} &= \frac{1}{4}(1-r)(1+s)(1-t)\\
S_5^{\mathrm{v}} &= t
\end{aligned}
\tag{4.188}
$$

The convergence of four vertices at a single point causes the Jacobian determinant at that point to approach zero, making it impossible to compute derivatives

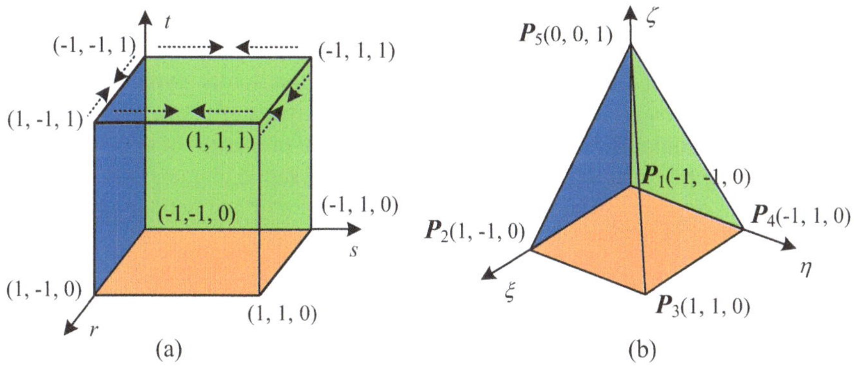

Fig. 4.36 Degenerated pyramid element: **a** a unit hexahedron, **b** a unit pyramid

there. In practice, although the errors introduced by this phenomenon may trigger warnings about unreliable local stress results (for example, in the degenerate form of the SOLID186 element in ANSYS), their impact remains highly localized. As a result, such elements still have certain applications. In 1973, Newton demonstrated that the shape functions of degenerate hexahedral isoparametric elements require fundamental improvements.

In 1992, Bedrosian discovered that it is impossible to achieve the compatibility between pyramid elements and adjacent elements by using only polynomial basis functions. To address this issue, he proposed the so-called "singular functions," that is, rational basis functions on the pyramid, which can be expressed as

$$
\begin{aligned}
S_1^{\mathrm{v}} &= \frac{1}{4}\left(1 - \xi - \eta - \zeta + \frac{\xi\eta}{1-\zeta}\right) \\
S_2^{\mathrm{v}} &= \frac{1}{4}\left(1 + \xi - \eta - \zeta - \frac{\xi\eta}{1-\zeta}\right) \\
S_3^{\mathrm{v}} &= \frac{1}{4}\left(1 + \xi + \eta - \zeta + \frac{\xi\eta}{1-\zeta}\right) \\
S_4^{\mathrm{v}} &= \frac{1}{4}\left(1 - \xi + \eta - \zeta - \frac{\xi\eta}{1-\zeta}\right) \\
S_5^{\mathrm{v}} &= \zeta
\end{aligned}
\tag{4.189}
$$

It has been verified that these shape functions satisfy the C^0 continuity requirements on all faces. For any face, it is sufficient that the surface shape functions can be reduced to two-dimensional triangular or quadrilateral shape functions, since adjacent tetrahedral, triangular prism, or hexahedral elements exhibit the same property. Furthermore, it is observed that for any pyramid element with a planar parallelogram base, its Jacobian determinant is constant; in the general case of curved elements, the faces are not necessarily planar, and the Jacobian determinant varies with position, but this variation is much less pronounced than in pyramid elements derived from degenerate hexahedra.

The rational "singular functions" of first- and second-order pyramid elements proposed by Bedrosian have been shown to be at least as effective as other classical elements. Coulomb and others have discussed the properties of these "singular functions" and their derivatives. At the apex of the pyramid element, since $\zeta = 1$, it follows that $\xi = \eta = 0$, resulting in both the numerator and denominator of the rational terms being zero. It is easy to verify the inequality

$$\begin{aligned} -(1-\zeta) \leq \xi \leq (1-\zeta) \\ -(1-\zeta) \leq \eta \leq (1-\zeta) \\ 0 \leq \zeta \leq 1 \end{aligned} \tag{4.190}$$

holds true. As $\zeta \rightarrow 1$, the rational term $\xi\eta/(1-\zeta) \rightarrow 0$, therefore the shape functions $S_1{}^{\mathrm{v}}, S_2{}^{\mathrm{v}}, S_3{}^{\mathrm{v}}$ and $S_4{}^{\mathrm{v}}$ will not have difficulty in the computation at the vertex. Inequality (4.190) can also be expressed in the following form in local coordinates:

$$\begin{aligned} -1 \leq \frac{\xi}{1-\zeta} \leq 1 \\ -1 \leq \frac{\eta}{1-\zeta} \leq 1 \end{aligned} \tag{4.191}$$

This indicates that the derivative is indeterminate. However, Bravo et al. have demonstrated that both the vertex basis functions themselves and their derivatives are zero at the apex of the pyramid, and therefore do not pose any computational difficulties.

In 2010, Bergot et al. constructed arbitrary-order efficient nodal pyramid elements using rational "singular functions" at the vertices of the pyramid element and Dubiner orthogonal expansion polynomials within the pyramid domain. Numerical tests using these elements (up to sixth order) demonstrated low phase error, good stability, and excellent performance in hybrid meshes.

In 2014, Bravo et al. constructed a set of p-th order rational hierarchical basis functions on the pyramid element. The concept of "edges" in the pyramid element was extended to include vertices, edges, faces, and the interior. The rational terms are restricted within the element domain and vanish on the element boundary. The edge, face, and volume functions are constructed using rational "singular functions" and Chebyshev polynomials. Elements of order 0 to 7 were tested, and rational monomials were considered.

Starting from the rational basis functions at the vertices given in Eq. (4.189), we derive the modified Dubiner orthogonal polynomials on the pyramid element. The orthogonal basis functions are classified into interior and boundary types (see Fig. 4.37). A unit hexahedral domain can be transformed into a unit pyramid domain via the Duffy transformation, that is,

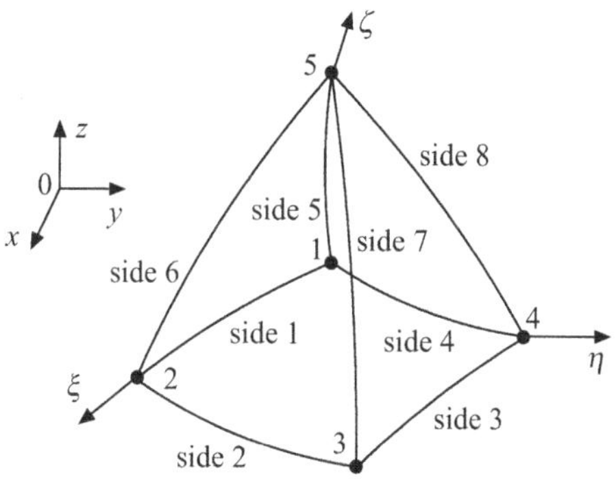

Fig. 4.37 A Curved Pyramid Element

$$\xi = r(1-t) \qquad \eta = s(1-t) \qquad \zeta = t \tag{4.192}$$

The inverse transformation is

$$r = \frac{\xi}{1-\zeta} \qquad s = \frac{\eta}{1-\zeta} \qquad t = \zeta \tag{4.193}$$

As shown in Fig. 4.38, the coordinate transformation given by Eq. (4.193) connects the parametric domains of the unit hexahedron and the unit pyramid. For example, the $r = -1$ plane of the unit hexahedron corresponds to the $\xi/(1-\zeta) = -1$ plane of the unit pyramid. It is straightforward to verify that Eq. (4.189) can be obtained by substituting Eqs. (4.193) into (4.188).

Based on the vertex shape functions, the shape functions for edges, faces, and volumes can be derived in a similar manner (see Fig. 4.39). To this end, Eq. (4.189) is presented in different forms as follows:

$$\begin{aligned}
S_1^{\mathrm{v}} &= \frac{1}{4}\frac{(1-\xi-\zeta)(1-\eta-\zeta)}{1-\zeta} \\
S_2^{\mathrm{v}} &= \frac{1}{4}\frac{(1+\xi-\zeta)(1-\eta-\zeta)}{1-\zeta} \\
S_3^{\mathrm{v}} &= \frac{1}{4}\frac{(1+\xi-\zeta)(1+\eta-\zeta)}{1-\zeta} \\
S_4^{\mathrm{v}} &= \frac{1}{4}\frac{(1-\xi-\zeta)(1+\eta-\zeta)}{1-\zeta} \\
S_5^{\mathrm{v}} &= \zeta
\end{aligned} \tag{4.194}$$

It is easy to see that a vertex shape function on the bottom square face vanishes on the two triangular faces opposite to the vertex. The top vertex function, denoted as $S_5{}^{\mathrm{v}}$, is zero on the bottom square face.

Given the properties of the vertex shape functions described above, the function $\mathrm{S}_i{}^{\mathrm{e},1}$ corresponding to the edge formed by vertices 1 and 2 (see E_1 in Fig. 4.39) can

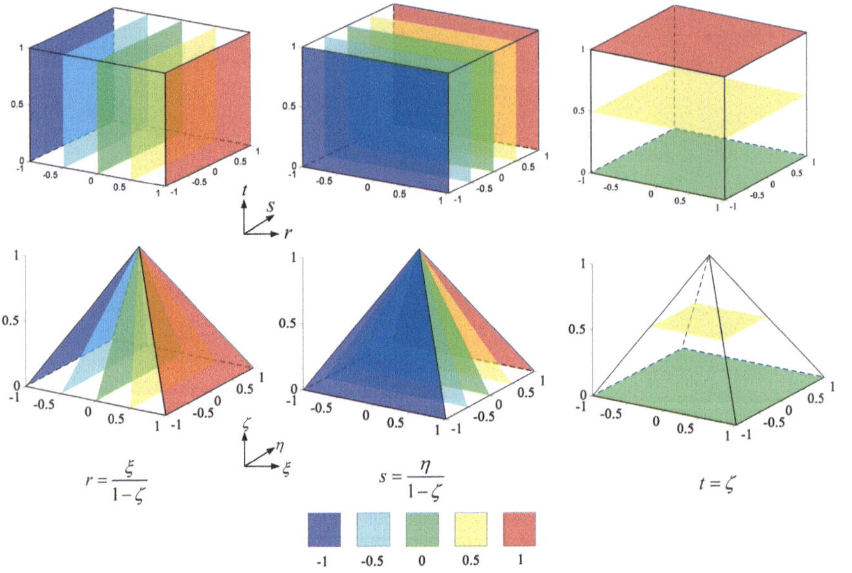

Fig. 4.38 Parametric relationship between a unit hexahedron (top) and a unit pyramid (bottom)

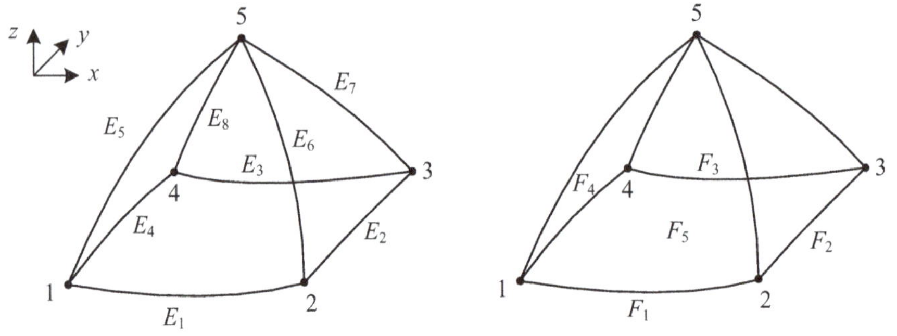

Fig. 4.39 Numbering of edges and faces of the pyramid element

be expressed as

$$S_i^{\mathrm{e},1} = S_1^{\mathrm{v}} S_2^{\mathrm{v}} H_i \tag{4.195}$$

Here, H_i denotes the undetermined polynomial, and i is an integer index. To improve the numerical properties of the basis functions, they are required to be orthogonal, that is, the polygonal functions must satisfy

$$\int_V S_i^{\mathrm{e},1} S_j^{\mathrm{e},1} \mathrm{d}V = \int_0^1 \int_{\zeta-1}^{1-\zeta} \int_{\zeta-1}^{1-\zeta} S_i^{\mathrm{e},1} S_j^{\mathrm{e},1} \mathrm{d}\xi \mathrm{d}\eta \mathrm{d}\zeta = \delta_i^j C_i \tag{4.196}$$

Here, δ_{ij} denotes the Kronecker delta. Substituting Eqs. (4.195) into (4.196) yields

$$\int_0^1\int_{\zeta-1}^{1-\zeta}\int_{\zeta-1}^{1-\zeta} S_i^{\mathrm{e},1}S_j^{\mathrm{e},1}\mathrm{d}\xi\mathrm{d}\eta\mathrm{d}\zeta = \int_0^1\int_{\zeta-1}^{1-\zeta}\int_{\zeta-1}^{1-\zeta}\left(S_1^{\mathrm{v}}S_2^{\mathrm{v}}\right)^2 H_iH_j\mathrm{d}\xi\mathrm{d}\eta\mathrm{d}\zeta \quad (4.197)$$

Let

$$r = \frac{\xi}{1-\zeta},\quad s = \frac{\eta}{1-\zeta},\quad t = 2\zeta - 1 \quad (4.198)$$

Compared with Eqs. (4.193), (4.198) extends the t-coordinate to the interval [–1, 1]. Therefore, we have

$$\xi = \frac{1}{2}r(1-t),\quad \eta = \frac{1}{2}s(1-t),\quad \zeta = \frac{1}{2}(1+t) \quad (4.199)$$

The Jacobian determinant of the parametric coordinates (ξ, η, ζ) with respect to the coordinates (r, s, t) is

$$|\mathbf{J}| = \left|\frac{\partial(\xi,\ \eta\ \zeta)}{\partial(r,\ s\ t)}\right| = \begin{vmatrix} \frac{1-t}{2} & 0 & -\frac{r}{2} \\ 0 & \frac{1-t}{2} & -\frac{s}{2} \\ 0 & 0 & \frac{1}{2} \end{vmatrix} = \frac{1}{2^3}(1-t)^2 \quad (4.200)$$

Therefore, substituting Eqs. (4.199) into (4.197) yields

$$\int_0^1\int_{\zeta-1}^{1-\zeta}\int_{\zeta-1}^{1-\zeta}\left(S_1^{\mathrm{v}}S_2^{\mathrm{v}}\right)^2 H_iH_j\mathrm{d}\xi\mathrm{d}\eta\mathrm{d}\zeta = \int_{-1}^1\int_{-1}^1\int_{-1}^1 \bar{J}H_iH_j\mathrm{d}r\mathrm{d}s\mathrm{d}t \quad (4.201)$$

where

$$\bar{J} = \frac{1}{2^{19}}(1-r)^2(1+r)^2(1-s)^4(1-t)^6 \quad (4.202)$$

The basis function $H_i(r, s, t)$ can be defined as

$$H_i = (1-r)^a(1+r)^b(1-s)^c(1+s)^d(1-t)^e(1+t)^f P_i^{(2a+2,\ 2b+2)}(r) \quad (4.203)$$

where a, b, c, d, e and f should be determined based on convenience for the application and the requirements of the Jacobi polynomials. To eliminate the term $(1-\zeta)$ in the denominators of r and s in Eq. (4.203), it is necessary to set $e = i$. For the C^0 problem considered here, set $a = b = c = d = f = 0$. Therefore, we have:

$$H_i(r, s, t) = (1-t)^i P_i^{(2,2)}(r) \quad (4.204)$$

Therefore, the edge function $S_i^{\mathrm{e},1}$ can be written as

$$S_i^{\mathrm{e},1} = S_1^{\mathrm{v}} S_2^{\mathrm{v}} 2^i (1-\zeta)^i P_i^{(2,2)} \left(\frac{\xi}{1-\zeta} \right) \tag{4.205}$$

The constant coefficient 2^i in Eq. (4.205) is typically omitted. By following the procedure outlined in Eqs. (4.124) to (4.129), one can avoid calculating the $(1-\zeta)$ term in the denominator.

Similarly, the edge functions for other edges can be obtained. The edge functions of the pyramid element can be expressed as

$$\begin{aligned}
S_i^{\mathrm{e},1} &= S_1^{\mathrm{v}} S_2^{\mathrm{v}} \tilde{P}_i^{(2,2)}(\xi,\ 1-\zeta) \\
S_j^{\mathrm{e},2} &= S_2^{\mathrm{v}} S_3^{\mathrm{v}} \tilde{P}_j^{(2,2)}(\eta,\ 1-\zeta) \\
S_i^{\mathrm{e},3} &= S_3^{\mathrm{v}} S_4^{\mathrm{v}} \tilde{P}_i^{(2,2)}(\xi,\ 1-\zeta) \\
S_j^{\mathrm{e},4} &= S_4^{\mathrm{v}} S_1^{\mathrm{v}} \tilde{P}_j^{(2,2)}(\eta,\ 1-\zeta) \\
S_k^{\mathrm{e},5} &= S_1^{\mathrm{v}} S_5^{\mathrm{v}} P_k^{(4,2)}(2\zeta-1) \\
S_k^{\mathrm{e},6} &= S_2^{\mathrm{v}} S_5^{\mathrm{v}} P_k^{(4,2)}(2\zeta-1) \\
S_k^{\mathrm{e},7} &= S_3^{\mathrm{v}} S_5^{\mathrm{v}} P_k^{(4,2)}(2\zeta-1) \\
S_k^{\mathrm{e},8} &= S_4^{\mathrm{v}} S_5^{\mathrm{v}} P_k^{(4,2)}(2\zeta-1)
\end{aligned} \tag{4.206}$$

Here, the indices i, j, and k represent the basis function indices in the ξ, η, and ζ parametric directions, respectively. It can be observed that the edge function can be expressed as $S_i^{\mathrm{e},k} = S_m^{\mathrm{v}} S_n^{\mathrm{v}} H_i$, where the vertex indices m and n determine edge k, as shown in Fig. 4.39.

The derivation process for surface and volume functions is similar to that of the edge functions described above and can be expressed as:

$$\begin{cases}
S_{ik}^{\mathrm{f},1} = S_1^{\mathrm{v}} S_2^{\mathrm{v}} S_5^{\mathrm{v}} \tilde{P}_i^{(2,2)}(\xi, 1-\zeta) P_k^{(2i+6,2)}(2\zeta-1), \\
S_{jk}^{\mathrm{f},2} = S_2^{\mathrm{v}} S_3^{\mathrm{v}} S_5^{\mathrm{v}} \tilde{P}_j^{(2,2)}(\eta, 1-\zeta) P_k^{(2j+6,2)}(2\zeta-1), \\
S_{ik}^{\mathrm{f},3} = S_3^{\mathrm{v}} S_4^{\mathrm{v}} S_5^{\mathrm{v}} \tilde{P}_i^{(2,2)}(\xi, 1-\zeta) P_k^{(2i+6,2)}(2\zeta-1), \\
S_{jk}^{\mathrm{f},4} = S_4^{\mathrm{v}} S_1^{\mathrm{v}} S_5^{\mathrm{v}} \tilde{P}_j^{(2,2)}(\eta, 1-\zeta) P_k^{(2j+6,2)}(2\zeta-1), \\
S_{ij}^{\mathrm{f},5} = S_1^{\mathrm{v}} S_3^{\mathrm{v}} \tilde{P}_i^{(2,2)}(\xi, 1-\zeta) \tilde{P}_j^{(2,2)}(\eta, 1-\zeta)
\end{cases} \tag{4.207}$$

$$S_{ijk}^{\mathrm{b}} = S_1^{\mathrm{v}} S_3^{\mathrm{v}} S_5^{\mathrm{v}} \tilde{P}_i^{(2,2)}(\xi,\ 1-\zeta) \tilde{P}_j^{(2,2)}(\eta,\ 1-\zeta) P_k^{(2i+2j+6,2)}(2\zeta-1) \tag{4.208}$$

Figure 4.40 presents graphical representations of some edge, face, and volume functions. As can be observed from Fig. 4.40, the orthogonal basis provided here exhibits symmetric or antisymmetric axes.

Based on the derivation above, edge, face, and volume functions can be expressed in the following general form.

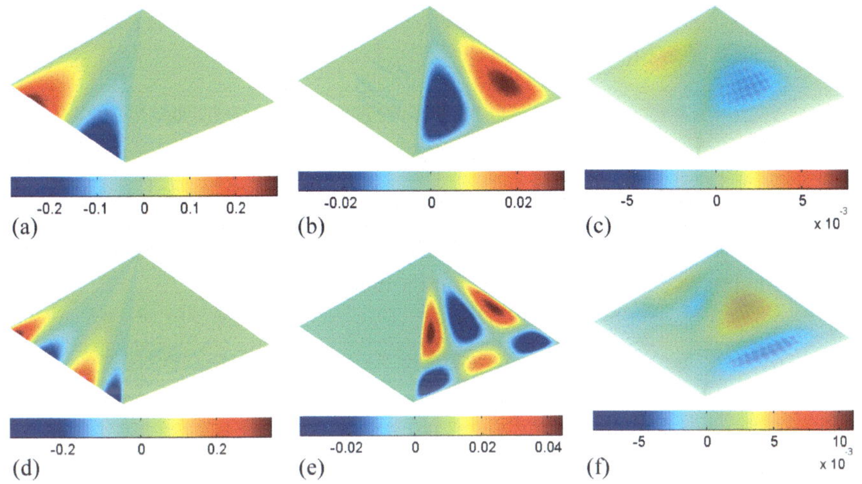

Fig. 4.40 Hierarchical shape functions: **a** $S_1^{\mathrm{e},1}$, **b** $S_{1,0}^{\mathrm{f},2}$, **c** $S_{0,1,0}^{\mathrm{b}}$, **d** $S_3^{\mathrm{e},1}$, **e** $S_{2,1}^{\mathrm{f},1}$, **f** $S_{0,2,1}^{\mathrm{b}}$

$$N_{ijk}(\xi,\ \eta,\ \zeta) = \left(S_1^{\mathrm{v}}\right)^g\left(S_2^{\mathrm{v}}\right)^h\left(S_3^{\mathrm{v}}\right)^l\left(S_4^{\mathrm{v}}\right)^m\left(S_5^{\mathrm{v}}\right)^n H_{ijk}(\xi,\ \eta,\ \zeta) \tag{4.209}$$

Satisfying the orthogonality condition

$$\int_0^1\int_{\zeta-1}^{1-\zeta}\int_{\zeta-1}^{1-\zeta} N_{ijk}N_{pqw}\mathrm{d}\xi\mathrm{d}\eta\mathrm{d}\zeta = \int_{-1}^1\int_{-1}^1\int_{-1}^1 \tilde{J}H_{ijk}H_{pqw}\mathrm{d}r\mathrm{d}s\mathrm{d}t = \delta_{ip}\delta_{jq}\delta_{kw}C_{ijk} \tag{4.210}$$

where

$$\tilde{J} = \frac{(1-r)^{2(g+m)}(1+r)^{2(h+l)}(1-s)^{2(g+h)}(1+s)^{2(l+m)}(1-t)^{2(g+h+l+m)+2}(1+t)^{2n}}{2^{6(g+h+l+m)+2n+3}} \tag{4.211}$$

Moreover, H_{ijk} can be determined in a manner similar to Eqs. (4.198) through (4.205) as follows:

$$\begin{aligned} H_{ijk}(r,s,t) = &\left[P_i^{(2g+2m,\ 2h+2l)}(r)\right]\left[P_j^{(2g+2h,\ 2l+2m)}(s)\right] \\ &\left[(1-t)^{i+j}P_k^{(2g+2h+2l+2m+2+2i+2j,\ 2n)}(t)\right] \end{aligned} \tag{4.212}$$

or

$$\begin{aligned} H_{ijk}(\xi,\ \eta,\ \zeta) = 2^{i+j}\tilde{P}_i^{(2g+2m,\ 2h+2l)}(\xi,\ 1-\zeta)\tilde{P}_j^{(2g+2h,\ 2l+2m)}(\eta,\ 1-\zeta)\times \\ P_k^{(2g+2h+2l+2m+2+2i+2j,\ 2n)}(2\zeta-1) \end{aligned} \tag{4.213}$$

Here, g, h, l, m and n are equal to 0 or 1. The constant coefficient 2^{i+j} is omitted here. The constant C_{ijk} in Eq. (4.209) are

$$C_{ijk} = \frac{C_i^{(2g+2m,\ 2h+2l)} C_j^{(2g+2h,\ 2l+2m)} C_k^{(2g+2h+2l+2m+2+2i+2j,\ 2n)}}{2^{6(g+h+l+m)+2n+3}} \tag{4.214}$$

It can be seen that when $k = l = n = 1$ and $h = m = 0$, Eq. (4.209) reduces to the volume function $S_{ijk}{}^{\mathrm{b}}$ in Eq. (4.207). The other edge and face shape functions mentioned above can be obtained in a similar manner.

4.3.2 Strain Matrix

Since the faces of tetrahedral, triangular prism (wedge), hexahedral, and pyramidal elements are either triangles or quadrilaterals, it is only necessary to use the method introduced in Sect. 4.2.4.1 to convert the basis functions on the boundary faces into Lagrange interpolation functions. Similar to Sect. 4.2.4.1, the displacement field given by Eq. (4.169) can be expressed as

$$u[x(\xi,\eta,\zeta), y(\xi,\eta,\zeta), z(\xi,\eta,\zeta)] = \begin{bmatrix} \mathbf{S}_1^{\mathrm{T}}(\xi,\eta,\zeta) & \mathbf{S}_2^{\mathrm{T}}(\xi,\eta,\zeta) \end{bmatrix} \begin{bmatrix} \tilde{\mathbf{u}}_1 \\ \tilde{\mathbf{u}}_2 \end{bmatrix} = \mathbf{S}^{\mathrm{T}}(\xi,\eta,\zeta)\tilde{\mathbf{u}} \tag{4.215}$$

Here, $\mathbf{S}_1{}^{\mathrm{T}}$ and $\mathbf{S}_2{}^{\mathrm{T}}$ represent the vectors of face (including edge and vertex) and internal functions, respectively; $\tilde{\mathbf{u}}_1$ and $\tilde{\mathbf{u}}_2$ are the coefficient vectors corresponding to the generalized displacements. After converting the edge functions into interpolation functions, Eq. (4.215) can be expressed as

$$u[x(\xi,\eta,\zeta), y(\xi,\eta,\zeta), z(\xi,\eta,\zeta)] = \mathbf{S}^{\mathrm{T}}(\xi,\eta,\zeta)\mathbf{u} \tag{4.216}$$

Substituting the computational points $(\bar{\boldsymbol{\xi}}, \bar{\boldsymbol{\eta}}, \bar{\boldsymbol{\zeta}})$ into Eq. (4.216) yields the following interpolation relationships:

$$u\left(\bar{\boldsymbol{\xi}}, \bar{\boldsymbol{\eta}}, \bar{\boldsymbol{\zeta}}\right) = \mathbf{Gu} \quad \text{or} \quad \bar{\mathbf{u}} = \mathbf{Gu} \tag{4.217}$$

Its first-order partial derivatives with respect to the coordinates are

$$\frac{\partial u\left(\bar{\boldsymbol{\xi}}, \bar{\boldsymbol{\eta}}, \bar{\boldsymbol{\zeta}}\right)}{\partial \xi} = \mathbf{A}^{(1)}\mathbf{u}, \quad \frac{\partial u\left(\bar{\boldsymbol{\xi}}, \bar{\boldsymbol{\eta}}, \bar{\boldsymbol{\zeta}}\right)}{\partial \eta} = \mathbf{B}^{(1)}\mathbf{u}, \quad \frac{\partial u\left(\bar{\boldsymbol{\xi}}, \bar{\boldsymbol{\eta}}, \bar{\boldsymbol{\zeta}}\right)}{\partial \zeta} = \mathbf{C}^{(1)}\mathbf{u} \tag{4.218}$$

Note: The sampling point $(\overline{\xi}, \overline{\eta}, \overline{\zeta})$ is typically integration nodes.

For a three-dimensional function $u(x, y, z)$, let ξ, η and ζ be the variables such that $u = u[x(\xi, \eta, \zeta), y(\xi, \eta, \zeta), z(\xi, \eta, \zeta)]$. Then, the first-order derivatives of u with respect to x, y and z can be obtained using the chain rule of differentiation as follows:

$$\begin{bmatrix} \frac{\partial u}{\partial x} \\ \frac{\partial u}{\partial y} \\ \frac{\partial u}{\partial z} \end{bmatrix} = \frac{1}{|\mathbf{J}|} \begin{bmatrix} \frac{\partial y}{\partial \eta}\frac{\partial z}{\partial \zeta} - \frac{\partial z}{\partial \eta}\frac{\partial y}{\partial \zeta} & \frac{\partial y}{\partial \zeta}\frac{\partial z}{\partial \xi} - \frac{\partial z}{\partial \zeta}\frac{\partial y}{\partial \xi} & \frac{\partial y}{\partial \xi}\frac{\partial z}{\partial \eta} - \frac{\partial z}{\partial \xi}\frac{\partial y}{\partial \eta} \\ \frac{\partial z}{\partial \eta}\frac{\partial x}{\partial \zeta} - \frac{\partial x}{\partial \eta}\frac{\partial z}{\partial \zeta} & \frac{\partial z}{\partial \zeta}\frac{\partial x}{\partial \xi} - \frac{\partial x}{\partial \zeta}\frac{\partial z}{\partial \xi} & \frac{\partial z}{\partial \xi}\frac{\partial x}{\partial \eta} - \frac{\partial x}{\partial \xi}\frac{\partial z}{\partial \eta} \\ \frac{\partial x}{\partial \eta}\frac{\partial y}{\partial \zeta} - \frac{\partial y}{\partial \eta}\frac{\partial x}{\partial \zeta} & \frac{\partial x}{\partial \zeta}\frac{\partial y}{\partial \xi} - \frac{\partial y}{\partial \zeta}\frac{\partial x}{\partial \xi} & \frac{\partial x}{\partial \xi}\frac{\partial y}{\partial \eta} - \frac{\partial y}{\partial \xi}\frac{\partial x}{\partial \eta} \end{bmatrix} \begin{bmatrix} \frac{\partial u}{\partial \xi} \\ \frac{\partial u}{\partial \eta} \\ \frac{\partial u}{\partial \zeta} \end{bmatrix} \tag{4.219}$$

The determinant of the Jacobian $\mathbf{J} = \partial(x, y, z)/\partial(\xi, \eta, \zeta)$ is

$$|\mathbf{J}| = \frac{\partial x}{\partial \xi}\left(\frac{\partial y}{\partial \eta}\frac{\partial z}{\partial \zeta} - \frac{\partial z}{\partial \eta}\frac{\partial y}{\partial \zeta}\right) + \frac{\partial y}{\partial \xi}\left(\frac{\partial z}{\partial \eta}\frac{\partial x}{\partial \zeta} - \frac{\partial x}{\partial \eta}\frac{\partial z}{\partial \zeta}\right) + \frac{\partial z}{\partial \xi}\left(\frac{\partial x}{\partial \eta}\frac{\partial y}{\partial \zeta} - \frac{\partial y}{\partial \eta}\frac{\partial x}{\partial \zeta}\right) \tag{4.220}$$

By applying the method described in Sect. 4.2.4.2 and substituting Eq. (4.218) into Eq. (4.219), we obtain

$$\overline{\mathbf{u}}_x = \mathbf{G}_x\mathbf{u}, \quad \overline{\mathbf{u}}_y = \mathbf{G}_y\mathbf{u}, \quad \overline{\mathbf{u}}_z = \mathbf{G}_z\mathbf{u} \tag{4.221}$$

This is the generalized differential quadrature (DQ) rule for three-dimensional solid elements.

According to the strain–displacement relationship given by Eq. (4.25), it can be discretized using the generalized differential quadrature (DQ) rule (4.221) as

$$\varepsilon = \begin{bmatrix} \mathbf{G}_x & \mathbf{0} & \mathbf{0} \\ \mathbf{0} & \mathbf{G}_y & \mathbf{0} \\ \mathbf{0} & \mathbf{0} & \mathbf{G}_z \\ \mathbf{G}_y & \mathbf{G}_x & \mathbf{0} \\ \mathbf{G}_z & \mathbf{0} & \mathbf{G}_x \\ \mathbf{0} & \mathbf{G}_z & \mathbf{G}_y \end{bmatrix} \begin{bmatrix} \mathbf{u} \\ \mathbf{v} \\ \mathbf{w} \end{bmatrix} = \mathbf{B}\mathbf{d}_e \tag{4.222}$$

It can be seen that this is similar to the lower-order case given in Eq. (4.60).

4.3.3 Element Matrices

The derivation method for the unit matrix is similar to that in Sect. 4.2.4.3. The stiffness matrix for a three-dimensional isotropic solid is given by

$$\mathbf{k}_e = \frac{G}{\upsilon_2}\begin{bmatrix} \mathbf{K}_{11} & & \text{sym} \\ \mathbf{K}_{21} & \mathbf{K}_{22} & \\ \mathbf{K}_{31} & \mathbf{K}_{32} & \mathbf{K}_{33} \end{bmatrix} \tag{4.223}$$

Here, E denotes Young's modulus; $G = E/2(1+\upsilon)$ is the shear modulus, υ is the Poisson's ratio, $\upsilon_1 = (1-\upsilon)/2$, $\upsilon_2 = 0.5 - \upsilon$,

$$\begin{aligned} \begin{bmatrix} \mathbf{K}_{11} \\ \mathbf{K}_{22} \\ \mathbf{K}_{33} \end{bmatrix} &= \begin{bmatrix} 2\upsilon_1 & \upsilon_2 & \upsilon_2 \\ \upsilon_2 & 2\upsilon_1 & \upsilon_2 \\ \upsilon_2 & \upsilon_2 & 2\upsilon_1 \end{bmatrix} \begin{bmatrix} \mathbf{G}_x^{\mathrm{T}}\mathbf{C}\mathbf{G}_x \\ \mathbf{G}_y^{\mathrm{T}}\mathbf{C}\mathbf{G}_y \\ \mathbf{G}_z^{\mathrm{T}}\mathbf{C}\mathbf{G}_z \end{bmatrix} \\ \mathbf{K}_{21} &= \upsilon\mathbf{G}_y^{\mathrm{T}}\mathbf{C}\mathbf{G}_x + \upsilon_2\mathbf{G}_x^{\mathrm{T}}\mathbf{C}\mathbf{G}_y \\ \mathbf{K}_{31} &= \upsilon\mathbf{G}_z^{\mathrm{T}}\mathbf{C}\mathbf{G}_x + \upsilon_2\mathbf{G}_x^{\mathrm{T}}\mathbf{C}\mathbf{G}_z \\ \mathbf{K}_{32} &= \upsilon\mathbf{G}_z^{\mathrm{T}}\mathbf{C}\mathbf{G}_y + \upsilon_2\mathbf{G}_y^{\mathrm{T}}\mathbf{C}\mathbf{G}_z \end{aligned} \tag{4.224}$$

The mass matrix and load vector are given by

$$\mathbf{m}_e = \rho\begin{bmatrix} \mathbf{G}^{\mathrm{T}}\mathbf{C}\mathbf{G} & \mathbf{0} & \mathbf{0} \\ \mathbf{0} & \mathbf{G}^{\mathrm{T}}\mathbf{C}\mathbf{G} & \mathbf{0} \\ \mathbf{0} & \mathbf{0} & \mathbf{G}^{\mathrm{T}}\mathbf{C}\mathbf{G} \end{bmatrix} \tag{4.225}$$

$$\mathbf{f}_e = \begin{bmatrix} \mathbf{G}^{\mathrm{T}}\mathbf{C}\mathbf{q}_u \\ \mathbf{G}^{\mathrm{T}}\mathbf{C}\mathbf{q}_v \\ \mathbf{G}^{\mathrm{T}}\mathbf{C}\mathbf{q}_w \end{bmatrix} \tag{4.226}$$

Here, ρ denotes the density, and $\boldsymbol{q}_u$, $\boldsymbol{q}_v$ and $\boldsymbol{q}_w$ are the distributed nodal load vectors in the x, y and z directions, respectively.

4.3.4 Three-Dimensional Vibration Analysis

This section will validate the performance of higher-order spectral quadrature elements through vibration analysis of three-dimensional solid structures. First, consider a triangular prism (wedge) element. Let the number of nodes on each edge of the two triangular faces be denoted as N_e; let the number of nodes in the two directions on the triangular faces be assigned the same value and denoted as N_f; let the number of nodes along the thickness direction on all quadrilateral edges and faces be assigned the same value and denoted as N_t (excluding both ends).

The hierarchical quadrature element method (HQEM) is used to analyze the free vibration of a clamped thick circular plate, with the frequency parameters listed in Table 4.5. The plate is discretized into eight prismatic (wedge-shaped) elements, as shown in Fig. 4.41. The prismatic (wedge-shaped) mesh is generated based on the triangular mesh of the mid-surface. In Table 4.5, the results obtained by the higher-order spectral quadrature element method are compared with those from the three-dimensional Chebyshev-Ritz solution and the three-dimensional polynomial Ritz (p Ritz) solution. It can be observed that the HQEM solutions using prismatic (wedge-shaped) elements closely resemble those obtained from higher-order shear deformation theory for plates and shells. In other words, the higher-order spectral quadrature element method can achieve the accuracy of three-dimensional elasticity theory at the computational cost of shear deformation theory.

Next, the hierarchical quadrature element method (HQEM) is used to analyze the free vibration of clamped square thin cylindrical shells, with the frequency parameters listed in Table 4.6. The HQEM mesh for this shell structure is shown in Fig. 4.42. The frequencies are presented according to the symmetry (S) and antisymmetry (A) of the modal shapes with respect to the coordinate axes. Once again, the rapid convergence characteristic of the hierarchical quadrature element method is demonstrated.

Finally, taking the three-dimensional structural vibration problem as an example, several cases of combined use of three-dimensional hierarchical quadrature elements are presented. As shown in Fig. 4.43, the model looks like a rocket, and is composed of

Table 4.5 Comparison of frequency parameters for clamped thick circular plates $\Omega = \omega R^2 \sqrt{\rho h/D}$, $\upsilon = 0.3$

h/b	(N_e, N_f, N_T)	Mode number							
		1	2	3	4	5	6	7	8
0.1	(6,3,1)	10.263	20.973	33.482	38.293	47.845	56.542	64.328	68.207
	(7,4,1)	10.079	20.552	33.075	37.611	47.228	55.763	62.815	68.104
	(8,5,2)	10.024	20.364	32.545	36.878	46.368	54.558	61.553	68.049
	(15,12,2)	9.9824	20.282	32.408	36.720	46.114	54.279	61.153	67.956
	(15,12,5)	9.9810	20.278	32.400	36.711	46.099	54.261	61.130	67.951
	3D Ritz	9.9755	20.267	32.383	36.692	46.076	54.234	61.113	67.941
	3D *p*-Ritz	9.9909	20.297	32.430	36.744	46.140	54.308	61.186	67.969
0.3	(6,3,1)	8.6966	15.976	22.771	23.502	26.162	26.186	31.284	35.460
	(8,5,2)	8.4897	15.497	22.739	22.741	25.237	26.176	30.209	34.387
	(15,12,2)	8.4804	15.485	22.725	22.727	25.221	26.176	30.185	34.360
	(15,12,3)	8.4690	15.458	22.678	22.722	25.163	26.176	30.110	34.261
	(17,14,5)	8.4636	15.447	22.661	22.717	25.142	26.176	30.086	34.232
	3D Ritz	8.4606	15.442	22.654	22.715	25.134	26.176	30.078	34.222
	3D *p*-Ritz	8.4676	15.453	22.667	22.721	25.150	–	30.093	34.239
0.5	(6,3,2)	6.8497	11.576	13.683	15.712	16.352	17.992	21.229	21.234
	(8,5,3)	6.8209	11.520	13.665	15.705	16.261	17.862	21.073	21.201
	(10,7,4)	6.8108	11.503	13.661	15.705	16.236	17.833	21.037	21.194
	(12,9,5)	6.8080	11.499	13.658	15.705	16.232	17.827	21.032	21.188
	(14,11,7)	6.8052	11.495	13.656	15.705	16.227	17.822	21.027	21.185
	3D Ritz	6.8027	11.492	13.654	15.705	16.224	17.819	21.024	21.183
	3D *p*-Ritz	6.8068	11.497	13.658	–	16.229	17.825	21.029	21.188

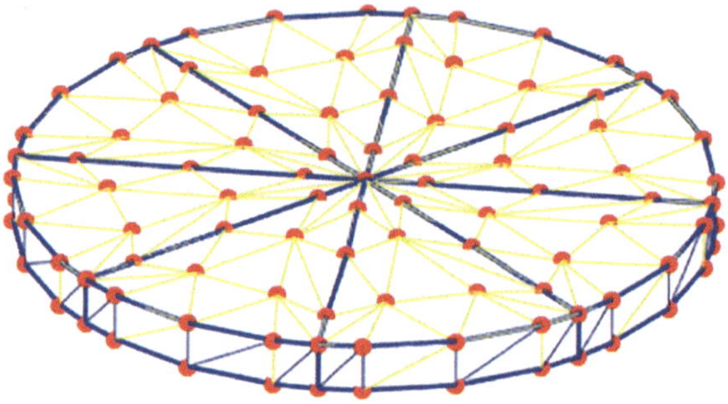

Fig. 4.41 Circular plate discretized into 8 triangular prism (wedge) elements

Table 4.6 Comparison of frequency parameters for clamped square thin cylindrical shells $\Omega = \omega a\sqrt{(\rho/E)}$, $\upsilon = 0.3$, $a/b = 1$, $h/b = 0.1$, $b/Ry = 0.5$

(N_e, N_f, N_T)	Mode number							
	SS-1	SS-2	SA-1	SA-2	AS-1	AS-2	AA-1	AA-2
(6, 3, 1)	1.0973	3.2408	1.9847	3.7402	1.9323	3.7566	2.7538	4.4477
(8, 5, 2)	1.0841	3.1466	1.9471	3.7252	1.8936	3.7492	2.6722	4.4463
(10, 7, 3)	1.0828	3.1419	1.9446	3.7229	1.8910	3.7476	2.6686	4.4461
(12, 9, 5)	1.0822	3.1395	1.9432	3.7219	1.8897	3.7470	2.6667	4.4460
HOSDT	1.0822	3.1337	1.9376	3.7110	1.8888	3.7355	2.6608	4.4441

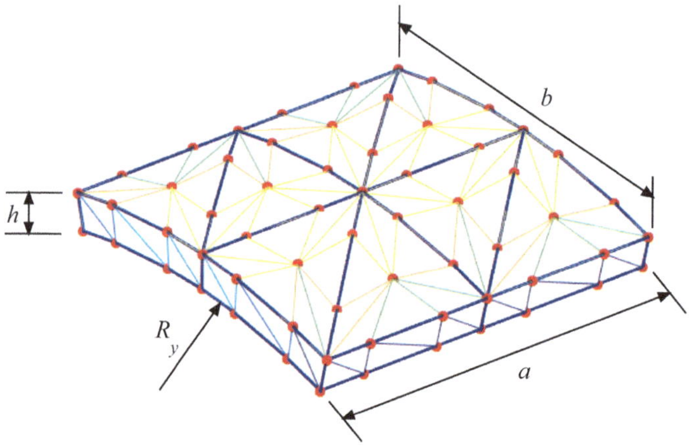

Fig. 4.42 Thin cylindrical shell discretized into 8 wedge-shaped elements

two triangular prism (wedge) elements, one pyramid element, and three hexahedral elements. The number of nodes on the edges, faces, and volume of each element are denoted by $[n_e, n_f, n_h]$. The material constants are density $\rho = 7722.7 \text{kg/m}^3$, Young's modulus $E = 190$ GPa, and Poisson's ratio $\upsilon = 0.3$. As can be seen from the results in Table 4.7, the hierarchical quadrature element method still exhibits rapid

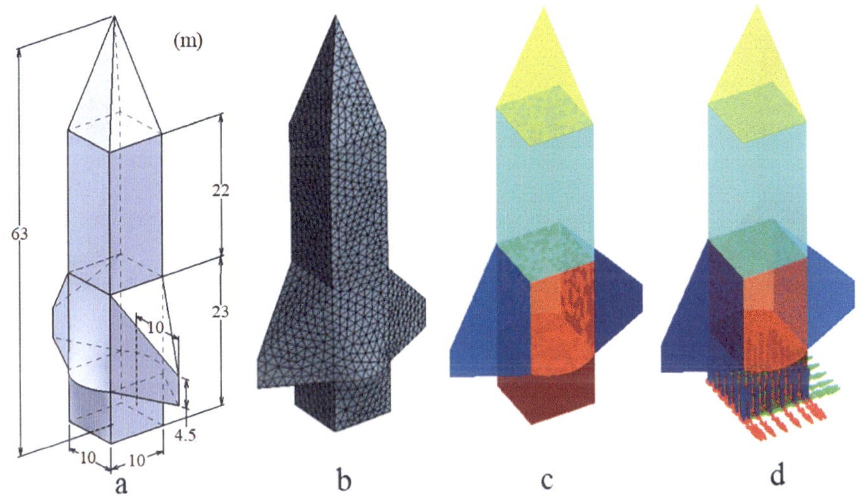

Fig. 4.43 Rocket model and its dimensions (**a**), its FEM mesh, HQEM mesh (**c**) and boundary conditions (**d**)

convergence and agrees well with the ANSYS results. Figure 4.44 shows the vibration modes obtained from the hierarchical quadrature element method and ANSYS. It can be observed that the modes obtained from the hierarchical quadrature element method align closely with those from ANSYS.

4.4 Discussion on Gaussian Integrals

When applying the Gaussian quadrature method, it is essential to determine the appropriate number of Gauss points to use. Theoretically, for one-dimensional integration, using m points yields exact integration for polynomials of up to degree $2m-1$. Generally, a greater number of points should be used for higher-order elements. It should also be noted that using fewer Gauss points can often compensate for the "overly stiff" behavior of the displacement-based finite element method.

This excessive stiffness in the displacement finite element method is primarily due to the use of shape functions. As discussed in previous sections, the displacement within a element is assumed by interpolating the nodal displacements using shape functions. This means that the deformation of the element is actually prescribed by the form of the shape functions. As a result, the element is subject to a constraint, making its behavior stiffer than in reality. It is generally observed that higher-order elements are "softer" than lower-order elements, as the use of more nodes reduces the constraints imposed on the element.

In most cases, linear elements require two Gauss points, while quadratic elements generally need about two or three Gauss points in each direction. For higher-order

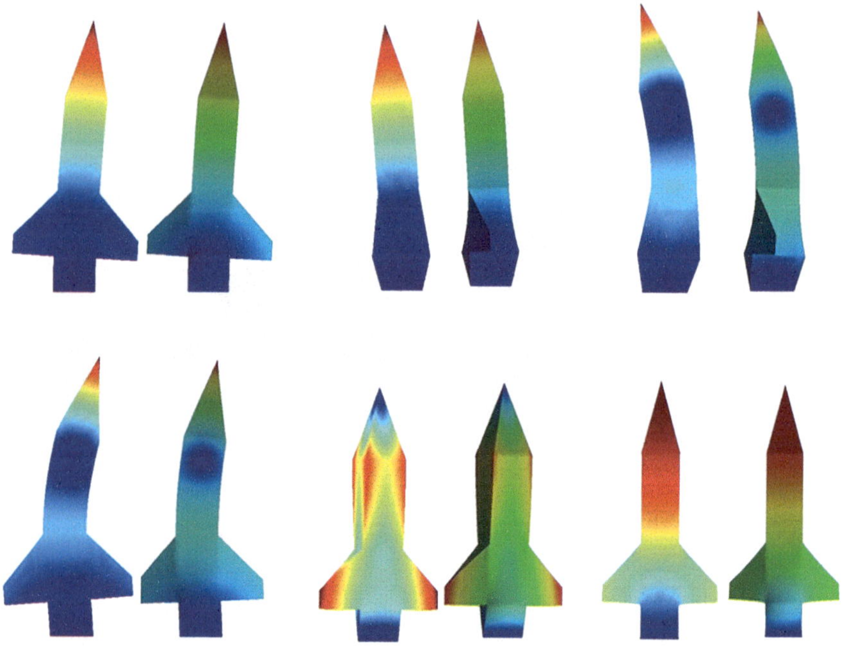

Fig. 4.44 Vibration modes obtained by the hierarchical quadrature element method (left) and the finite element method (right)

Table 4.7 Frequencies of the rocket model with a clamped boundary condition at the base

Method	Element order	Frequency order					
		1	2	3	4	5	6
HQEM	(2,2,1)	3.5784	3.7206	15.6509	15.6810	16.4131	25.9542
	(3,3,2)	3.4761	3.6069	15.0136	15.1583	15.6175	25.7387
	(4,4,3)	3.4463	3.5848	14.9414	15.0896	15.5484	25.6247
	(5,5,4)	3.4338	3.5760	14.9059	15.0647	15.4911	25.5906
	(6,6,5)	3.4279	3.5724	14.8935	15.0531	15.4787	25.5643
	(9,9,8)	3.4244	3.5700	14.8843	15.0465	15.4664	25.5545
	(10,10,9)	3.4222	3.5688	14.8802	15.0425	15.4626	25.5446
	(13,13,12)	3.4179	3.5664	14.8712	15.0352	15.4530	25.5301
	(14,14,13)	3.4176	3.5662	14.8707	15.0347	15.4525	25.5286
	Number of elements						
Finite element	17,683	3.4195	3.5061	14.8782	15.0839	15.4604	25.4155
	24,060	3.419	3.505	14.8737	15.0832	15.4594	25.413

elements, the number of Gauss points depends on the order of the basis functions used. Many explicit dynamic finite element codes tend to use single-point integration to maximize computational efficiency.

4.5 Summary

This chapter presents the finite element analysis of two-dimensional and three-dimensional solid elements. All elements are C^{0} continuous, and therefore share certain similarities. The construction of linear and higher-order triangular elements is discussed in detail, as these procedures form the foundation for developing other two- and three-dimensional elements. Theoretically, quadrilateral and hexahedral elements are the most efficient, accurate, and straightforward elements for two- and three-dimensional analyses, respectively. Therefore, these two types of elements should be used whenever possible for 2D and 3D analyses. However, to date, the technology for generating such meshes has not been fully developed. Commercial software currently only supports the automatic generation of triangular and tetrahedral meshes. As a result, to improve computational efficiency, it may be necessary to use hexahedron-dominated meshes supported by some software. For such meshes, in addition to tetrahedral and hexahedral elements, transition elements such as triangular prisms (wedges) and pyramids are also essential. Unlike other elements with many symmetries, the pyramid element is quite unique and requires rational vertex shape functions. Once the shape functions are constructed, the remaining procedures of the finite element method are more or less similar. In addition, to demonstrate the performance of the elements, several numerical examples for various elements are also provided.

Chapter 5
Finite Element Analysis of Plates and Solid Shells

Plates and shells are merely special forms of three-dimensional solids, and there are no theoretical difficulties in treating them within the framework of elasticity. However, since the thickness of such structures is very small compared to their other dimensions, employing a full three-dimensional numerical approach is not only costly but also frequently leads to severe numerical ill-conditioning. To simplify the analysis, classical assumptions regarding the behavior of these structures were introduced even before the advent of numerical methods. Clearly, such assumptions result in a series of approximations. Therefore, in general, numerical analysis concerns itself with approximating an already simplified theory (or mathematical model), and the validity of these approximations is inherently limited. Sometimes, we point out the shortcomings of the original assumptions and modify them as needed or as convenient. This is no longer difficult, as we now have much greater flexibility than in the pre-computer era.

Plate theory is based on the assumptions proposed by Kirchhoff in 1850. In fact, his name is often associated with this theory, although its earliest form was introduced by Sophie Germain in 1811. Reissner extended these assumptions in 1945, and Mindlin presented a slightly different formulation of the theory in 1951. These revised theories broadened the applicability of plate theory to thick plates, and as a result, the theory for thick plates is often associated with the Reissner–Mindlin assumptions.

It has been shown that, although thick plate theory posed considerable challenges in early analytical treatments, its implementation within the finite element method is relatively straightforward. While it may be more convenient to first introduce thick plate theory and then, through additional assumptions, restrict it to thin plate theory, for the sake of discussing numerical solutions, we will reverse this process. We will begin by presenting methods for analyzing thin plates, then extend these methods to thick plates, and finally introduce mixed methods.

In thick plate theory, deformation is determined by the transverse displacement w of the mid-surface and the rotational variable φ. The strains resulting from displacement and rotational variables involve only first-order derivatives, allowing the use

B. Liu and C. Liu, *Introduction to Linear and Nonlinear Finite Element Analysis*,
https://doi.org/10.1007/978-981-95-9670-6_5

of the previously introduced C^0 elements. However, it is found that thick plates are prone to locking issues, which necessitates careful selection of shape functions and unknown variables.

Similar to the Euler–Bernoulli beam element, the Kirchhoff plate element requires the deflection to possess C^1 continuity due to the presence of second-order derivatives of deflection in the virtual work expression. However, unlike beam elements, it is more challenging for Kirchhoff plate elements to satisfy the continuity requirements between elements. This has led to the development of "nonconforming" elements, some of which are still used in practical applications today. Kirchhoff plate elements are included in most commercial codes and continue to evolve with the introduction of new concepts.

Essentially, a shell can be regarded as a plate structure whose mid-surface is curved. The assumptions regarding the transverse distribution of strain and stress used for thin plates remain valid for shells. However, the manner in which shells support external loads differs significantly from that of flat plates. In shells, the resultant stresses acting on the mid-surface include both tangential and normal components, both of which bear the majority of the load. This load-bearing mechanism and its efficiency have led to the widespread application of shells in engineering structures.

There are numerous challenges in deriving the detailed governing equations for shell problems. As a result, the use of different approximation methods has led to the existence of various shell formulations. In early approaches to shell elements, these difficulties were addressed by introducing further approximations. However, these approximations are physical in nature rather than mathematical. This method assumes that the properties of a continuous curved surface can be adequately approximated by the properties of many small planar elements. Intuitively, as the subdivision becomes finer, convergence seems inevitable, and this has indeed been confirmed by experience.

This chapter presents the energy function of curved shells in orthogonal curvilinear coordinate systems. Based on this energy function, methods similar to those used for plate elements can be applied to obtain the shell element in the corresponding coordinate system. This chapter will not discuss general thin shell elements, but will introduce a general thick curved shell element (directly based on three-dimensional elasticity theory and avoiding the abstraction of shell equations).

As discussed in Chap. 4, iso-parametric curved two-dimensional and three-dimensional elements are particularly effective in solid analysis. Clearly, in shell analysis, these elements can be utilized by simply reducing their dimensions in the thickness direction. However, directly applying the aforementioned three-dimensional concepts presents certain difficulties.

First, retaining three displacement degrees of freedom at each node results in excessive stiffness corresponding to strains in the shell thickness direction. This may lead to numerical issues, and when the shell thickness is much smaller than the other dimensions of the element, it can cause the resulting algebraic equations to become ill-conditioned.

Secondly, there is the issue of computational efficiency. Using multiple nodes for shell thickness overlooks an important fact: even for thick shells, the "normal" to the mid-surface remains essentially straight after deformation. Therefore, employing thick shell elements introduces a large number of unnecessary degrees of freedom, resulting in additional computational effort.

In this chapter, we present a thick shell element that overcomes these two challenges. By neglecting the strain components corresponding to the stress components perpendicular to the mid-surface, numerical performance is improved, and the introduction of the "straight normal" constraint enhances computational efficiency. Through these modifications, an effective thick shell element is achieved.

5.1 Element Energy Functional

5.1.1 *Thin Plate Bending Element*

As shown in Fig. 5.1, a thin plate with uniform thickness h is subjected to a surface distributed load, with the load acting perpendicular to the mid-plane ($z = 0$).

In deriving the energy function for thin plates, it is assumed that the transverse normal stress σ_z is zero; the normal to the mid-plane of the undeformed plate remains straight, perpendicular to the mid-plane, and inextensible during deformation. Therefore, the displacement parallel to the undeformed mid-plane can be expressed as

$$\begin{aligned} u(x, y, z) &= -z\frac{\partial w}{\partial x} \\ v(x, y, z) &= -z\frac{\partial w}{\partial y} \end{aligned} \tag{5.1}$$

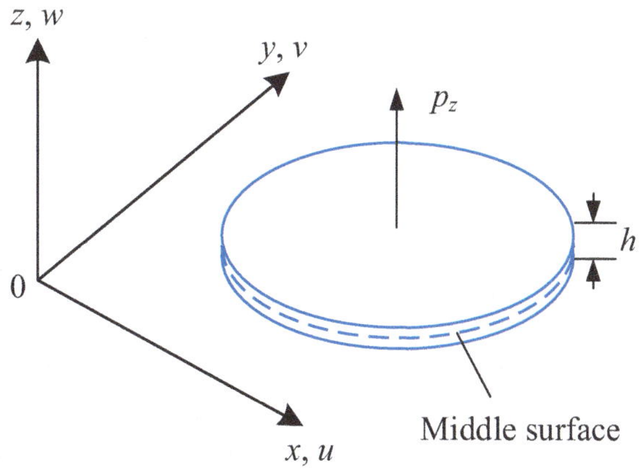

Fig. 5.1 Plate bending element

Here, $w(x, y)$ is the displacement function of the mid-surface in the z-direction. The strain components can be expressed as

$$\begin{aligned}
\varepsilon_x &= \frac{\partial u}{\partial x} = -z\frac{\partial^2 w}{\partial x^2} \\
\varepsilon_y &= \frac{\partial v}{\partial y} = -z\frac{\partial^2 w}{\partial y^2} \\
\gamma_{xy} &= \frac{\partial u}{\partial y} + \frac{\partial v}{\partial x} = -2z\frac{\partial^2 w}{\partial x \partial y} \\
\gamma_{xz} &= \frac{\partial u}{\partial z} + \frac{\partial w}{\partial x} = 0, \\
\gamma_{yz} &= \frac{\partial v}{\partial z} + \frac{\partial w}{\partial y} = 0
\end{aligned} \tag{5.2}$$

Since σ_z, γ_{xz} and γ_{yz} are all zero, the strain energy stored in the element can be expressed as

$$U = \frac{1}{2}\int_V \left(\sigma_x\varepsilon_x + \sigma_y\varepsilon_y + \tau_{xy}\gamma_{xy}+\right)\mathrm{d}V \tag{5.3}$$

This is identical to the strain energy Eq. (4.1) for the membrane element. Equation (5.3) can be expressed in matrix form as

$$U = \frac{1}{2}\int_V \{\boldsymbol{\sigma}\}^{\mathrm{T}}\{\boldsymbol{\varepsilon}\}\mathrm{d}V \tag{5.4}$$

Here, $\{\boldsymbol{\sigma}\}$ and $\{\boldsymbol{\varepsilon}\}$ are defined by Eq. (4.3).

Since $\sigma_z = 0$, the stress–strain relationship takes the following form.

$$\{\boldsymbol{\sigma}\} = [\mathbf{D}]\{\boldsymbol{\varepsilon}\} \tag{5.5}$$

Here, $[\mathbf{D}]$ is defined by Eq. (4.5).

Substituting Eqs. (5.5) into (5.4) yields

$$U = \frac{1}{2}\int_V \{\boldsymbol{\varepsilon}\}^{\mathrm{T}}[\mathbf{D}]\{\boldsymbol{\varepsilon}\}\mathrm{d}V \tag{5.6}$$

According to Eq. (5.2), the strain matrix can be expressed as

$$\{\boldsymbol{\varepsilon}\} = -z\{\boldsymbol{\chi}\} \tag{5.7}$$

where

$$\{\chi\} = \begin{Bmatrix} \frac{\partial^2 w}{\partial x^2} \\ \frac{\partial^2 w}{\partial y^2} \\ 2\frac{\partial^2 w}{\partial x \partial y} \end{Bmatrix} = \boldsymbol{L}w \tag{5.8}$$

Substituting Eqs. (5.7) into (5.6) and integrating with respect to z yields:

$$U = \frac{1}{2}\int_A \frac{h^3}{12}\{\chi\}^{\mathrm{T}}[\mathbf{D}]\{\chi\}\mathrm{d}A \tag{5.9}$$

The kinetic energy of the plate is

$$T = \frac{1}{2}\int_A \rho h \dot{w}^2 \mathrm{d}A \tag{5.10}$$

The virtual work of the transverse load is

$$\delta W = \int_A p_z \delta w \mathrm{d}A \tag{5.11}$$

As shown in Fig. 5.2, a representative plate differential element with thickness h and dimensions $\mathrm{d}x \times \mathrm{d}y$ is subjected to an external force f_z and an inertial force $\rho h\, \ddot{w}$. Here, ρ denotes the material density. As shown in Fig. 5.3, the internal moments M_x, M_y, M_z and M_{xy}, as well as the shear forces Q_x and Q_y, act on the plate element. The moments and shear forces are generated by the distributed normal and shear stresses σ_{xx}, σ_{yy} and σ_{xy}, as illustrated in Fig. 5.2. The normal stress on the cross-section varies linearly along the thickness direction. The moments on the cross-section can be calculated by the following integrals, which is similar to the method used for beams.

$$\{\mathbf{M}_p\} = \begin{Bmatrix} M_x \\ M_y \\ M_{xy} \end{Bmatrix} = \int_A \sigma z \mathrm{d}A = -\frac{h^3}{12}[\mathbf{D}]\{\chi\} \tag{5.12}$$

First, consider the equilibrium of the plate differential element in the z-direction. Noting that $Q_x = (\partial Q_x/\partial x)\mathrm{d}x$ and $\mathrm{d}Q_y = (\partial Q_y/\partial y)\mathrm{d}y$, we obtain:

$$\left(\frac{\partial Q_x}{\partial x}\mathrm{d}x\right)\mathrm{d}y + \left(\frac{\partial Q_y}{\partial y}\mathrm{d}y\right)\mathrm{d}x + (p_z - \rho h \ddot{w})\mathrm{d}x\mathrm{d}y = 0 \tag{5.13}$$

or

$$\frac{\partial Q_x}{\partial x} + \frac{\partial Q_y}{\partial y} + p_z = \rho h \ddot{w} \tag{5.14}$$

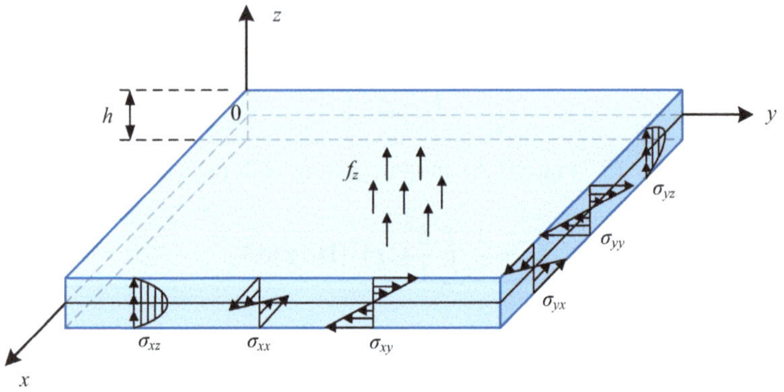

Fig. 5.2 Stresses on a plate differential element. Integration of these stresses yields the corresponding moments and shear forces

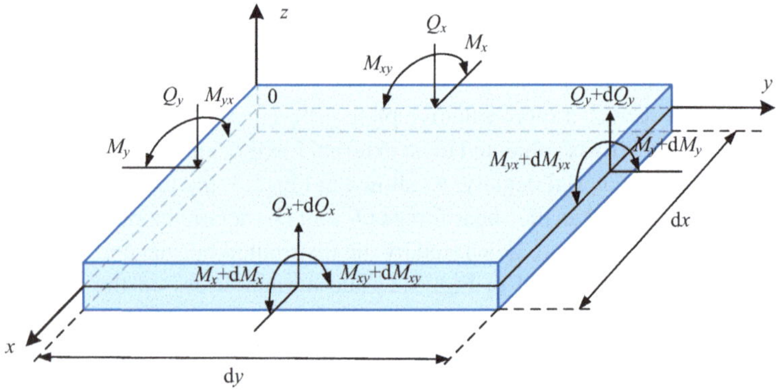

Fig. 5.3 Shear forces and moments on the plate differential element dx × dy. The system equilibrium equations are established based on the above force and moment states

Then, consider the moment equilibrium of the plate element about the x-axis. By neglecting the second-order infinitesimal terms, the formula for the shear force Q_x is derived:

$$Q_x = \frac{\partial M_x}{\partial x} + \frac{\partial M_{xy}}{\partial y} \tag{5.15}$$

Finally, considering the moment equilibrium of the plate element about the y-axis and neglecting second-order infinitesimals, we obtain

$$Q_y = \frac{\partial M_{xy}}{\partial x} + \frac{\partial M_y}{\partial y} \tag{5.16}$$

Here, $M_{yx} = M_{xy}$.

The dynamic equilibrium equation for the microelement is obtained by first substituting Eqs. (5.12) into (5.15) and (5.16), respectively. Substituting the obtained Q_x and Q_y into Eq. (5.14), we have

$$D\left(\frac{\partial^4 w}{\partial x^4} + \frac{\partial^4 w}{\partial x^2 \partial y^2} + \frac{\partial^4 w}{\partial y^4}\right) + \rho h\ddot{w} = p_z \tag{5.17}$$

Here, $D = Eh^3/(12(1 - \upsilon_2))$ denotes the flexural rigidity of the plate.

The irreducible weak form of the thin plate expressed in terms of w is

$$\begin{aligned} \mathrm{G}(\delta w, w) = & \int_\Omega \delta w\left[\rho h\ddot{w} - q\right]\mathrm{d}\Omega + \int_\Omega (\boldsymbol{L}\delta w)^{\mathrm{T}}\mathbf{D}\boldsymbol{L}w\mathrm{d}\Omega \\ & + \int_{\Gamma_n} \frac{\partial \delta w}{\partial n} M_n \mathrm{d}\Gamma - \int_{\Gamma_q} \delta w V_n \mathrm{d}\Gamma - \sum_i \delta w_i R_i \end{aligned} \tag{5.18}$$

Here, n denotes the normal direction, s denotes the tangential direction, and

$$V_n = Q_n + \frac{\partial M_{ns}}{\partial s} \tag{5.19}$$

represents the "equivalent boundary shear force", and R_i denotes the concentrated force at the corner position. The total potential energy for a static problem is

$$\begin{aligned} \Pi = & \frac{1}{2}\int_\Omega (\boldsymbol{L}w)^{\mathrm{T}}\mathbf{D}\boldsymbol{L}w\mathrm{d}\Omega - \int_\Omega wq\mathrm{d}\Omega \\ & + \int_{\Gamma_n} \frac{\partial w}{\partial n} M_n \mathrm{d}\Gamma - \int_{\Gamma_q} wV_n \mathrm{d}\Gamma - \sum_i w_i R_i \end{aligned} \tag{5.20}$$

5.1.2 Thick Plate Bending Element

Similar to the case of shear beams, when the local bending wavelength is less than ten times the plate thickness, shear deformation and rotary inertia effects must be taken into account. When shear deformation is significant, it can no longer be assumed that the normal to the mid-surface remains perpendicular to the mid-surface. In this case, the displacement parallel to the mid-surface can be expressed as

$$u_x = z\phi_x(x, y, t) \quad u_y = z\phi_y(x, y, t) \quad u_z = w(x, y, t) \tag{5.21}$$

Note: the displacement parameters are now functions of x and y.

It is sometimes convenient to replace the rotations θ_x and θ_y about the x and y coordinates with ϕ_x and ϕ_y themselves. This transformation is accomplished by the following relations:

$$\begin{Bmatrix} \phi_x \\ \phi_y \end{Bmatrix} = \begin{bmatrix} 0 & 1 \\ -1 & 0 \end{bmatrix} \begin{Bmatrix} \theta_x \\ \theta_y \end{Bmatrix} \; or \; \begin{Bmatrix} \theta_x \\ \theta_y \end{Bmatrix} = \begin{bmatrix} 0 & -1 \\ 1 & 0 \end{bmatrix} \begin{Bmatrix} \phi_x \\ \phi_y \end{Bmatrix} \tag{5.22}$$

or

$$\boldsymbol{\phi} = \mathbf{T}\boldsymbol{\theta} \tag{5.23}$$

Equation (5.22) can be substituted at any point during the derivation. However, this substitution is generally not performed during the derivation, as it would making the expression of the control equation more complex in form will only lead to the substitution of nodal parameters after the shape functions have been constructed.

The strain can now be decomposed into bending and transverse shear components. The in-plane bending strain can be expressed as

$$\{\boldsymbol{\varepsilon}\} = \begin{Bmatrix} \varepsilon_x \\ \varepsilon_y \\ \gamma_{xy} \end{Bmatrix} = z \begin{bmatrix} \frac{\partial}{\partial x} & 0 \\ 0 & \frac{\partial}{\partial y} \\ \frac{\partial}{\partial y} & \frac{\partial}{\partial x} \end{bmatrix} \begin{Bmatrix} \phi_x \\ \phi_y \end{Bmatrix} \equiv z\boldsymbol{S}\boldsymbol{\phi} = z\{\boldsymbol{\chi}\} \tag{5.24}$$

where

$$\{\boldsymbol{\chi}\} = \begin{Bmatrix} \frac{\partial \phi_x}{\partial x} \\ \frac{\partial \phi_y}{\partial y} \\ \frac{\partial \phi_x}{\partial y} + \frac{\partial \phi_y}{\partial x} \end{Bmatrix} \tag{5.25}$$

The relationships between the transverse shear strains γ_{xz}, γ_{yz} and displacement are given by

$$\boldsymbol{\gamma} = \begin{Bmatrix} \gamma_{xz} \\ \gamma_{yz} \end{Bmatrix} = \begin{Bmatrix} \frac{\partial w}{\partial x} \\ \frac{\partial w}{\partial y} \end{Bmatrix} + \begin{Bmatrix} \phi_x \\ \phi_y \end{Bmatrix} = \nabla w + \boldsymbol{\phi} \tag{5.26}$$

It should be noted that when transverse shear can be neglected, Eq. (5.26) yields:

$$\nabla w + \boldsymbol{\phi} = \mathbf{0} \quad or \quad \boldsymbol{\phi} = -\nabla w \tag{5.27}$$

With these relationships, it can be seen that Eq. (5.24) can be simplified to Eq. (5.7).

The equilibrium equations for Mindlin plates can also be derived using a method similar to that for thin plate equilibrium equations. For isotropic elastic materials, the moments M_x, M_y and M_{xy}, as well as the shear forces Q_x and Q_y, can be expressed in terms of displacements as follows:

$$
\begin{aligned}
M_x &= \int_{-\frac{h}{2}}^{\frac{h}{2}} z\sigma_x \mathrm{d}z = D\left(\frac{\partial \phi_x}{\partial x} + \upsilon\frac{\partial \phi_y}{\partial y}\right) \\
M_y &= \int_{-\frac{h}{2}}^{\frac{h}{2}} z\sigma_y \mathrm{d}z = D\left(\frac{\partial \phi_y}{\partial y} + \upsilon\frac{\partial \phi_x}{\partial x}\right) \\
M_{xy} &= \int_{-\frac{h}{2}}^{\frac{h}{2}} z\tau_{xy} \mathrm{d}z = \frac{1}{2}(1-\upsilon)D\left(\frac{\partial \phi_x}{\partial y} + \frac{\partial \phi_y}{\partial x}\right)
\end{aligned} \tag{5.28}
$$

$$
\begin{aligned}
Q_x &= \int_{-\frac{h}{2}}^{\frac{h}{2}} \tau_{xz} \mathrm{d}z = C\left(\frac{\partial w}{\partial x} + \phi_x\right) \\
Q_y &= \int_{-\frac{h}{2}}^{\frac{h}{2}} \tau_{yz} \mathrm{d}z = C\left(\frac{\partial w}{\partial y} + \phi_y\right)
\end{aligned} \tag{5.29}
$$

Here, $D = Eh^3/12(1\text{-}\upsilon^2)$ and $C = \kappa Gh$ represent the bending stiffness and shear stiffness of the plate, respectively. Equation (5.28) and (5.29) can also be expressed as

$$
\mathbf{M} = \begin{Bmatrix} M_x \\ M_y \\ M_{xy} \end{Bmatrix} = \mathbf{DS}\boldsymbol{\phi} \tag{5.30}
$$

$$
\mathbf{Q} = \begin{Bmatrix} Q_x \\ Q_y \end{Bmatrix} = \boldsymbol{\alpha}(\nabla w + \boldsymbol{\phi}) \tag{5.31}
$$

where

$$
\boldsymbol{\alpha} = \kappa Gh\mathbf{I} \tag{5.32}
$$

Here, I is a 2 × 2 identity matrix (G is not associated with E or υ here, allowing for different possible shear stiffnesses). The introduction of the constant κ is to account for the variation of shear stress and strain in the thickness direction. Typically, κ is taken as $\pi^2/12$ or 5/6.

Of course, the constitutive relations described above can also be readily extended to anisotropic or non-homogeneous materials. For example, several layers of materials can be symmetrically assembled to form a composite material. The only difference introduced in this case lies in the structure of the **D** matrix and the $\boldsymbol{\alpha}$ matrix, both of which can be obtained through straightforward integration.

The dynamic governing equations for thick and thin plates can be derived from equilibrium relations. By neglecting in-plane properties, Eq. (5.14) can be rewritten as

$$\begin{bmatrix} \frac{\partial}{\partial x} & \frac{\partial}{\partial y} \end{bmatrix} \begin{Bmatrix} Q_x \\ Q_y \end{Bmatrix} + q \equiv \nabla^{\mathrm{T}} \mathbf{Q} + q = \rho h \ddot{w} \tag{5.33}$$

Use Eq. (5.33) to replace Eqs. (5.15) and (5.16), we obtain

$$\begin{bmatrix} \frac{\partial}{\partial x} & 0 & \frac{\partial}{\partial y} \\ 0 & \frac{\partial}{\partial y} & \frac{\partial}{\partial x} \end{bmatrix} \begin{Bmatrix} M_x \\ M_y \\ M_{xy} \end{Bmatrix} - \begin{Bmatrix} Q_x \\ Q_y \end{Bmatrix} = \frac{1}{12} \rho h^3 \begin{Bmatrix} \ddot{\phi}_x \\ \ddot{\phi}_y \end{Bmatrix} \tag{5.34}$$

or

$$\mathbf{S}^{\mathrm{T}} \mathbf{M} - \mathbf{Q} = \frac{1}{12} \rho h^3 \ddot{\boldsymbol{\phi}} \tag{5.35}$$

Equations (5.30) to (5.35) form the basis for solving thick and thin plates. For thick plates, any (or all) independent variables can be solved independently, which leads to mixed methods that will be discussed in detail later.

Substituting Eqs. (5.28) and (5.29) into (5.33) and (5.34), we obtain

$$\begin{aligned} &\frac{\partial^2 \phi_x}{\partial x^2} + \frac{1-\upsilon}{2} \frac{\partial^2 \phi_x}{\partial y^2} + \frac{1+\upsilon}{2} \frac{\partial^2 \phi_y}{\partial x \partial y} - \frac{C}{D} \left(\frac{\partial w}{\partial x} + \phi_x \right) - \frac{\rho h^3}{12D} \frac{\partial^2 \phi_x}{\partial t^2} = 0 \\ &\frac{\partial^2 \phi_y}{\partial y^2} + \frac{1-\upsilon}{2} \frac{\partial^2 \phi_y}{\partial x^2} + \frac{1+\upsilon}{2} \frac{\partial^2 \theta_x}{\partial x \partial y} - \frac{C}{D} \left(\frac{\partial w}{\partial y} + \phi_y \right) - \frac{\rho h^3}{12D} \frac{\partial^2 \phi_y}{\partial t^2} = 0 \\ &\frac{\partial^2 w}{\partial x^2} + \frac{\partial^2 w}{\partial y^2} - \left(\frac{\partial \phi_x}{\partial x} + \frac{\partial \phi_y}{\partial y} \right) + q - \frac{\rho h}{C} \frac{\partial^2 w}{\partial t^2} = 0 \end{aligned} \tag{5.36}$$

The exact natural frequency parameter for a Mindlin plate is

$$\frac{\omega^2 \rho h}{D} = \frac{\pi^4 \left[\left(\frac{m}{a} \right)^2 + \left(\frac{n}{b} \right)^2 \right]^2}{1 + \left(\frac{D}{C} + \frac{h^2}{12} \right) \left[\left(\frac{m\pi}{a} \right)^2 + \left(\frac{n\pi}{b} \right)^2 \right] - \frac{D}{C} \frac{h^2}{12} \left(\frac{\omega^2 \rho h}{D} \right)} \tag{5.37}$$

The exact natural frequency parameter for a Kirchhoff thin plate is

$$\frac{\omega^2 \rho h}{D} = \pi^4 \left[\left(\frac{m}{a} \right)^2 + \left(\frac{n}{b} \right)^2 \right]^2 \tag{5.38}$$

A precise comparison with the Mindlin plate theory reveals that the Kirchhoff thin plate theory may overestimate the natural frequencies.

By combining Eqs. (5.33), (5.35), and (5.31), the moment **M** can be eliminated, yielding

$$\nabla^{\mathrm{T}} \mathbf{Q} + q = \rho h \ddot{w}$$

$$\begin{aligned} & \boldsymbol{S}^{\mathrm{T}}\mathbf{D}\boldsymbol{S}\boldsymbol{\phi} - \mathbf{Q} = \frac{1}{12}\rho h^3\ddot{\boldsymbol{\phi}} \\ & \boldsymbol{\alpha}^{-1}\mathbf{Q} - \boldsymbol{\phi} - \nabla w = \mathbf{0} \end{aligned} \tag{5.39}$$

This system of equations can serve as the foundation for constructing mixed methods, or it can be further simplified to obtain an irreducible form. In Sect. 5.2.1, we will discuss the fourth-order irreducible equation expressed solely in terms of w, which can only be used to solve thin plate problems, i.e., when $\boldsymbol{\alpha} = \infty$. Additionally, another irreducible form can be readily derived, which holds only when $\boldsymbol{\alpha} \neq \infty$. Therefore, eliminating the shear force yields:

$$\begin{aligned} & \nabla^{\mathrm{T}}[\boldsymbol{\alpha}(\nabla w + \boldsymbol{\phi})] + q = \rho h\ddot{w} \\ & \boldsymbol{S}^{\mathrm{T}}\mathbf{D}\boldsymbol{S}\boldsymbol{\phi} - \boldsymbol{\alpha}(\nabla w + \boldsymbol{\phi}) = \frac{1}{12}\rho h^3\ddot{\boldsymbol{\phi}} \end{aligned} \tag{5.40}$$

This is an irreducible system. For static problems, it is easy to verify the corresponding minimum potential energy principle, namely

$$\begin{aligned} \Pi = & \frac{1}{2}\int_{\Omega}(\boldsymbol{S}\boldsymbol{\phi})^{\mathrm{T}}\mathbf{D}\boldsymbol{S}\boldsymbol{\phi}\mathrm{d}\Omega + \frac{1}{2}\int_{\Omega}(\nabla w + \boldsymbol{\phi})^{\mathrm{T}}\boldsymbol{\alpha}(\nabla w + \boldsymbol{\phi})\mathrm{d}\Omega \\ & - \int_{\Omega} wq\mathrm{d}\Omega - \Pi_{bt} = \text{Stationary value} \end{aligned} \tag{5.41}$$

In the equation, the first term represents the bending strain energy, the second term is the shear strain energy, the third term is the work done by the transverse load, and the last term is the work done by the boundary forces, moments, and shear forces, that is,

$$\Pi_{bt} = \int_{\Gamma_n} \phi_n \overline{M}_n \mathrm{d}\Gamma + \int_{\Gamma_s} \phi_s \overline{M}_{ns} \mathrm{d}\Gamma + \int_{\Gamma_q} w\overline{Q}_n \mathrm{d}\Gamma \tag{5.42}$$

The kinetic energy of the plate can be expressed as

$$T = \frac{1}{2}\int_A \rho h\left(\dot{u}^2 + \dot{v}^2 + \dot{w}^2\right)\mathrm{d}A \tag{5.43}$$

Substituting u and v from Eq. (5.21), as well as $\mathrm{d}V = \mathrm{d}z{\cdot}\mathrm{d}A$ into Eq. (5.43), and integrating with respect to z, we obtain

$$T = \frac{1}{2}\int_A \rho\left(h\dot{w}^2 + \frac{h^3}{12}\dot{\phi}_x^2 + \frac{h^3}{12}\dot{\phi}_y^2\right)\mathrm{d}A \tag{5.44}$$

Clearly, the irreducible system given by Eq. (5.41) is valid only when $\boldsymbol{\alpha} \neq \infty$. Treating $\boldsymbol{\alpha}$ as a penalty factor, Eq. (5.41) can be interpreted as the potential energy solution of a "thin" plate, as given by Eq. (5.20), with the constraints specified in

Eq. (5.27) incorporated via a penalty function. As is intuitively evident in physics, the thin plate formulation represents a limiting case of the thick plate formulation.

Therefore, only when the discretized mixed formulation satisfies the necessary convergence conditions can the penalty function method yield satisfactory solutions.

5.1.3 Deep Shell Elements

For ease of understanding, this section will briefly introduce the concept of deep shell elements based on orthogonal curvilinear coordinate systems. Practical shell elements used as special cases in 3D analysis will be discussed in Sect. 5.4. Assuming that the normal to the midsurface remains straight during deformation but is not necessarily perpendicular to the midsurface, the displacement can be expressed as follows:

$$\begin{aligned} u_1(\alpha, \beta, \gamma) &= u(\alpha, \beta) + \gamma \psi_\alpha(\alpha, \beta) \\ u_2(\alpha, \beta, \gamma) &= v(\alpha, \beta) + \gamma \psi_\beta(\alpha, \beta) \\ u_3(\alpha, \beta, \gamma) &= w(\alpha, \beta) \end{aligned} \tag{5.45}$$

Here, u, v and w represent the mid-surface displacements of the shell, while ψ_α and ψ_β are the two mid-surface rotations $e\gamma = 0$. The third equation in Eq. (5.45) yields as

$$H_1 = A(1 + k_1\gamma), H_2 = B(1 + k_2\gamma) \tag{5.46}$$

The numbers k_1 and k_2 represent the principal curvatures of the mid-surface with respect to α and β, respectively; A and B denote the stretching of the mid-surface ($\gamma = 0$) with respect to α and β, respectively.

$$\begin{aligned} A &= \left[\left(\frac{\partial x}{\partial \alpha}\right)^2 + \left(\frac{\partial y}{\partial \alpha}\right)^2 + \left(\frac{\partial z}{\partial \alpha}\right)^2\right]^{\frac{1}{2}} \\ B &= \left[\left(\frac{\partial x}{\partial \beta}\right)^2 + \left(\frac{\partial y}{\partial \beta}\right)^2 + \left(\frac{\partial z}{\partial \beta}\right)^2\right]^{\frac{1}{2}} \end{aligned} \tag{5.47}$$

The strain at any point in the shell can be expressed in terms of the mid-surface strain and the change in curvature as follows:

$$e_1 = e_1^0 + \gamma\chi_1, \quad e_2 = e_2^0 + \gamma\chi_2, \quad e_{12} = e_{12}^0 + \gamma\chi_{12} \tag{5.48}$$

$$e_{23} = \frac{1}{H_2}\left(B\psi_\beta + \frac{\partial w}{\partial \beta} - B\frac{v}{R_2}\right)$$

$$e_{31} = \frac{1}{H_1}\left(A\psi_\alpha + \frac{\partial w}{\partial \alpha} - A\frac{u}{R_1}\right) \tag{5.49}$$

where

$$\left.\begin{aligned} e_1^0 &= \frac{1}{H_1}\left(\frac{\partial u}{\partial \alpha} + \frac{v}{H_2}\frac{\partial H_1}{\partial \beta} + A\frac{w}{R_1}\right) \\ e_2^0 &= \frac{1}{H_2}\left(\frac{\partial v}{\partial \beta} + \frac{u}{H_1}\frac{\partial H_2}{\partial \alpha} + B\frac{w}{R_2}\right) \\ e_{12}^0 &= \frac{H_2}{H_1}\frac{\partial}{\partial \alpha}\left(\frac{v}{H_2}\right) + \frac{H_1}{H_2}\frac{\partial}{\partial \beta}\left(\frac{u}{H_1}\right) \end{aligned}\right\} \tag{5.50}$$

$$\left.\begin{aligned} \chi_1 &= \frac{1}{H_1}\left(\frac{\partial \psi_\alpha}{\partial \alpha} + \frac{1}{H_2}\frac{\partial H_1}{\partial \beta}\psi_\beta\right) \\ \chi_2 &= \frac{1}{H_2}\left(\frac{\partial \psi_\beta}{\partial \beta} + \frac{1}{H_1}\frac{\partial H_2}{\partial \alpha}\psi_\alpha\right) \\ \chi_{12} &= \frac{H_2}{H_1}\frac{\partial}{\partial \alpha}\left(\frac{\psi_\beta}{H_2}\right) + \frac{H_1}{H_2}\frac{\partial}{\partial \beta}\left(\frac{\psi_\alpha}{H_1}\right) \end{aligned}\right\} \tag{5.51}$$

For the Kirchhoff–Love thin shell theory, we have $e_{23} = e_{31} = 0$. Then, from Eq. (5.49), we obtain:

$$\begin{aligned} \psi_\alpha &= \frac{u}{R_1} - \frac{1}{A}\frac{\partial w}{\partial \alpha} \\ \psi_\beta &= \frac{v}{R_2} - \frac{1}{B}\frac{\partial w}{\partial \beta} \end{aligned} \tag{5.52}$$

Here, $R_1 = 1/k_1$ and $R_2 = 1/k_2$ are the radii of curvature. Considering that

$$\frac{\gamma}{R_1} << 1, \quad \frac{\gamma}{R_2} << 1 \tag{5.53}$$

Replacing H_1 and H_2 with A and B respectively, we have

$$\left.\begin{aligned} e_1^0 &= \frac{1}{A}\frac{\partial u}{\partial \alpha} + \frac{v}{AB}\frac{\partial A}{\partial \beta} + \frac{w}{R_1} \\ e_2^0 &= \frac{1}{B}\frac{\partial v}{\partial \beta} + \frac{u}{AB}\frac{\partial B}{\partial \alpha} + \frac{w}{R_2} \\ e_{12}^0 &= \frac{B}{A}\frac{\partial}{\partial \alpha}\left(\frac{v}{B}\right) + \frac{A}{B}\frac{\partial}{\partial \beta}\left(\frac{u}{A}\right) \end{aligned}\right\} \tag{5.54}$$

$$\left.\begin{aligned}\chi_1 &= \frac{1}{A}\left(\frac{\partial \psi_\alpha}{\partial \alpha} + \frac{1}{B}\frac{\partial A}{\partial \beta}\psi_\beta\right)\\ \chi_2 &= \frac{1}{B}\left(\frac{\partial \psi_\beta}{\partial \beta} + \frac{1}{A}\frac{\partial B}{\partial \alpha}\psi_\alpha\right)\\ \chi_{12} &= \frac{B}{A}\frac{\partial}{\partial \alpha}\left(\frac{\psi_\beta}{B}\right) + \frac{A}{B}\frac{\partial}{\partial \beta}\left(\frac{\psi_\alpha}{A}\right)\end{aligned}\right\} \tag{5.55}$$

If we take Eq. (5.52) as

$$\psi_\alpha = -\frac{1}{A}\frac{\partial w}{\partial \alpha}, \quad \psi_\beta = -\frac{1}{B}\frac{\partial w}{\partial \beta} \tag{5.56}$$

This yields the Donnell thin shell theory, also known as the engineering shell theory.

The stress–strain relationship for isotropic materials is given by

$$\sigma_1 = \frac{E}{1-\upsilon^2}(e_1 + \upsilon e_2), \quad \sigma_2 = \frac{E}{1-\upsilon^2}(e_2 + \upsilon e_1), \quad \tau_{12} = G e_{12} \tag{5.57}$$

$$\tau_{13} = G e_{13}, \quad \tau_{23} = G e_{23} \tag{5.58}$$

Substituting Eqs. (5.48) into(5.57) yields

$$\begin{aligned}\sigma_1 &= \frac{E}{1-\upsilon^2}\left[\left(e_1^0 + \upsilon e_2^0\right) + \gamma(\chi_1 + \upsilon\chi_2)\right]\\ \sigma_2 &= \frac{E}{1-\upsilon^2}\left[\left(e_2^0 + \upsilon e_1^0\right) + \gamma(\chi_2 + \upsilon\chi_1)\right]\\ \tau_{12} &= \frac{E}{2(1+\upsilon)}\left(e_{12}^0 + \gamma\chi_{12}\right)\end{aligned} \tag{5.59}$$

The moments M_1, M_2, M_{21} and M_{12}, the shear forces Q_1 and Q_2, as well as the membrane forces N_1, N_2, N_{21} and N_{12}, can be expressed in terms of stresses as follows:

$$\begin{aligned}\begin{bmatrix} N_1 \\ N_{12} \\ Q_1 \end{bmatrix} &= \int_{-h/2}^{h/2} \begin{bmatrix} \sigma_1 \\ \tau_{12} \\ \tau_{13} \end{bmatrix}(1+\gamma/R_2)\mathrm{d}\gamma,\\ \begin{bmatrix} N_2 \\ N_{21} \\ Q_2 \end{bmatrix} &= \int_{-h/2}^{h/2} \begin{bmatrix} \sigma_2 \\ \tau_{12} \\ \tau_{23} \end{bmatrix}(1+\gamma/R_1)\mathrm{d}\gamma\\ \begin{bmatrix} M_1 \\ M_{12} \end{bmatrix} &= \int_{-h/2}^{h/2} \begin{bmatrix} \sigma_1 \\ \tau_{12} \end{bmatrix}(1+\gamma/R_2)\gamma\mathrm{d}\gamma,\end{aligned} \tag{5.60}$$

$$\begin{bmatrix} M_2 \\ M_{21} \end{bmatrix} = \int_{-h/2}^{h/2} \begin{bmatrix} \sigma_2 \\ \tau_{12} \end{bmatrix} (1 + \gamma / R_1) \gamma \mathrm{d}\gamma \tag{5.61}$$

The forces and moments above for the shell are shown in Figs. 5.4 and 5.5, respectively.

Substituting Eqs. (5.59) and (5.58) into (5.60) and (5.61), we obtain

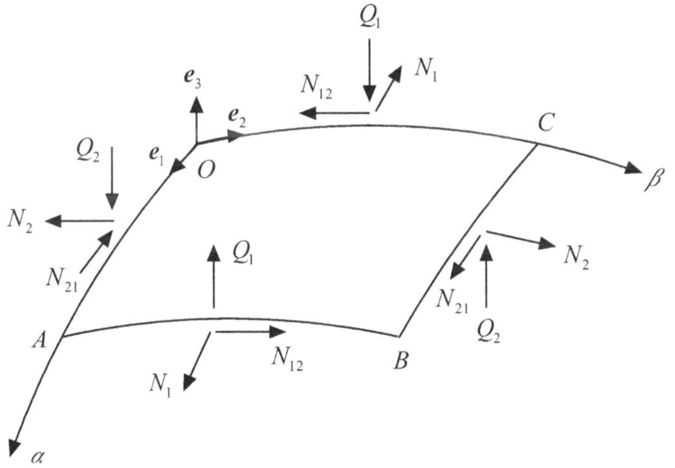

Fig. 5.4 Internal forces of the shell

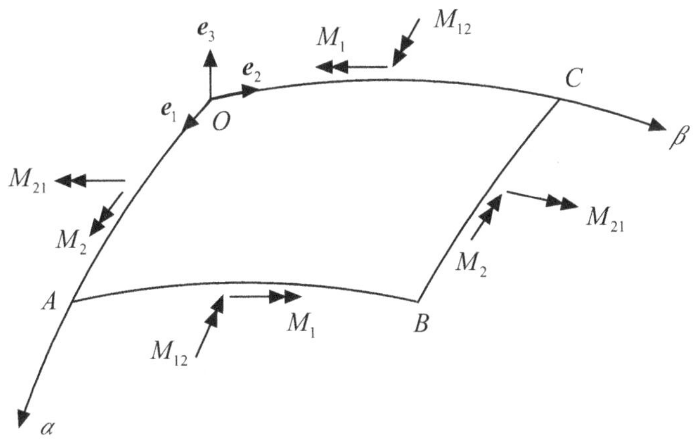

Fig. 5.5 Internal moment of the shell

$$\left.\begin{aligned} N_1 &= \frac{Eh}{1-\upsilon^2}\left[\left(e_1^0 + \upsilon e_2^0\right) + \frac{h^2}{12R_2}(\chi_1 + \upsilon\chi_2)\right] \\ N_{12} &= \frac{Eh}{2(1+\upsilon)}\left(e_{12}^0 + \frac{h^2}{12R_2}\chi_{12}\right) \\ N_2 &= \frac{Eh}{1-\upsilon^2}\left[\left(e_2^0 + \upsilon e_1^0\right) + \frac{h^2}{12R_1}(\chi_2 + \upsilon\chi_1)\right] \\ N_{21} &= \frac{Eh}{2(1+\upsilon)}\left(e_{12}^0 + \frac{h^2}{12R_1}\chi_{12}\right) \end{aligned}\right\} \tag{5.62}$$

$$\left.\begin{aligned} M_1 &= \frac{Eh^3}{12(1-\upsilon^2)}\left[(\chi_1 + \upsilon\chi_2) + \frac{1}{R_2}\left(e_1^0 + \upsilon e_2^0\right)\right] \\ M_{12} &= \frac{Eh^3}{24(1+\upsilon)}\left(\chi_{12} + \frac{1}{R_2}e_{12}^0\right) \\ M_2 &= \frac{Eh^3}{12(1-\upsilon^2)}\left[(\chi_2 + \upsilon\chi_1) + \frac{1}{R_1}\left(e_2^0 + \upsilon e_1^0\right)\right] \\ M_{21} &= \frac{Eh^3}{24(1+\upsilon)}\left(\chi_{12} + \frac{1}{R_1}e_{12}^0\right) \end{aligned}\right\} \tag{5.63}$$

$$Q_1 = \frac{Eh}{2(1+\upsilon)}e_{13}, \quad Q_2 = \frac{Eh}{2(1+\upsilon)}e_{23} \tag{5.64}$$

For the Kirchhoff–Love and Donnell thin shell theories, the shear forces Q_1 and Q_2 are calculated using the equilibrium equations. The $1 + \gamma/R_1$ and $1 + \gamma/R_2$ terms should be omitted, thus we have

$$\begin{aligned} N_1 &= C\left(e_1^0 + \upsilon e_2^0\right), \quad N_{12} = \upsilon_1 C e_{12}^0 \\ N_2 &= C\left(e_2^0 + \upsilon e_1^0\right), \quad N_{21} = \upsilon_1 C e_{12}^0 \\ M_1 &= D(\chi_1 + \upsilon\chi_2), \quad M_{12} = \upsilon_1 D\chi_{12} \\ M_2 &= D(\chi_2 + \upsilon\chi_1), \quad M_{21} = \upsilon_1 D\chi_{12} \end{aligned} \tag{5.65}$$

where

$$C = \frac{Eh}{1-\upsilon^2}, \quad D = \frac{Eh^3}{12(1-\upsilon^2)}, \quad \upsilon_1 = \frac{1-\upsilon}{2} \tag{5.66}$$

The strain energy stored in the element can be expressed as

$$\begin{aligned} U =& \frac{1}{2}\int_\alpha\int_\beta\int_{-h/2}^{+h/2} \{\sigma_1 e_1 + \sigma_2 e_2 + \tau_{12}e_{12} \\ &+ \tau_{13}e_{13} + \tau_{23}e_{23}\}AB\mathrm{d}\alpha\mathrm{d}\beta\mathrm{d}\gamma \end{aligned} \tag{5.67}$$

It can also be expressed in terms of mid-surface variables as

$$U = \frac{1}{2}\int_\alpha\int_\beta \{N_1 e_1^0 + N_2 e_2^0 + N_{12} e_{12}^0 + M_1\chi_1 + M_2\chi_2 + M_{12}\chi_{12} + M_{21}\chi_{21} + Q_1 e_{13}^0 + Q_2 e_{23}^0\} AB\mathrm{d}\alpha\mathrm{d}\beta \tag{5.68}$$

The kinetic energy of the shell can be expressed as

$$\begin{aligned} T &= \frac{1}{2}\int_V \{\dot{u}_1^2 + \dot{u}_2^2 + \dot{u}_3^2\}\mathrm{d}V \\ &= \frac{1}{2}\int_\alpha\int_\beta\int_{-h/2}^{+h/2} (\dot{u}^2 + \dot{v}^2 + \dot{w}^2 + \gamma^2\dot{\psi}_\alpha^2 + \gamma^2\dot{\psi}_\beta^2 + 2\gamma\dot{u}\dot{\psi}_\alpha + 2\gamma\dot{v}\dot{\psi}_\beta)\times \\ &(1 + \frac{\gamma}{R_1})(1 + \frac{\gamma}{R_2})AB\mathrm{d}\alpha\mathrm{d}\beta\mathrm{d}\gamma \end{aligned} \tag{5.69}$$

The virtual work of the shell can be expressed as

$$\delta W = \int_\alpha\int_\beta \{q_1\delta u + q_2\delta v + q_3\delta w + m_1\delta\psi_\alpha + m_2\delta\psi_\beta\} AB\mathrm{d}\alpha\mathrm{d}\beta \tag{5.70}$$

Here, m_1 and m_2 represent the distributed bending moments.

The equilibrium equation of the shell is given by

$$\left.\begin{aligned} &\frac{\partial(BN_1)}{\partial\alpha} + \frac{\partial(AN_{21})}{\partial\beta} + N_{12}\frac{\partial A}{\partial\beta} - N_2\frac{\partial B}{\partial\alpha} + AB\frac{Q_1}{R_1} + ABq_1 = AB\rho h\frac{\partial^2 u}{\partial t^2} \\ &\frac{\partial(BN_{12})}{\partial\alpha} + \frac{\partial(AN_2)}{\partial\beta} + N_{21}\frac{\partial B}{\partial\alpha} - N_1\frac{\partial A}{\partial\beta} + AB\frac{Q_2}{R_2} + ABq_2 = AB\rho h\frac{\partial^2 v}{\partial t^2} \\ &\frac{\partial(BQ_1)}{\partial\alpha} + \frac{\partial(AQ_2)}{\partial\beta} - AB\left(\frac{N_1}{R_1} + \frac{N_2}{R_2}\right) + ABq_3 = AB\rho h\frac{\partial^2 w}{\partial t^2} \end{aligned}\right\} \tag{5.71}$$

$$\left.\begin{aligned} &\frac{\partial(BM_1)}{\partial\alpha} + \frac{\partial(AM_{21})}{\partial\beta} + M_{12}\frac{\partial A}{\partial\beta} - M_2\frac{\partial B}{\partial\alpha} - ABQ_1 + ABm_1 = AB\rho\frac{h^3}{12}\frac{\partial^2\psi_\alpha}{\partial t^2} \\ &\frac{\partial(BM_{12})}{\partial\alpha} + \frac{\partial(AM_2)}{\partial\beta} + M_{21}\frac{B}{\partial\alpha} - M_1\frac{\partial A}{\partial\beta} - ABQ_2 + ABm_2 = AB\rho\frac{h^3}{12}\frac{\partial^2\psi_\beta}{\partial t^2} \end{aligned}\right\} \tag{5.72}$$

5.2 Thin Plate Finite Elements

5.2.1 Finite Element Approximation

Assume the displacement field of the thin plate is

$$w = \mathbf{N}(x, y)\tilde{\mathbf{u}}(t) \quad \text{and} \quad \delta w = \mathbf{N}(x, y)\delta\tilde{\mathbf{u}} \tag{5.73}$$

Here, $\mathbf{N}$ and $\tilde{\mathbf{u}}$ represent the shape function and the undetermined coefficients, respectively. According to the standard displacement finite element method, the following ordinary differential equation can be obtained:

$$\mathbf{M}\ddot{\tilde{\mathbf{u}}} + \mathbf{K}\tilde{\mathbf{u}} = \mathbf{f} \tag{5.74}$$

The mass matrix, stiffness matrix, and load vector are given by

$$\begin{aligned}\mathbf{M} &= \int_\Omega \mathbf{N}^{\mathrm{T}}\rho h\mathbf{N}\mathrm{d}\Omega \\ \mathbf{K} &= \int_\Omega \mathbf{B}^{\mathrm{T}}\mathbf{D}\mathbf{B}\mathrm{d}\Omega \\ \mathbf{f} &= \int_\Omega \mathbf{N}^{\mathrm{T}}q\mathrm{d}\Omega + \mathbf{f}_b\end{aligned} \tag{5.75}$$

where, $\mathbf{f}_b$ is the boundary load

$$\begin{aligned}\mathbf{f}_b = &\int_{\Gamma_s} \mathbf{N}V_n\mathrm{d}\Gamma - \int_{\Gamma_s} \frac{\partial\mathbf{N}}{\partial n}M_n\mathrm{d}\Gamma \\ &+ \sum_i \mathbf{N}(x_i, y_i)R_i\end{aligned} \tag{5.76}$$

and

$$\mathbf{B} = -\boldsymbol{L}\mathbf{N} = -\begin{Bmatrix} \frac{\partial^2\mathbf{N}}{\partial x^2} \\ \frac{\partial^2\mathbf{N}}{\partial y^2} \\ 2\frac{\partial^2\mathbf{N}}{\partial x\partial y} \end{Bmatrix} \tag{5.77}$$

The reader can once again confirm the key points of the displacement finite element method from the above process: once the shape function vector $\mathbf{N}$ is determined, the subsequent computational procedure is entirely analogous.

5.2.2 Continuity Requirements for Shape Functions (C^1 Continuity)

The weak form presented in Sect. 5.1.1 involves all second-order derivatives of the displacement function w. Therefore, it is evident that at least all first-order derivatives of w must be continuous within the domain Ω. As mentioned above, such functions are referred to as C^1 continuous. If the first-order derivatives are not continuous, the plate exhibits "kinks," resulting in singular second-order derivatives, and the square of a singular function is not integrable over the domain Ω.

As demonstrated in Sect. 3.3.1, constructing a piecewise C^1 continuous polynomial along a straight line using Hermite interpolation is not difficult. It should also be noted, however, that Hermite interpolation is not iso-parametric. When attempting to extend shape functions based on Hermite interpolation to arbitrarily shaped thin plate elements, certain difficulties arise.

To ensure the continuity of the displacement function w and its normal slope across an interface, it is necessary to guarantee that both w and $\partial w/\partial n$ are uniquely determined by the nodal parameter values along that interface. For example, for the side 1–2 of the rectangular element shown in Fig. 5.6, the normal direction n is actually the y direction. Therefore, it must be ensured that both w and $\partial w/\partial y$ are uniquely determined by the values of w, $\partial w/\partial x$, and $\partial w/\partial y$ at the nodes along this line.

To ensure C^1continuity along edge 1–2, we have

$$w = A_1 + A_2 x + A_3 y + \cdots \tag{5.78}$$

and

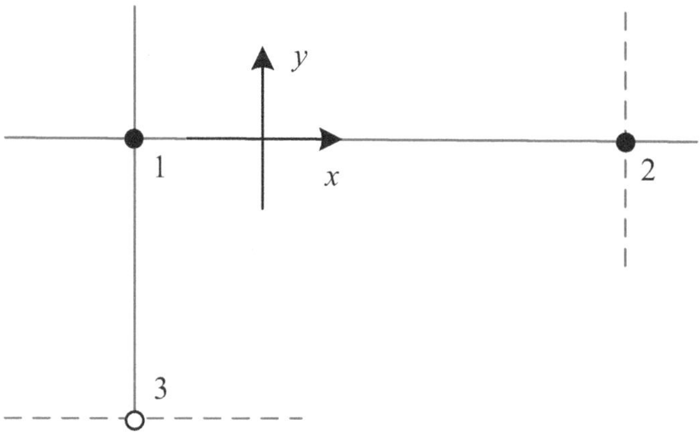

Fig. 5.6 Continuity Requirement for the Slope of the Normal

$$\frac{\partial w}{\partial y} = B_1 + B_2 x + B_3 y + \cdots \tag{5.79}$$

Each expression contains certain constant coefficients, which are sufficient to uniquely determine the parameters of the nodes on the line.

Therefore, if there are only two nodes and considering that both $\partial w/\partial x$ and w are specified at each node, w must be a cubic function along the boundary. Similarly, for the case of two nodes, only a linear variation of $\partial w/\partial y$ along the boundary is permitted. However, note that a similar procedure can be applied along the edge in the y-direction to maintain the continuity of $\partial w/\partial x$ along this edge. Thus, the value of $\partial w/\partial y$ along edge 1–2 depends solely on the nodal parameters of edge 1–2; similarly, the value of $\partial w/\partial x$ along edge 1–3 depends solely on the nodal parameters of edge 1–3. Differentiating $\partial w/\partial y$ along edge 1–2 with respect to x yields $\partial^2 w/\partial x \partial y$ on edge 1–2, which also depends only on the nodal parameters of edge 1–2. Likewise, $\partial^2 w/\partial y \partial x$ on edge 1–3 depends only on the nodal parameters of edge 1–3.

An immediate inconsistency arises at common point 1, because it is difficult to automatically satisfy the necessary identities for continuity when arbitrary parameter values are chosen at nodes 2 and 3.

$$\frac{\partial^2 w}{\partial x \partial y} \equiv \frac{\partial^2 w}{\partial y \partial x} \tag{5.80}$$

Therefore, when only the values of w and its derivatives are specified at the corner points, it is not possible to ensure complete compatibility of the shape functions using simple polynomials.

If any shape function with three nodal variables that satisfies the compatibility requirements is found, then such a shape function will necessarily not be continuously differentiable at the corner nodes, and the mixed partial derivatives will not be unique. There already exists a proof of this conclusion for rectangular elements. Clearly, this result can be extended to any two arbitrary directions at a corner. Currently, there are some thin plate elements with discontinuities at the corners.

A way to overcome the aforementioned difficulties is to specify the mixed partial derivatives as one of the nodal parameters. For rectangular elements, this approach is both convenient and permissible. Simple functions of this type have already been constructed and successfully applied to problems that can be solved using square elements. For the rectangular element shown in Fig. 5.7, its shape functions can be regarded as combinations of Hermite interpolation functions and can be expressed as follows:

$$\hat{w}(\xi, \eta) = \sum_{a=1}^{4} \mathbf{N}_a(\xi, \eta)\, \tilde{\mathbf{u}}_a \tag{5.81}$$

where

$$\mathbf{N}_a = \left[H_a^w(\xi) H_a^w(\eta) \;\; H_a^\theta(\xi) H_a^w(\eta) \;\; H_a^w(\xi) H_a^\theta(\eta) \;\; H_a^\theta(\xi) H_a^\theta(\eta) \right] \tag{5.82}$$

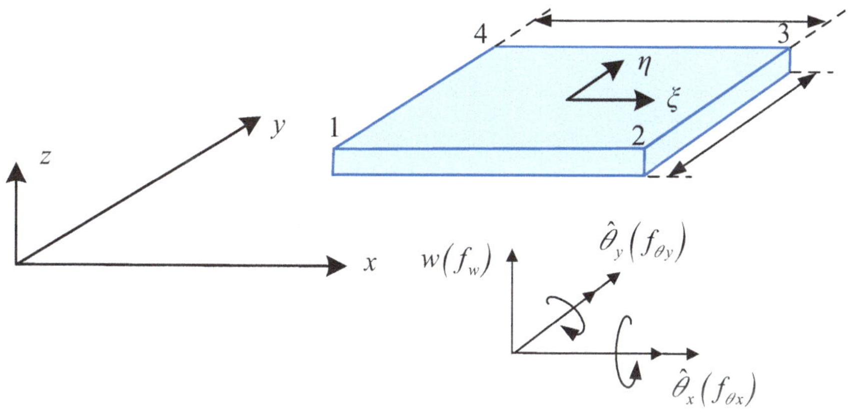

Fig. 5.7 Rectangular plate element

and

$$\tilde{\mathbf{u}}_a = \begin{bmatrix} \tilde{w}_a & \frac{\partial \tilde{w}_a}{\partial x} & \frac{\partial \tilde{w}_a}{\partial y} & \frac{\partial^2 \tilde{w}_a}{\partial x \partial y} \end{bmatrix}^{\mathrm{T}} \tag{5.83}$$

The geometry of the element can be represented as

$$\mathbf{x} = \sum_{a=1}^{4} N_a(\xi, \eta)\, \mathbf{x}_a \tag{5.84}$$

Here, basis N_{a} can be the first term of $\mathbf{N}_a$ and $\mathbf{x}_{\mathrm{a}}$ represents the coordinate vector of the node.

In fact, this element is the most accurate rectangular element for thin plate problems. However, like all rectangular elements, its applicability is very limited. Moreover, since this element is not of the iso-parametric type, it generally cannot be used in situations where multiple elements intersect at different angles (see Fig. 5.8). Continuity of cross-derivatives at a point along several orthogonal directions implies that all second-order derivatives are continuous at that point.

However, when the second derivative is continuous, if the stiffness of the plate changes abruptly from one element to another, then, it cannot be guaranteed that the moments perpendicular to the interface are equal, which violates actual physical requirements. Although this excessive continuity may seem unreasonable, this element has achieved some success for isotropic homogeneous plates. In cases involving thickness or material variations, the nodal parameters can be decomposed into components normal and tangential to the interface, and the continuity of the parameter $\partial 2w_a/\partial n2$ can be relaxed.

Due to the difficulty of finding compatible displacement functions, many have attempted to forgo continuity along the entire boundary and instead satisfy continuity under other necessary criteria. If compatibility requirements are met at the nodes,

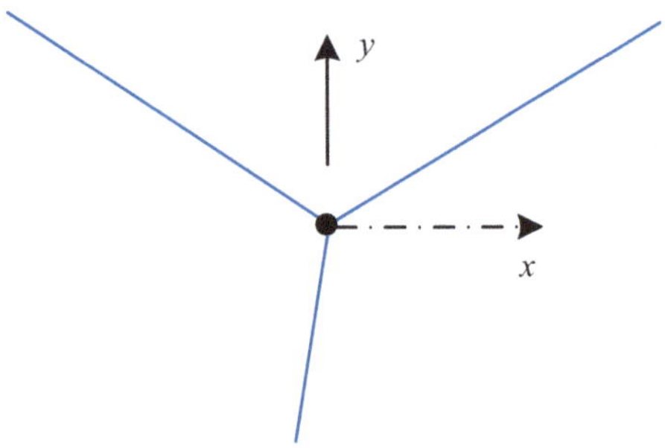

Fig. 5.8 A node where elements intersect at an arbitrary angle

then, in the limiting case, compatibility is satisfied along the entire edge. For plate problems, patch tests are particularly important in both the design and testing of elements, and therefore should never be neglected.

5.2.3 *Patch Test: Theoretical Requirements*

Different forms of patch tests are typically applied numerically to the final form of the tested element. However, if certain conditions are met, it is possible to accurately predict whether the shape functions violate the fundamental requirements of compatibility when selecting them. These conditions are based on the requirement that the internal virtual work at discontinuities under constant strain must be zero. Therefore, as shown in Fig. 5.9, if the tractions acting on the interface of the plate element are M_n, M_{ns}, and Q_n, and the corresponding virtual displacements are

$$\Delta\phi_n \equiv \Delta\left(\frac{\partial w}{\partial n}\right), \quad \Delta\phi_s \equiv \Delta\left(\frac{\partial w}{\partial s}\right)$$

which do not match Δw, then, ideally, it is desired that the integral given below equals zero for all constant stress states:

$$\int_{\Gamma_e} M_n \Delta\phi_n \mathrm{d}\Gamma + \int_{\Gamma_e} M_{ns} \Delta\phi_s \mathrm{d}\Gamma + \int_{\Gamma_e} Q_n \Delta w \,\mathrm{d}\Gamma = 0 \tag{5.85}$$

The last term will always be zero for constant M_x, M_y, M_{xy}, therefore $Q_x = Q_y = 0$ (without considering the force couple, see Eqs. (5.15) and (5.16)). The remaining conditions that need to be satisfied are then

$$\int_{\Gamma_e} \Delta\phi_n \mathrm{d}\Gamma = 0 \quad \text{and} \quad \int_{\Gamma_e} \Delta\phi_s \mathrm{d}\Gamma = 0 \tag{5.86}$$

If this condition is satisfied on each straight edge Γ_e of the element, the remaining conditions are automatically fulfilled.

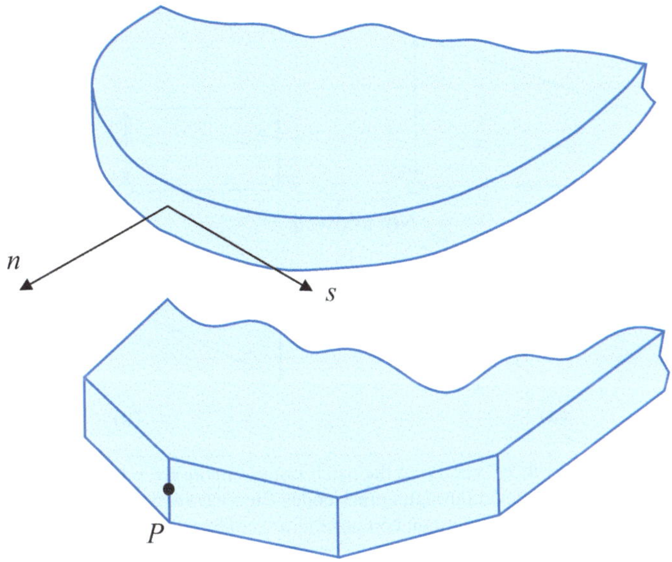

Fig. 5.9 Boundary forces (M_n, M_{ns} and Q_n) and the corresponding displacements (θ_n, θ_s and w)

Note: On the boundary, the simply supported boundary conditions $M_n = 0$, $\varphi_s = 0$ and $w = 0$ are equivalent to the clamped condition $\varphi_n = \varphi_s = w = 0$ at the corner points. Contradictions may arise if a polygonal shape (bottom) is used to approximate a curved boundary (top).

For elements connected at nodes where $\partial w / \partial n$ is specified at the nodes, if the polynomial contains only linear and antisymmetric cubic terms, and there are no quadratic terms involving the normal gradient, these integrals will be zero, as shown in Fig. 5.10. If the function w is determined solely by the parameters of the corresponding edge, the second condition in Eq. (5.86) is always satisfied.

5.2.4 Quadrilateral Elements

In this section, we introduce hierarchical basis functions for quadrilateral elements. The boundary shape functions (vertex and edge functions) are derived using mixed function interpolation, while the face functions are constructed as tensor products of one-dimensional basis functions in the ξ and η directions.

5.2.4.1 Vertex Form Functions

In fact, the continuity of cross derivatives in several orthogonal directions implies that all second-order derivatives at that node are continuous. Therefore, each vertex

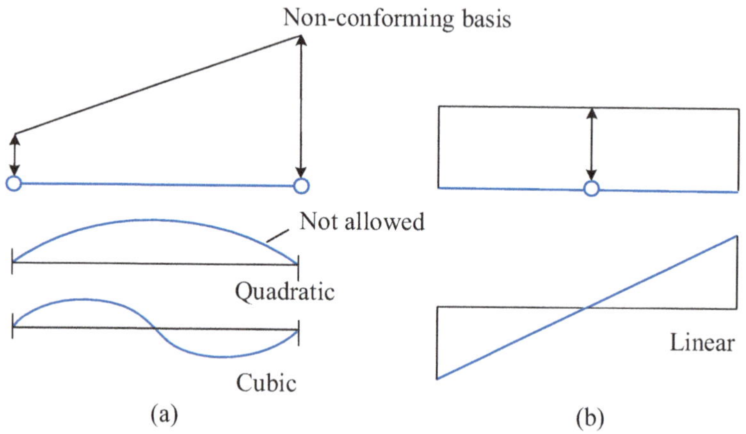

Fig. 5.10 Continuity conditions satisfying the patch test $[\int(\partial w/\partial n)\mathrm{d}s = 0]$, and variation of $\partial w/\partial n$ along the boundary: **a** defined through corner nodes (linear components are conforming) and **b** defined through a central node (constant components are conforming)

is assigned the following degrees of freedom: w, $\partial w/\partial \xi$, $\partial w/\partial \eta$, $\partial 2w/\partial \eta 2$, $\partial 2w/\partial \xi 2$, and $\partial 2w/\partial \xi \partial \eta$, resulting in a quadrilateral element with a total of 24 vertex degrees of freedom (see Fig. 5.11). The basis functions are derived using a mixed interpolation method based on third- and fifth-order Hermite functions. The third-order Hermite interpolation polynomial is given by

$$
\begin{aligned}
h_1(\xi) &= \frac{1}{4}(\xi + 1)(\xi - 1)^2,\ h_2(\xi) = \frac{1}{4}(\xi + 2)(\xi - 1)^2 \\
h_3(\xi) &= \frac{1}{4}(2 - \xi)(\xi + 1)^2,\ h_4(\xi) = \frac{1}{4}(\xi - 1)(\xi + 1)^2
\end{aligned}
\tag{5.87}
$$

The fifth-order Hermite interpolation polynomial is

$$
\begin{aligned}
H_1^{(2)} &= (1 - \xi)^3(1 + \xi)^2/16, \\
H_2^{(2)} &= (1 + \xi)^3(1 - \xi)^2/16 \\
H_1^{(1)} &= (1 - \xi)^3(1 + \xi)(5 + 3\xi)/16, \\
H_2^{(1)} &= (1 + \xi)^3(\xi - 1)(5 - 3\xi)/16 \\
H_1 &= (1 - \xi)^3\left(3\xi^2 + 9\xi + 8\right)/16, \\
H_2 &= (1 + \xi)^3\left(3\xi^2 - 9\xi + 8\right)/16
\end{aligned}
\tag{5.88}
$$

The vertex shape functions expressed by the 3rd-order and 5th-order Hermite interpolation polynomials are.

Vertex 1:

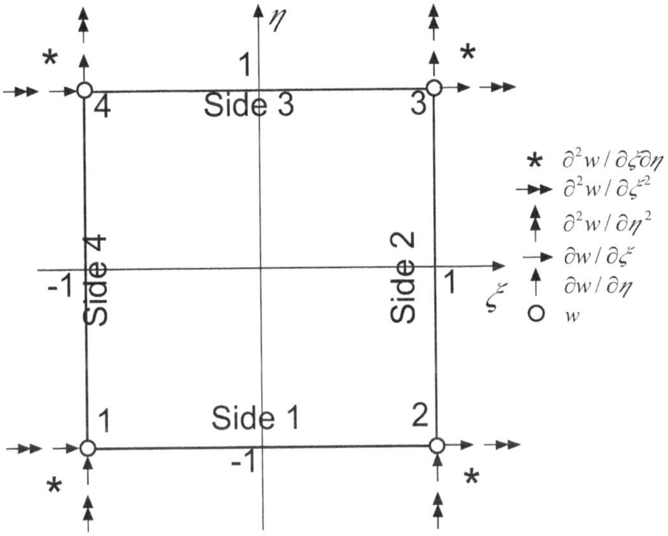

Fig. 5.11 Vertex degrees of freedom configuration for the unit quadrilateral element

$$
\begin{aligned}
S_{w}^{V_1} &= h_2(\xi)H_1(\eta) + H_1(\xi)h_2(\eta) - h_2(\xi)h_2(\eta) \\
S_{w,\xi}^{V_1} &= H_1^{(1)}(\xi)h_2(\eta) \\
S_{w,\eta}^{V_1} &= h_2(\xi)H_1^{(1)}(\eta) \\
S_{w,\xi\xi}^{V_1} &= H_1^{(2)}(\xi)h_2(\eta) \\
S_{w,\eta\eta}^{V_1} &= h_2(\xi)H_1^{(2)}(\eta) \\
S_{w,\xi\eta}^{V_1} &= h_1(\xi)h_1(\eta)
\end{aligned}
\tag{5.89}
$$

Vertex 2:

$$
\begin{aligned}
S_{w}^{V_2} &= h_3(\xi)H_1(\eta) + H_2(\xi)h_2(\eta) - h_3(\xi)h_2(\eta) \\
S_{w_\xi}^{V_2} &= H_2^{(1)}(\xi)h_2(\eta) \\
S_{w_\eta}^{V_2} &= h_3(\xi)\ H_1^{(1)}(\eta) \\
S_{w_{\xi\xi}}^{V_2} &= H_2^{(2)}(\xi)h_2(\eta) \\
S_{w_{\eta\eta}}^{V_2} &= h_3(\xi)H_1^{(2)}(\eta) \\
S_{w_{\xi\eta}}^{V_2} &= h_4(\xi)h_1(\eta)
\end{aligned}
\tag{5.90}
$$

Vertex 3:

$$
S_{w}^{V_3} = h_3(\xi)H_2(\eta) + H_2(\xi)h_3(\eta) - h_3(\xi)h_3(\eta)
$$

$$
\begin{aligned}
S_{w_\xi}^{V_3} &= H_2^{(1)}(\xi)h_3(\eta) \\
S_{w_\eta}^{V_3} &= h_3(\xi)H_2^{(1)}(\eta) \\
S_{w_{\xi\xi}}^{V_3} &= H_2^{(2)}(\xi)h_3(\eta) \\
S_{w,\eta\eta}^{V_3} &= h_3(\xi)H_2^{(2)}(\eta) \\
S_{w,\xi\eta}^{V_3} &= h_4(\xi)h_4(\eta)
\end{aligned}
\tag{5.91}
$$

Vertex 4:

$$
\begin{aligned}
S_w^{V_4} &= h_2(\xi)H_2(\eta) + H_1(\xi)h_3(\eta) - h_2(\xi)h_3(\eta) \\
S_{w_\xi}^{V_4} &= H_1^{(1)}(\xi)h_3(\eta) \\
S_{w_\eta}^{V_4} &= h_2(\xi)H_2^{(1)}(\eta) \\
S_{w_{\xi\xi}}^{V_4} &= H_1^{(2)}(\xi)h_3(\eta) \\
S_{w_{\eta\eta}}^{V_4} &= h_2(\xi)H_2^{(2)}(\eta) \\
S_{w_{\xi\eta}}^{V_4} &= h_1(\xi)h_4(\eta)
\end{aligned}
\tag{5.92}
$$

The shape function for vertex 1 is plotted in Fig. 5.12.

5.2.4.2 Boundary Shape Functions

The boundary shape functions for the C^1 element can be divided into two parts: first, the shape functions associated with the boundary value (i.e., w), which have zero normal derivative; second, the shape functions associated with the boundary normal derivative (i.e., w_n), which have zero function value. As described earlier, the boundary functions for w and w_n reach the 5th and 3rd orders, respectively. Therefore, the higher-order spectral basis functions for w and w_n should start from the 6th and 4th orders, respectively. The boundary shape functions are as follows:

Boundary 1:

$$
\begin{aligned}
S_{w,i}^{S_1} &= (1+\xi)^3(1-\xi)^3 J_i^{(4,4)}(\xi)h_2(\eta);\ i = 0 \sim M-1 \\
S_{w_n,i}^{S_1} &= (1+\xi)^2(1-\xi)^2 J_i^{(4,4)}(\xi)h_1(\eta);\ i = 0 \sim M
\end{aligned}
\tag{5.93}
$$

Boundary 2:

$$
\begin{aligned}
S_{w,i}^{S_2} &= (1-\eta)^3(1+\eta)^3 J_i^{(4,4)}(\eta)h_3(\xi);\ i = 0 \sim N-1 \\
S_{w_n,i}^{S_2} &= (1-\eta)^2(1+\eta)^2 J_i^{(4,4)}(\eta)h_4(\xi);\ i = 0 \sim N
\end{aligned}
\tag{5.94}
$$

Boundary 3:

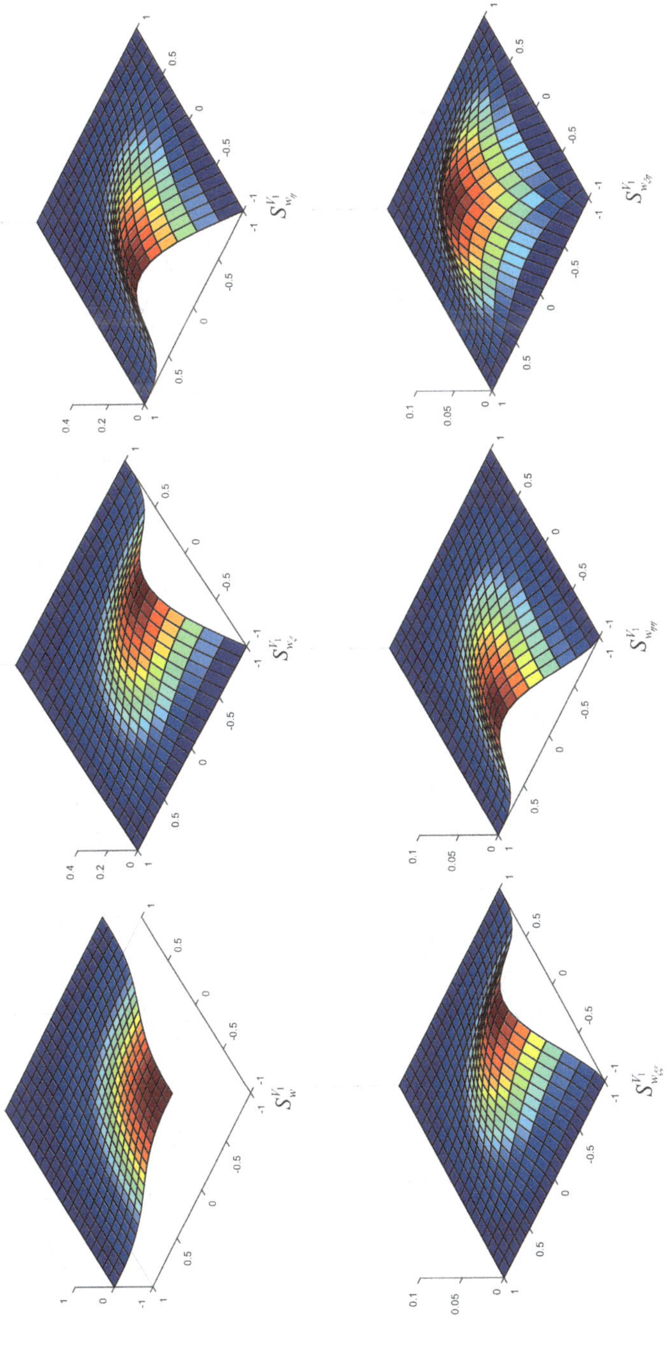

Fig. 5.12 Shape function of vertex 1 on a unit quadrilateral element

$$S_{w,i}^{S_3} = (1-\xi)^3(1+\xi)^3 J_i^{(4,4)}(\xi)h_3(\eta);\ i = 0 \sim P-1$$
$$S_{w_n,i}^{S_3} = (1-\xi)^2(1+\xi)^2 J_i^{(4,4)}(\xi)h_4(\eta);\ i = 0 \sim P \tag{5.95}$$

Boundary 4:

$$S_{w,i}^{S_4} = (1-\eta)^3(1+\eta)^3 J_i^{(4,4)}(\eta)h_2(\xi);\ i = 0 \sim Q-1$$
$$S_{w_n,i}^{S_4} = (1-\eta)^2(1+\eta)^2 J_i^{(4,4)}(\eta)h_1(\xi);\ i = 0 \sim Q \tag{5.96}$$

The number of edge functions on each side of a quadrilateral element can vary. Therefore, it is possible to connect elements with different polynomial orders. In addition, the number of edge functions associated with w and w_n on the same side can also differ (the number of edge functions for w_n should be greater than that for w). As a result, the approximation orders for w and w_n on the boundary are relatively independent. This feature is advantageous for enhancing C^1 continuity along curved boundaries. Further details will be provided later. As shown in Fig. 5.13, the first three shape functions for w and w_n on the boundary of a unit quadrilateral element are illustrated.

5.2.4.3 Surface Functions

The shape functions for quadrilateral elements can be readily constructed as the tensor product of one-dimensional C^1 hierarchical basis functions, that is,

$$S_{mn}^{\mathrm{F}} = \left(1-\xi^2\right)^2\left(1-\eta^2\right)^2 J_{m-1}^{(4,4)}(\xi)J_{n-1}^{(4,4)}(\eta),$$
$$m = 1, 2, \cdots, N_\xi,\ n = 1, 2, \cdots, N_\eta \tag{5.97}$$

Here, m and n are indices; N_ξ and N_η denote the number of basis functions along the two natural coordinate directions. The surface functions satisfy the following orthogonality conditions

$$\int_{-1}^{1}\int_{-1}^{1} S_{pq}^{\mathrm{F}} S_{mn}^{\mathrm{F}} \mathrm{d}\xi \mathrm{d}\eta = \delta_{pm}\delta_{qn}C_{mn} \tag{5.98}$$

where

$$C_{mn} = 2^9\sqrt{\left[\prod_{i=0}^{3}(m+i)\prod_{j=0}^{3}(n+i)\right]\Big/\left[(2m+7)(2n+7)\prod_{i=4}^{7}(m+i)\prod_{j=4}^{7}(n+j)\right]} \tag{5.99}$$

Here, δ_{pm} and δ_{qn} are the Kronecker delta. The first three face functions are shown in Fig. 5.14. According to Eq. (5.99), both the function values and the normal derivatives of the face functions are zero on the element boundaries.

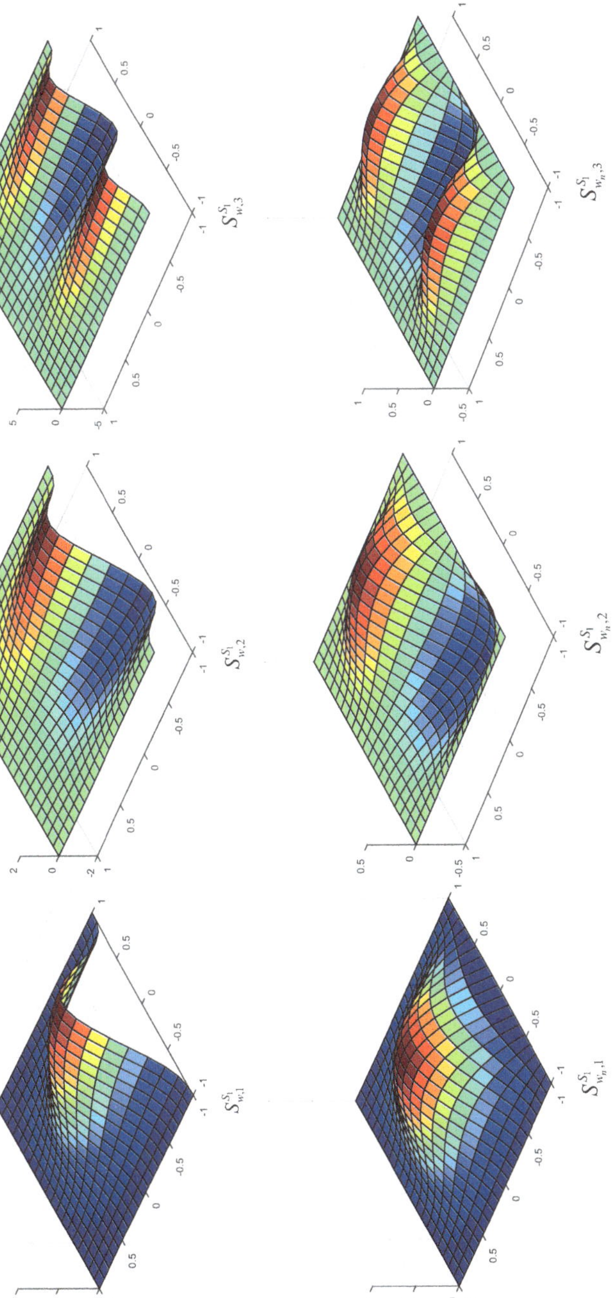

Fig. 5.13 Shape functions of boundary 1 on a unit quadrilateral element

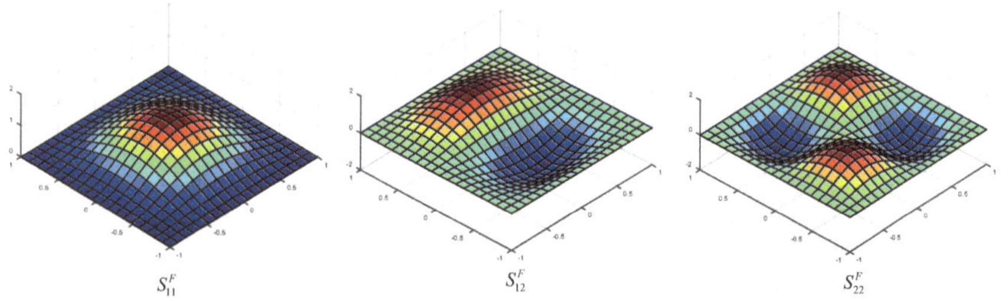

Fig. 5.14 The first three face functions of a unit quadrilateral element

5.2.5 *Triangular Elements*

5.2.5.1 Vertex Shape Functions

Compared to quadrilateral elements, the presence of a hypotenuse in triangular elements results in Hermite interpolation functions that are rational. For example, the interpolation function in the ξ direction (see Fig. 5.15b), namely ϕ_i, can be expressed as

$$\begin{aligned} \phi_1(\xi,\eta) &= \frac{\xi(\xi+\eta-1)^2}{(1-\eta)^2}, \quad \phi_2(\xi,\eta) = \frac{(\xi+\eta-1)^2(2\xi+1-\eta)}{(1-\eta)^3} \\ \phi_3(\xi,\eta) &= \frac{\xi^2(2\xi-3+3\eta)}{(\eta-1)^3}, \quad \phi_4(\xi,\eta) = \frac{(\xi+\eta-1)\xi^2}{(1-\eta)^2} \end{aligned} \tag{5.100}$$

Here, ϕ_1 and ϕ_4 are interpolation functions for the derivatives at the two ends of the segment alongξ direction, while ϕ_2 and ϕ_3 are interpolation functions for the function values at the two ends of the segment along ξ direction. Equation (5.100) is derived using cubic Hermite interpolation in combination with the inverse Duffy transformation (see Sect. 4.2.4.1). Similarly, the interpolation functions in η direction (see Fig. 5.15c), denoted as ψi, can be obtained from Eq. (5.100) through the following

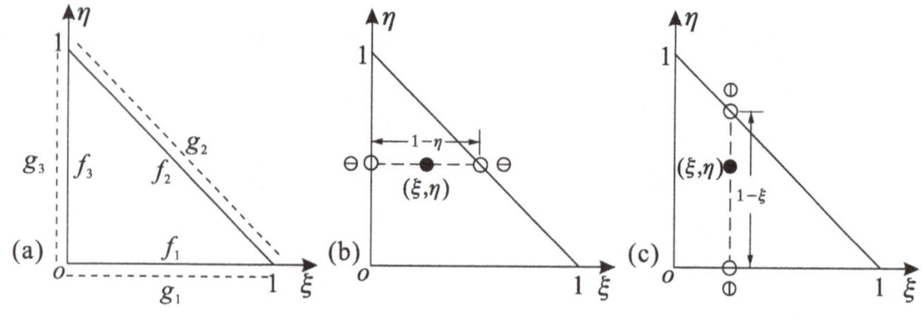

Fig. 5.15 Mixed function interpolation on a unit triangular element: **a** function and its derivatives on the boundary, **b** interpolation along ξ-direction, **c** interpolation along η-direction

transformation:

$$
\begin{aligned}
\psi_1(\xi,\eta) &= \phi_1(\eta,\xi),\ \psi_2(\xi,\eta) = \phi_2(\eta,\xi) \\
\psi_3(\xi,\eta) &= \phi_3(\eta,\xi),\ \psi_4(\xi,\eta) = \phi_4(\eta,\xi)
\end{aligned} \tag{5.101}
$$

Among them, ψ_1 and ψ_4 are related to the derivatives at the two ends of the segment along η direction, while ψ_2 and ψ_3 are related to the function values at the two ends of the segment along η direction.

A process similar to that in the previous section can be used to derive the shape functions on the unit triangular element. However, special care must be taken when determining the shape functions at vertices 2 and 3, since the angles at these points are no longer right angles as in the quadrilateral element (see Fig. 5.16). To obtain the boundary shape functions, the fifth-order Hermite polynomials defined on [−1, 1] in Eq. (5.88) are shifted to the interval [0, 1], that is,

$$
\begin{aligned}
\tilde{H}_1^{(2)}(\eta) &= (1-\eta)^3\eta^2/2, \\
\tilde{H}_2^{(2)}(\eta) &= \eta^3(1-\eta)^2/2 \\
\tilde{H}_1^{(1)}(\eta) &= (1-\eta)^3\eta(3\eta+1), \\
\tilde{H}_2^{(1)}(\eta) &= \eta^3(\eta-1)(4-3\eta) \\
\tilde{H}_1(\eta) &= (1-\eta)^3\left(6\eta^2+3\eta+1\right), \\
\tilde{H}_2(\eta) &= \eta^3\left(6\eta^2-15\eta+10\right)
\end{aligned} \tag{5.102}
$$

Using the blended function interpolation method, the shape function of vertex 3 with respect to $w_\eta{}^{V3}$ is

$$
S_{w_\eta}^{V_3} = H_2^{(1)}(\eta)[\phi_2(\xi,\eta)+\phi_3(\xi,\eta)] \tag{5.103}
$$

which can be simplified to

$$
S_{w_\eta}^{V_3} = -\eta^3\left(3\eta^2-7\eta+4\right) \tag{5.104}
$$

Similarly, the shape functions for the other vertices can be derived. For simplicity, the area coordinate relation $\zeta = 1-\xi-\eta$ will be used below.

Vertex 1:

$$
\begin{aligned}
S_w^{V_1} &= 6\zeta^5 - 5\zeta^2\left(3\zeta^2 - 2\zeta - 6\xi\eta + 6\xi\eta\zeta\right) \\
S_{w_\xi}^{V_1} &= \xi(3\xi+1)\zeta^2(3\eta+\zeta) \\
S_{w_\eta}^{V_1} &= \eta(3\eta+1)\zeta^2(3\xi+\zeta) \\
S_{w_{\xi\xi}}^{V_1} &= \xi^2\zeta^2(3\eta+\zeta)/2 \\
S_{w_{\eta\eta}}^{V_1} &= \eta^2\zeta^2(3\xi+\zeta)/2
\end{aligned}
$$

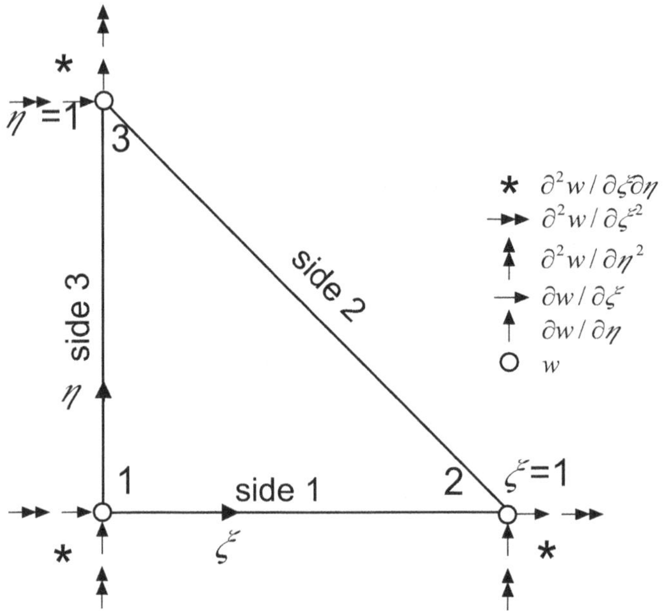

Fig. 5.16 Configuration of vertex degrees of freedom for a unit triangular element

$$S^{V_1}_{w_{\xi\eta}} = \xi\eta\zeta^2 \tag{5.105}$$

Vertex 2:

$$\begin{aligned}
S^{V_2}_{w} &= \xi^3\left(6\xi^2 - 15\xi + 10\right)\\
S^{V_2}_{w_\xi} &= -\xi^3\left(3\xi^2 - 7\xi + 4\right)\\
S^{V_2}_{w_\eta} &= \xi^2\eta\left(9\xi\eta - 9\eta - 2\xi + 6\eta^2 + 3\right)\\
S^{V_2}_{w_{\xi\xi}} &= \xi^3(\xi - 1)^2/2\\
S^{V_2}_{w_{\eta\eta}} &= \xi^2\eta^2(3\xi + 2\eta - 2)/2\\
S^{V_2}_{w_{\xi\eta}} &= -\xi^2\eta\left(3\xi\eta - 3\eta - \xi + 2\eta^2 + 1\right)
\end{aligned} \tag{5.106}$$

Vertex 3:

$$\begin{aligned}
S^{V_3}_{w} &= \eta^3\left(6\eta^2 - 15\eta + 10\right)\\
S^{V_3}_{w_\xi} &= \eta^2\xi\left(9\xi\eta - 9\xi - 2\eta + 6\xi^2 + 3\right)\\
S^{V_3}_{w_\eta} &= -\eta^3\left(3\eta^2 - 7\eta + 4\right)\\
S^{V_3}_{w_{\xi\xi}} &= \eta^2\xi^2(3\eta + 2\xi - 2)/2\\
S^{V_3}_{w_{\eta\eta}} &= \eta^3(1 - \eta)^2/2
\end{aligned}$$

$$S_{w_{\xi\eta}}^{V_3} = -\eta^2\xi\left(3\xi\eta - 3\xi - \eta + 2\xi^2 + 1\right) \tag{5.107}$$

It can be seen that the final vertex shape function is not rational. The shape function for node 3 is plotted in Fig. 5.17.

5.2.5.2 Edge Shape Functions

The derivation of the edge shape function is similar. For example, the shape function on edge 1 with respect to w can be expressed as

$$S_{w,i}^{S_1}(\xi) = \xi^3(1-\xi)^3 P_i(\xi)\psi_2(\xi,\eta), \quad i = 0, 1\cdots \tag{5.108}$$

Theoretically, P_i can be any complete set of polynomials. To improve the numerical properties during the basis function transformation process (see the next section), P_i must satisfy the following orthogonality condition.

$$\int_A S_{w,i}^{S_1} S_{w,j}^{S_1} \mathrm{d}A = C\delta_{ij} \tag{5.109}$$

Here, C is a constant and δ_{ij} denotes the Kronecker delta. Thus, P_i can be expressed as

$$P_i(\xi) = J_i^{(7,6)}(2\xi - 1) \tag{5.110}$$

Therefore, the shape function for w on edge 1 can be expressed as:

$$S_{w,i}^{S_1}(\xi) = \xi^3(\xi + \eta - 1)^2(2\eta + 1 - \xi)J_i^{(7,6)}(2\xi - 1), \\ i = 0,\ 1,\ 2,\ \cdots \tag{5.111}$$

Similarly, the shape function for w_η on edge 1 is

$$S_{w_\eta,i}^{S_1}(\xi) = \xi^2\eta(\xi + \eta - 1)^2 J_i^{(7,4)}(2\xi - 1), \quad i = 0,\ 1,\ 2,\ \cdots \tag{5.112}$$

The shape functions for the other edges can be derived in a similar manner. The edge shape function for the three edges of a triangular element is:

Edge1:

$$\begin{aligned} S_{w,i}^{S_1}(\xi) &= \xi^3\zeta^2(3\eta + \zeta)J_i^{(7,6)}(2\xi - 1),\ i = 0 \sim M - 1 \\ S_{w_\eta,i}^{S_1}(\xi) &= \xi^2\eta\zeta^2 J_i^{(7,4)}(2\xi - 1),\ i = 0 \sim M \end{aligned} \tag{5.113}$$

Edge 2:

$$S_{w,i}^{S_2}(\eta) = -\xi^2\eta^3(\xi + 3\zeta)J_i^{(7,6)}(2\eta - 1),\ i = 0 \sim N - 1$$

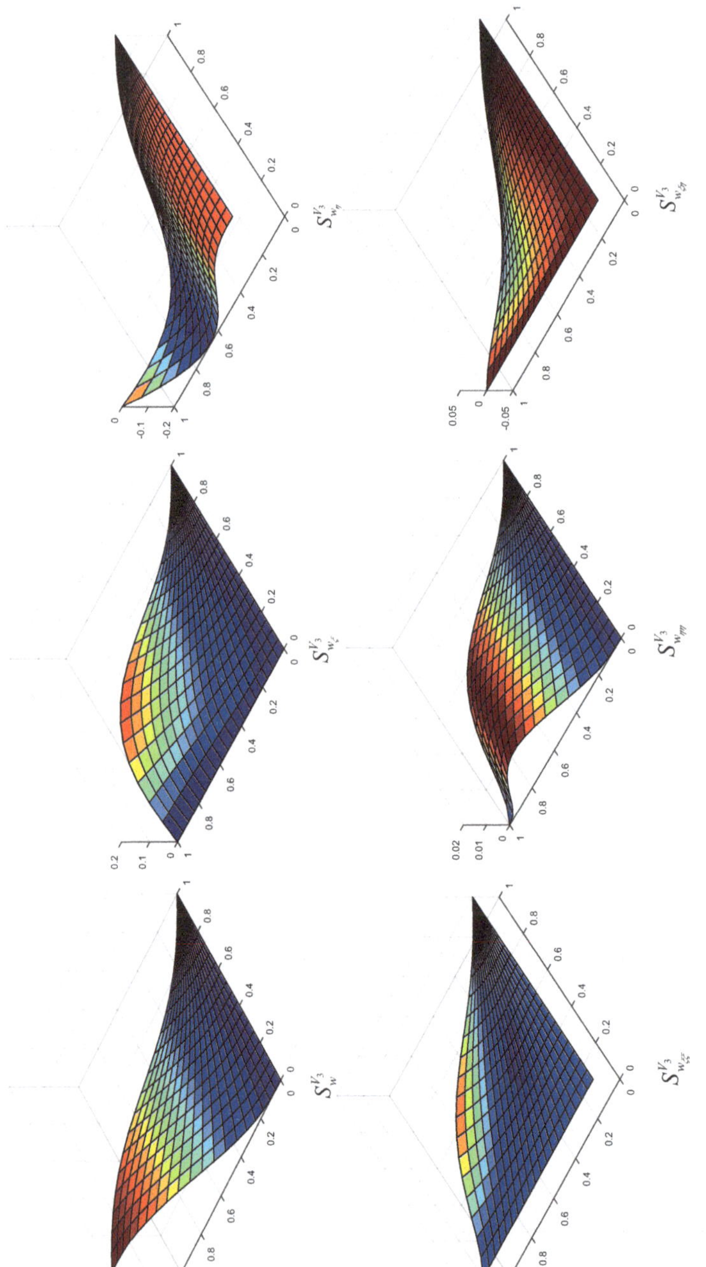

Fig. 5.17 Shape function of vertex 3 in a unit triangular element

$$S^{S_1}_{w_\xi,i}(\eta) = -\xi^2\eta^2\varsigma J_i^{(7,4)}(2\eta - 1),\ i = 0 \sim N \tag{5.114}$$

Edge 3:

$$\begin{aligned} S^{S_2}_{w,i}(\eta) &= -\xi^2\eta^3(\xi + 3\varsigma)J_i^{(7,6)}(2\eta - 1),\ i = 0 \sim N - 1 \\ S^{S_1}_{w_\xi,i}(\eta) &= -\xi^2\eta^2\varsigma J_i^{(7,4)}(2\eta - 1),\ i = 0 \sim N \end{aligned} \tag{5.115}$$

As shown in Fig. 5.18, the first three edge functions with respect to w and w_η on edge 1 of the unit triangular element are presented.

5.2.5.3 Face Functions

The orthogonal face functions on the C^1 triangular element are derived using the method described in Sect. 4.2.4.1. The derivation steps are omitted in this section, and the results are as follows:

$$\begin{aligned} S^F_{pn}(\xi, \eta) &= (1 - \xi - \eta)^2\xi^2\eta^2(1 - \eta)^{p-n-6} \\ &J^{(4,4)}_{p-n-6}\left(\frac{2\xi}{1-\eta} - 1\right)J_n^{(2p-2n-3,4)}(2\eta - 1) \end{aligned} \tag{5.116}$$

Here, p ($\geq$6) is the order of the polynomial, and $n = 0, 1, \bullet\bullet\bullet, p-6$ is the index of the p-th order face function. Equation (5.116) satisfies the orthogonal condition of the Eq. (5.98), and the constant C_{pn} is determined as

$$C_{pn} = \frac{\sqrt{\left[\prod_{i=1}^{4}(p - n - 6 + i)(n + i)\right]}}{\sqrt{\left[(2p - 2n - 3)(2p + 2)\prod_{i=1}^{4}(p - n - 2 + i)(2p - n - 3 + i)\right]}} \tag{5.117}$$

For a given order $p \geq 6$, the number of complete face functions is $N_f = (p-4)(p-5)/2$. The rational term in Eq. (5.116) does not exist in the computational process. The method for calculating the rational term is detailed in Sect. 4.2.4.1. The first three face functions can be expressed as:

$$\begin{aligned} S^F_{60} &= C_{60}\xi^2\eta^2(1 - \xi - \eta)^2 \\ S^F_{70} &= C_{70}5\xi^2\eta^2(1 - \xi - \eta)^2(2\xi + \eta - 1) \\ S^F_{71} &= C_{71}5\xi^2\eta^2(1 - \xi - \eta)^2(3\eta - 1) \end{aligned} \tag{5.118}$$

They are plotted in Fig. 5.19, where both their function values and normal derivatives vanish on the boundaries.

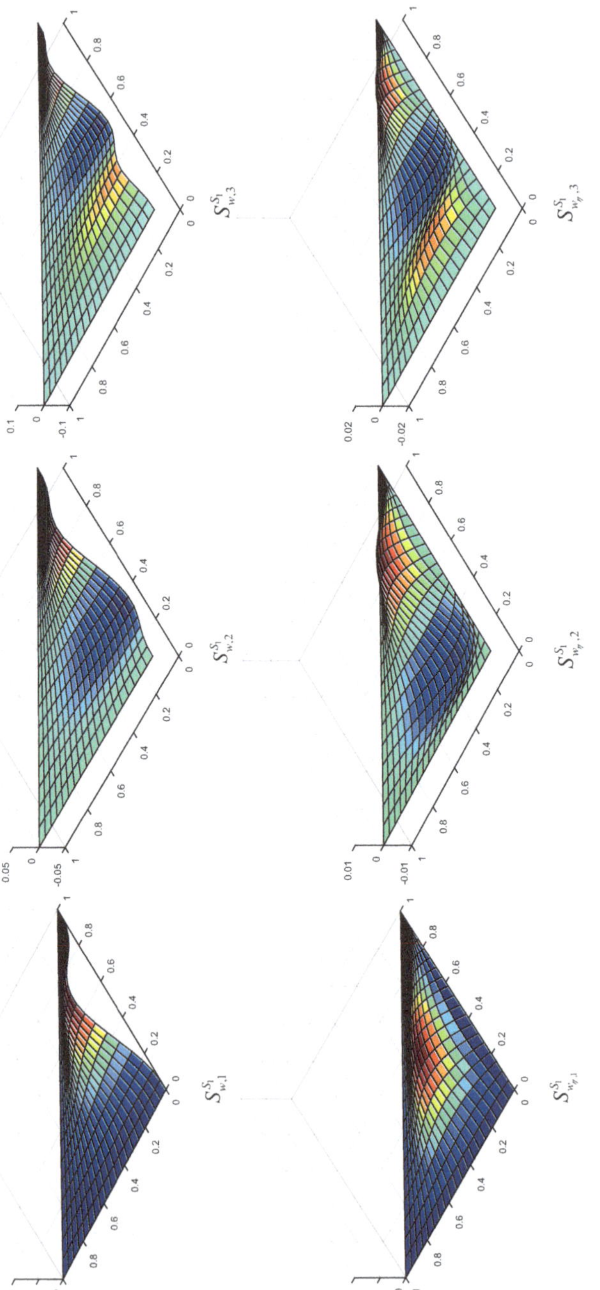

Fig. 5.18 Shape functions on edge 1 of the unit triangular element

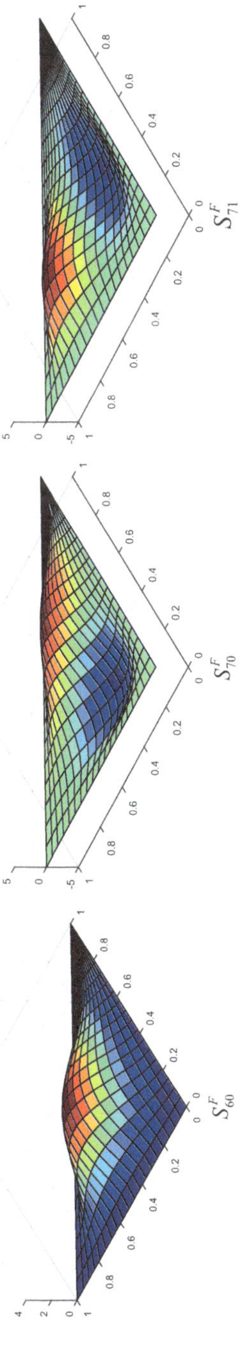

Fig. 5.19 The first three face functions of the unit triangular element

5.2.6 Shape Function Transformation

The hierarchical basis functions described above are defined in natural coordinates. However, due to the difficulty in achieving C^1 continuity, they cannot be directly applied to curved elements. To address this issue, it is necessary to transform them into nodal shape functions in the Cartesian coordinate system. For the quadrilateral element shown in Fig. 5.20a, the displacement field w on the unit square domain can be expressed as follows:

$$\begin{aligned} w(\xi,\eta) &= \sum_{i=1}^{4}\left(S_{w}^{V_i}w^{V_i} + S_{w_\xi}^{V_i}w_\xi^{V_i} + S_{w_\eta}^{V_i}w_\eta^{V_i} + S_{w_{\xi\xi}}^{V_i}w_{\xi\xi}^{V_i} + S_{w_{\eta\eta}}^{V_i}w_{\eta\eta}^{V_i} + S_{w_{\xi\eta}}^{V_i}w_{\xi\eta}^{V_i}\right) \\ &+ \sum_{i=1}^{4}\left[\sum_{j=0}^{N_i-1}\left(S_{w,j}^{S_i}a_j^{S_i}\right) + \sum_{j=0}^{N_i+\Delta-1}\left(S_{w_n,j}^{S_i}b_j^{S_i}\right)\right] + \sum_{i=1}^{N_\xi}\sum_{j=1}^{N_\eta}\left(S_{ij}^{F}c_{ij}^{F}\right) \\ &= \mathbf{N}^{\mathrm{T}}\mathbf{a} \end{aligned} \tag{5.119}$$

In the above description, $N_1 = M$, $N_2 = N$, $N_3 = P$, and $N_4 = Q$ represent the numbers of edge functions for w on each respective edge; V_i $(i = 1, 2, 3, 4)$ denotes vertex i; S_i $(i = 1, 2, 3, 4)$ denotes edge i; and F denotes the face. The number of edge functions related to the normal derivative w_n on each edge is set as $N_i + \Delta$. For straight-edged elements, the increment Δ is typically set to 1 to ensure C^1 continuity. For curve-edged elements, additional nodes can be employed to improve the interpolation accuracy of the normal derivative (i.e., C^1 continuity). Thus, for w, the total number of edge functions is $N_w = \sum N_i$; for w_n, the total number of edge functions is $N_d = 4\Delta + \sum N_i$; and the number of face functions is $N_f = N_\xi N_\eta$.

By using an appropriate geometric mapping function, the natural coordinate unit domain shown in Fig. 5.20a can be mapped to the global coordinate curved domain shown in Fig. 5.20b. In this book, the C^0 blending function (see Sect. 4.2.5.1) is adopted as the exact mapping function for planar quadrilateral and triangular

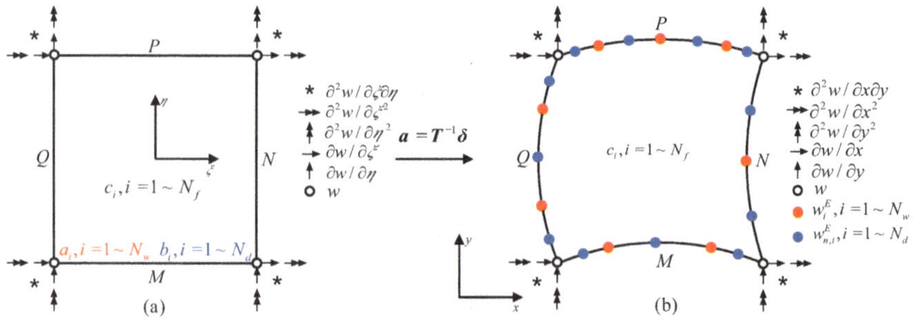

Fig. 5.20 Transformation of shape functions for quadrilateral elements: **a** coefficients of the hierarchical basis functions, **b** coefficients of the nodal basis functions, where $M = 2$, $N = 1$, $P = 3$, $Q = 2$, and $\Delta = 1$

elements, converting nodal variables in natural coordinates to variables in global coordinates. The transformed nodal variables are illustrated in Fig. 5.20b. Along each edge in the x–y coordinate system, a certain number of nodal variables w and w_n (deflection and normal derivative) are assigned. The number of these variables should match the number of higher-order spectral edge functions for w and w_n. The nodal values in Eq. (5.119) can thus be obtained as follows:

$$\delta = \mathbf{Ta} \tag{5.120}$$

Here, $\boldsymbol{\delta}$ includes all transformed boundary nodal variables and untransformed face function coefficients c_i, and $\mathbf{T}$ is the transformation matrix. Therefore, the displacement field can be rewritten as

$$w(\xi, \eta) = \mathbf{N}^{\mathrm{T}}\mathbf{T}^{-1}\boldsymbol{\delta} = \tilde{\mathbf{N}}^{\mathrm{T}}\boldsymbol{\delta} \tag{5.121}$$

Here, $\tilde{\mathbf{N}}$ includes both nodal interpolation shape functions and face functions. The transformation of shape functions for triangular elements is shown in Fig. 5.21. Due to the interpolation properties of $\tilde{\mathbf{N}}$, element assembly and the application of boundary conditions can be carried out as conveniently as in the finite element method. For parallelogram elements and straight-edged triangular elements, since the Jacobian matrix is constant, the C^1 continuity of these elements can be fully satisfied.

In traditional h-version Lagrange elements, higher-order elements commonly use equally spaced nodes. For example, quadratic and cubic elements use 3 and 4 nodes per edge, respectively. However, for p-type elements, when the polynomial order is high, more nodes may be required along the element boundaries. In such cases, equally spaced nodes often result in poorly conditioned matrices, and the basis functions exhibit significant oscillations when the number of nodes increases (see Fig. 5.22a, b). For C^0 problems, it is known that Lagrange functions based on Gauss–Lobatto nodes provide optimal interpolation accuracy (see Fig. 5.22c). In this scenario, each basis function attains its maximum at the corresponding node, and the function values vary within a small range, approximately [−0.2, 1]. For the current

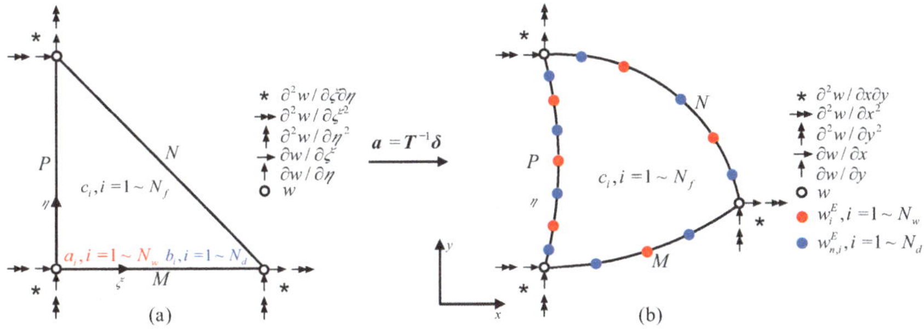

Fig. 5.21 Basis function transformation for triangular elements: **a** coefficients of the hierarchical basis functions, **b** coefficients of the nodal basis functions, where, $M = 1, N = 2, P = 3$, and $\Delta = 1$

thin plate elements, the transformed basis functions along the element boundaries for w and w_n are one-dimensional C^2 and C^1 Hermite basis functions, respectively (see Fig. 5.23). The corresponding node optimization can be achieved by ensuring that each basis function reaches its maximum at its associated node. The following takes the node optimization of the C^1 Hermite basis functions as an example.

As shown in Fig. 5.24, let $f(\xi)$ be a one-dimensional function defined on the interval $[-1, 1]$, given at the nodes $[\xi_1 = -1, \xi_2, \bullet\bullet\bullet, \xi_{N-1}, \xi_N = 1]$, as well as the values of its first derivatives at both endpoints. The C^1 Hermite interpolation of the function f can be expressed as

$$\tilde{f}(\xi) = h_1^{(1)}(\xi) f'(-1) + h_N^{(1)}(\xi) f'(1) + \sum_{i=1}^{N} h_i(\xi) f(\xi_i) \tag{5.122}$$

where $h_1{}^{(1)}(\xi)$ and $h_N{}^{(1)}(\xi)$ are basis functions associated with derivatives, while $h_i(\xi)$ are basis functions associated with the function values at node ξ_i. The definitions of

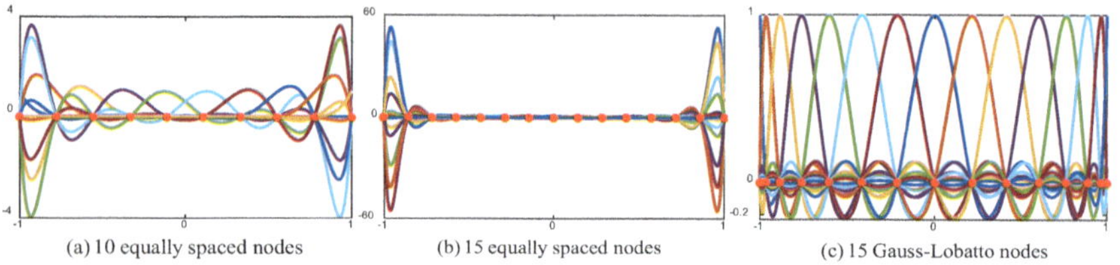

Fig. 5.22 Comparison of Lagrange functions with different nodal distributions

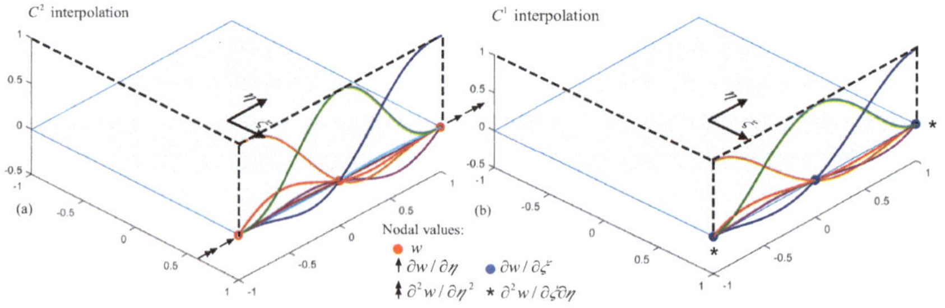

Fig. 5.23 Interpolation basis functions on the boundary with 3 boundary nodes (Edge 2)

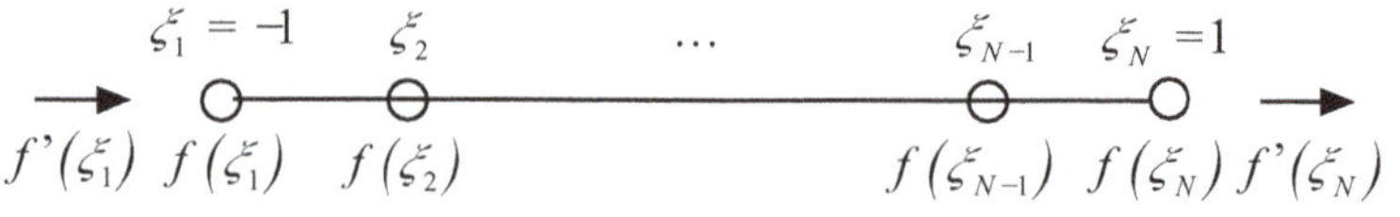

Fig. 5.24 C^1 Hermite interpolation with N nodes

these basis functions are as follows:

$$\begin{aligned}
h_1(\xi) &= (c_1\xi + c_2)\frac{1-\xi}{2}L_1(\xi); \\
h_N(\xi) &= (c_3\xi + c_4)\frac{1+\xi}{2}L_N(\xi) \\
h_1^{(1)}(\xi) &= \frac{1-\xi^2}{2}L_1(\xi); \\
h_N^{(1)}(\xi) &= \frac{\xi^2-1}{2}L_N(\xi) \\
h_i(\xi) &= \frac{1-\xi^2}{1-\xi_i^2}L_i(\xi),\ j = 2 \sim N-1
\end{aligned} \tag{5.123}$$

where $L_i(\xi)$, $i = 1, 2, \bullet\bullet\bullet, N$ are the Lagrange functions, defined as:

$$L_i(\xi) = \prod_{k=1,\ k\neq i}^{N} \frac{\xi - \xi_k}{\xi_i - \xi_k} \tag{5.124}$$

The coefficient c_i ($i = 1, 2, 3, 4$) is expressed as,

$$\begin{aligned}
c_1 &= \frac{1}{2} - L_1'(\xi_1),\ c_2 = \frac{3}{2} - L_1'(\xi_1) \\
c_3 &= -\frac{1}{2} - L_N'(\xi_N),\ c_4 = \frac{3}{2} + L_N'(\xi_N)
\end{aligned} \tag{5.125}$$

In order to obtain optimized nodes, the basis functions of the intermediate nodes, i.e., $h_i(\xi)$, $i = 2, 3, \bullet\bullet\bullet\ N-1$, are forced to reach their maximum values at the nodes ξ_i, similar to the Lagrange functions based on Gauss–Lobatto nodes. This yields:

$$\begin{aligned}
&g_i(\boldsymbol{\xi}) = \frac{\mathrm{d}h_i(\xi_i)}{\mathrm{d}\xi} = \frac{2\xi_i}{\xi_i^2 - 1} + \frac{\mathrm{d}L_i(\xi_i)}{\mathrm{d}\xi} = 0; \\
&(\boldsymbol{\xi} = [\xi_2, \cdots, \xi_{N-1}]\ i = 2,\ 3,\ \cdots,\ N-1)
\end{aligned} \tag{5.126}$$

These $N-2$ intermediate nodes can be determined by the aforementioned $N-2$ nonlinear equations and solved using the Newton–Raphson method. The required Jacobian matrix for the solution is

$$\frac{\partial g_i(\boldsymbol{\xi})}{\partial \xi_j} = \begin{cases} -\dfrac{2\left(1+\xi_j^2\right)}{\left(1-\xi_j^2\right)^2} - \displaystyle\sum_{k=1,\ k\neq j}^{N} \frac{1}{\left(\xi_j - \xi_k\right)^2}, & (i = j) \\ \dfrac{1}{\left(\xi_j - \xi_i\right)^2}, & (i \neq j) \end{cases} \tag{5.127}$$

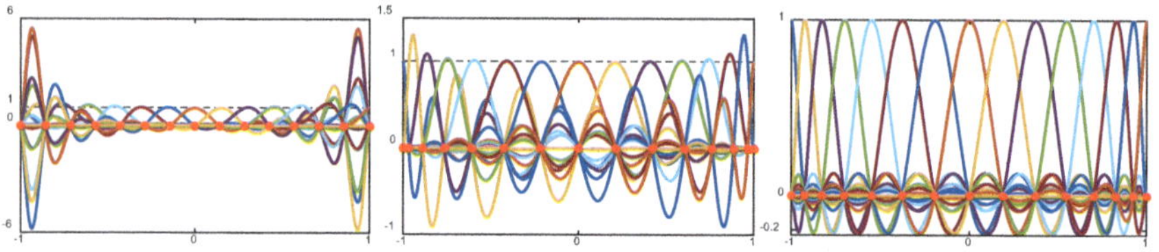

(a) 15 equally spaced nodes (b) 15 Gauss-Lobatto nodes (c) 15 Gauss-Jacobi (3, 3) nodes

Fig. 5.25 Comparison of C^1 Hermite basis functions obtained using different types of nodes

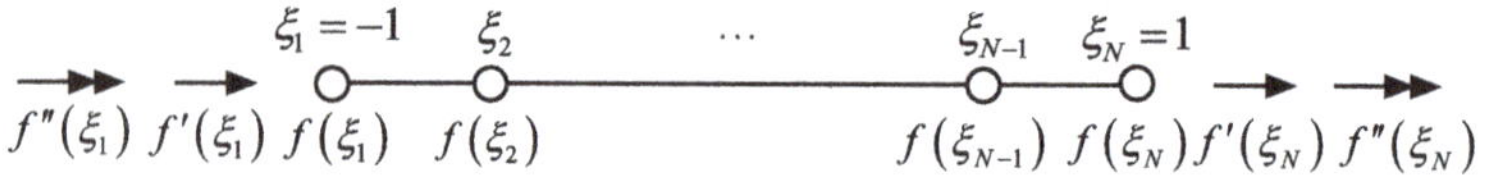

Fig. 5.26 C^2 Hermite interpolation with N nodes

The Gauss–Lobatto nodes can be used as initial guesses. The solution to Eq. (6.126) corresponds to the zeros of the Jacobi polynomial $J^{(3,3)}{}_{N-2}(\xi)$; hence, they are referred to here as Gauss-Jacobi (GJ) nodes. Figure 5.25 compares the C^1 Hermite basis functions obtained using different types of nodes. The basis functions derived from equally spaced nodes oscillate more intensely than those derived from Gauss-Jacobi nodes.

Figure 5.26 illustrates C^2 Hermite interpolation, where the endpoints include second-order derivatives (the fifth-order Hermite basis in the Eq. (5.104) corresponds to the special case of $N = 2$. The derivation of the optimized C^2 Gauss-Jacobi nodes follows a procedure similar to the one described above. They are the zeros of the Jacobi polynomial $J^{(5,5)}{}_{N-2}(\xi)$.

5.2.7 *Finite Element Discretization*

For an isotropic Kirchhoff plate, its potential energy functional and maximum kinetic energy can be expressed as

$$\begin{aligned}\Pi =& \frac{D}{2}\iint_{\Omega}\left(\frac{\partial^2 w}{\partial x^2}\right)^2+\left(\frac{\partial^2 w}{\partial y^2}\right)^2+2\upsilon\frac{\partial^2 w}{\partial x^2}\frac{\partial^2 w}{\partial y^2}\\&+2(1-\upsilon)\left(\frac{\partial^2 w}{\partial x\partial y}\right)^2\mathrm{d}x\mathrm{d}y-\iint_{\Omega}\boldsymbol{qw}\,\mathrm{d}x\mathrm{d}y\\ T_{\max} =& \iint_{\Omega}\frac{1}{2}\rho h\omega^2 w^2\mathrm{d}x\mathrm{d}y\end{aligned}\tag{5.128}$$

where υ is Poisson's ratio, $D = Eh^3/(12(1-\upsilon^2))$ is the bending stiffness, ρ is the volume density, and h is the thickness. For the convenience of numerical discretization, we define the following differential weighting coefficient matrices for the Differential Quadrature (DQ) method:

$$\mathbf{D}_0 = \begin{bmatrix} \mathbf{N}^{\mathrm{T}}(\xi_1, \eta_1) \\ \vdots \\ \mathbf{N}^{\mathrm{T}}\left(\xi_{N_\xi}, \eta_{N_\eta}\right) \end{bmatrix}, \mathbf{D}_{xx} = \begin{bmatrix} \mathbf{N}//^{\mathrm{T}}_{,xx}(\xi_1, \eta_1) \\ \vdots \\ \mathbf{N}//^{\mathrm{T}}_{,xx}\left(\xi_{N_\xi}, \eta_{N_\eta}\right) \end{bmatrix}$$

$$\mathbf{D}_{yy} = \begin{bmatrix} \mathbf{N}//^{\mathrm{T}}_{,yy}(\xi_1, \eta_1) \\ \vdots \\ \mathbf{N}//^{\mathrm{T}}_{,yy}\left(\xi_{N_\xi}, \eta_{N_\eta}\right) \end{bmatrix}, \mathbf{D}_{xy} = \begin{bmatrix} \mathbf{N}//^{\mathrm{T}}_{,xy}(\xi_1, \eta_1) \\ \vdots \\ \mathbf{N}//^{\mathrm{T}}_{,xy}\left(\xi_{N_\xi}, \eta_{N_\eta}\right) \end{bmatrix} \quad (5.129)$$

where (ξ_i, η_j) are the integration points, and N_ξ and N_η are the numbers of nodes in the two directions. To simplify post-processing, it is recommended to use Gauss–Lobatto nodes. When computing the equation, the following chain rule of differentiation is required:

$$\begin{bmatrix} \frac{\partial w}{\partial x} \\ \frac{\partial w}{\partial y} \end{bmatrix} = \frac{1}{|\mathbf{J}|}\begin{bmatrix} J_{22} & -J_{12} \\ -J_{21} & J_{11} \end{bmatrix}\begin{bmatrix} \frac{\partial w}{\partial \xi} \\ \frac{\partial w}{\partial \eta} \end{bmatrix} \quad (5.130)$$

$$\begin{bmatrix} \frac{\partial^2 w}{\partial x^2} \\ \frac{\partial^2 w}{\partial y^2} \\ \frac{\partial^2 w}{\partial x \partial y} \end{bmatrix} = \frac{1}{|\mathbf{J}|^2}\begin{bmatrix} J_{22}^2 & J_{12}^2 & -2J_{12}J_{22} \\ J_{21}^2 & J_{11}^2 & -2J_{11}J_{21} \\ -J_{21}J_{22} & -J_{11}J_{12} & J_{11}J_{22} + J_{12}J_{21} \end{bmatrix} \begin{bmatrix} \frac{\partial^2 w}{\partial \xi^2} - \frac{\partial w}{\partial x}\frac{\partial^2 x}{\partial \xi^2} - \frac{\partial w}{\partial y}\frac{\partial^2 y}{\partial \xi^2} \\ \frac{\partial^2 w}{\partial \eta^2} - \frac{\partial w}{\partial x}\frac{\partial^2 x}{\partial \eta^2} - \frac{\partial w}{\partial y}\frac{\partial^2 y}{\partial \eta^2} \\ \frac{\partial^2 w}{\partial \eta \partial \xi} - \frac{\partial w}{\partial x}\frac{\partial^2 x}{\partial \eta \partial \xi} - \frac{\partial w}{\partial y}\frac{\partial^2 y}{\partial \eta \partial \xi} \end{bmatrix} \quad (5.131)$$

where $x(\xi, \eta)$ and $y(\xi, \eta)$ are the geometric mapping functions, and $\mathbf{J}$ is the Jacobian matrix, defined as:

$$\mathbf{J} = \begin{bmatrix} J_{11} & J_{12} \\ J_{21} & J_{22} \end{bmatrix} = \begin{bmatrix} \frac{\partial x}{\partial \xi} & \frac{\partial y}{\partial \xi} \\ \frac{\partial x}{\partial \eta} & \frac{\partial y}{\partial \eta} \end{bmatrix} \quad (5.132)$$

Using Eqs. (5.121) and (5.129), the functional in Eq. (5.129) can be discretized as:

$$\Pi = \frac{1}{2}\boldsymbol{\delta}^{\mathrm{T}}\mathbf{T}^{-\mathrm{T}}\mathbf{H}^{\mathrm{T}}\mathbf{B}\mathbf{H}\mathbf{T}^{-1}\boldsymbol{\delta} - \mathbf{q}^{\mathrm{T}}\mathbf{C}\mathbf{D}_0\mathbf{T}^{-1}\boldsymbol{\delta}$$
$$T_{\max} = \frac{1}{2}\omega^2 \rho h \boldsymbol{\delta}^{\mathrm{T}}\mathbf{T}^{-\mathrm{T}}\mathbf{D}_0^{\mathrm{T}}\mathbf{C}\mathbf{D}_0\mathbf{T}^{-1}\boldsymbol{\delta} \tag{5.133}$$

where

$$\mathbf{B} = D\begin{bmatrix} \mathbf{C} & \upsilon\mathbf{C} & \mathbf{0} \\ \upsilon\mathbf{C} & \mathbf{C} & \mathbf{0} \\ \mathbf{0} & \mathbf{0} & 2(1-\upsilon)\mathbf{C} \end{bmatrix}, \mathbf{H} = \begin{bmatrix} \mathbf{D}_{xx} \\ \mathbf{D}_{yy} \\ \mathbf{D}_{xy} \end{bmatrix}$$
$$\mathbf{q} = \left[q(\xi_1, \eta_1), \cdots\cdots, q\left(\xi_{N_\xi}, \eta_{N_\eta}\right)\right]^{\mathrm{T}}$$
$$\mathbf{C} = \mathrm{diag}\left\{J_{ij} C_i^{\xi} C_j^{\eta}\right\}, J_{ij} = \left|\mathbf{J}\left(\xi_i, \eta_j\right)\right| \tag{5.134}$$

Here, C_i^{ξ} and C_j^{η} are the integration weighting coefficients, and J_{ij} is the value of the Jacobian matrix at the integration point. The stiffness matrix, mass matrix, and load vector can be expressed as follows:

$$\begin{aligned} &\mathbf{K} = D\mathbf{T}^{-\mathrm{T}}\left[\mathbf{D}_{xx}^{\mathrm{T}}\mathbf{C}\mathbf{D}_{xx} + \mathbf{D}_{yy}^{\mathrm{T}}\mathbf{C}\mathbf{D}_{yy} + \upsilon\left(\mathbf{D}_{xx}^{\mathrm{T}}\mathbf{C}\mathbf{D}_{yy} + \mathbf{D}_{yy}^{\mathrm{T}}\mathbf{C}\mathbf{D}_{yy}\right) + 2(1-\upsilon)\mathbf{D}_{xy}^{\mathrm{T}}\mathbf{C}\mathbf{D}_{xy}\right]\mathbf{T}^{-1} \\ &\mathbf{M} = \rho h\mathbf{T}^{-\mathrm{T}}\mathbf{D}_0^{\mathrm{T}}\mathbf{C}\mathbf{D}_0\mathbf{T}^{-1}, \quad \mathbf{F} = \mathbf{T}^{-\mathrm{T}}\mathbf{D}_0^{\mathrm{T}}\mathbf{C}\mathbf{q} \end{aligned} \tag{5.135}$$

Plate elements have multiple degrees of freedom at each node, so boundary conditions must be applied with care.

5.2.8 *Thin Plate Bending*

This section will validate the computational performance of quadrilateral (abbreviated as *Quad.*) and triangular (*Tri.*) elements through some examples of Kirchhoff plate bending deformation. For simplicity, unless otherwise specified, the numbers of intermediate nodes for w and w_n for all elements are taken as $N \geq 0$ and $N + 1$, respectively. The numbers of face functions for quadrilateral and triangular elements are set to $N_f = N_\xi N_\eta = (N + 1)^2$ and $N_f = (N + 1)N/2$, respectively. Thus, for a given value of N, a quadrilateral element have $N^2 + 10N + 29$ basis functions, and a triangular element have $(N + 6)(N + 7)/2$ basis functions. Gauss-Jacobi points are used as nodes for w and w_n on the element boundaries, which can improve numerical performance.

First, consider a simply supported square plate under a sinusoidal load (Fig. 5.27), analyzed using both h- and p-refinement methods. Figure 5.28 shows two sets of grids for h-refinement using non-distorted quadrilateral and triangular elements. The distorted quadrilateral and triangular grids for h-refinement are shown in Figs. 5.29 and 5.30, respectively, while the grids for p-refinement are shown in Fig. 5.31. The quadrilateral elements in h-refinement are generated by forcing the node counts to satisfy $M = N = P = Q = N_f = 0$; these elements are identical to Watkins' 28-degree-of-freedom rectangular elements. The triangular elements in h-refinement are generated by forcing the node counts to satisfy $M = N = P = N_f = 0$; they are identical to TUBA-21 elements. The energy norm of the displacement error and the relative error of the central moment M_x are defined as follows, both of which will be used to study the convergence rates.

$$\|e\|_E = \left[D \iint_\Omega \left(\frac{\partial^2 e}{\partial x^2}\right)^2 + \left(\frac{\partial^2 e}{\partial y^2}\right)^2 + 2\upsilon \frac{\partial^2 e}{\partial x^2}\frac{\partial^2 e}{\partial y^2} + 2(1-\upsilon)\left(\frac{\partial^2 e}{\partial x \partial y}\right)^2 \mathrm{d}x\mathrm{d}y \right]^{1/2} \tag{5.136}$$

$$\|e_r\|_{M_x} = \frac{|M_{x,app} - M_{x,Exact}|}{M_{x,Exact}}, \quad M_x = -D\left(\frac{\partial^2 w}{\partial x^2} + \upsilon \frac{\partial^2 w}{\partial y^2}\right) \tag{5.137}$$

Here, $e = w^* - w_{app}$ represents the difference between the exact solution w^* and the approximate numerical solution w_{app}, and $M_{x,app}$ denotes the approximate numerical value of the exact moment $M_{x,Exact}$.

In Fig. 5.31a, c, d, N is set to 0, 2, 4, 6, and 8 for p-refinement, while in Fig. 5.31b, N is set to 1, 3, 5, 7, and 9 to ensure a comparable number of degrees of freedom for comparison. In the error plot shown in Fig. 5.32, the error presented in logarithmic form. As shown in Fig. 5.32a, the convergence rate of h-refinement on distorted triangular meshes is slightly lower than that on undistorted triangular meshes; the convergence of h-refinement on distorted quadrilateral meshes is relatively poor, which is due to the loss of C^1 continuity—a common issue in h-version finite elements. All

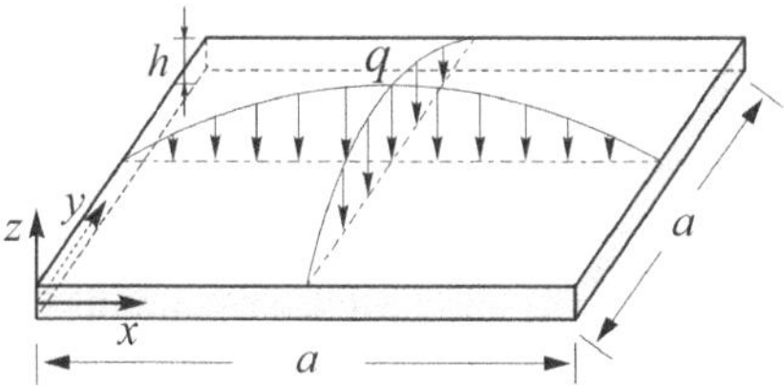

$q = q_0 \sin(\pi x / a)\sin(\pi y / a)$, $q_0 = -2\text{KN/m}^2$
$E = 206\text{GPa}$; $\upsilon = 0.3$;
$h = 0.01\text{m}$; $a = 1\text{m}$
Exact solution: $w = \dfrac{q_0 a^4}{4\pi^4 D} \sin(\pi x / a)\sin(\pi y / a)$

Fig. 5.27 Simply supported rectangular plate under sinusoidal distributed load

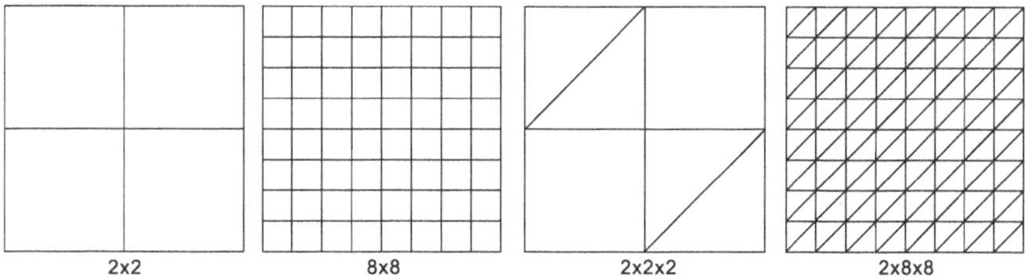

Fig. 5.28 Non-distorted mesh for h-refinement

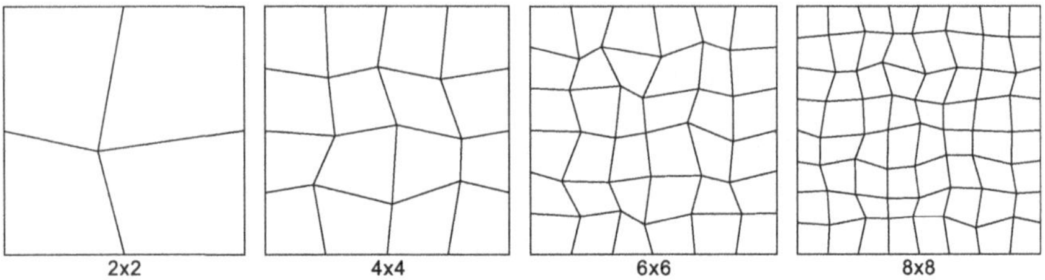

Fig. 5.29 Twisted quadrilateral mesh for h-refinement

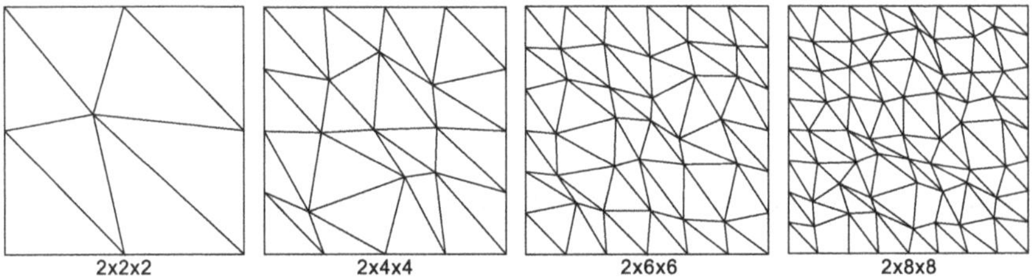

Fig. 5.30 Twisted triangular mesh for h-refinement

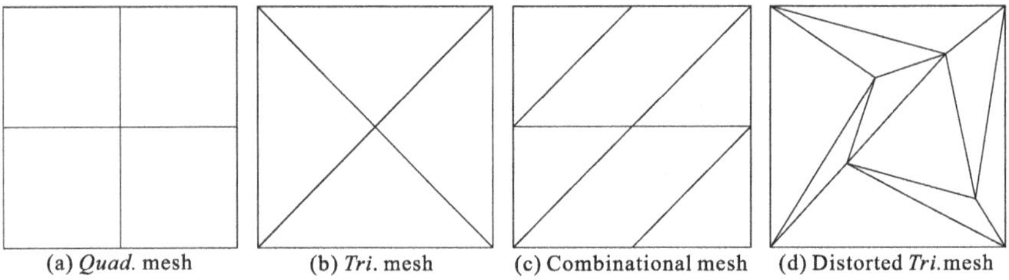

Fig. 5.31 Mesh for p-refinement

p-refinement examples exhibit exponential convergence rates. In p-refinement, the convergence rate for undistorted triangular meshes is higher than that for undistorted quadrilateral meshes, because for the same number of degrees of freedom, the completeness order of triangular meshes is higher than that of quadrilateral meshes (see Fig. 5.33). When using distorted triangular meshes, satisfactory results can still be obtained, with convergence rates very close to those of composite meshes with only mild element distortion. As shown in Fig. 5.32b, rapid convergence of the midpoint moment can also be observed.

For curved-edge elements (including distorted straight-edged quadrilateral elements), the continuity of the displacement function w is fully satisfied along the boundary, while the normal derivative function w_n is only continuous at the nodes. Although full C^1 compatibility is not achieved, it will be shown below that these elements still exhibit a high rate of convergence. As illustrated in Fig. 5.34, a square plate is discretized using a curved-edge mesh. Let N denote the number of nodal

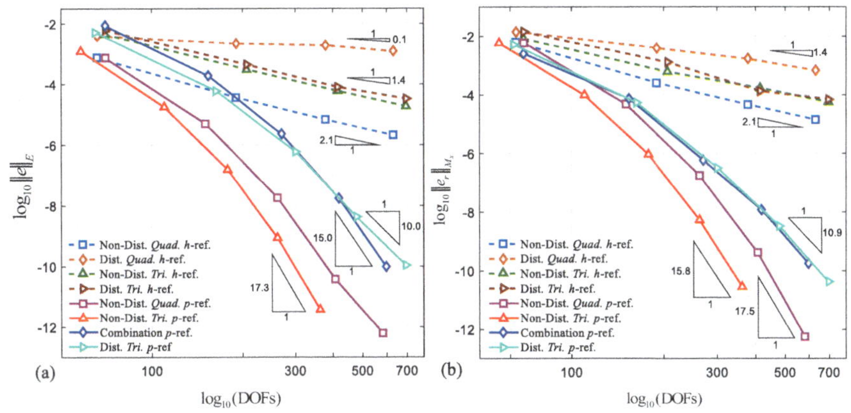

Fig. 5.32 Comparison of convergence rates

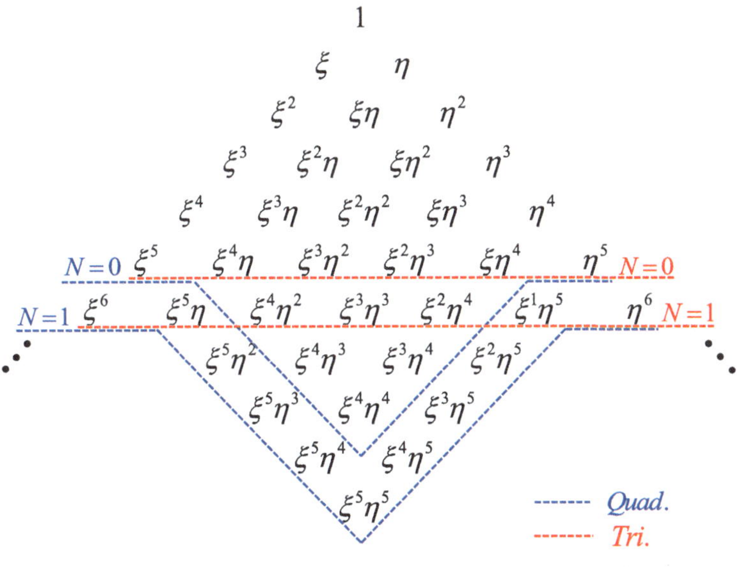

Fig. 5.33 Complete orders of quadrilateral and triangular elements

points for w along each edge, and set the number of nodal points for w_n as $N + \Delta$. The number of surface functions is as previously described.

The convergence curves of the error energy norm for distorted quadrilateral meshes with $N = 0, 2, 4, 6$ and 8 (corresponding to p ranging from 5 to 13) are shown in Fig. 5.35a. Gauss–Lobatto points are used as nodes for w_n (while Gauss-Jacobi nodes with weights (5,5) remain the nodes for w, with Δ ranging from 1 to 3. Results for rectangular conforming elements (Fig. 5.31a) are also included in the figure. Exponential convergence rates are achieved for different increments in the

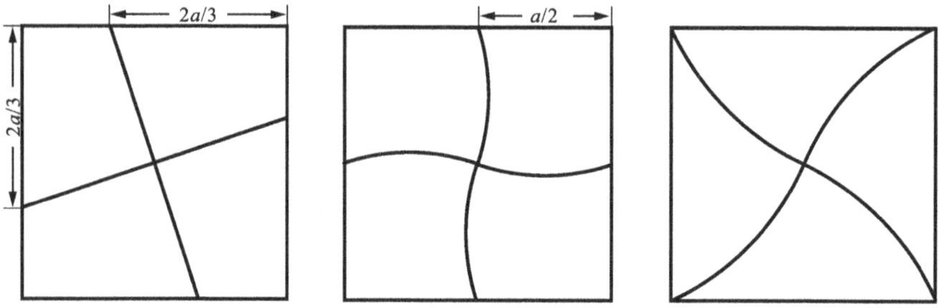

Fig. 5.34 Meshes of three square plates using quasi-conforming elements

number of normal rotation nodes, denoted as Δ. As Δ increases, the convergence speed improves due to the refinement of C^1 continuity as the number of nodes for w_n increases. Surprisingly, when $\Delta = 3$, the convergence curve obtained from distorted elements closely matches the results of conforming elements using non-distorted meshes. To examine the influence of node type on convergence speed, Fig. 5.35b shows convergence curves for different nodal types in distorted quadrilateral meshes with $\Delta = 3$. Exponential convergence is achieved only when Gauss–Lobatto nodes are used. Figure 5.35c and d compare the convergence curves of the error energy norm and the central moment for various meshes with $\Delta = 3$. Rapid convergence is observed in all cases.

5.3 Thick Plate Finite Elements

5.3.1 Irreducible Forms: Reduced Integration

The discretization process of Eq. (5.40) is relatively straightforward, but it is particularly sensitive to certain parameters. The two displacement variables can be approximated using standard isoparametric interpolation, expressed in terms of shape functions and displacement parameters as follows:

$$\boldsymbol{\phi} = \boldsymbol{N}_{\phi}\tilde{\boldsymbol{\phi}} \text{ and } w = \boldsymbol{N}_{w}\tilde{\boldsymbol{w}} \tag{5.138}$$

where the rotation $\boldsymbol{\phi}$ can be transformed into the physical rotation angle $\boldsymbol{\theta}$ about the coordinate axis via Eq. (5.23). The parameter $\boldsymbol{\theta}$ is often more convenient for computation in shell element. The approximate equations can now be obtained directly by applying the principle of total potential energy (Eq. (5.41)), the Galerkin method based on the weak form, or the virtual work expression. Here, the generalized strain components corresponding to the bending moment $\mathbf{M}$ and the shear force $\mathbf{Q}$ are

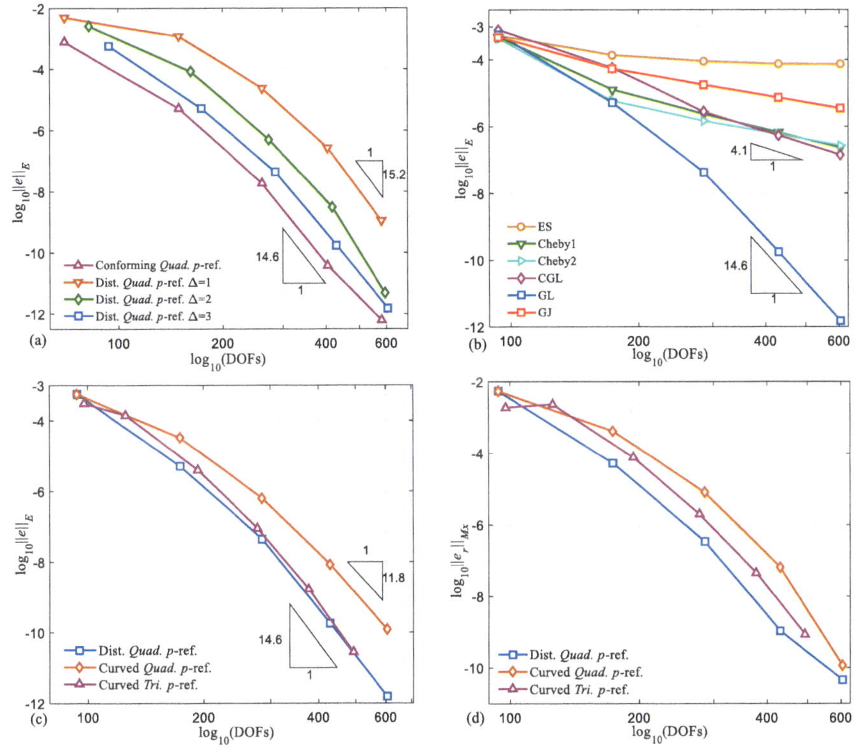

Fig. 5.35 Convergence Comparison: **a** using different meshes with $\Delta = 1, 2$, and 3; **b** with different node types for distorted quadrilateral meshes and $\Delta = 3$; **c** convergence of the energy norm of the error; **d** Convergence of the relative error of the central moment M_x

$$\boldsymbol{\varepsilon}_m = \boldsymbol{S\phi} = \left(\mathbf{SN}_\phi\right)\tilde{\boldsymbol{\phi}} \tag{5.139}$$

and

$$\boldsymbol{\varepsilon}_s = \nabla w + \boldsymbol{\phi} = (\nabla \mathbf{N}_w)\tilde{\mathbf{w}} + \mathbf{N}_\phi \tilde{\boldsymbol{\phi}} \tag{5.140}$$

Therefore, the discrete problem can be obtained as follows.

$$\left(\int_\Omega \left(\mathbf{SN}_\phi\right)^{\mathrm{T}} \mathbf{DSN}_\phi \mathrm{d}\Omega + \int_\Omega \mathbf{N}_\phi^{\mathrm{T}} \boldsymbol{\alpha} \mathbf{N}_\phi \mathrm{d}\Omega\right)\tilde{\boldsymbol{\phi}} + \left(\int_\Omega \mathbf{N}_\phi^{\mathrm{T}} \boldsymbol{\alpha} \nabla \mathbf{N}_w \mathrm{d}\Omega\right)\tilde{\mathbf{w}} = \mathbf{f}_\phi \tag{5.141}$$

and

$$\left(\int_\Omega (\nabla \mathbf{N}_w)^{\mathrm{T}} \boldsymbol{\alpha} \mathbf{N}_\phi \mathrm{d}\Omega\right)\tilde{\boldsymbol{\phi}} + \left(\int_\Omega (\nabla \mathbf{N}_w)^{\mathrm{T}} \boldsymbol{\alpha} \nabla \mathbf{N}_w \mathrm{d}\Omega\right)\tilde{\mathbf{w}} = \mathbf{f}_w \tag{5.142}$$

or, in simplified form,

$$\begin{bmatrix} \mathbf{K}_{ww} & \mathbf{K}_{w\phi} \\ \mathbf{K}_{\phi w} & \mathbf{K}_{\phi\phi} \end{bmatrix} \begin{Bmatrix} \tilde{\mathbf{w}} \\ \tilde{\boldsymbol{\phi}} \end{Bmatrix} = \mathbf{K}\tilde{\mathbf{u}} = (\mathbf{K}_b + \mathbf{K}_s)\tilde{\mathbf{u}} = \begin{Bmatrix} \mathbf{f}_w \\ \mathbf{f}_\phi \end{Bmatrix} = \mathbf{f} \tag{5.143}$$

where

$$\tilde{\mathbf{u}} = \begin{Bmatrix} \tilde{\mathbf{w}} \\ \tilde{\boldsymbol{\phi}} \end{Bmatrix}, \quad \tilde{\boldsymbol{\phi}} = \begin{Bmatrix} \tilde{\boldsymbol{\phi}}_x \\ \tilde{\boldsymbol{\phi}}_y \end{Bmatrix}, \quad \mathbf{K}_b = \begin{bmatrix} \mathbf{0} & \mathbf{0} \\ \mathbf{0} & \mathbf{K}^b_{\phi\phi} \end{bmatrix}, \quad \mathbf{K}_s = \begin{bmatrix} \mathbf{K}^s_{ww} & \mathbf{K}^s_{w\phi} \\ \mathbf{K}^s_{\phi w} & \mathbf{K}^s_{\phi\phi} \end{bmatrix} \tag{5.144}$$

The matrices in the above equation are

$$\begin{aligned} &\mathbf{K}^b_{\phi\phi} = \int_\Omega (\mathbf{S}\mathbf{N}_\phi)^{\mathrm{T}} \mathbf{D}\mathbf{S}\mathbf{N}_\phi \mathrm{d}\Omega \ \ \mathbf{K}^s_{ww} = \int_\Omega (\nabla \mathbf{N}_w)^{\mathrm{T}} \boldsymbol{\alpha} \nabla \mathbf{N}_w \mathrm{d}\Omega \\ &\mathbf{K}^s_{\phi\phi} \int_\Omega \mathbf{N}_\phi^{\mathrm{T}} \boldsymbol{\alpha} \mathbf{N}_\phi \mathrm{d}\Omega \ \ \mathbf{K}^s_{\phi w} = \int_\Omega \mathbf{N}_\phi^{\mathrm{T}} \boldsymbol{\alpha} \nabla \mathbf{N}_w \mathrm{d}\Omega = \left(\mathbf{K}^s_{w\phi}\right)^{\mathrm{T}} \end{aligned} \tag{5.145}$$

The load vector is

$$\begin{aligned} \mathbf{f}_w &= \int_\Omega \mathbf{N}_w^{\mathrm{T}} q \mathrm{d}\Omega + \int_{\Gamma_s} \mathbf{N}_w^{\mathrm{T}} \overline{Q}_n \mathrm{d}\Gamma \\ \mathbf{f}_\phi &= \int_{\Gamma_m} \mathbf{N}_\phi^{\mathrm{T}} \overline{\mathbf{M}} \mathrm{d}\Gamma \end{aligned} \tag{5.146}$$

Here, $\overline{Q}_n$ denotes the shear force on the boundary Γ_s, and $\overline{\mathbf{M}}$ ($\overline{M}_n$, $\overline{M}_{ns}$) represents the moment on the boundary Γ_m (for simplicity, Γ_m and Γ_s are combined).

Since this formulation involves only first-order derivatives, it is evident that any two-dimensional C^0 function can be used to interpolate the two rotations and the transverse displacement. However, it should be noted that when the span-to-thickness ratio L/h is very small, shear locking is likely to occur. Although techniques such as reduced integration have been developed, these methods may fail in certain cases, either resulting in persistent locking or exhibiting anomalous behavior. Therefore, these elements are not considered "robust" and should not be used universally.

It is necessary to better explain the reasons for their failure in order to understand how to design new elements. In the next section, this issue will be addressed using a mixed formulation.

5.3.2 *Hybrid Forms of Thick Plates*

5.3.2.1 Approximation Methods

The thick plate problem can certainly be solved using the mixed method, directly and independently approximating the variables w, $\boldsymbol{\phi}$ and $\mathbf{Q}$ according to Eq. (5.39). In the following derivation, only the stiffness behavior is considered, as this is the source of locking and singularities. Inertial effects are temporarily neglected. Based on Eq. (5.39), a static weak form can be constructed as follows:

$$\begin{aligned}
&\int_{\Omega} \delta w\left[\nabla^{\mathrm{T}}\mathbf{Q}+q\right]\mathrm{d}\Omega=0\\
&\int_{\Omega}\left[(\boldsymbol{S}\delta\boldsymbol{\phi})^{\mathrm{T}}\mathbf{D}\boldsymbol{S}\boldsymbol{\phi}-\delta\boldsymbol{\phi}^{\mathrm{T}}\mathbf{Q}\right]\mathrm{d}\Omega=0\\
&\int_{\Omega}\delta\mathbf{Q}^{\mathrm{T}}\left[-\boldsymbol{\alpha}^{-1}\mathbf{Q}+\boldsymbol{\phi}+\nabla w\right]\mathrm{d}\Omega=0
\end{aligned} \tag{5.147}$$

Now, the standard Galerkin method is used to express the approximate forms of the variables as:

$$\begin{aligned}
&w=\mathbf{N}_w\tilde{\mathbf{w}},\quad \boldsymbol{\phi}=\mathbf{N}_\phi\tilde{\boldsymbol{\phi}}\quad\text{and}\quad \mathbf{Q}=\mathbf{N}_q\tilde{\mathbf{Q}}\\
&\delta w=\mathbf{N}_w\delta\tilde{\mathbf{w}},\quad \delta\boldsymbol{\phi}=\mathbf{N}_\phi\delta\tilde{\boldsymbol{\phi}}\quad\text{and}\quad \delta\mathbf{Q}=\mathbf{N}_q\delta\tilde{\mathbf{Q}}
\end{aligned} \tag{5.148}$$

Other forms of interpolation can also be used, which will be discussed later.

After integrating the first two parts of Eq. (5.147) appropriately, the discrete symmetric system of equations can be obtained.

$$\begin{bmatrix}\mathbf{0} & \mathbf{0} & \mathbf{E}^{\mathrm{T}}\\ \mathbf{0} & \mathbf{K}_b & \mathbf{C}^{\mathrm{T}}\\ \mathbf{E} & \mathbf{C} & \mathbf{H}\end{bmatrix}\begin{Bmatrix}\tilde{\mathbf{w}}\\ \tilde{\boldsymbol{\phi}}\\ \tilde{\mathbf{Q}}\end{Bmatrix}=\begin{Bmatrix}\mathbf{f}_w\\ \mathbf{f}_\phi\\ \mathbf{0}\end{Bmatrix} \tag{5.149}$$

where

$$\begin{aligned}
\mathbf{K}_b&=\int_{\Omega}(\boldsymbol{S}\mathbf{N}_\phi)^{\mathrm{T}}\mathbf{D}\boldsymbol{S}\mathbf{N}_\phi\mathrm{d}\Omega\\
\mathbf{E}&=\int_{\Omega}\mathbf{N}_q^{\mathrm{T}}\nabla\mathbf{N}_w\mathrm{d}\Omega\\
\mathbf{C}&=\int_{\Omega}\mathbf{N}_q^{\mathrm{T}}\boldsymbol{\alpha}^{-1}\mathbf{N}_\phi\mathrm{d}\Omega\\
\mathbf{H}&=-\int_{\Omega}\mathbf{N}_q^{\mathrm{T}}\mathbf{N}_q\mathrm{d}\Omega
\end{aligned} \tag{5.150}$$

The definitions of $\mathbf{f}_w$ and $\mathbf{f}\boldsymbol{\phi}$ are given in Eq. (5.146).

The above formulations represent a typical three-variable mixed problem. As the thickness of the plate approaches zero (which can now be solved exactly), the problem must satisfy certain approximate stability criteria in order to be solvable. In the limiting case, we have

$$\boldsymbol{\alpha}=\infty\quad\text{and}\quad\mathbf{H}=\mathbf{0} \tag{5.151}$$

In this case, a necessary condition for stable solutions under any element assembly and boundary conditions is

$$n_\phi + n_w \geq n_Q \quad \text{or} \quad \alpha_P \equiv \frac{n_\phi + n_w}{n_Q} \geq 1 \tag{5.152}$$

and

$$n_Q \geq n_w \quad \text{or} \quad \beta_P \equiv \frac{n_Q}{n_q} \geq 1 \tag{5.153}$$

Here, $n\phi$, n_Q and n_w represent the number of variables in $\tilde{\phi}$, $\tilde{Q}$ and $\tilde{w}$, respectively, as defined in Eq. (5.149).

Equations (5.152) and (5.153) are conditions that must be satisfied by the entire system. When the above counting conditions are not met, the system of equations is either singular or exhibits locking. Furthermore, to avoid local instability or oscillations, each element must also satisfy these conditions.

Note: Eqs. (5.152) and (5.153) are necessary conditions, but not sufficient conditions. Therefore, it is essential to always perform compatibility and stability tests to ensure that the developed hybrid elements can pass the patch test.

As will be shown later, the above necessary conditions will assist us in designing appropriate thick plate elements that not only exhibit good convergence but can also be used to modify thin plate elements.

5.3.2.2 Continuity Requirements

The approximate forms given in Eqs. (5.149) and (5.150) imply certain continuity requirements. Clearly, the rotation shape function $\mathbf{N}\boldsymbol{\phi}$ requires C^0 continuity (since the approximate form involves the product of first derivatives), whereas $\mathbf{N}_w$ or $\mathbf{N}_q$ may be discontinuous. In the form presented in Eq. (5.150), C^0 continuity of w is implied; however, after integration by parts, a C^0 approximate form for $\mathbf{Q}$ can be obtained. Of course, from a physical standpoint, only the component of $\mathbf{Q}$ normal to the boundary needs to be continuous.

In all the approximations discussed in the previous section, it was assumed that the variables $\boldsymbol{\phi}$ and w are C^0 continuous. While this continuity requirement cannot be considered excessive (as it does not violate any physical conditions), in subsequent chapters, successful elements can also be constructed using discontinuous w interpolation (for reasons unrelated to physical considerations).

It is clearly more convenient to use completely discontinuous interpolation for $\mathbf{Q}$, as this allows for the elimination of shear at the element level, thereby making the final element stiffness matrix can be simply expressed using the standard variables $\tilde{\boldsymbol{\phi}}$ and $\tilde{w}$. In the following sections, we will discuss several specific cases. Here, $\boldsymbol{\alpha}^{-1}$ is allowed to be zero, whereas in other cases it must not be zero.

As mentioned above, in the absence of line or point loads, the continuous interpolation of the normal component of **Q** is physically correct.

However, using this type of interpolation generally does not eliminate $\tilde{\mathbf{Q}}$, and retaining these additional variables incurs significant computational cost. As a result, it has not yet been adopted in practice. Nevertheless, employing an iterative solution process in mixed form can greatly reduce the computational cost associated with these additional variables.

In practice, explicitly employing mixed interpolation and eliminating **Q** variables through static condensation is more convenient (and often more effective) than using special integration schemes. However, in problems where coupling between bending and shear responses arises due to material properties, selective reduced integration is not convenient. Additionally, if $\boldsymbol{\alpha}$ varies within the element, or if isoparametric mapping leads to different interpolation schemes, the condensation method may not be effective.

5.3.3 Higher-Order Elements

5.3.3.1 Element Matrices

The displacement field of a Mindlin plate is identical to that of a two-dimensional solid. Therefore, the shape functions constructed in Sect. 4.2 can also be applied to Mindlin plates. Higher-order elements are not sensitive to shear locking, which provides certain advantages in this regard. Using the weighting coefficient matrix given in Sect. 4.2.4.2, the displacement field can be expressed as follows:

$$w\left(\overline{\xi}, \overline{\eta}\right) = \mathbf{G}w(\xi, \eta) \quad \text{or} \quad \overline{\mathbf{w}} = \mathbf{G}\mathbf{w} \tag{5.154}$$

$$\overline{\theta}_x = \mathbf{G}\theta_x, \quad \overline{\theta}_y = \mathbf{G}\theta_y \tag{5.155}$$

For isotropic Mindlin plates, the stiffness and mass matrices, as well as the load vector, can be expressed as follows:

$$\mathbf{K} = D\begin{bmatrix} \mathbf{K}_{11} & & \text{sym} \\ \mathbf{K}_{21} & \mathbf{K}_{22} & \\ \mathbf{K}_{31} & \mathbf{K}_{32} & \mathbf{K}_{33} \end{bmatrix},$$

$$\mathbf{M} = \rho\begin{bmatrix} J\mathbf{G}^{\mathrm{T}}\mathbf{C}\mathbf{G} & \mathbf{0} & \mathbf{0} \\ \mathbf{0} & J\mathbf{G}^{\mathrm{T}}\mathbf{C}\mathbf{G} & \mathbf{0} \\ \mathbf{0} & \mathbf{0} & h\mathbf{G}^{\mathrm{T}}\mathbf{C}\mathbf{G} \end{bmatrix},$$

$$\mathbf{f} = \begin{bmatrix} \mathbf{G}^{\mathrm{T}}\mathbf{C}\mathbf{m}_x \\ \mathbf{G}^{\mathrm{T}}\mathbf{C}\mathbf{m}_y \\ \mathbf{G}^{\mathrm{T}}\mathbf{C}\mathbf{q}_w \end{bmatrix} \tag{5.156}$$

$$\begin{bmatrix} \mathbf{K}_{11} \\ \mathbf{K}_{22} \\ \mathbf{K}_{33} \end{bmatrix} = \begin{bmatrix} 1 & \upsilon_1 & \upsilon_s \\ \upsilon_1 & 1 & \upsilon_s \\ \upsilon_s & \upsilon_s & 0 \end{bmatrix} \begin{bmatrix} \mathbf{G}_x^{\mathrm{T}}\mathbf{C}\mathbf{G}_x \\ \mathbf{G}_y^{\mathrm{T}}\mathbf{C}\mathbf{G}_y \\ \mathbf{G}^{\mathrm{T}}\mathbf{C}\mathbf{G} \end{bmatrix}, \quad \begin{aligned} \mathbf{K}_{21} &= \upsilon \mathbf{G}_y^{\mathrm{T}}\mathbf{C}\mathbf{G}_x + \upsilon_1 \mathbf{G}_x^{\mathrm{T}}\mathbf{C}\mathbf{G}_y \\ \mathbf{K}_{31} &= -\upsilon_s \mathbf{G}_x^{\mathrm{T}}\mathbf{C}\mathbf{G} \\ \mathbf{K}_{32} &= -\upsilon_s \mathbf{G}_y^{\mathrm{T}}\mathbf{C}\mathbf{G} \end{aligned} \tag{5.157}$$

In the formula, h is the plate thickness, $\upsilon_s = 6\kappa(1-\upsilon)/h^2$, $J = h^3/12$, $D = Eh^3/12(1-\upsilon^2)$ is the bending stiffness of the plate, $\mathbf{C}$ is the integral weight coefficient matrix, $\boldsymbol{m}_x$ and $\boldsymbol{m}_y$ are the nodal moment vectors in the x and y directions, respectively, $\boldsymbol{q}_w$ is the nodal load vector in the z direction, $\mathbf{w}$ is the discretized deflection vector, and $\boldsymbol{\theta}_x$ and $\boldsymbol{\theta}_y$ are the discretized rotation vectors of the mid-plane normal along the x and y directions, respectively. The shear stiffness of the Mindlin plate is $C = \kappa Gh = \upsilon_s D$, where κ is the shear correction factor and G is the shear modulus.

5.3.3.2 Vibration and Bending of Mindlin Plates

In this section, the hierarchical quadrature element method (HQEM) is applied to the vibration and bending analysis of Mindlin plates to validate its accuracy and convergence. For vibration analysis, frequencies are presented in a dimensionless form, denoted as Ω. Results for various geometric shapes and boundary conditions are provided, and the convergence of the method is examined through a series of nodal variations. In all cases, Poisson's ratio is set to 0.3. For convenience in numerical computations, the same number of nodes is adopted in the ξ and η directions along the element boundaries, denoted as N_e, and the same number of hierarchical shape functions is used in both directions, denoted as N_h.

Table 5.1 presents a comparison of bending moments for elliptical plates (Fig. 5.36) under uniformly distributed loads, with both clamped and simply supported boundary conditions. The geometric and material parameters used in the calculation are: $a = 0.50$ m, $b = 0.33333$ m, $h = 0.01$ m, $E = 1$ MPa, $q = 1.0$ Pa and $\kappa = \pi^2/12$. As shown in Fig. 5.36, the HQEM results at points O, A, and B are provided, along with results based on the p-version finite element method for Mindlin plate theory and the Differential Quadrature Finite Element Method (DQFEM). The geometric mapping in Table 5.1 is achieved using blending functions. It can be observed that the HQEM results converge very rapidly, requiring approximately 10 nodes in each direction, which is significantly faster than the convergence rate of the DQFEM reported in the literature, where convergence is achieved around 15 nodes. For both clamped and simply supported boundary conditions, excellent agreement is found among the four sets of results.

In Tables 5.2 and 5.3, the hierarchical quadrature element method (HQEM) is applied to analyze the vibration of square thin plates with four combinations of boundary conditions ¾ simply supported, clamped, and free. Since the results in the literature are based on classical thin plate theory, a relative thickness h/a of 0.0001 is adopted here. The results show that, to all available significant digits, the HQEM solutions are very close to the Kantorovich solutions and are identical to the differential quadrature finite element method (DQFEM) solutions. This verifies

Table 5.1 Computation of clamped and simply supported elliptical plates under uniform load

(N_e, N_h)	$100(w)_0$	$100(M_x)_0$	$100(M_y)_0$	$100(M_x)_a$	$100(M_y)_b$	$100(w)_0$	$100(M_x)_0$	$100(M_y)_0$
	Fixed (Clamped) boundary conditions					Simply supported boundary conditions		
Exact	−0.3759	−0.9227	−1.4048	1.1016	2.4791	−1.5549	−2.4662	−3.5660
FEM	−0.3776	−0.9238	−1.4064	1.1012	2.5005	−1.5546	−2.4100	−3.5065
DQFEM	−0.3773	−0.9232	−1.4048	1.1034	2.4779	−1.5462	−2.4195	−3.5583
	Hierarchical Quadrature Element Method (HQEM) solution based on Mindlin plate Theory ($h = 0.01m$)							
(6, 6)	−0.3763	−0.9119	−1.3872	1.0675	2.4052	−1.5447	−2.4057	−3.5397
(6, 8)	−0.3773	−0.9228	−1.4042	1.1045	2.4814	−1.5459	−2.4193	−3.5578
(6, 10)	−0.3773	−0.9232	−1.4048	1.1036	2.4786	−1.5459	−2.4192	−3.5585
(8, 10)	−0.3773	−0.9232	−1.4048	1.1036	2.4786	−1.5459	−2.4198	−3.5582
(10, 10)	−0.3773	−0.9232	−1.4048	1.1036	2.4786	−1.5459	−2.4198	−3.5581
(12, 12)	−0.3773	−0.9232	−1.4049	1.1037	2.4788	−1.5459	−2.4198	−3.5582

Fig. 5.36 A elliptical plate

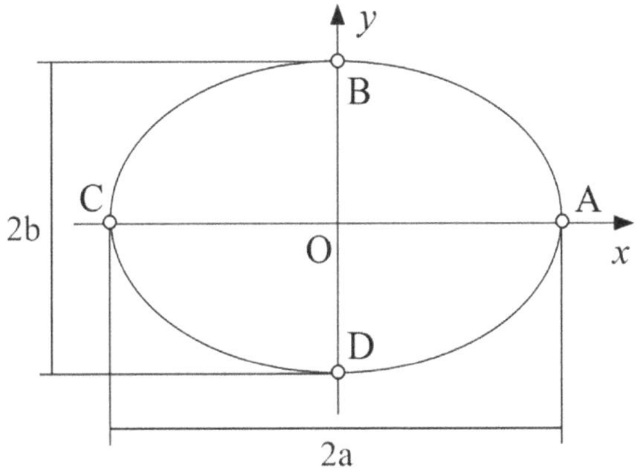

the high accuracy of HQEM under various boundary conditions. Furthermore, it is evident that the HQEM does not suffer from shear locking. It can also be observed that the HQEM can yield highly accurate results with only a small number of sampling points.

Table 5.4 analyzes a square Mindlin plate with SFSF boundary conditions. For this plate, only two elements are used, as shown in Fig. 5.37. Here, the hierarchical quadrature element method (HQEM) results using triangular elements are compared with the results from Liew's p-version Ritz method and Shufrin's semi-analytical solution. As can be seen from Table 5.4, the hierarchical quadrature element method (HQEM) requires only a few nodes to achieve fast convergence and high accuracy.

Table 5.2 Comparison of vibration frequency parameters $\Omega_{ij} = \omega_{ij}a^2\sqrt{\rho h/D}$ for square thin plates with two combinations of simply supported, clamped, and free boundary conditions (CSCS Plate and FSFS Plate)

(N_e, N_h)	Modal number							
	(1,1)	(2,1)	(1,2)	(2,2)	(1,1)	(2,1)	(1,2)	(2,2)
	CSCS Plate				*FSFS* Plate			
Kantorovich	28.951	69.327	54.743	94.585	9.631	16.135	38.945	46.738
DQFEM	28.951	69.327	54.743	94.585	9.631	16.135	38.945	46.738
	Hierarchical Quadrature Element Method (HQEM) solution based on Mindlin plate theory (h/a = 0.0001)							
(6,6)	28.951	69.364	54.765	94.620	9.633	16.136	39.006	46.828
(6,8)	28.951	69.348	54.759	94.607	9.633	16.136	39.003	46.826
(6,10)	28.951	69.348	54.759	94.607	9.633	16.136	39.002	46.826
(8,10)	28.951	69.327	54.743	94.586	9.631	16.135	38.945	46.738
(10,10)	28.951	69.327	54.743	94.585	9.631	16.135	38.945	46.738

Table 5.3 Comparison of vibration frequency parameters $\Omega_{ij} = \omega_{ij}a^2\sqrt{\rho h/D}$ for square thin plates with two combinations of simply supported, clamped, and free boundary conditions (CCCC Plate and FFFF Plate)

(N_e, N_h)	Modal number							
	(1,1)	(2,1)	(1,2)	(2,2)	(1,1)	(2,1)	(1,2)	(2,2)
	CCCC Plate				*FFFF* Plate			
Kantorovich	35.999	73.405	73.405	108.24	–	–	–	–
DQFEM	35.985	73.394	73.394	108.22	13.468	19.596	24.270	34.801
	Hierarchical Quadrature Element Method (HQEM) solution based on Mindlin plate theory (h/a = 0.0001)							
(6,6)	35.990	73.420	73.420	108.26	13.469	19.641	24.374	35.111
(6, 8)	35.986	73.395	73.395	108.22	13.469	19.641	24.369	35.105
(6,10)	35.985	73.394	73.394	108.22	13.469	19.641	24.368	35.105
(8,10)	35.985	73.394	73.394	108.22	13.468	19.596	24.270	34.801
(10,10)	35.985	73.394	73.394	108.22	13.468	19.596	24.270	34.801

5.4 Three-Dimensional Degenerated Shell Elements

This section introduces shell elements with displacement and rotational parameters. These two parameters correspond precisely to the Reissner–Mindlin assumptions used to describe the behavior of thick plates, as discussed in Sect. 5.3. By relaxing the assumptions associated with thin plate theory, three constraints (the deformed normal remains perpendicular to the mid-surface) allow the shell to undergo transverse shear deformation—an important characteristic of thick shells.

Table 5.4 Comparison of vibration frequency parameters $\Omega = \left(\omega a^2/\pi^2\right)\sqrt{\rho h/D}$ ($\upsilon = 0.3$, $h/a = 0.1$, $\kappa = 0.82305$) for a square thin plate with SFSF boundary condition

(N_h, N_e)	Modal number							
	(1, 1)	(2, 1)	(3, 1)	(1, 2)	(2, 2)	(3, 2)	(1, 3)	(2, 3)
(2,5)	0.9587	4.0651	–	1.5916	–	–	3.5913	–
(3,6)	0.9572	3.7029	–	1.5683	4.4869	–	3.5178	6.6371
(4,7)	0.9568	3.6885	7.8620	1.5628	4.3529	8.7432	3.4429	6.4870
(5, 8)	0.9565	3.6831	7.7926	1.5607	4.3418	8.4067	3.4344	6.3200
(6, 9)	0.9564	3.6821	7.7601	1.5594	4.3354	8.3646	3.4308	6.3057
(7,10)	0.9564	3.6817	7.7572	1.5590	4.3336	8.3441	3.4294	6.2935
(8,11)	0.9564	3.6816	7.7561	1.5588	4.3330	8.3411	3.4291	6.2915
(9,12)	0.9564	3.6815	7.7558	1.5588	4.3328	8.3400	3.4289	6.2910
(10,13)	0.9564	3.6815	7.7557	1.5587	4.3327	8.3397	3.4289	6.2908
(12,15)	0.9564	3.6815	7.7557	1.5587	4.3327	8.3396	3.4289	6.2908
(15,18)	0.9564	3.6815	7.7557	1.5587	4.3327	8.3396	3.4289	6.2908
Shufrin	0.9564	3.6815	7.7557	1.5587	4.3327	8.3396	3.4289	6.2908
Liew	0.9564	3.6815	7.7558	1.5588	4.3329	–	3.4290	6.2910

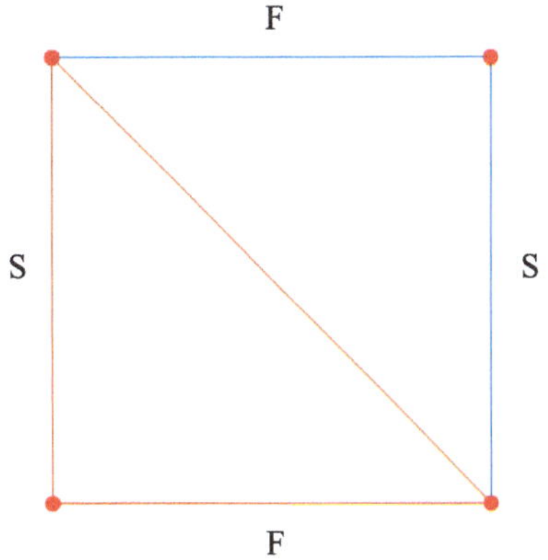

Fig. 5.37 A square plate with two elements and SFSF boundary conditions

Compared to directly using three-dimensional elements, the formulas proposed here are somewhat more complex. The elements constructed in this section are essentially an alternative form of the plate elements previously discussed, which employ independent interpolation for deflection and rotation. When dealing with thin shells, reduced integration is still required. Some of the earlier discussions regarding the robustness of plate elements are also applicable to shells. In general, elements that perform well for plates tend to perform well for shells as well.

5.4.1 Geometric Definition of an Element

As shown in Fig. 5.38, the typical shell element has a curved outer surface, while the lines in the thickness direction of the cross-section are straight. The shell element's shape is defined by the Cartesian coordinates of two given points on the upper and lower surfaces, $\tilde{\mathbf{x}}_{a_{\text{top}}}$ and $\tilde{\mathbf{x}}_{a_{\text{bottom}}}$.

Let ξ and η be the two curvilinear coordinates on the mid-surface of the shell, and ζ be the linear coordinate in the thickness direction. If we further assume that ξ, η, and ζ are defined on each face of the element and their values range from -1 to 1, then the relationship between the Cartesian coordinates and the curvilinear coordinates at any point in the shell can be expressed as follows:

$$\begin{Bmatrix} x \\ y \\ z \end{Bmatrix} = \sum_a N_a(\xi, \eta) \left(\frac{1+\zeta}{2} \begin{Bmatrix} \tilde{x}_a \\ \tilde{y}_a \\ \tilde{z}_a \end{Bmatrix}_{\text{top}} + \frac{1-\zeta}{2} \begin{Bmatrix} \tilde{x}_a \\ \tilde{y}_a \\ \tilde{z}_a \end{Bmatrix}_{\text{bottom}} \right) \tag{5.158}$$

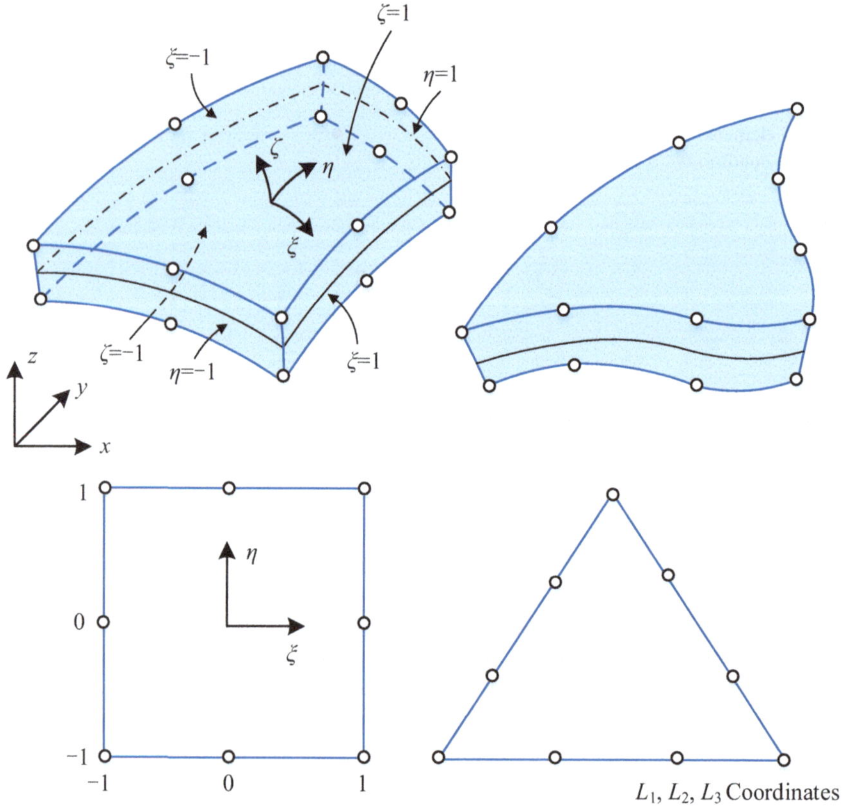

Fig. 5.38 Typical curved thick shell element

Here, $N_a(\xi, \eta)$ denotes the standard two-dimensional shape function; it takes the value 1 at node a on the top and bottom, and 0 at all other nodes. If the basis function N_a is the shape function for a unit square or triangular domain and satisfies compatibility across element interfaces, it can also be applied to curved elements. By employing quadratic or higher-order shape functions, elements of arbitrary curved shapes can be accommodated. In practice, all two-dimensional shape functions introduced in Chap. 4 are applicable here.

After establishing the relationship between Cartesian coordinates and curvilinear coordinates, these shape functions are also applicable in the curvilinear coordinate system. It should be noted that the coordinate direction ζ is generally only approximately perpendicular to the mid-surface.

The relationship in Eq. (5.158) can be rewritten in terms of a "vector" connecting the nodes on the upper and lower surfaces (i.e., a vector of length equal to the shell thickness t) and the coordinates of the shell mid-surface (see Fig. 5.39), as follows:

$$\begin{Bmatrix} x \\ y \\ z \end{Bmatrix} = \sum_a N_a(\xi, \eta)\left(\begin{Bmatrix} \tilde{x}_a \\ \tilde{y}_a \\ \tilde{z}_a \end{Bmatrix} + \frac{\zeta}{2}\mathbf{V}_{3a}\right) \tag{5.159}$$

where

$$\begin{Bmatrix} \tilde{x}_a \\ \tilde{y}_a \\ \tilde{z}_a \end{Bmatrix} = \frac{1}{2}\left(\begin{Bmatrix} \tilde{x}_a \\ \tilde{y}_a \\ \tilde{z}_a \end{Bmatrix}_{\text{top}} + \begin{Bmatrix} \tilde{x}_a \\ \tilde{y}_a \\ \tilde{z}_a \end{Bmatrix}_{\text{bottom}}\right),$$

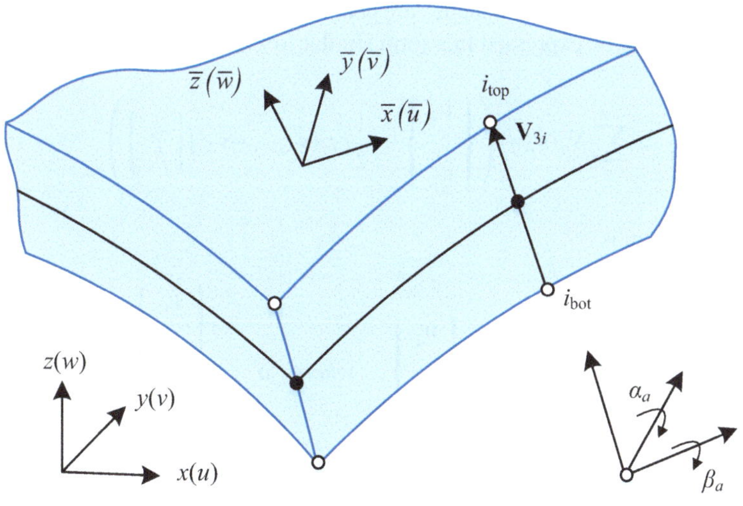

Fig. 5.39 Local and global coordinates

$$\mathbf{V}_{3a} = \begin{Bmatrix} \tilde{x}_a \\ \tilde{y}_a \\ \tilde{z}_a \end{Bmatrix}_{\text{top}} - \begin{Bmatrix} \tilde{x}_a \\ \tilde{y}_a \\ \tilde{z}_a \end{Bmatrix}_{\text{bottom}} \tag{5.160}$$

The vector $\mathbf{x}_a$ defines the coordinates of the midsurface node, while the direction of vector $\mathbf{V}_{3a}$ corresponds to the normal direction of the shell, and its magnitude equals the shell thickness.

For thin shells, the vector $\mathbf{V}_{3\text{a}}$ perpendicular to the mid-surface can be replaced by the unit vector $\mathbf{v}_{3\text{a}}$. Thus, Eq. (5.159) can be written as

$$\begin{Bmatrix} x \\ y \\ z \end{Bmatrix} = \sum_a N_a(\xi, \eta) \left(\begin{Bmatrix} \tilde{x}_a \\ \tilde{y}_a \\ \tilde{z}_a \end{Bmatrix} + \frac{1}{2} \zeta t_a \mathbf{v}_{3a} \right) \tag{5.161}$$

Here, t_a denotes the shell thickness at node a. Constructing a vector perpendicular to the mid-surface is relatively straightforward.

5.4.2 Displacement Field

Since the strain in the direction perpendicular to the mid-surface is assumed to be negligible, the displacement of the entire element can be uniquely determined by the three Cartesian components of the mid-surface nodal displacements, as well as the rotations about two orthogonal directions perpendicular to the nodal vector $\mathbf{V}_{3a}$ (or $\mathbf{v}_{3a}$). If these two orthogonal directions are represented by unit vectors $\mathbf{v}_{1a}$ and $\mathbf{v}_{2a}$, and the corresponding rotation angles are denoted by $\tilde{\alpha}_a$ and $\tilde{\beta}_a$ (see Fig. 5.39), then displacements can be expressed in a form similar to Eq. (5.159), namely,

$$\begin{Bmatrix} u \\ v \\ w \end{Bmatrix} = \sum_a N_a(\xi, \eta) \left(\begin{Bmatrix} \tilde{u}_a \\ \tilde{v}_a \\ \tilde{w}_a \end{Bmatrix} + \frac{1}{2} \zeta t_a [\mathbf{v}_{1a}, \ -\mathbf{v}_{2a}] \begin{Bmatrix} \tilde{\alpha}_a \\ \tilde{\beta}_a \end{Bmatrix} \right) \tag{5.162}$$

Its vector form is

$$\begin{Bmatrix} u \\ v \\ w \end{Bmatrix} = \mathbf{N}\tilde{\mathbf{u}}^e, \quad \tilde{\mathbf{u}}^e = \begin{Bmatrix} \tilde{\mathbf{u}}_1^e \\ \vdots \\ \tilde{\mathbf{u}}_m^2 \end{Bmatrix}, \quad where \quad \tilde{\mathbf{u}}_a^e = \begin{Bmatrix} \tilde{u}_a \\ \tilde{v}_a \\ \tilde{w}_a \\ \tilde{\alpha}_a \\ \tilde{\beta}_a \end{Bmatrix} \tag{5.163}$$

where u, v, and w are the displacements in the global x, y, and z directions, respectively, and

$$\mathbf{N}_a(\xi, \eta, \zeta) = N_a(\xi, \eta)\left[\mathbf{I} \ \tfrac{1}{2}\zeta t_a \mathbf{v}_{1a} \ -\tfrac{1}{2}\zeta t_a \mathbf{v}_{2a}\right] \quad (5.164)$$

Since it is possible to generate infinitely many vectors perpendicular to a given direction, a specific scheme must be designed to ensure uniqueness. A unique method is presented below: let $\mathbf{V}_{3a}$ be the vector whose normal direction is to be constructed. In Cartesian coordinates, the coordinate vector can be defined as

$$\mathbf{x} = x\mathbf{e}_x + y\mathbf{e}_y + z\mathbf{e}_z \quad (5.165)$$

Here, $\mathbf{e}_x$, $\mathbf{e}_y$ and $\mathbf{e}_z$ are three (orthogonal) basis vectors. To construct the first normal vector, identify the smallest component of $\mathbf{V}_{3a}$, and then define vector $\mathbf{V}_{1a}$ as the cross product with the unit vector in that direction. For example, if the x-component of $\mathbf{V}_{3a}$ is the smallest one, then

$$\mathbf{V}_{1a} = \mathbf{e}_x \times \mathbf{V}_{3a} \quad (5.166)$$

where

$$\mathbf{e}_x = \left[1\,0\,0\right]^{\mathrm{T}} \quad (5.167)$$

is the unit vector in x-direction. Then,

$$\mathbf{v}_{1a} = \frac{\mathbf{V}_{1a}}{|\mathbf{V}_{1a}|}, \text{ where } |\mathbf{V}_{1a}| = \sqrt{\mathbf{V}_{1a}^{\mathrm{T}}\mathbf{V}_{1a}} \quad (5.168)$$

is defined as the first unit vector.

The second normal vector can be expressed as

$$\mathbf{V}_{2a} = \mathbf{V}_{3a} \times \mathbf{V}_{1a} \quad (5.169)$$

and normalized using the form given in Eq. (5.168). Therefore, there are three local orthogonal coordinate systems defined by unit vectors $\mathbf{v}_{1a}$, $\mathbf{v}_{2a}$ and $\mathbf{v}_{3a}$. If N_a is a C^0 function, then according to Eq. (5.162), displacement compatibility between adjacent elements can be ensured.

The definition of element coordinates is given by Eq. (5.159), and includes more degrees of freedom than those in the displacement definition. Therefore, this element belongs to the category of "super-parametric" elements, and the constant strain criterion is not automatically satisfied. However, as can be seen from the definition of the strain components involved, this "super-parametric" element is applicable to both rigid body motion and constant strain conditions.

Physically, the definition in Eq. (5.162) assumes that no strain occurs in the thickness direction ζ. Although this direction is not always perfectly perpendicular to the mid-surface, it remains a good approximation commonly used in shell element assumptions.

At each mid-surface node a, as shown in Fig. 5.39, there are five fundamental degrees of freedom, and the connectivity of the elements still follows the standard element assembly procedure.

5.4.3 Definition of Strain (E) and Stress

To derive the finite element matrices, it is necessary to define the strain and stress of the element. According to the fundamental assumptions of shell theory, it is essential to consider the components in the local orthogonal coordinate system associated with the ζ (constant) surface. Therefore, at any point on this surface, a normal axis z- is established, along with two additional orthogonal axes x- and y- that are tangent to the surface (see Fig. 5.39). The strain components can then be expressed using the three-dimensional relationships presented in Sect. 3.1.

$$\overline{\boldsymbol{\varepsilon}} = \begin{Bmatrix} \varepsilon_{\overline{x}} \\ \varepsilon_{\overline{y}} \\ \gamma_{\overline{xy}} \\ \gamma_{\overline{yz}} \\ \gamma_{\overline{zx}} \end{Bmatrix} = \begin{Bmatrix} \overline{u}_{,\overline{x}} \\ \overline{v}_{,\overline{y}} \\ \overline{u}_{,\overline{y}} + \overline{v}_{,\overline{x}} \\ \overline{v}_{,\overline{z}} + \overline{w}_{,\overline{y}} \\ \overline{w}_{,\overline{x}} + \overline{u}_{,\overline{z}} \end{Bmatrix} \tag{5.170}$$

To be consistent with the fundamental assumptions of conventional shells, the strain in the z-direction is neglected. It should be noted that these directions do not generally coincide with the directions of the curvilinear coordinates ξ, η, ζ, even though the $\overline{x}$, $\overline{y}$ directions lie within the ξ-η plane (ζ = constant).

The stress corresponding to the above strain is defined by the matrix $\overline{\boldsymbol{\sigma}}$, and in the case of elasticity, the stress is related to the elasticity matrix $\overline{\mathbf{D}}$, that is,

$$\overline{\boldsymbol{\sigma}} = \begin{Bmatrix} \sigma_{\overline{x}} \\ \sigma_{\overline{y}} \\ \tau_{\overline{xy}} \\ \tau_{\overline{yz}} \\ \tau_{\overline{zx}} \end{Bmatrix} = \overline{\boldsymbol{D}}(\overline{\boldsymbol{\varepsilon}} - \overline{\boldsymbol{\varepsilon}}_0) + \overline{\boldsymbol{\sigma}}_0 \tag{5.171}$$

Here, $\overline{\varepsilon}_0$ and $\overline{\sigma}_0$ denote the "initial" strain and stress, respectively.

The 5×5 matrix $\overline{\mathbf{D}}$ can encompass any anisotropic properties. For laminated (or composite) structures, it can in fact be regarded as a function of ζ.

For isotropic materials, the matrix $\overline{\mathbf{D}}$ is defined as follows:

$$\overline{\mathbf{D}} = \frac{E}{1-\upsilon^2}\begin{bmatrix} 1 & \upsilon & 0 & 0 & 0 \\ \upsilon & 1 & 0 & 0 & 0 \\ 0 & 0 & (1-\upsilon)/2 & 0 & 0 \\ 0 & 0 & 0 & \kappa(1-\upsilon)/2 & 0 \\ 0 & 0 & 0 & 0 & \kappa(1-\upsilon)/2 \end{bmatrix} \tag{5.172}$$

Here, E and υ are Young's modulus and Poisson's ratio, respectively. The coefficient κ included in the last two shear terms is taken as 5/6, which is a shear correction factor (please refer to Sect. 5.1.2). From the definition of displacement, it can be seen that the shear distribution is approximated as constant across the entire thickness direction, whereas in reality, the shear distribution for elastic behavior is approximately parabolic. The coefficient $\kappa = 5/6$ scales the relevant strain energy.

It should be noted that this matrix is not defined by removing appropriate terms from the equivalent three-dimensional stress matrix. To satisfy this important shell assumption, it is necessary to substitute $\sigma_{\overline{z}} = 0$ into the three-dimensional constitutive equations and perform appropriate elimination. This process is analogous to deriving plane stress behavior in two-dimensional analysis.

5.4.4 Element Properties and Necessary Transformations

The element stiffness matrix, as well as other matrices and vectors, all involve integration over the element volume or its boundary. These integrals typically take the following form:

$$\int_{\Omega^e} \mathbf{H}\mathrm{d}x\mathrm{d}y\mathrm{d}z \tag{5.173}$$

Here, the matrix $\mathbf{H}$ is a function of the coordinates. For example, in the stiffness matrix,

$$\mathbf{H} = \overline{\mathbf{B}}^{\mathrm{T}}\overline{\mathbf{D}}\overline{\mathbf{B}} \tag{5.174}$$

Usually, the strain is defined as

$$\overline{\boldsymbol{\varepsilon}} = \overline{\mathbf{B}}\tilde{\mathbf{u}}^e \tag{5.175}$$

Here, $\overline{\mathbf{B}}$ is defined by Eq. (5.173), that is, it is defined in terms of the derivatives of displacement with respect to the local Cartesian coordinates $\overline{x}$, $\overline{y}$, $\overline{z}$. Therefore, two sets of transformations are required before the element can be integrated with respect to the curvilinear coordinates ξ, η, and ζ.

First, by following the same procedure as for iso-parametric elements, the derivatives with respect to the x, y, z directions are obtained. As shown in Eq. (5.162), the global displacements u, v, w are related to the curvilinear coordinates. The derivatives of these displacements with respect to the global x, y, z coordinates can be expressed by the following matrix relationship.

$$\begin{bmatrix} \frac{\partial u}{\partial x} & \frac{\partial v}{\partial x} & \frac{\partial w}{\partial x} \\ \frac{\partial u}{\partial y} & \frac{\partial v}{\partial y} & \frac{\partial w}{\partial y} \\ \frac{\partial u}{\partial z} & \frac{\partial v}{\partial z} & \frac{\partial w}{\partial z} \end{bmatrix} = \mathbf{J}^{-1} \begin{bmatrix} \frac{\partial u}{\partial \xi} & \frac{\partial v}{\partial \xi} & \frac{\partial w}{\partial \xi} \\ \frac{\partial u}{\partial \eta} & \frac{\partial v}{\partial \eta} & \frac{\partial w}{\partial \eta} \\ \frac{\partial u}{\partial \zeta} & \frac{\partial v}{\partial \zeta} & \frac{\partial w}{\partial \zeta} \end{bmatrix} \tag{5.176}$$

where the Jacobian matrix

$$\mathbf{J} = \begin{bmatrix} \frac{\partial x}{\partial \xi} & \frac{\partial y}{\partial \xi} & \frac{\partial z}{\partial \xi} \\ \frac{\partial x}{\partial \eta} & \frac{\partial y}{\partial \eta} & \frac{\partial z}{\partial \eta} \\ \frac{\partial x}{\partial \zeta} & \frac{\partial y}{\partial \zeta} & \frac{\partial z}{\partial \zeta} \end{bmatrix} \tag{5.177}$$

Equation (5.177) is derived based on the coordinate definitions provided in Eq. (5.161). Therefore, for any curvilinear coordinate, the displacement with respect to the global derivatives of the coordinates can all be obtained numerically.

Next, we further expand the coordinate transformation for the local displacements in the $\overline{x}$, $\overline{y}$, $\overline{z}$ directions to compute the strain matrix $\overline{\mathbf{B}}$. The orientation of the local coordinate axes can be established based on the normal vector perpendicular to the ξ-η mid-surface ($\zeta = 0$), and this normal vector can be obtained from two tangent vectors to the mid-surface, $\partial \mathbf{x}/\partial \xi$ and $\partial \mathbf{x}/\partial \eta$. Therefore,

$$\mathbf{V}_3 = \begin{bmatrix} \frac{\partial x}{\partial \xi} \\ \frac{\partial y}{\partial \xi} \\ \frac{\partial z}{\partial \xi} \end{bmatrix} \times \begin{bmatrix} \frac{\partial x}{\partial \eta} \\ \frac{\partial y}{\partial \eta} \\ \frac{\partial z}{\partial \eta} \end{bmatrix} = \begin{bmatrix} \frac{\partial y}{\partial \xi}\frac{\partial z}{\partial \eta} - \frac{\partial y}{\partial \eta}\frac{\partial z}{\partial \xi} \\ \frac{\partial z}{\partial \xi}\frac{\partial x}{\partial \eta} - \frac{\partial z}{\partial \eta}\frac{\partial x}{\partial \xi} \\ \frac{\partial x}{\partial \xi}\frac{\partial y}{\partial \eta} - \frac{\partial x}{\partial \eta}\frac{\partial y}{\partial \xi} \end{bmatrix} \tag{5.178}$$

Following the procedure outlined in Sect. 5.4.2, two orthogonal vectors $\mathbf{V}_1$ and $\mathbf{V}_2$ can be constructed to represent the x- and y-directions, respectively. Three orthogonal vectors can be normalized to unit vectors, thereby yielding a vector matrix in the $\overline{x}$, $\overline{y}$, $\overline{z}$ directions (in fact, a direction cosine matrix), namely

$$\boldsymbol{\theta} = [\mathbf{v}_1,\ \mathbf{v}_2,\ \mathbf{v}_3] \tag{5.179}$$

Thus, the derivatives of the displacements u, v, and w with respect to the global coordinates can be transformed into derivatives with respect to the local orthogonal coordinate system through standard operations.

$$\begin{bmatrix} \frac{\partial \overline{u}}{\partial \overline{x}} & \frac{\partial \overline{v}}{\partial \overline{x}} & \frac{\partial \overline{w}}{\partial \overline{x}} \\ \frac{\partial \overline{u}}{\partial \overline{y}} & \frac{\partial \overline{v}}{\partial \overline{y}} & \frac{\partial \overline{w}}{\partial \overline{y}} \\ \frac{\partial \overline{u}}{\partial \overline{z}} & \frac{\partial \overline{v}}{\partial \overline{z}} & \frac{\partial \overline{w}}{\partial \overline{z}} \end{bmatrix} = \boldsymbol{\theta}^{\mathrm{T}} \begin{bmatrix} \frac{\partial u}{\partial x} & \frac{\partial v}{\partial x} & \frac{\partial w}{\partial x} \\ \frac{\partial u}{\partial y} & \frac{\partial v}{\partial y} & \frac{\partial w}{\partial y} \\ \frac{\partial u}{\partial z} & \frac{\partial v}{\partial z} & \frac{\partial w}{\partial z} \end{bmatrix} \boldsymbol{\theta} \tag{5.180}$$

Therefore, the components of matrix $\overline{\mathbf{B}}$ can be explicitly determined (each node has 5 degrees of freedom).

$$\overline{\boldsymbol{\varepsilon}} = \overline{\mathbf{B}}\tilde{\mathbf{u}}^e \tag{5.181}$$

The expression for $\tilde{\mathbf{u}}^e$ is given by Eq. (5.163).

According to the infinitesimal volume element in curvilinear coordinates, we have

$$\mathrm{d}x\mathrm{d}y\mathrm{d}z = j\,\mathrm{d}\xi\mathrm{d}\eta\mathrm{d}\zeta \tag{5.182}$$

where $j = \det \mathbf{J}$.

The value of the integral is exactly the same as the Gaussian integration in three-dimensional elements. A similar procedure can be used to compute all other related element matrices, such as the inertia matrix, volume and surface load vectors, etc.

Since the dependent variable varies linearly in the thickness (or ζ) direction, only two Gauss points are sufficient in this direction for a homogeneous elastic cross-section. However, for quadratic and cubic shape functions $N_a(\xi, \eta)$, about 2 to 4 points are required in the ξ and η directions.

5.4.5 Some Discussions on the Representation of Stress

The computation method for the element matrix has been provided earlier; the assembly and solution of elements can be carried out using the standard procedure. However, it is still necessary to further discuss the representation of stress. The stress $\overline{\boldsymbol{\sigma}}$ defined in the local coordinate system is readily obtained, and in some cases, these stress components are directly required. Nevertheless, since the local coordinate system is not easily visualized (and may, in fact, be discontinuous between adjacent elements), it is often more convenient to use the following standard transformation to convert them to the global coordinate system, namely,

$$\begin{bmatrix} \sigma_x & \tau_{xy} & \tau_{xz} \\ \tau_{yx} & \sigma_y & \tau_{yz} \\ \tau_{zx} & \tau_{zy} & \sigma_z \end{bmatrix} = \boldsymbol{\theta} \begin{bmatrix} \sigma_{\overline{x}} & \tau_{\overline{xy}} & \tau_{\overline{xz}} \\ \tau_{\overline{yx}} & \sigma_{\overline{y}} & \tau_{\overline{yz}} \\ \tau_{\overline{zx}} & \tau_{\overline{zy}} & \sigma_{\overline{z}} \end{bmatrix} \boldsymbol{\theta}^{\mathrm{T}} \tag{5.183}$$

The above transformation is only applicable to relatively smooth surfaces or when the materials are identical.

However, in general shell structures, the stresses in the global coordinate system do not clearly represent the surface stresses of the shell. Therefore, calculating the principal stresses (or stress invariants) through appropriate transformations is often advantageous. Some of these stress components, such as the shear stresses $\tau_{\overline{xz}}$ and $\tau_{\overline{yz}}$ on the upper and lower surfaces of the shell are typically zero. When transforming the stress components to the global coordinate system according to Eq. (5.186), this point should be noted before establishing the coordinate system, and it is also necessary to ensure that the principal stress components of the shell are located on the shell's surface. The directly obtained shear stress components are actually the average values over the entire cross-section. The maximum transverse shear stress on a solid cross-section is located at the mid-surface. Assuming the variation of shear stress through the thickness is approximately parabolic, the maximum value is about 1.5 times the average value.

5.4.6 Special Cases of Axisymmetric Thick Shells

For axisymmetric shells, the corresponding displacement equations can be further simplified. Since the surface within the element is defined by only two coordinates, ξ and η, the computational effort is significantly reduced.

As shown in Fig. 5.40, the axisymmetric thick shell element is defined in two dimensions.

By modifying Eqs. (5.158) and (5.159) into their two-dimensional forms, the relationship between the global and local coordinates can be defined as follows:

$$\begin{Bmatrix} r \\ z \end{Bmatrix} = \sum_a N_a(\xi,\eta)\left(\frac{1+\eta}{2}\begin{Bmatrix} \tilde{r}_a \\ \tilde{z}_a \end{Bmatrix}_{\text{top}} + \frac{1-\eta}{2}\begin{Bmatrix} \tilde{r}_a \\ \tilde{z}_a \end{Bmatrix}_{\text{bottom}}\right)$$
$$= \sum_a N_a(\xi,\eta)\left(\begin{Bmatrix} \tilde{r}_a \\ \tilde{z}_a \end{Bmatrix} + \frac{1}{2}\eta t_a \mathbf{v}_{3a}\right) \tag{5.184}$$

where

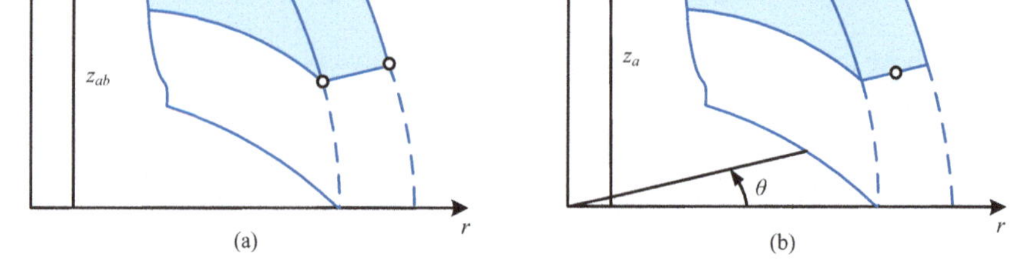

Fig. 5.40 Coordinate system of an axisymmetric thick shell

$$\begin{bmatrix} \tilde{r}_a \\ \tilde{z}_a \end{bmatrix} = \frac{1}{2}\left(\begin{bmatrix} \tilde{r}_a \\ \tilde{z}_a \end{bmatrix}_{\text{top}} + \begin{bmatrix} \tilde{r}_a \\ \tilde{z}_a \end{bmatrix}_{\text{bottom}}\right), \quad \mathbf{v}_{3a} = \begin{bmatrix} \cos\psi_a \\ \sin\psi_a \end{bmatrix} \tag{5.185}$$

In the above equation, ψ_a is the angle defined in Fig. 5.40b, and t_a is the shell thickness. The displacement is defined similarly to Eq. (5.162).

In this case, only axisymmetric loads in the r and z directions (without torsion) need to be considered. Therefore, the two displacement components are specified as follows:

$$\begin{Bmatrix} u \\ w \end{Bmatrix} = \sum_a N_a(\xi,\eta)\left(\begin{Bmatrix} \tilde{u}_a \\ \tilde{w}_a \end{Bmatrix} + \frac{\eta t_a}{2}\begin{Bmatrix} -\sin\tilde{\psi}_a \\ \cos\tilde{\psi}_a \end{Bmatrix}\tilde{\phi}_a\right) \tag{5.186}$$

Here, $\tilde{\phi}_a$ is the rotation angle as shown in Fig. 5.41; $\tilde{u}_a$, $\tilde{w}_a$ are the displacements of the mid-surface nodes.

All transformations follow the methods introduced in the previous sections; it should be noted that they are only applied between the directions ξ, η, r, z and $\bar{r}$, $\bar{z}$, thus involving only two variables.

Similarly, the integration of element properties is performed numerically only with respect to ξ and η. However, it should be noted that the element is a volumetric element.

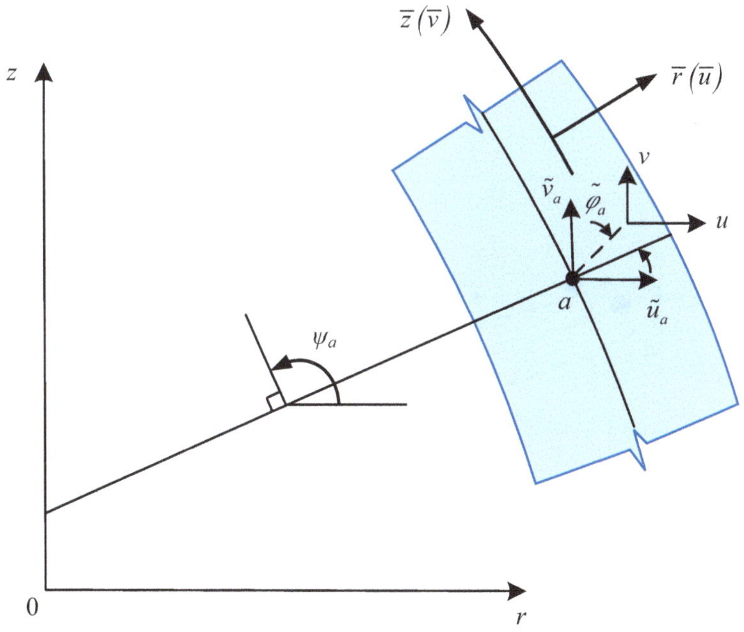

Fig. 5.41 Global displacement in an axisymmetric thick shell

$$\mathrm{d}x\mathrm{d}y\mathrm{d}z = 2\pi r \det \mathbf{J}\, \mathrm{d}\xi \mathrm{d}\eta \mathrm{d}\theta = 2\pi rj\, \mathrm{d}\xi \mathrm{d}\eta \mathrm{d}\theta \tag{5.187}$$

By selecting appropriate shape functions $N_a(\xi)$, linear, parabolic, or cubic variable-thickness shell elements can be used, as shown in Fig. 5.42.

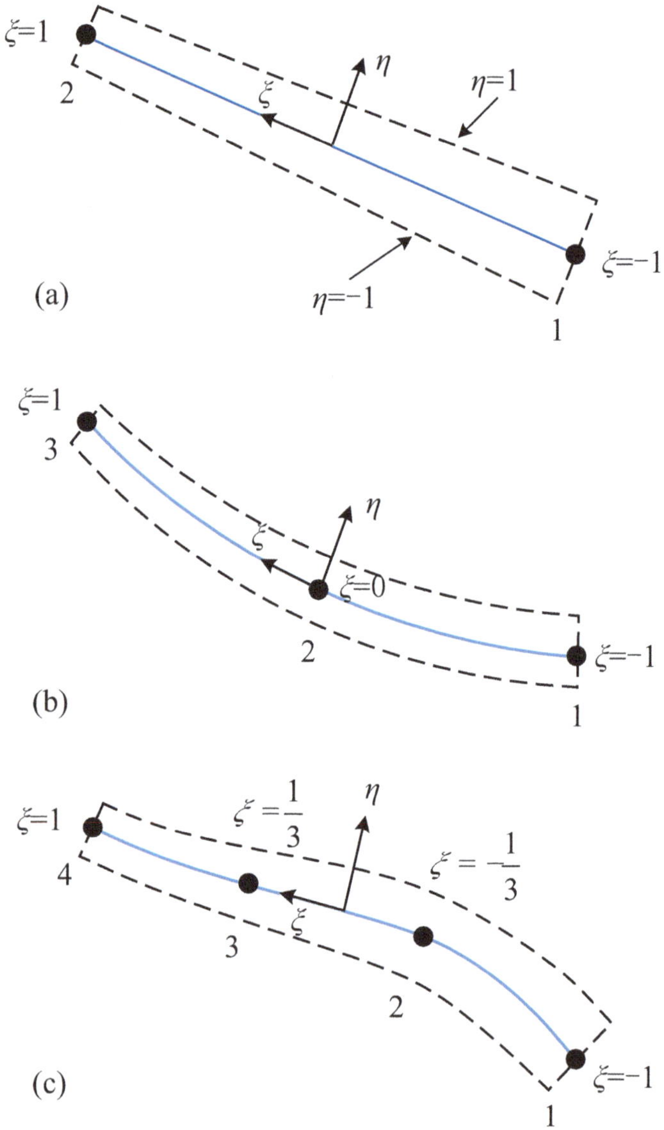

Fig. 5.42 An axisymmetric thick shell element

5.4.7 Special Cases of Thick Plates

The transformations described in this chapter are relatively complex, as are the programming steps involved. However, the methods discussed are equally applicable to thick plates. It is recommended to first gain an understanding through simpler problems such as thick plates.

In the case of thick plates, the following significant simplifications arise:

- $\zeta = 2z/t$, the unit vectors $\boldsymbol{v}_1$, $\boldsymbol{v}_2$ and $\boldsymbol{v}_3$ can be represented by $\boldsymbol{e}_x$, $\boldsymbol{e}_y$ and $\boldsymbol{e}_z$, respectively.
- $\tilde{\alpha}_a$ and $\tilde{\beta}_a$ represent the rotation angles $\tilde{\theta}_y$ and $\tilde{\theta}_x$, respectively (see Sect. 5.1.2).
- It is no longer necessary to transform the stress and strain components to the local $\bar{x}$, $\bar{y}$, $\bar{z}$ coordinate systems; the global coordinate system x, y, z can be used throughout. For this type of element, numerical integration in the thickness direction can be avoided. One can practice deriving the stiffness matrix for linear and rectangular elements, among others. The results obtained are identical to those derived in Sect. 5.3.1, including independent displacement and rotation interpolation, and employing shear constraints. This demonstrates the equivalence of the two methods.

5.5 Summary

This chapter begins with an introduction to thin plate elements, which represent one of the most complex problems in finite element analysis. Nonconforming elements are not discussed in this chapter; although they still have some applications, their accuracy is limited. The chapter presents conforming thin plate elements based on the hierarchical finite element method (HFEM). By removing or reducing the number of basis functions on the surface and edges, lower-order elements can be obtained. The convergence of both the lower-order methods and the hierarchical finite element method (HFEM) is verified through plate bending analysis. Since thick plate elements are very similar to planar problems, only reduced integration, mixed methods, and the hierarchical quadrature element method (HQEM) are discussed in detail. The hierarchical quadrature element method (HQEM) is not sensitive to shear locking. Shell elements in orthogonal curvilinear coordinate systems are similar to plate elements and are therefore only briefly introduced.

The degenerated shell elements introduced in this chapter are almost identical to thick plate elements when applied to plate problems, as thick plate elements also interpolate rotations and displacements independently on the mid-surface. Applying shell elements to axisymmetric problems yields equally accurate results. For general curved shells, this analogy is less apparent but still holds. Therefore, we expect that the discussion on the robustness of plate elements remains valid for shell elements as well.

Although the importance of transverse shear strain constraints is now well understood, the constraints introduced by in-plane (membrane) stresses are still not easily addressed (even though the associated elastic modulus E_t and shear modulus G_t are of the same order of magnitude). Membrane locking may occur in bending deformations involving tension. Although this type of locking has been studied, it has not yet been satisfactorily resolved.

Chapter 6
Cartesian Tensors

In mathematics, a tensor is a type of geometric entity that provides a concise mathematical framework for formulating equations in physical and engineering problems. As such, tensors are indispensable in nonlinear analysis. This book focuses exclusively on nonlinear problems described in Cartesian coordinates, and therefore only introduces Cartesian tensors. The most important characteristic of a tensor is its coordinate invariance, meaning that a tensor is independent of the choice of coordinate system. This property is analogous to the invariance of length and direction in geometric figures, regardless of the coordinate system used in algebraic expressions. In contrast, the components of a tensor in structural analysis are coordinate-dependent. In this chapter, we will discuss how the choice of coordinate system affects the components of a tensor.

6.1 Rotation of Coordinate Axes

6.1.1 Tensors and Coordinate Transformations

A tensor is a natural generalization of vectors or scalars encountered in elementary vector operations. In fact, both are special cases of n-th order tensors, which are determined by 3^n variables in three-dimensional space, known as the components of the tensor. In this context, a scalar is a zeroth order tensor with $3^\circ = 1$ component, and a vector is a first order tensor with $3^1 = 3$ components.

It is important to note that an n-th order tensor is not merely a set of 3^n numbers. The key characteristic of a tensor is that it adheres to the transformation law of its components under changes in the coordinate system (e.g., from Cartesian coordinates to elliptic, polar, or other curvilinear coordinates). If the coordinate system is changed to a new one, the components of the tensor transform according to a specific transformation law. We will see that this transformation law elucidates the physical

B. Liu and C. Liu, *Introduction to Linear and Nonlinear Finite Element Analysis*,
https://doi.org/10.1007/978-981-95-9670-6_6

(or geometric) meaning of tensors, which remain invariant under coordinate transformations. This coordinate-invariant property of tensors satisfies the requirement that the correct formulation of physical laws should be independent of the choice of coordinate systems.

It is evident that physical processes must be independent of coordinate systems. However, what is crucial is the requirement imposed by the coordinate-independent nature of physical processes on the transformation rules of mathematical objects, i.e., tensors. The study of these implications and the classification of physical quantities through transformation laws constitute the main content of this chapter. It should be noted that various tensors are geometric objects whose representations (i.e., the values of their components) adhere to specific transformation laws under coordinate changes.

6.1.2 Summation Convention

To simplify the notation in the derivation process, the following two conventions will be introduced:

Summation convention: When the same index appears more than once in a single term, summation is implied over that index. The range of summation is from 1 to n, where n is the dimension of the space.

Range convention: All distinct indices are assumed to range from 1 to n.

Unless otherwise specified, the above two conventions remain in effect throughout this chapter.

Example The new notation introduced by the summation convention is

$$a_i b_i = \sum_{i=1}^{n} a_i b_i = a_1 b_1 + a_2 b_2 + \cdots + a_n b_n$$

Similarly, if the ranges of i and j are 1 to 2, then

$$a_{ij} b_{ij} = a_{1j} b_{1j} + a_{2j} b_{2j} = a_{11} b_{11} + a_{12} b_{12} + a_{21} b_{21} + a_{22} b_{22}$$

It does not matter whether the summation is performed over i or j first.

Note: Repeated indices are called dummy indices. Due to the implied summation, any such pair of indices can be replaced by any other pair of repeated indices without changing the meaning of the mathematical expression.

6.1.3 Cartesian Coordinate System

A tensor is a mathematical object composed of several components. The values of these components depend on the choice of coordinate system, so even though the tensor itself remains unchanged, its components will vary under a change of coordinates. Among the many possible coordinate transformations, **rigid rotations** of the **Cartesian coordinate system** are the simplest. The fundamental properties of this simplest coordinate transformation will be explained below, serving as a preliminary foundation for studying tensors in more general coordinate systems.

We begin with the following two formal definitions:

Cartesian coordinate system: Associates each point in a given n-dimensional space with a unique ordered set of real numbers (coordinates) $(x_1, x_2, \ldots, x_n)$ by means of a set of oriented straight lines $(Ox_1, Ox_2, \ldots, Ox_n)$ (coordinate axes) intersecting at the origin O.

Rectangular Cartesian Coordinate System: If the three directed lines (coordinate axes) of a Cartesian coordinate system are mutually perpendicular, the system is called a rectangular Cartesian coordinate system.

As shown in Fig. 6.1, this is a schematic diagram of the three-dimensional Cartesian coordinate system. With reference to this coordinate system, let $\boldsymbol{e}_1, \boldsymbol{e}_2, \boldsymbol{e}_3$ denote the vectors (1, 0, 0), (0, 1, 0), and (0, 0, 1), respectively, which are represented geometrically by mutually perpendicular unit arrows.

The Cartesian coordinate axes Ox_1, Ox_2 and Ox_3 form a right-handed system if and only if the angle required to rotate the x_1 axis toward the x_2 axis, $\angle x_1Ox_2 < \pi$, is less than π, and the rotation follows the right-hand rule along the positive direction of the x_3 axis. Conversely, if rotating in the positive direction of the x_3 axis follows the left-hand rule, the axes constitute a left-handed system. In this section, we will use only right-handed Cartesian coordinate systems.

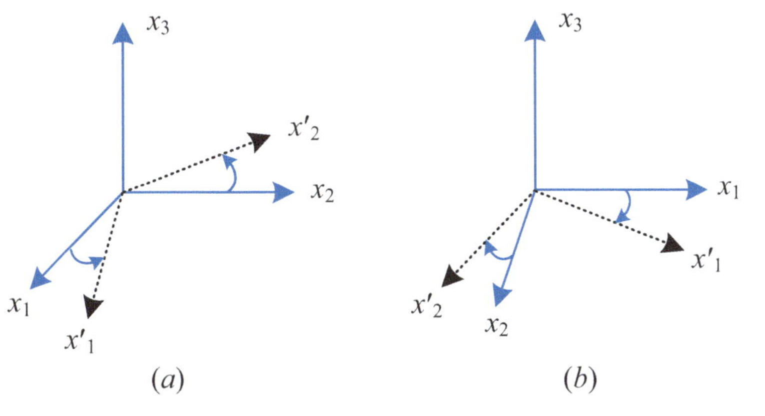

Fig. 6.1 Cartesian coordinate system

6.1.4 Rotation of Coordinate Axes

Next, we establish the rigid rotation of Cartesian axes. Suppose the position vector $\boldsymbol{r}$ has components (x_1, x_2, x_3) and (x'_1, x'_2, x'_3) in two different Cartesian coordinate systems sharing a common origin. We denote the sets of orthonormal basis vectors associated with the original and new coordinate systems by $\{\boldsymbol{e}_i\}$ and $\{\boldsymbol{e}'_i\}$, respectively. The transformation from one Cartesian coordinate system to another is called a rigid rotation of Cartesian axes and possesses the following properties:

The transformation equation for x_k is given by

$$x'_j = R_{jk}x_k \quad \text{(summed over } k\text{)} \tag{6.1}$$

$$x_k = R_{jk}x'_j \quad \text{(summed over } j\text{)} \tag{6.2}$$

Here, $R_{jk} = \boldsymbol{e}'_j \,\Box\, \boldsymbol{e}_k$ represents the direction cosine associated with $\boldsymbol{e}'_j$ and $\boldsymbol{e}_k$.

Note:

- Each of the two subscripts j and k in R_{jk} corresponds to a different basis: the first subscript j refers to the new set $\{\boldsymbol{e}'_j\}$, while the second subscript k refers to the original set $\{\boldsymbol{e}_k\}$. In general, $R_{jk} \neq R_{kj}$.
- The transformation coefficients R_{jk} do not form a tensor; they are simply a set of real numbers. (See the second note in Sect. 6.2.3).

Proof The geometric vector $\boldsymbol{r}$ connecting the origin O and point P can be expressed as

$$\boldsymbol{r} = x_k\boldsymbol{e}_k = x'_j\boldsymbol{e}'_j \tag{6.3}$$

We use $\{\boldsymbol{e}'_j\}$ to express $\boldsymbol{e}_k$, that is,

$$\boldsymbol{e}_k = \left(\boldsymbol{e}'_j \cdot \boldsymbol{e}_k\right)\boldsymbol{e}'_j = R_{jk}\mathbf{e}'_j \tag{6.4}$$

By simultaneously applying the range convention and the summation convention, and substituting Eqs. (6.4) into (6.3), we obtain

$$x_kR_{jk}\mathbf{e}'_j = x'_j\mathbf{e}'_j$$

Therefore

$$\left(x_kR_{jk} - x'_j\right)\mathbf{e}'_j = \mathbf{0}$$

Since the set of vectors $\{\boldsymbol{e}'_j\}$ is linearly independent, the expression in the parentheses above equals zero, thus Eq. (6.1) is proven.

Similarly, by expressing $\{e'_j\}$ in terms of $\{e_k\}$, we obtain

$$e'_j = \left(e_k \cdot e'_j\right)e_k = R_{jk}e_k \tag{6.5}$$

Substituting it into Eqs. (6.3), (6.2) is thus proven.

Note: In the transformation Eq. (6.1) and the expansion (6.5), R_{jk} acts on quantities in the original coordinate system (i.e., x_k or e_k) to produce the corresponding quantities in the new coordinate system (i.e., x'_j or e'_j). However, in Eqs. (6.2) and (6.4), R_{kj} acts on quantities in the new coordinate system to yield the corresponding quantities in the original coordinate system. In all of the above cases, it is essential to ensure that the order of the indices j and k in the coefficient R_{jk} remains unchanged: the first index j always refers to quantities in the original coordinate system, while the second index k always refers to quantities in the new coordinate system.

6.1.5 Orthogonality Relations of Transformation Coefficients

The following theorem describes an important property of the transformation coefficient R_{jk}, which defines the rigid rotation of Cartesian axes.

Orthogonality: The transformation coefficients R_{jk} for a rigid rotation of Cartesian axes satisfy the following condition

$$R_{ik}R_{jk} = \delta_{ij}, \quad R_{ik}R_{il} = \delta_{kl} \tag{6.6}$$

Proof: The first relation in Eq. (6.6) is derived from the geometric formula for the angle θ between the two basis vectors e'_i and e'_j. Let $e'_i = R_{ik}e_k e_i$' and $\mathbf{e}'_j = R_{jl}\mathbf{e}_l$

$$\cos\theta = \mathbf{e}'_i \cdot \mathbf{e}'_j = R_{ik}R_{jl}(\mathbf{e}_k \cdot \mathbf{e}_l) = R_{ik}R_{jl}\delta_{kl} = R_{ik}R_{jk} \tag{6.7}$$

If $i = j$, then $\mathbf{e}'_i$ and $\mathbf{e}'_j$ coincide, i.e., $\theta = 0$; if $i \neq j$, then e'_i and e'_j are orthogonal, i.e., $\theta = \pi/2$. Therefore, we have

$$R_{ik}R_{jk} = \begin{cases} 1, & i = j \\ 0, & i \neq j \end{cases}$$

Similarly, the second equation in Eq. (6.6) can also be verified using the angle between e_k and e_l.

The physical meaning of Eq. (6.6) is quite clear: it ensures that the coordinate axes of each set, $\{e'_i\}$ or $\{e_k\}$, are mutually orthogonal, that is,

$$\mathbf{e}'_i \cdot \mathbf{e}'_j = \delta_{ij}, \mathbf{e}_k \cdot \mathbf{e}_l = \delta_{kl}$$

6.1.6 Matrix Representation of Transformation Coefficients

Since the transformation coefficients R_{jk} have two subscripts, their values can be represented in matrix form. The notation $[R_{jk}]$ denotes the matrix whose element in the j-th row and k-th column is R_{jk}. Furthermore, when $\boldsymbol{R}$ is used alone, it refers to the linear operator of axis rotation, without reference to any specific values of its coefficients R_{jk}.

Example In the two-dimensional case, the rotation angle θ of a rigidly rotating Cartesian coordinate system is given by the following formula

$$\mathbf{e}'_i = R_{ij}\mathbf{e}_j, \quad [R_{ij}] = \begin{pmatrix} R_{11} & R_{12} \\ R_{21} & R_{22} \end{pmatrix} = \begin{pmatrix} \cos\theta & \sin\theta \\ -\sin\theta & \cos\theta \end{pmatrix} \tag{6.8}$$

The coordinate transformation corresponding to this rotation of the coordinate system is

$$x'_i = R_{ij}x_j$$

or equivalently expressed as

$$\begin{pmatrix} x'_1 \\ x'_2 \end{pmatrix} = \begin{pmatrix} R_{11} & R_{12} \\ R_{21} & R_{22} \end{pmatrix}\begin{pmatrix} x_1 \\ x_2 \end{pmatrix} = \begin{pmatrix} \cos\theta & \sin\theta \\ -\sin\theta & \cos\theta \end{pmatrix}\begin{pmatrix} x_1 \\ x_2 \end{pmatrix} \tag{6.9}$$

Note: The following is a brief discussion of the distinction between active and passive transformations, as this is a common source of confusion. In this chapter, we focus exclusively on passive transformations. In a passive transformation, the physical entity under consideration (such as the mass of a particle or a geometric vector) remains unchanged, while only the coordinate system changes from $\{\boldsymbol{e}_i\}$ to $\{\boldsymbol{e}'_i\}$, as shown in Eq. (6.8). In contrast, an active transformation alters the position or orientation of the physical entity itself, while the basis vectors $\{\boldsymbol{e}_i\}$ remain fixed. For an active transformation, the relationship between a two-dimensional geometric vector $\mathbf{x}$ and the rotation angle θ is given by

$$\boldsymbol{e}'_i = \boldsymbol{e}_i \quad \text{and} \quad \begin{pmatrix} x'_1 \\ x'_2 \end{pmatrix} = \begin{pmatrix} \cos\theta & -\sin\theta \\ \sin\theta & \cos\theta \end{pmatrix}\begin{pmatrix} x_1 \\ x_2 \end{pmatrix}$$

This is clearly different from passive transformation. Figure 6.2 illustrates the distinction between the two types of transformations.

It can be shown that the determinant of the matrix $[R_{kl}]$ is

$$\det[R_{kl}] = \pm 1 \tag{6.10}$$

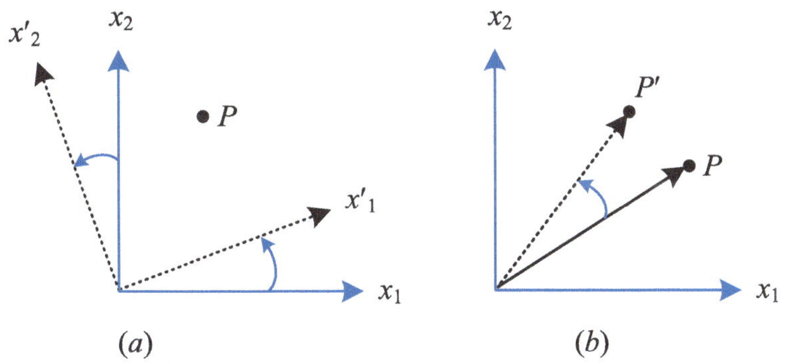

Fig. 6.2 Differences between passive and active transformation

This means there are two types of right-handed Cartesian coordinate systems, corresponding to the positive and negative signs in the equation. In this chapter, only the case where $\det[R_{kl}] = +1$ is considered, which corresponds to the earlier restriction to a single type of coordinate system, namely the right-handed system. As we will see in Sect. 6.3.1, rotations of coordinate axes with a transformation coefficient $\det[R_{kl}] = -1$ result in a left-handed system.

6.1.7 *Determinant of a Matrix*

The following discusses the formal definition of the determinant of a matrix and related information.

- The determinant of a square matrix $[a_{ij}]$ with $n2$ elements

$$D = \det\left[a_{ij}\right] = \begin{bmatrix} a_{11} & a_{12} & \cdots & a_{1n} \\ a_{21} & a_{22} & \cdots & a_{2n} \\ \vdots & \vdots & \ddots & \vdots \\ a_{n1} & a_{n2} & \cdots & a_{nn} \end{bmatrix} \tag{6.11}$$

can be expressed as the sum of $n!$ terms

$$(-1)^r a_{1k_1} a_{2k_2} \cdots a_{nk_n} \tag{6.12}$$

Each term corresponds to $n!$ distinct ordered sets $k_1, k_2, \ldots, k_n$, which are obtained by performing r permutations on the 1st, 2nd, …, n^{th} element of the ordered set.

- The algebraic complement minor M_{ij} of the element a_{ij} in $D = \det[a_{ij}]$ is obtained by deleting the i-th row and the j-th column from expression (6.11).

Example Given a 3rd-order determinant.

$$D = \begin{vmatrix} a_{11} & a_{12} & a_{13} \\ a_{21} & a_{22} & a_{23} \\ a_{31} & a_{32} & a_{33} \end{vmatrix}$$

The algebraic complement minor of element a_{12} is obtained by deleting the first row and the second column from D, that is,

$$M_{12} = \begin{vmatrix} a_{21} & a_{23} \\ a_{31} & a_{33} \end{vmatrix}.$$

- The cofactor C_{ij} of the element a_{ij} is defined as

$$C_{ij} = \frac{\partial D}{\partial a_{ij}} \tag{6.13}$$

Alternatively, it can be equivalently expressed as

$$C_{ij} = (-1)^{i+j} M_{ij}$$

- The determinant D can be expressed in terms of the elements of any row or column and their corresponding algebraic complement minor, that is,

$$D = \det\left[a_{ij}\right] = \sum_{i=1}^{n} a_{ij} C_{ij} = \sum_{k=1}^{n} a_{jk} C_{jk} \quad (Fix\, j) \tag{6.14}$$

This is known as the simple Laplace expansion of the determinant D. Regardless of which row or column is chosen for the expansion, Eq. (6.14) yields the same value for D (that is, the value of D is independent of the choice of j in Eq. (6.14)). Additionally, note that for $j \neq h$,

$$\sum_{i=1}^{n} a_{ij} C_{ih} = \sum_{k=1}^{n} a_{jk} C_{hk} = 0$$

Example Expanding along the first row gives.

$$D = \begin{vmatrix} 1 & 3 & 0 \\ 2 & 6 & 4 \\ -1 & 0 & 2 \end{vmatrix} = 1\begin{vmatrix} 6 & 4 \\ 0 & 2 \end{vmatrix} - 3\begin{vmatrix} 2 & 4 \\ -1 & 2 \end{vmatrix} + 0\begin{vmatrix} 2 & 6 \\ -1 & 0 \end{vmatrix} = -12$$

Note: In expansion (6.13), an n-th order determinant D is expressed as a linear combination of n determinants of order $n-1$. Similarly, each of the subsequent $n-1$ order determinants is in turn expressed as a combination of $n-1$ determinants of order $n-2$. Continuing in this manner, one ultimately obtains $n!$ first-order determinants (that is, $n!$ real numbers), each represented by Eq. (6.12).

6.2 Cartesian Tensors

6.2.1 Cartesian Vectors

After addressing the rotation of coordinate axes, the concept of tensors and their transformation laws in Cartesian coordinate systems will now be introduced. Consider an ordered set of three quantities, v_i $(i = 1, 2, 3)$, which are explicit or implicit functions of x_j. Let us examine how the values of $v_i(x_j)$ change under rigid rotations of the Cartesian axes. If they transform according to the law given below, then v_i is referred to as a component of a specific type of tensor—namely, a Cartesian vector (or a first-order Cartesian tensor).

A Cartesian vector **v** is represented in the x-coordinate system by an ordered set of n functions $v_i(x_j)$, and in the x'-coordinate system by another set of n functions $v'_i(x'_j)$. The functions v'_i and v_i are related at each point by the following transformation law

$$v'_i = R_{ij} v_j \quad \text{and} \quad v_i = R_{ki} v'_k \tag{6.15}$$

where $R_{ij} = \boldsymbol{e}'_i \square\, \boldsymbol{e}_j$.

Clearly, the vector **v** is a geometric object, and thus it is uniquely determined independently of the coordinate system. On the other hand, the functional forms of the components *vi*(*xj*) depend on the choice of the coordinate system, even when considering the same vector **v**. This is why the concepts of a vector and its components are inherently distinct from each other. (For more on this, refer to Sect. 6.2.2).

We reiterate that in the equation above, the subscript i in *Rij* refers to the transformed function v'_i, while j refers to the original function v_j.

6.2.2 Cartesian Vectors and Geometric Vectors

Equation (6.15) may be confusing to some readers: the function v_i appears to represent a geometric vector. To clarify this point, it is necessary to distinguish between the definition of a vector as a first-order Cartesian tensor and its definition as a geometric vector.

In elementary calculations, a vector is simply defined as a geometric entity with a certain magnitude and direction, and is typically denoted by a boldface letter, such as

v. By definition, **v** is uniquely determined by specifying its magnitude and direction, both of which are independent of the choice of coordinate system. However, if a vector is defined algebraically by specifying its components v_k with respect to a given set of coordinate axes, this uniqueness is lost. Even for the same vector **v**, the values of its components v_k depend on the choice of coordinate system. Therefore, when a coordinate transformation is performed according to Eq. (6.15), the values of v_k change in such a way that the magnitude and direction of the vector **v** remain unchanged. This is why the set of n functions v_k is not referred to as a vector itself, but rather as the components of a vector.

In short, we should always keep in mind that vectors are geometric objects independent of any coordinate system, while the components of a vector are merely its mathematical representation with respect to a particular coordinate system. This principle applies to all categories of tensors introduced in this section.

Note: Despite the above considerations, a set of components of a tensor is sometimes also referred to as a "tensor." However, it is important to recognize the fundamental distinction between a tensor (i.e., a coordinate-independent object) and the components of a tensor (i.e., quantities that depend on the choice of coordinates).

6.2.3 Cartesian Tensors

For a second-order Cartesian tensor, two subscripts are required to specify a particular element of the set.

A second-order Cartesian tensor **T** is represented by a set of ordered double subscript components Tij in the x coordinate system and another set of double subscript components $T'kl$ in the x' coordinate system, where at each point, Tij and $T'kl$ are related by the following expression

$$T'_{ij} = R_{ik}R_{jl}T_{kl}, \quad T_{kl} = R_{mk}R_{nl}T'_{mn} \tag{6.16}$$

Here, the double subscript quantity Tij is called a tensor.

In a similar manner, a general-rank Cartesian tensor can be defined, with components $Tij\ldots k$. For any rotation of the axes, the components $T'lm\ldots n$ in the new coordinate system are given by the following expression:

$$T'_{ij\cdots k} = R_{il}R_{jm}\cdots R_{kn}T_{lm\cdots n},$$

$$T_{lm\cdots n} = R_{pl}R_{qm}\cdots R_{rn}T'_{pq\cdots r}.$$

It is clear that an n-th order Cartesian tensor in three dimensions has 3^n components.

Example Suppose there are two Cartesian vectors $\boldsymbol{a}$ and $\boldsymbol{b}$, each represented by components a_j and b_k in the same coordinate system. Then, nine components can be formed as expressed by the following equation.

$$a_j b_k \quad (j, k = 1, 2, 3)$$

This is known as the outer product (or direct product) of vectors $\boldsymbol{a}$ and $\boldsymbol{b}$ (see Sect. 6.4.3 for details). The outer product forms a second-order Cartesian tensor, since the transformation formulas for each a_i and b_j are given by

$$a_i' = R_{ik} a_k \quad \text{and} \quad b_j' = R_{jl} b_l$$

Then we have

$$T_{ij}' = a_i' b_j' = R_{ik} R_{jl} a_k b_l = R_{ik} R_{jl} T_{kl}$$

Note: We emphasize that transformation coefficients, such as *Rij*, do not form a tensor. It is important to note that in the tensor *Tij*, both indices *i* and *j* refer to the same coordinate system, whereas in the coefficients *Rij*, the indices *i* and *j* correspond to different coordinate systems. Therefore, *Tij* and *Rij* are fundamentally different, even though both require two indices.

6.2.4 Scalars

In contrast to the case of finite-order tensors, quantities that remain unchanged under rotation of the coordinate axes are called scalars or zero-order tensors, and they contain only a single component. For example, the square of the distance from a point to the origin is necessarily invariant under any rotation of the coordinate axes. Other examples of scalars include the dot product $\boldsymbol{u} \cdot \boldsymbol{v}$, which also remains unchanged under rotation. In the following divergence is used as an example.

Example If v_i is a component of the vector, then the divergence ∇□v = ∂vi/∂xi is a scalar. In a rotating coordinate system, ∇□**v** is given by the following expression:

$$\frac{\partial v_i'}{\partial x_i'} = \frac{\partial}{\partial x_i'}(R_{ik} v_k) = R_{ik} \frac{\partial v_k}{\partial x_i'}$$

Here, $R_{ik} = \boldsymbol{e}_k$□$\boldsymbol{e}'_i$ is not a function of position. According to the relation $\partial x_j / \partial x'_i = R_{ij}$, we have

$$R_{ik} \frac{\partial v_k}{\partial x_i'} = R_{ik} \frac{\partial x_j}{\partial x_i'} \frac{\partial v_k}{\partial x_j} = R_{ik} R_{ij} \frac{\partial v_k}{\partial x_j} = \delta_{jk} \frac{\partial v_k}{\partial x_j} = \frac{\partial v_j}{\partial x_j}$$

That is,

$$\frac{\partial v_i'}{\partial x_i'} = \frac{\partial v_j}{\partial x_j}$$

6.3 Pseudotensors

6.3.1 Abnormal Rotation

So far, the coordinate transformations discussed have been limited to rigid rotations described by orthogonal matrices [Rij], which possess the following properties:

$$|\mathbf{R}| \equiv \det\left[R_{ij}\right] = 1$$

This transformation is called a proper rotation. If the coordinate transformation includes a rigid rotation described by an orthogonal matrix [Rij], its properties are as follows:

$$|\mathbf{R}| = -1$$

This type of transformation is called an improper rotation (or reflection rotation). Below are two examples of improper rotations.

Example (Inversion) Inversion of the coordinate axis through the origin, i.e.,

$$\mathbf{e}'_i = -\mathbf{e}_i \quad i = 1, 2, 3$$

In this case, the position vector **x** is described with respect to the bases $\mathbf{e}_i$’ and $\mathbf{e}_i$ as follows:

$$\mathbf{x} = x_i\mathbf{e}_i \quad and \quad \mathbf{x} = x_i'\mathbf{e}'_i = x_i'(-\mathbf{e}_i)$$

That is,

$$x_i = -x_i' = -\delta_{ij}x_j'$$

This indicates that axis inversion is represented by $Rij = -\delta ij$. In fact, its determinant becomes

$$|\mathbf{R}| = \begin{vmatrix} -1 & 0 & 0 \\ 0 & -1 & 0 \\ 0 & 0 & -1 \end{vmatrix} = -1$$

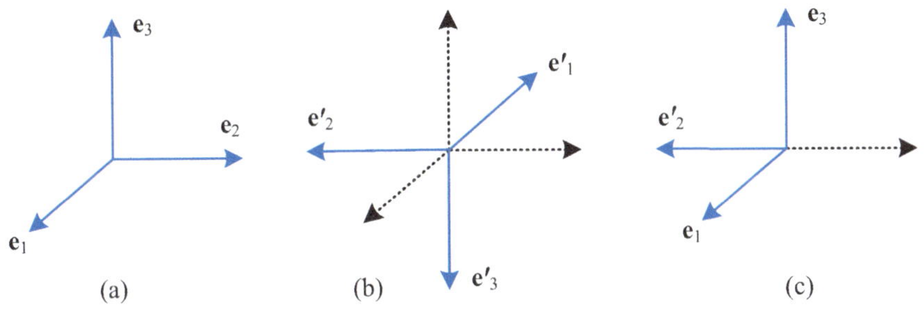

Fig. 6.3 Abnormal rotation: **a** Right-handed Cartesian coordinate system, **b** inversion of the coordinate system, **c** Reflection of the coordinate system

Example (Reflection) A reflection that reverses the orientation of a basis, i.e.,

$$\mathbf{e}'_i = -\mathbf{e}_i \text{ for a certain } i$$

For reflection about the x-axis, we have

$$|\mathbf{R}| = \begin{vmatrix} -1 & 0 & 0 \\ 0 & 1 & 0 \\ 0 & 0 & 1 \end{vmatrix} = -1$$

Note: Inversion differs from standard rotation, which applies only to odd-dimensional spaces. For example, in two or four dimensions, inversion is equivalent to standard rotation. In contrast, reflection, which changes the sign of only one coordinate, is always distinct from standard rotation, regardless of the dimension.

Through an improper rotation, the original right-handed coordinate system is transformed into a left-handed coordinate system, as shown in Fig. 6.3.

Note: This transformation cannot be achieved through any standard rotation.

6.3.2 Pseudovectors

Regardless of whether the coordinate system is orthonormal or non-orthonormal, any rotation described by Rij will transform the components v_i of the vector $\mathbf{v}$ into

$$v'_i = R_{ij} v_j$$

This is because any real physical vector $\mathbf{v}$ can be regarded as a geometric object (i.e., an arrow in space), whose direction and magnitude cannot be altered simply by choosing a different coordinate system to describe it.

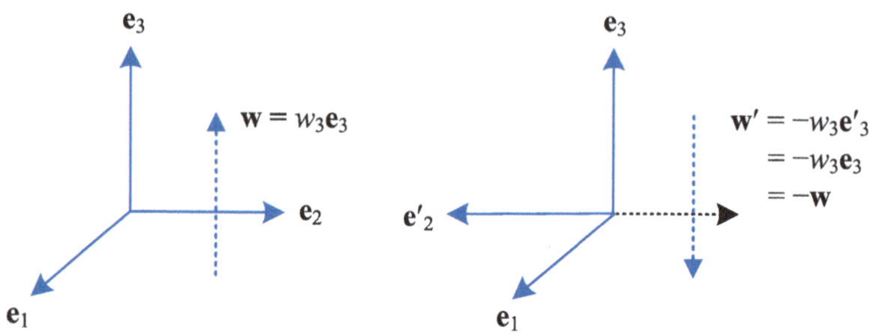

Fig. 6.4 Inversion behavior of the pseudovector **w** (reflection through the $\mathbf{e}_2$ axis)

However, another type of vector **w** can be defined, whose components w_i transform as follows:

$$w_i' = R_{ij} w_j \text{ Normal rotation}$$
$$w_i' = -R_{ij} w_j \text{ Abnormal rotation}$$

or equivalently expressed as

$$w_i' = |\mathbf{R}| R_{ij} w_j$$

In this case, w_i are no longer components of a true Cartesian vector in the strict sense; instead, they are referred to as components forming a pseudovector or a first-order Cartesian pseudotensor. Pseudovectors are also known as axial vectors; correspondingly, true vectors are referred to as polar vectors.

Note: Pseudovectors should not be regarded as true geometric objects (arrows) in space, as their direction reverses under improper coordinate transformations. As shown in Fig. 6.4, the pseudovector **w** is depicted with a dashed line to indicate that it is not a genuine physical vector.

The components v_i of a vector **v** transform under rigid rotations of the Cartesian axes as:

$$v_i' = R_{ij} v_j$$

The components w_i of a pseudovector **w** transform under rigid rotations of the Cartesian axes as:

$$w_i' = |\mathbf{R}| R_{ij} w_j$$

where $|\mathbf{R}|$ is the determinant of the transformation matrix $[Rij]$.

Thus, the distinction between vectors and pseudovectors becomes evident when improper rotations resulting in $|\mathbf{R}| = -1$ are applied.

Pseudovectors frequently arise in physics, although this fact is often not explicitly stated. Below are physical examples of pseudovectors.

Examples The following three physical quantities are all pseudovectors:

1. The angular momentum of a moving particle, $\mathbf{L} = \mathbf{r} \times \mathbf{p}$, where $\mathbf{r}$ is the position vector of the particle and $\mathbf{p}$ is its momentum vector.
2. The torque on a particle, $\mathbf{N} = \mathbf{r} \times \mathbf{F}$, where $\mathbf{r}$ is the position vector of the particle and $\mathbf{F}$ is the force acting on it.
3. The magnetic field, $\mathbf{B} = \nabla \times \mathbf{A}$, defined as the curl of the vector potential $\mathbf{A}$.

It is worth noting that each of these pseudovectors is composed of the vector product of two vectors.

6.3.3 Pseudotensors

The concepts of vectors and pseudovectors can be extended to objects with two or more indices. For example, suppose a component transforms under a proper rotation as follows:

$$T'_{ij} = R_{ik}R_{jl}T_{kl}$$

However, under non-standard rotation, it becomes

$$T'_{ij} = -R_{ik}R_{jl}T_{kl}$$

Thus, Tij represents the components of a second-order Cartesian pseudotensor. Similarly, a Cartesian pseudotensor of any order is defined such that its components transform as follows:

$$T'_{ij\cdots k} = |\mathbf{R}|R_{il}R_{jm}\cdots R_{kn}T_{lm\cdots n}$$

Here, $|\mathbf{R}|$ denotes the determinant of the transformation matrix $[Rij]$. Correspondingly, zeroth-order objects can also be classified as tensors and pseudotensors; the latter remain invariant under proper rotations but change sign under improper rotations.

6.3.4 Levi–Civita Symbol

A typical example of a third-order pseudotensor is the Levi–Civita symbol ε_{ijk}.

If the ordered sets i, j, k are obtained from sets $1, 2, 3$ by an even or odd permutation, then the Levi–Civita symbol (or permutation symbol) ε_{ijk} takes the value $+1$ or -1 accordingly.

In fact, under the following conditions, the value of ε_{ijk} is

$$\varepsilon_{123} = \varepsilon_{231} = \varepsilon_{312} = +1$$
$$\varepsilon_{213} = \varepsilon_{321} = \varepsilon_{132} = -1$$

If any two indices i, j, k are equal, then $\varepsilon_{ijk} = 0$.

The determinant $|\mathbf{A}|$ of a general 3×3 matrix $[Aij]$ is expressed as

$$|\mathbf{A}|\varepsilon_{lmn} = A_{il}A_{jm}A_{kn}\varepsilon_{ijk}$$

Of course, the above equation applies to the transformation matrix $[Rji]$ for rigid-body rotation. Consequently, we obtain

$$|\mathbf{R}|\varepsilon_{lmn} = R_{il}R_{jm}R_{kn}\varepsilon_{ijk}$$

or equivalently expressed as

$$\varepsilon_{ijk} = |\mathbf{R}|R_{il}R_{jm}R_{kn}\varepsilon_{lmn} \tag{6.17}$$

This indicates that ε_{ijk} is a third-order Cartesian pseudotensor.

Equation (6.17) not only demonstrates the pseudotensor nature of ε_{ijk}, but also clearly shows that all components of ε_{ijk} remain unchanged under any rotation of the axes. Tensors with this property are called isotropic tensors (invariant tensors or fundamental tensors). It is known that there exists no first-order isotropic tensors, and that the only second- and third-order isotropic tensors are scalar multiples of δij and ε_{ijk}, respectively. Furthermore, the most general fourth-order isotropic tensor can be expressed as

$$\lambda\delta_{ik}\delta_{mp} + \mu\delta_{im}\delta_{kp} + \upsilon\delta_{ip}\delta_{km}$$

Here, λ, μ, and υ are arbitrary constants (such fourth-order isotropic tensors appear in the theory of solid elasticity; see Sect. 6.5.2 for details). All of the isotropic tensors mentioned above are related to the description of the physical properties of isotropic media (that is, media that exhibit the same properties regardless of their orientation).

The following notations are commonly used in applications

$$\varepsilon_{ijk}\varepsilon_{klm} = \delta_{il}\delta_{jm} - \delta_{im}\delta_{jl} \tag{6.18}$$

This can be verified using the properties of the Levi–Civita symbol.

6.4 Tensor Algebra

6.4.1 Addition and Subtraction

Addition and subtraction of tensors form the foundation of tensor algebra, providing methods to construct new tensors from existing ones. For convenience, we may refer to *Tij* simply as a tensor, but it should always be remembered that *Tij* represents the components of $\mathbf{T}$ in a specific coordinate system.

Addition and subtraction of tensors are defined in an obvious manner. If $A_{ij\ldots k}$ and $B_{ij\ldots k}$ are tensors (or components) of the same order, then for each set of indices $i, j, \ldots, k$, their sum and difference, $S_{ij\ldots k}$ and $D_{ij\ldots k}$, are given by

$$\begin{aligned} S_{ij\cdots k} &= A_{ij\cdots k} + B_{ij\cdots k} \\ D_{ij\cdots k} &= A_{ij\cdots k} - B_{ij\cdots k} \end{aligned}$$

Furthermore, the linearity of coordinate rotation yields

$$\begin{aligned} R_{ip}S_{pq\cdots r} &= R_{ip}\left(A_{pq\cdots r} + B_{pq\cdots r}\right) = R_{ip}A_{pq\cdots r} + R_{ip}B_{pq\cdots r} \\ &= A_{iq\cdots r} + B_{iq\cdots r} = S_{iq\cdots r} \end{aligned}$$

6.4.2 Contraction

Contraction is the operation of setting two indices equal and summing over all possible values of the repeated index. It is a unique operation in tensor algebra and is crucial in certain calculations.

For example, the transformation law of a third-order tensor T_{ijk} can be expressed as

$$T'_{ijk} = R_{il}R_{jm}R_{kn}T_{lmn} \tag{6.19}$$

If we perform contraction on the indices j and k of this tensor, and set $j = k$ in Eq. (6.19) and sum over k, we obtain

$$T'_{ikk} = R_{il}R_{km}R_{kn}T_{lmn} = R_{il}\delta_{mn}T_{lmn} = R_{il}T_{lnn}$$

In the above equation, the orthogonality condition is applied to $R_{km}R_{kn}$. The result shows that T_{ikk} transforms as the components of a first-order tensor ($1 = 3 - 2$).

In general, contraction reduces the order of a tensor by 2. An N-th order tensor $T_{ij\ldots l\ldots m\ldots k}$, upon contraction by setting the indices l and m equal, yields another tensor of order $N-2$. In particular, if contraction is applied to a second-order tensor, the result is a scalar.

6.4.3 Cross Product and Dot Product

The outer product and inner product are tensor multiplication operations. For example, given two tensors A_{ij} and B_{klm}, writing them side by side in different orders simply yields

$$C_{ijklm} = A_{ij} B_{klm} \tag{6.20}$$

where C_{ijklm} are the components of a fifth-order tensor, which directly follow the transformation law of tensors. A product of this form, as shown in Eq. (6.20), is called the outer product of the components of tensors A_{ij} and B_{klm}, where all indices are mutually distinct.

Another important product derived from the outer product by contraction is called the inner product of tensors. For example, setting $k = j$ in Eq. (6.20) yields

$$C_{ijjlm} = A_{ij} B_{jlm} \tag{6.21}$$

It is thus a component of a third-order tensor, as described in Sect. 6.4.2. The product in Eq. (6.21) is referred to as the inner product of the components of tensors A_{ij} and B_{klm}.

Example The process of taking the scalar product of two vectors $\boldsymbol{u}$ and $\boldsymbol{v}$ can be represented as $u_i v_i$, or it can first be expressed in the form of a tensor outer product as follows.

$$T_{ij} = u_i v_j$$

Then rewrite as

$$T_{ii} = u_i v_i$$

By employing the concepts of the tensor outer (and inner) product, many familiar expressions from vector algebra can be reformulated as contracted tensors. For example, the vector product $\boldsymbol{a} = \boldsymbol{b} \times \boldsymbol{c}$ can be written as

$$a_i = \varepsilon_{ijk} b_j c_k \tag{6.22}$$

As its i-th component, where ε_{ijk} is the Levi–Civita symbol, this illustrates the distinction between the pseudovector composed of components $\varepsilon_{ijk}b_jc_k$ and the second-order tensor formed by the outer product b_ic_j.

Note: The outer product of two vectors without reference to any coordinate system is typically expressed as

$$\mathbf{T} = \mathbf{u} \otimes \mathbf{v} \tag{6.23}$$

It should not be confused with the cross product of two vectors, which itself is a pseudovector. According to Eq. (6.23), the components Tij of a second-order tensor can be interpreted as follows: since $\boldsymbol{u} = u_i\boldsymbol{e}_i$ and $\boldsymbol{v} = v_j\boldsymbol{e}_j$, the tensor $\boldsymbol{T}$ can be expressed as

$$\mathbf{T} = \mathbf{u} \otimes \mathbf{v} = u_i\mathbf{e}_i \otimes v_j\mathbf{e}_j = u_iv_j\mathbf{e}_i \otimes \mathbf{e}_j = T_{ij}\mathbf{e}_i \otimes \mathbf{e}_j$$

In addition, we have

$$\mathbf{T} = u_i\mathbf{e}_i \otimes v_j\mathbf{e}_j = u'_i\mathbf{e}'_i \otimes v'_j\mathbf{e}'_j$$

This indicates that the variables T'_{ij} are components of the same tensor $\boldsymbol{T}$, but are referenced in different coordinate systems. These concepts can be extended to higher-order tensors.

The expression $\boldsymbol{e}_{i\otimes}\boldsymbol{e}_j$ is sometimes simply written as $\mathbf{e}_i\mathbf{e}_j$, and is also referred to as the tensor product. This notation (without $_\otimes$) emphasizes that, for example, just as $\boldsymbol{e}_1$ is an element of the set of vectors, $\mathbf{e}_1\mathbf{e}_2$ is an element of the set of second-order tensors. The term "tensor product" indicates that this operation shares certain properties with the product in elementary algebra.

Gradient: In structural mechanics, many governing equations involve derivatives of field variables with respect to spatial coordinates. Here, a "field" refers to a function defined in space, such as the temperature or displacement function of a structure. Field variables may be scalars, vectors, or tensors. Therefore, it is convenient to define the gradient operator using tensor notation. The gradient operator is defined as a vector (or a first-order tensor), namely,

$$\nabla = \frac{\partial}{\partial \mathbf{x}} = \mathbf{e}_i\frac{\partial}{\partial x_i} \tag{6.24}$$

For example, the gradient of a scalar field $\phi(x)$ can be expressed as

$$\nabla\phi = \text{grad}\phi = \mathbf{e}_i\frac{\partial \phi}{\partial x_i} \tag{6.25}$$

This is a vector. The gradient of a vector field $\mathbf{u}(x)$ can be defined as

$$\nabla \mathbf{u} = \left(\mathbf{e}_i \frac{\partial}{\partial x_i}\right) \otimes \left(u_j \mathbf{e}_j\right) = \frac{\partial u_j}{\partial x_i} \mathbf{e}_i \otimes \mathbf{e}_j \tag{6.26}$$

Note: The gradient of a vector is a second-order tensor.

For convenience in notation, the gradient is often represented using a comma, i.e., $v_{i,j} = \partial v_i / \partial x_j$. The divergence can be defined by constructing a scalar from the gradient of a vector.

$$\nabla \cdot \mathbf{u} = \left(\mathbf{e}_i \frac{\partial}{\partial x_i}\right) \cdot \left(u_j \mathbf{e}_j\right) = \frac{\partial u_i}{\partial x_i} = \frac{\partial u_1}{\partial x_1} + \frac{\partial u_2}{\partial x_2} + \frac{\partial u_3}{\partial x_3} \tag{6.27}$$

The Laplace operator can be defined as the inner product of two gradient operators, that is,

$$\nabla^2 = \nabla \cdot \nabla = \left(\mathbf{e}_i \frac{\partial}{\partial x_i}\right) \cdot \left(\mathbf{e}_j \frac{\partial}{\partial x_j}\right) = \frac{\partial}{\partial x_i} \frac{\partial}{\partial x_i} = \frac{\partial^2}{\partial x_1^2} + \frac{\partial^2}{\partial x_2^2} + \frac{\partial^2}{\partial x_3^2} \tag{6.28}$$

The following expressions in vector algebra are presented as contracted Cartesian tensors (where $[\boldsymbol{a}]_i$ denotes the i-th component of the vector (or tensor) $\boldsymbol{a}$):

1. $\mathbf{a} \cdot \mathbf{b} = a_i b_i = \delta_{ij} a_i b_j$.
2. $[\mathbf{a} \cdot (\mathbf{b} \times \mathbf{c})]_i = \delta_{il} a_i [\mathbf{b} \times \mathbf{c}]_l = \delta_{il} a_i \left(\varepsilon_{ljk} b_j c_k\right) = \varepsilon_{ijk} a_i b_j c_k$.
3. $(\mathbf{a}, \mathbf{b}, \mathbf{c}) = [\mathbf{a}, \mathbf{b}, \mathbf{c}] = (\mathbf{a} \times \mathbf{b}) \cdot \mathbf{c} = \mathbf{a} \cdot (\mathbf{b} \times \mathbf{c}) = \begin{vmatrix} a_1 & a_2 & a_3 \\ b_1 & b_2 & b_3 \\ c_1 & c_2 & c_3 \end{vmatrix}$
4. $(\mathbf{a} \times \mathbf{b}) \times \mathbf{c} = (\mathbf{a} \cdot \mathbf{c})\mathbf{b} - (\mathbf{b} \cdot \mathbf{c})\mathbf{a}$.
5. $(\mathbf{a} \times \mathbf{b}) \cdot (\mathbf{c} \times \mathbf{d}) = (\mathbf{a} \cdot \mathbf{c})(\mathbf{b} \cdot \mathbf{d}) - (\mathbf{a} \cdot \mathbf{d})(\mathbf{b} \cdot \mathbf{c}) = \begin{vmatrix} \mathbf{a} \cdot \mathbf{c} & \mathbf{a} \cdot \mathbf{d} \\ \mathbf{b} \cdot \mathbf{c} & \mathbf{b} \cdot \mathbf{d} \end{vmatrix}$.
6. $(\mathbf{a} \times \mathbf{b}) \times (\mathbf{c} \times \mathbf{d}) = (\mathbf{a}, \mathbf{c}, \mathbf{d})\mathbf{b} - (\mathbf{b}, \mathbf{c}, \mathbf{d})\mathbf{a} = (\mathbf{a}, \mathbf{b}, \mathbf{d})\mathbf{c} - (\mathbf{a}, \mathbf{b}, \mathbf{c})\mathbf{d}$.
7. $(\mathbf{a} \otimes \mathbf{b}) \cdot \mathbf{c} = \mathbf{a}(\mathbf{b} \cdot \mathbf{c})$, $\mathbf{c} \cdot (\mathbf{a} \otimes \mathbf{b}) = \mathbf{b}(\mathbf{c} \cdot \mathbf{a})$, $(\mathbf{a} \otimes \mathbf{b}) \cdot (\mathbf{c} \otimes \mathbf{d}) = (\mathbf{b} \cdot \mathbf{c})(\mathbf{a} \otimes \mathbf{d})$
8. $(\nabla \cdot \boldsymbol{\sigma})_j = \frac{\partial \sigma_{ij}}{\partial x_i}$

This is the divergence of the stress tensor. By expanding all the components in the above equation and setting the divergence to zero, the equilibrium differential equations can be obtained.

$$\begin{cases} \dfrac{\partial \sigma_{11}}{\partial x_1} + \dfrac{\partial \sigma_{21}}{\partial x_2} + \dfrac{\partial \sigma_{31}}{\partial x_3} = 0 \\ \dfrac{\partial \sigma_{12}}{\partial x_1} + \dfrac{\partial \sigma_{22}}{\partial x_2} + \dfrac{\partial \sigma_{32}}{\partial x_3} = 0 \\ \dfrac{\partial \sigma_{13}}{\partial x_1} + \dfrac{\partial \sigma_{23}}{\partial x_2} + \dfrac{\partial \sigma_{33}}{\partial x_3} = 0 \end{cases}$$

Since the stress tensor is symmetric (see Sect. 6.4.4 for details), we have $\sigma_{12} = \sigma_{21}, \sigma_{23} = \sigma_{32}$, and $\sigma_{13} = \sigma_{31}$.

9. $[\nabla \times \mathbf{v}]_i = \varepsilon_{ijk} \frac{\partial v_k}{\partial x_j}$
10. $[\nabla(\nabla \cdot \mathbf{v})]_i = \frac{\partial}{\partial x_i}(\nabla \cdot \mathbf{v}) = \frac{\partial}{\partial x_i}\left(\frac{\partial v_j}{\partial x_j}\right) = \delta_{jk} \frac{\partial^2 v_j}{\partial x_i \partial x_k}$
11. $[\nabla \times (\nabla \times \mathbf{v})]_i = \varepsilon_{ijk} \frac{\partial}{\partial x_j}[\nabla \times \mathbf{v}]_k = \varepsilon_{ijk}\varepsilon_{klm} \frac{\partial^2 v_m}{\partial x_j \partial x_l}$

6.4.4 Symmetric and Antisymmetric Tensors

The order of tensor indices is important; in general, *Tij* and *Tji* are not the same. If

$$T_{ij} = T_{ji}$$

holds for all indices i and j, then the tensor composed of *Tij* is called a symmetric tensor. If

$$T_{ij} = -T_{ji} \tag{6.29}$$

then the tensor is said to be an antisymmetric (or skew-symmetric) tensor.

A tensor that is symmetric (or antisymmetric) in one coordinate system remains symmetric (or antisymmetric) in any other coordinate system. If *Tij* is symmetric in a given system, that is, *Tij* = *Tji*, then

$$T'_{ij} = R_{ik}R_{jl}T_{kl} = R_{jl}R_{ik}T_{lk} = T'_{ji}$$

The above conclusion also applies to antisymmetric and higher-order tensors.

It is noteworthy that every tensor can be decomposed into symmetric and antisymmetric parts by means of an identity

$$T_{ij} = S_{ij} + A_{ij} \tag{6.30}$$

where

$$S_{ij} = \frac{1}{2}(T_{ij} + T_{ji})$$
$$A_{ij} = \frac{1}{2}(T_{ij} - T_{ji})$$

Clearly, S_{ij} is a symmetric tensor, as it remains unchanged even when the indices i and j are interchanged; A_{ij} is an antisymmetric tensor, since swapping the indices i and j changes the sign of all its components. Therefore, S_{ij} and A_{ij} are referred to as the symmetric and antisymmetric parts of T_{ij}, respectively.

6.4.5 Equivalence Between Antisymmetric Second-Order Tensors and Pseudo-Vectors

It is noteworthy that in three-dimensional space, a second-order antisymmetric tensor $\boldsymbol{W}$ is associated with a pseudovector $\boldsymbol{w}$. To illustrate this, let Wi be the components of an antisymmetric second-order tensor ($W11 = W22 = W33 = 0$), whose transformation law is given by

$$\begin{aligned} W'_{ij} =& R_{il}R_{jm}W_{lm} = R_{i1}R_{j2}W_{12} + R_{i1}R_{j3}W_{13} + R_{i2}R_{j1}W_{21} \\ &+ R_{i2}R_{j3}W_{23} + R_{i3}R_{j1}W_{31} + R_{i3}R_{j2}W_{32} \end{aligned} \tag{6.31}$$

Furthermore, since $Wlm = -Wml$, Eq. (6.31) can be simplified as follows:

$$W'_{ij} = \sum_{(l,m)} \left(R_{il}R_{jm} - R_{im}R_{jl}\right)W_{lm} \tag{6.32}$$

Here, the sum $\sum_{(l,\,m)}$ restricts the values of (l, m) to (1, 2), (2, 3), or (3, 1).

Introduction of notation

$$w_1 \equiv W_{23} = -W_{32}, \quad w_2 \equiv W_{31} = -W_{13}, \quad w_3 \equiv W_{12} = -W_{21},$$

or more concisely expressed as

$$w_n = W_{lm}$$

Here, l, m, and n represent a cyclic permutation of the numbers 1, 2, and 3, that is,

$$(l, m, n) = (1, 2, 3), \; (2, 3, 1), \; (3, 1, 2)$$

Thus, Eq. (6.32) can be written as

$$w'_k = \sum_{(l,m,n)} \left(R_{il}R_{jm} - R_{im}R_{jl}\right)W_n \tag{6.33}$$

Here, i, j, k and l, m, n are all cyclic permutations of 1, 2, 3.

It is noteworthy that Eq. (6.33) is equivalent to the transformation law for the components w_k of the pseudovector $\boldsymbol{w}$. After some algebraic manipulation, Eq. (6.33) can be simplified into a more compact form.

$$w'_k = |\mathbf{R}|R_{kn}w_n \tag{6.34}$$

This is precisely the transformation law for pseudovectors.

Therefore, the following theorem can be established:

Suppose there is a second-order antisymmetric tensor $\boldsymbol{W}$ in three dimensions, whose components Wij are given by

$$[W_{ij}] = \begin{pmatrix} 0 & W_{12} & -W_{31} \\ -W_{12} & 0 & W_{23} \\ W_{31} & -W_{23} & 0 \end{pmatrix}$$

The three components $W12$, $W31$ and $W23$ can be associated with the pseudovector $\boldsymbol{w}$, whose components are given by the following expression:

$$(w_1, w_2, w_3) = (W_{23}, W_{31}, W_{12})$$

$$w_i = \frac{1}{2}\varepsilon_{ijk} W_{jk} \tag{6.35}$$

The right-hand side of Eq. (6.35) is a double contraction product of the third-order pseudotensor ε_{ijk} and the second-order tensor W_{ij}. Therefore, $\boldsymbol{w}$ is a pseudovector.

In physical applications, we often use the vector representation (6.35) of a second-order antisymmetric tensor. For example, for a mass of the angular momentum equation for a particle of mass m, assuming a force $\mathbf{F}$ acts on a particle located at $\boldsymbol{x}$, and letting i and j take values 1, 2, and 3, yields nine equations.

$$m(\ddot{x}_j x_k - \ddot{x}_k x_j) = F_j x_k - F_k x_j \tag{6.36}$$

Note: Both sides of Eq. (6.36) are antisymmetric tensors. Therefore, among the 9 equations, only 3 are independent, specifically when $(j, k) = (1,2)$, $(2,3)$, and $(3,1)$, Eq. (6.36) can be transformed into a more concise vector form as

$$mw_i = N_i$$

where

$$w_i = \varepsilon_{ijk}(\ddot{x}_j x_k - \ddot{x}_k x_j)$$
$$N_i = \varepsilon_{ijk}(F_j x_k - F_k x_j)$$

6.4.6 Quotient Theorem

Sometimes it is necessary to determine whether a set of functions constitutes the components of a vector, for example, $\{a_i(x_j)\}$. The direct approach is to check whether the functions satisfy the required transformation rules under a rotation of the coordinate axes, but this can be cumbersome in practice. In this section, we introduce a simple and more efficient indirect method, known as the quotient rule, which is used to determine whether a given set of quantities forms the components of a tensor.

Quotient Theorem: For any vector $\boldsymbol{v}$ in a rotating coordinate system, if $a_i v_i$ is a scalar, then a_i constitutes the components of vector $\boldsymbol{a}$.

Proof: Suppose we are given a set of n quantities a_i. In any rotated coordinate system, $a_i v_i$ is a scalar for any vector $\boldsymbol{v}$ and its component v_i. This can be expressed as

$$a_j v_j = \phi \tag{6.37}$$

Here, ϕ denotes a scalar. By expressing the transformation of a_i as a_j' (currently unknown), in the x' coordinate system Eq. (6.37) becomes

$$a_i' v_i' = \phi' \tag{6.38}$$

Since ϕ is a scalar, we have $\phi = \phi$'. Furthermore, as v_i is a component of a vector, it follows that

$$v_i' = R_{ij} v_j \tag{6.39}$$

Subtracting Eqs. (6.38) from (6.37) yields

$$\left(a_j - a_i' R_{ij}\right) v_j = 0 \tag{6.40}$$

The left-hand side of the above equation implies summation over the index j. Therefore, one cannot simply assume that the coefficient of v_j is zero. However, since Eq. (6.40) must hold in any coordinate system, we can specifically choose a coordinate system in which the components of $\boldsymbol{v}$ are $v_1 = 1$ and $v_{(i \neq 1)} = 0$. Equation (6.40) then reduces to

$$a_1 - R_{i1} a_i' = 0$$

Similarly, an appropriately rotated coordinate system can be chosen so that the components of $\boldsymbol{v}$ satisfy $v_2 = 1$ and $v_{(i\, \neq 2)} = 0$, thus

$$a_2 - R_{i2} a_i' = 0$$

Continuing in this manner, we observe that

$$a_j = R_{ij} a_i' \quad \textit{for all}\, j$$

Multiplying both sides of the above equation by Rkj yields

$$R_{kj} a_j = R_{kj} R_{ij} a_i' = \delta_{ki} a_i' = a_k'$$

That is,

$$a_i' = R_{ij} a_j$$

This is the transformation law for vector components. Therefore, it can be concluded that a_i constitutes the components of the vector, which is denoted by $\boldsymbol{a}$.

Note: In applying the above theorem, it is essential to ensure that the coordinate system used can be arbitrarily rotated. This assumption represents a very stringent condition and is generally not satisfied.

6.4.7 Quotient Theorem for Double-Indexed Quantities

Another common scenario is as follows: Suppose a set of $n2$ quantities a_{ij} is given, and for any vector $\boldsymbol{v}$ in a rotated coordinate system, if $a_{ij} v_i v_j$ is a scalar ϕ, then the quantities a_{ij} with two subscripts are considered the components of a second-order tensor. In fact, based on this assumption alone, we cannot conclude that a_{ij} possesses the properties of a tensor. This indicates that the quotient theorem for quantities with two subscripts needs to be revised.

To establish the modified quotient theorem, a discussion similar to that in Sect. 6.4.6 is required. Suppose in the given x-coordinate system, let

$$a_{ij} v_i v_j = \phi$$

Similarly, in the x' coordinate system, let

$$a_{kl}' v_k' v_l' = \phi' \tag{6.41}$$

In Eq. (6.41), let a_{ij}' denote the unknown transformation of a_{ij}. Using the transformation law of v_i and $\phi = \phi$', we have

$$\left(a_{ij} - R_{ki} R_{lj} a_{kl}'\right) v_i v_j = 0 \tag{6.42}$$

In the above expression, summation over i and j is implied, and it cannot be directly inferred that the coefficient of $v_i v_j$ is zero. However, by choosing the components $(v_1, v_2, v_3, \ldots)$ successively as $(1, 0, 0, \ldots)$ and $(0, 1, 0, \ldots)$, we obtain

$$a_{11} - R_{k1} R_{l1} a_{kl}' = 0, \quad a_{22} - R_{k2} R_{l2} a_{kl}' = 0, \quad \cdots \tag{6.43}$$

The above result implies that a_{ij} obeys the transformation law of a second-order tensor when $i = j$. However, it does not indicate the behavior of a_{ij} when $i\ ^{1}\ j$. To examine this further, let the components be $v_1\ ^{1}\ 0$, $v_2\ ^{1}\ 0$, and $v_i = 0$ for all other i.

Then Eq. (6.42) becomes.

$$\begin{aligned}&\left(a_{11}-R_{k1}R_{l1}a'_{kl}\right)v_1v_1+\left(a_{12}-R_{k1}R_{l2}a'_{kl}\right)v_1v_2\\&+\left(a_{21}-R_{k2}R_{l1}a'_{kl}\right)v_2v_1+\left(a_{22}-R_{k2}R_{l2}a'_{kl}\right)v_2v_2=0.\end{aligned}\tag{6.44}$$

According to Eq. (6.43), the coefficients of v_1v_1 and v_2v_2 are zero. Furthermore, since

$$R_{k1}R_{l2}a'_{kl}=R_{k2}R_{l1}a'_{lk}$$

is merely a relabeling of the subscripts k and l, Eq. (6.44) can be rewritten as

$$\left[(a_{12}+a_{21})-\left(a'_{kl}+a'_{lk}\right)R_{k2}R_{l1}\right]v_1v_2=0$$

By selecting the components $v_1=1$ and $v_2=1$ in the above equation, we obtain

$$a_{12}+a_{21}=\left(a'_{kl}+a'_{lk}\right)R_{k2}R_{l1}$$

Similarly, by repeatedly applying the above process, we obtain

$$a_{ij}+a_{ji}=\left(a'_{kl}+a'_{lk}\right)R_{kj}R_{li}$$

i.e.

$$a'_{kl}+a'_{lk}=R_{kj}R_{li}\left(a_{ij}+a_{ji}\right)$$

This is indeed the transformation law for a second-order tensor; however, the corresponding quantity is $a_{ij}+a_{ji}$, that is, the symmetric part $2a_{ij}$, rather than a_{ij} itself. Therefore, the quotient theorem for double-subscript quantities states: Suppose there is a set of n^2 quantities a_{ij} such that for any vector $\boldsymbol{v}$ and any rotation of the coordinate system, the sum $a_iv_iv_j$ is a scalar; then the symmetric part of a_{ij}, namely $(a_{ij}+a_{ji})/2$, constitutes the components of a second-order tensor.

Note:

- In addition to the assumptions above, if a_{ij} is symmetric, then a_{ij} itself constitutes the components of a second-order tensor.
- The tensor properties of the antisymmetric part of a_{ij} cannot be inferred from the above assumptions, as this part does not contribute to the scalar ϕ in any way.

  This can be seen from the following equation.

$$\left(a_{ij}+a_{ji}\right)v_iv_j=a_{ij}v_iv_j-a_{ji}v_iv_j=a_{ij}v_iv_j-a_{ij}v_jv_i=0.$$

  The indices i and j can be interchanged.

Example Using the multiplication theorem, it is proven that the double subscript quantity a_{ij} given by the following expression.

$$\left[a_{ij}\right] = \begin{bmatrix} (x_2)^2 & -x_1 x_2 \\ -x_1 x_2 & (x_1)^2 \end{bmatrix}$$

is the components of a second-order tensor.

Since $a_{ij} = a_{ji}$ and the outer product $x_k x_l$ is a second-order tensor, contracting a_{ij} with the outer product $x_k x_l$ yields

$$a_{ij} x_i x_j = (x_2)^2 (x_1)^2 - x_1 x_2 x_1 x_2 - x_1 x_2 x_2 x_1 + (x_1)^2 (x_2)^2 = 0 \tag{6.45}$$

In the above equation, the last term 0 is a zero-order tensor, so Eq. (6.45) holds for any rotated coordinate system. Therefore, we can conclude that a_{ij} are the components of a second-order tensor.

6.5 Applications in Physics

This section introduces the physical applications of second-order and higher-order Cartesian tensors in mechanics, elasticity, and related fields.

6.5.1 *Moment of Inertia Tensor*

For a set of rigidly connected particles, let the mass of the α-th particle be $m^{(\alpha)}$ and its position relative to the origin O be $\boldsymbol{r}^{(\alpha)}$. Assuming the rigid body assembly rotates with angular velocity $\boldsymbol{\omega}$ about an axis passing through the origin O, the angular momentum is given by

$$\boldsymbol{J} = \sum_{\alpha} \left(\boldsymbol{r}^{(\alpha)} \times \boldsymbol{p}^{(\alpha)} \right)$$

For any α, we have $p^{(\alpha)} = m^{(\alpha)} \dot{\boldsymbol{r}}^{(\alpha)}$ and $\dot{\boldsymbol{r}}^{(\alpha)} = \omega \times r^{(\alpha)}$, with their components expressed in subscript notation as follows:

$$p_k^{(\alpha)} = m^{(\alpha)} \dot{x}_k^{(\alpha)} \quad \text{and} \quad \dot{x}_k^{(\alpha)} = \varepsilon_{klm} \omega_l x_m^{(\alpha)}$$

Therefore, we obtain

$$J_i = \sum_{\alpha} \sum_{jk} \varepsilon_{ijk} x_j^{(\alpha)} p_k^{(\alpha)} = \sum_{\alpha} \sum_{jk} m^{(\alpha)} \varepsilon_{ijk} x_j^{(\alpha)} \varepsilon_{klm} \omega_l x_m^{(\alpha)}$$

$$= \sum_{\alpha}\sum_{jk} m^{(\alpha)}\left(\delta_{il}\delta_{jm} - \delta_{im}\delta_{jl}\right)x_j^{(\alpha)}x_m^{(\alpha)}\omega_l$$

$$= \sum_{\alpha}\sum_{l} m^{(\alpha)}\left[\left(x_m^{(\alpha)}\right)^2\delta_{il} - x_i^{(\alpha)}x_l^{(\alpha)}\right]\omega_l \equiv \sum_{l} I_{il}\omega_l \tag{6.46}$$

where

$$I_{il} = \sum_{\alpha} m^{(\alpha)}\left[\left(x_m^{(\alpha)}\right)^2\delta_{il} - x_i^{(\alpha)}x_l^{(\alpha)}\right] \tag{6.47}$$

I_{il} forms a symmetric second-order Cartesian tensor: the symmetry property represented by $I_{il} = I_{li}$ can be derived from Eq. (6.46); applying the quotient rule is called the moment of inertia tensor of the rigid body with respect to the origin O. Clearly, from Eq. (6.46), it can be seen that I_{il} depends only on the mass distribution of the rigid body, and it does not depend on the direction or magnitude of the rigid body's angular velocity $\boldsymbol{\omega}$.

For a continuous rigid body, $m^{(\alpha)}$ is replaced by the mass distribution $\rho(\boldsymbol{r})$, and the sum $\sum\alpha$ is replaced by the integral of $\int \mathrm{d}V$ over the entire volume of the object. When expressed in Cartesian coordinates, the inertia tensor of the continuum can be written as

$$\mathbf{I} = \left[I_{ij}\right] = \begin{pmatrix} \int \left(y^2 + z^2\right)\rho \mathrm{d}V & -\int xy\rho \mathrm{d}V & -\int zx\rho \mathrm{d}V \\ -\int xy\rho \mathrm{d}V & \int \left(z^2 + x^2\right)\rho \mathrm{d}V & -\int yz\rho \mathrm{d}V \\ -\int zx\rho \mathrm{d}V & -\int yz\rho \mathrm{d}V & -\int \left(x^2 + y^2\right)\rho \mathrm{d}V \end{pmatrix}$$

The diagonal elements of this tensor are called moments of inertia, while the off-diagonal elements without the negative sign are referred to as products of inertia.

It can be shown that the kinetic energy of a rotating system $K = (1/2)I_{jl}\omega_j\omega_l$ is a scalar obtained by contracting the vector ω_j twice with the inertia tensor I_{jl}, that is (similar to Eq. (6.45))

$$K = \frac{1}{2}\sum_{\alpha} m^{(\alpha)}\dot{\mathbf{r}}^{(\alpha)} \cdot \dot{\mathbf{r}}^{(\alpha)} = \frac{1}{2}\sum_{\alpha} m^{(\alpha)} \sum_{j,k,l,m} \varepsilon_{ijk}\omega_j x_k^{(\alpha)} \varepsilon_{ilm}\omega_l x_m^{(\alpha)} = \frac{1}{2}\sum_{j,l} I_{jl}\omega_j\omega_l$$

This indicates that the kinetic energy of a rotating body can be expressed as a scalar obtained by contracting the angular velocity vector ω_j with the inertia tensor I_{jl} twice. Alternatively, since $J_j = I_{jl}\omega_l$, the kinetic energy can also be written as $K = (1/2)J_j\omega_j$.

6.5.2 Elastic Tensor

The above discussion focused only on the physical applications of second-order tensors, which are associated with two vectors. Now, we extend this concept to fourth-order tensors, which are associated with two second-order physical tensors. This situation commonly arises in elasticity theory, where the local deformation at any point P in an elastic body can be described by a second-order symmetric tensor *eij*, known as the strain tensor, which is given by the following expression:

$$e_{ij} = \frac{1}{2}\left(\frac{\partial u_i}{\partial x_j} + \frac{\partial u_j}{\partial x_i}\right)$$

Here, $\boldsymbol{u}$ is the displacement vector describing the strain of an infinitesimal volume element. Similarly, the second-order symmetric stress tensor σ_{ij} can be used to describe the stress at point P within the body. σ_{ij} represents the x_j component of the stress vector acting on the plane passing through point P, whose normal is in the x_i direction. The general form of Hooke's law, which relates the stress and strain tensors, is given by

$$\sigma_{ij} = \sum_{k,l} C_{ijkl} e_{kl} \tag{6.48}$$

Here, C_{ijkl} is a fourth-order Cartesian tensor.

For isotropic media, C_{ijkl} must be an isotropic tensor, which can be expressed as

$$C_{ijkl} = \lambda\delta_{ij}\delta_{kl} + \eta\delta_{ik}\delta_{jl} + \upsilon\delta_{il}\delta_{jk}$$

Substituting the above equation into Eq. (6.48) yields

$$\sigma_{ij} = \lambda\delta_{ij}\sum_{k} e_{kk} + \eta e_{ij} + \upsilon e_{ji} \tag{6.49}$$

Therefore, if we let $\eta + \upsilon = 2\mu$, and since e_{ij} is symmetric, Eq. (6.49) can be rewritten in its commonly used form.

$$\sigma_{ij} = \lambda\sum_{k} e_{kk}\delta_{ij} + 2\mu e_{ij}$$

Here, λ and μ are referred to as Lamé constants.

Chapter 7
Finite Element Analysis of Elastoplastic Problems

The distinctive property of elastic materials is the existence of a strain energy density, from which stress can be defined as the derivative of the strain energy density with respect to strain. This establishes a one-to-one correspondence between strain and stress, a property of elastic materials known as path independence. For example, suppose a given elastic rod is subjected to a current strain ε. This state can be achieved either by gradually stretching the rod until the strain reaches ε, or by first stretching the rod beyond ε and then compressing it back to ε. Although these two processes involve different loading histories, the final strain state of the rod is the same, and the resulting stress is also identical. Utilizing the path independence of elastic materials, it follows that when the stress vanishes, the strain also disappears. Therefore, when the applied load is removed, the structure will always (reversibly) return to its original geometric shape, leaving no permanent deformation.

Unlike elastic materials, some materials exhibit permanent deformation when subjected to forces exceeding a certain threshold (the elastic limit), even after the force is removed. For these materials, if the applied force is relatively small, the structure will return to its original geometry once the force is removed. However, if the force exceeds the elastic limit (irreversible), the structure will undergo permanent deformation. This behavior is referred to as plasticity. Since these materials initially behave elastically and then transition to plastic behavior when the force exceeds the elastic limit, this material response is called elastoplasticity. This chapter primarily discusses the fundamental theory of elastoplasticity and related numerical solution algorithms.

Elastoplastic behavior is a type of material nonlinearity that originates from the constitutive relationship, namely the stress–strain relationship. In elastoplastic problems, there is no one-to-one correspondence between stress and strain. For example, if the bar in the previous example is made of an elastoplastic material, for a given strain value ε, the stress may take different values depending on whether the bar is in tension or compression. In fact, according to the loading history, the stress can assume any value less than the elastic limit. Therefore, the stress–strain relationship

B. Liu and C. Liu, *Introduction to Linear and Nonlinear Finite Element Analysis*,
https://doi.org/10.1007/978-981-95-9670-6_7

cannot be expressed solely in terms of total strain; the constitutive relationship can only be defined in terms of stress and strain rate. Here, the rate does not refer to the rate with respect to time, but should be understood as an increment in static analysis. Since the stress–strain relationship is given in rate form, the stress can only be determined by integrating the stress rate over the loading history. As a result, stress calculation depends on the loading history or path.

A key step in elastoplastic analysis is to separate the elastic strain from the total strain. Once the elastic strain is determined, the stress can be readily calculated, as stress depends solely on the elastic strain, while the plastic strain does not contribute to the stress. When the total strain is small (infinitesimal deformation), it is assumed that the total strain can be decomposed into elastic and plastic components. In this case, geometric nonlinearity is neglected, meaning the displacement–strain relationship is linear, and integration can be performed on the initially undeformed geometry. Considering that plastic deformation in metals typically occurs at strains around 0.2%, plastic deformation generally satisfies the small strain condition. All discussions of fundamental elastoplastic theory and related numerical solution algorithms in this chapter are based on the assumptions of small deformation and small strain.

Due to the path-dependent nature of elastoplastic problems, obtaining analytical solutions is virtually impossible. Therefore, numerical methods are indispensable for solving elastoplastic problems, and the finite element method has naturally become the primary approach for their analysis. Prior to the 1990s, nearly all research on the numerical solutions of elastoplastic problems was based on the *h*-version finite element method. However, as the research on *h*-version finite elements matured, progress largely stagnated, and issues such as preprocessing bottlenecks emerged. As a result, the *p*-version finite element method gradually gained prominence. This chapter will first introduce low-order (*h*-version) elements, but will focus on examples of high-order (*p*-version) elements, since the former are special cases of the latter at lower polynomial orders.

In 1992, Holzer was the first to apply the *p*-version finite element method to incremental elastoplastic analysis. Szabo noted in 1995 that, for nonlinear problems, error control during the adaptive analysis process is particularly important, as errors resulting from imprecise initial mesh discretization can accumulate continuously during nonlinear iterative computations. High-order schemes can reduce errors caused by mesh discretization through adaptive analysis, more accurate results can be achieved, and its rapid convergence characteristics can also shorten the iteration cycle. Building on this, Szabo systematically applied the *p*-version finite element method to the numerical solution of small deformation elastoplastic problems under plane stress, plane strain, and axisymmetry conditions, providing a comprehensive demonstration of the effectiveness of the *p*-version finite element method in solving small deformation elastoplastic problems. Holzer further used incremental plasticity theory to solve an example with an analytical solution, first comparing the accuracy differences between the *p*-version and *h*-version finite element methods, and then performing loading and unloading calculations with both methods. The results showed that the *p*-version finite element method maintained excellent performance even under cyclic elastoplastic loading and unloading. Düster conducted calculations on the same example as Holzer, comparing the efficiency of p-version and h-version finite element methods in adaptive analysis. The results indicated that the p-version finite element method is more efficient than the h-version finite element method.

The Hierarchical Finite Element Method (HFEM), as a classical high-order approach, offers advantages such as rapid convergence, ease of adaptive analysis implementation, and insensitivity to locking phenomena, and as previously mentioned, has been widely applied in elastoplastic analysis. However, its development has been limited by issues of numerical stability and difficulties in imposing boundary conditions. The Differential Quadrature Method (DQM) has also achieved significant success in solving elastoplastic problems. Nevertheless, since DQM is based on the strong form of differential equations, it is mainly suitable for regular domains, involves complex element assembly, and does not support local order elevation. The Hierarchical Quadrature Element Method (HQEM) combines the strengths of HFEM and DQM, employing the weak form of the governing equations, and is applicable to problems in complex domains. HQEM utilizes non-uniformly distributed Gauss–Lobatto integration points along element boundaries to facilitate the imposition of boundary conditions, thereby overcoming the limitation in DQM and its weak form, where the number of boundary basis functions is determined by the number of internal basis functions. This allows for high accuracy with very few degrees of freedom, a feature that has been validated in the analysis of linear problems in previous chapters. Furthermore, elements of different orders can be easily assembled. These characteristics make the hierarchical quadrature element method particularly well-suited for elastoplastic analysis.

This chapter primarily covers the fundamental theory and solution methods of elastoplasticity, as well as elastoplastic analysis examples based on the hierarchical quadrature element method. The chapter begins by introducing the general form of the elastoplastic constitutive relationship. For practical application to specific problems, its explicit form is provided. On this basis, the element matrix and global stiffness matrix are derived, and the finite element equations are established for solution. Subsequently, for the specific elastoplastic problems under investigation and their corresponding finite element equations, appropriate nonlinear numerical solution methods are presented. Building on the above, suitable hierarchical quadrature elements are selected to formulate the finite element equations for elastoplastic problems. Following the iterative procedure of the nonlinear numerical solution method, specific two-dimensional and three-dimensional elastoplastic problems are solved. The obtained results are compared with those from ABAQUS simulations and other sources to verify that the hierarchical quadrature element method can efficiently and accurately solve elastoplastic problems.

When a material undergoes both elastic and plastic deformation simultaneously, elastoplasticity occurs. The following section first introduces the basic concepts of elastoplastic finite element analysis using a one-dimensional bar, and then discusses the extension to multidimensional elastoplastic problems. Since the small deformation assumption is still adopted, geometric nonlinear effects can be neglected, and it is unnecessary to distinguish between different forms of stress.

7.1 One-Dimensional Elastoplastic Problems

7.1.1 Elastoplastic Material Behavior

In a one-dimensional tensile test of a rod, once the material deforms beyond its elastic limit, it exhibits a complex stress–strain relationship. For example, tensile tests on metals show that the initial stress is proportional to strain (elastic region); after reaching the elastic limit (yield stress), the material begins to undergo plastic deformation. In the first stage of plastic deformation, stress continues to increase proportionally with strain, but with a much lower slope (strain hardening), until the ultimate strength is reached. Thereafter, the stress gradually decreases (strain softening) until the material fractures. In addition, if the applied load is reduced (unloaded) after the material begins to undergo plastic deformation, the material will no longer follow the original stress–strain curve and will immediately exhibit elastic behavior. If cyclic loading is applied, the material response becomes increasingly complex.

The simulation of material behavior depends on the purpose of the analysis. For example, if the objective is to determine the behavior of a material up to fracture, it is necessary to model all stages of the stress–strain response in detail. However, if the goal is to examine the response of the material under small deformations, the material behavior can be simplified by considering only the elastic deformation and strain hardening regions. Of course, this model is not suitable for predicting material fracture. Figure 7.1 illustrates the ideal elastic–plastic stress–strain behavior for a uniaxial tension/compression test. When a tensile load is applied, the mechanical behavior is initially elastic until the yield stress σ_Y is reached (along line *o-a*). The elastic modulus is the slope of line *o-a* and is denoted by E. If the applied tensile load is removed before reaching the yield stress, the stress–strain relationship follows the same curve (along line *a-o*). Point *a* is referred to as the yield point; beyond this point, the material no longer exhibits purely elastic behavior.

After reaching the yield stress, if the load continues to increase, the plastic stage begins, and the stress increases further with a slope of E_t, known as the tangent modulus (along line *ab*). In this stage, the strain consists of both elastic and plastic components. In this simplified model, it is assumed that strain hardening is linear. If the tensile load is reduced after the material has undergone plastic deformation, it returns to elastic behavior, and the stress decreases linearly with the same initial elastic slope E (along line *bc*). If the applied load is completely removed, a permanent plastic strain remains (the strain at point *c*); if the tensile load is reapplied, the stress follows line *cb* and exhibits a new yield stress (the stress at point *b*). Thereafter, the stress increases with the same slope as the tangent modulus (along line *bd*). In elastoplastic materials, the yield stress changes due to the effect of strain hardening. On the other hand, if a compressive load is applied at the stress-free point *c*, the stress becomes negative and increases with a slope of E. If the compressive load continues to increase, the material will eventually yield again in compression (point *e*).

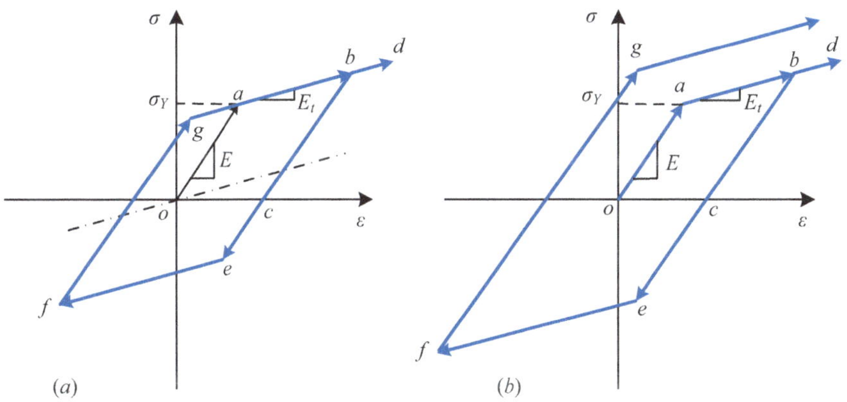

Fig. 7.1 Elastoplastic hardening model

If the applied load is proportional, meaning the load increases monotonically, strain hardening can be simply described using the tangent modulus E_t. However, when combined cyclic loading and unloading are applied, the situation can become complex, depending on when yielding occurs in the reverse direction. Several different hardening models have been proposed to determine the yield stress under cyclic loading. Figure 7.1 illustrates the two most commonly used models: the kinematic hardening model and the isotropic hardening model. In the kinematic hardening model, it is assumed that the elastic range (twice the initial yield stress) remains unchanged, and the center of the elastic region moves along a dashed line passing through the origin, parallel to the strain hardening line. Therefore, segments *be* and *fg* are equal and each is twice the length of *oa*. In the isotropic hardening model, the magnitude of the yield stress upon reverse loading is equal to the original yield stress, that is, the stress at points *b* and *e* is the same. As a result, the elastic range in this model increases.

In nonlinear finite element analysis, the Newton–Raphson iterative method can be used to reduce the displacement increment of the residuals of nonlinear equations. In addition, by using the finite element interpolation scheme, the strain increment can be computed from a given displacement increment. Therefore, a key task in elastoplastic analysis is to determine the stress increment corresponding to a given strain increment. However, it is not known a priori how much of the given strain increment is elastic and how much is plastic. Once this decomposition is achieved, the elastic strain increment can be used to calculate the stress increment. Note that plastic strain does not contribute to an increase in stress. When the material behaves elastically—whether it is in the initial elastic stage or has returned to elastic behavior due to unloading—the strain increment is purely elastic, and there is no plastic strain increment. Thus, the stress increment can be obtained by multiplying the strain increment by the elastic modulus. Since this process is essentially the same as that for linear elastic systems, only the case where the material is in the plastic stage will be discussed here.

Figure 7.2 presents the stress–strain curve for a one-dimensional elastoplastic material. When the applied load is proportional, both the isotropic hardening and kinematic hardening models yield identical results. In the previous load increment, it is assumed that the material has already entered the plastic stage. For the current load increment, the strain increment is obtained via the Newton–Raphson method. Since the material is in the plastic stage and the load continues to increase, the strain increment $\Delta\varepsilon$ can be decomposed into elastic strain and plastic strain, that is,

$$\Delta\varepsilon = \Delta\varepsilon_{\mathrm{e}} + \Delta\varepsilon_{\mathrm{p}}, \tag{7.1}$$

Here, the subscripts *e* and *p* denote elastic and plastic components, respectively. When the applied load is reduced or its direction is reversed (unloading), the elastic portion of the strain is removed, while the plastic portion remains unchanged during elastic unloading and increases again when the material yields. Even if the material yields in the opposite direction, the plastic strain only increases and never decreases, as it represents the accumulation of plastic deformation. Based on the small deformation assumption and Eq. (7.1), we have

$$\varepsilon = \varepsilon_{\mathrm{e}} + \varepsilon_{\mathrm{p}} \tag{7.2}$$

Here, ε_{e} and ε_{p} represent the total elastic and plastic strain increments, respectively, accumulated from all previous load increments.

Equation (7.2) highlights a crucial distinction between nonlinear elastic and elastoplastic materials. In the former, the unique stress is determined by the magnitude of the given total strain, which is essentially the total elastic strain. However, in the latter, for a given total strain, different values of elastic strain can be obtained by varying the plastic strain. Therefore, in elastoplastic materials, it is not possible to determine the stress corresponding to a given total strain. Since plastic strain accumulates each time the material yields, it is necessary to compute the plastic strain incrementally for each load step. This characteristic of elastoplasticity is referred

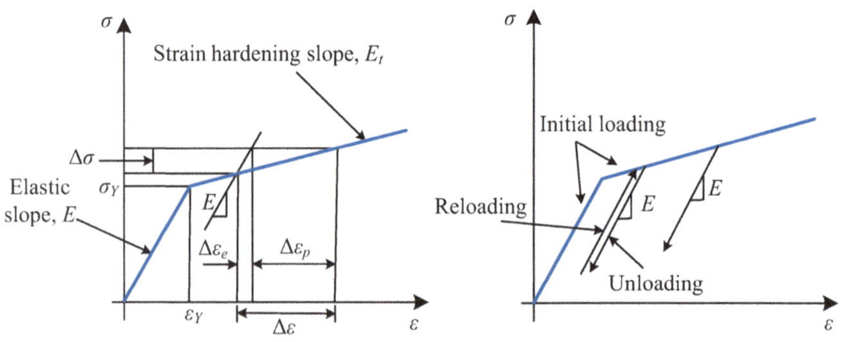

Fig. 7.2 One-dimensional elastoplasticity with strain hardening

to as path dependence or history dependence. To determine the stress, the complete loading history must be considered, while the accumulated plastic strain ε_p accounts for this loading history. Furthermore, the yield stress of the material is determined by the magnitude of the plastic strain.

Although the goal is to separate the elastic strain increment from the plastic strain increment, it is currently assumed that the elastic strain increment is known. Thus, the stress increment $\Delta\sigma$ can be calculated using the elastic strain increment, that is,

$$\Delta\sigma = E\Delta\varepsilon_e. \tag{7.3}$$

In addition to the elastic modulus and the tangent modulus, the plastic modulus H is also defined. It is the slope of the strain-hardened part of the stress–strain curve after the elastic strain component is removed, that is,

$$H = \frac{\Delta\sigma}{\Delta\varepsilon_p}. \tag{7.4}$$

Even though the stress–strain curve in Fig. 7.2 is given in terms of E and E_t, material properties are typically expressed using E and H. Since the stress–strain curve under proportional loading is identical for both isotropic hardening and kinematic hardening models, the plastic modulus for both hardening models is also the same.

Elastic modulus, plastic modulus, and tangent modulus are interrelated. The stress increment in the plastic stage can be expressed using any one of these three moduli as follows:

$$\Delta\sigma = E\Delta\varepsilon_e = H\Delta\varepsilon_p = E_t\Delta\varepsilon. \tag{7.5}$$

The expression $\Delta\sigma = H\Delta\varepsilon_p$ might be misinterpreted as implying that plastic strain directly causes an increase in stress, but this is actually caused by the strain hardening effect. Substituting Eq. (7.5) into the strain-increment term of Eq. (7.1) gives,

$$\frac{\Delta\sigma}{E_t} = \frac{\Delta\sigma}{E} + \frac{\Delta\sigma}{H} \Rightarrow \frac{1}{E_t} = \frac{1}{E} + \frac{1}{H}. \tag{7.6}$$

Therefore, for given values of E and E_t, the plastic modulus

$$H = \frac{EE_t}{E - E_t}. \tag{7.7}$$

Similarly, for given E and H, the tangent modulus

$$E_t = \frac{EH}{E + H} = E\left(1 - \frac{E}{E + H}\right). \tag{7.8}$$

When a material is subjected to elastoplastic strain (along line *ab* in Fig. 7.1), the plastic component of a given total strain increment can be determined.

$$\begin{aligned}\Delta\varepsilon = \Delta\varepsilon_{\mathrm{e}} + \Delta\varepsilon_{\mathrm{p}} = \frac{\Delta\sigma}{E} + \Delta\varepsilon_{\mathrm{p}} = \frac{H\Delta\varepsilon_{\mathrm{p}}}{E} + \Delta\varepsilon_{\mathrm{p}} = \Delta\varepsilon_{\mathrm{p}}\left(\frac{H}{E}+1\right).\\ \Rightarrow \Delta\varepsilon_{\mathrm{p}} = \frac{\Delta\varepsilon}{1+H/E}\end{aligned} \tag{7.9}$$

Therefore, for a given strain increment, the ratio between the plastic modulus and the elastic modulus can be used to calculate the plastic strain increment. As a special case, when there is no strain hardening—i.e., when the yield stress remains constant during plastic deformation—the strain increment becomes purely plastic.

Note: Eq. (7.9) does not hold when the material is initially in an elastic state.

7.1.2 Elastoplastic Finite Element Formulation

This section introduces the finite element formulation for elastoplastic bar elements. Since the small deformation assumption is adopted, only material nonlinearity is considered.

For greater generality, the load increment method is adopted, dividing the applied load into N load increments, denoted as $[t_1, t_2, \ldots, t_N]$. Although the analysis is static, these load increments are also referred to as time increments. It is assumed that the analysis has been completed up to load increment t_n, and the Newton–Raphson method is used to seek a new solution at load increment t_{n+1}. In the following discussion, a left superscript n or $n+1$ on a variable indicates the load increment, while a right superscript k or $k+1$ denotes the iteration count. Since all variables at t_n have already converged, the iteration count appears only for those variables at t_{n+1}. When a variable does not explicitly indicate an incremental load, it should be regarded as the variable or constant at t_{n+1}. It is also assumed that iteration step k has been completed and the current iteration is step $k+1$. The iteration index will be omitted whenever possible, unless this would cause confusion. In most cases, variables at t_{n+1} represent those at the $(k+1)-th$ iteration, that is, the current iteration.

As shown in Fig. 7.3, suppose the entire structure is modeled using a single bar element. The solution for the element is approximated by the nodal displacements $\boldsymbol{d} = \{u_1, u_2\}^{\mathrm{T}}$ and their increment vector. From the perspectives of load increment and iteration count, two different types of displacement increments can be considered

$$\begin{aligned}\Delta\boldsymbol{d}^k &= {}^{n+1}\boldsymbol{d}^k - {}^{n}\boldsymbol{d}\\ \delta\boldsymbol{d}^k &= {}^{n+1}\boldsymbol{d}^{k+1} - {}^{n+1}\boldsymbol{d}^k\end{aligned} \tag{7.10}$$

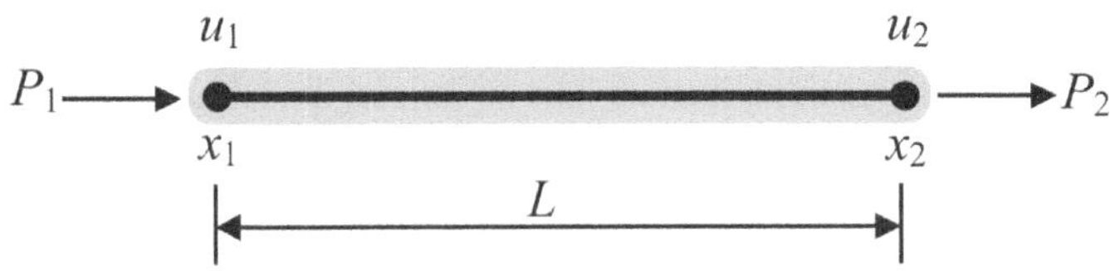

Fig. 7.3 One-dimensional elastoplastic rod element

Here, $\Delta \boldsymbol{d}$ represents the load increment from the last converged load step to the previous iteration, while $\delta \boldsymbol{d}$ denotes the increment from the previous iteration. The former is used to calculate the stress increment, whereas the latter is the displacement increment computed using the Newton–Raphson iteration. Therefore, during the Newton–Raphson iterative process, $\delta \boldsymbol{d}$ is accumulated into $\Delta \boldsymbol{d}$, and $\Delta \boldsymbol{d}$ is reset to zero before starting a new load increment.

As in the finite element method, for a bar element, there is an appropriate interpolation function vector $\boldsymbol{N}(\boldsymbol{x}) = [N_1, N_2]$,

$$\Delta u(x) = [N_1 \ N_2] \begin{Bmatrix} \Delta u_1 \\ \Delta u_2 \end{Bmatrix} = \mathbf{N} \cdot \Delta \mathbf{d}. \tag{7.11}$$

The corresponding strain increment can also be calculated by differentiating the displacement described above.

$$\Delta \varepsilon = \frac{\mathrm{d}}{\mathrm{d}x}(\Delta u) = \begin{bmatrix} -\frac{1}{L} & \frac{1}{L} \end{bmatrix} \begin{Bmatrix} \Delta u_1 \\ \Delta u_2 \end{Bmatrix} = \mathbf{B} \cdot \Delta \mathbf{d}, \tag{7.12}$$

Here, $\boldsymbol{B}$ is the displacement–strain matrix (in the case of a bar element, it is a row vector). Similarly, the strain increment refers to the increment from the last converged load increment to the previous iteration. Using the same interpolation scheme, the increments from the previous iteration can also be calculated via $\delta \mathbf{u}(\mathbf{x}) = \boldsymbol{N} \cdot \delta \mathbf{d}$ and $\delta \varepsilon = \boldsymbol{B} \cdot \delta \mathbf{d}$.

When the bar element in Fig. 7.3 is in equilibrium, the nodal forces induced by internal stress must be equal in magnitude and opposite in direction to the applied nodal forces. More specifically, the weak form of structural equilibrium can be expressed as

$$\overline{\boldsymbol{d}}^{\mathrm{T}} \int_0^L \boldsymbol{B}^{\mathrm{T}\ n+1} \sigma^{k+1} A \mathrm{d}x = \overline{\boldsymbol{d}}^{\mathrm{T}\ n+1} \boldsymbol{F}, \quad \forall \overline{\boldsymbol{d}} \in \mathbb{R}^2, \tag{7.13}$$

Here, $\overline{\boldsymbol{d}} = [\overline{u}_1, \overline{u}_2]^{\mathrm{T}}$ represents the displacement vector of the virtual nodes; A is the cross-sectional area of the bar; ${}^{n+1}\boldsymbol{F} = \left[{}^{n+1}F_1, {}^{n+1}F_2\right]^{\mathrm{T}}$ denotes the load.

Since the relationship between stress and strain is nonlinear, the above variational equation is also nonlinear with respect to strain or equivalent displacement. To solve the nonlinear equation, the Newton–Raphson method can be employed. Assuming that the applied load is constant or independent of displacement, only the left-hand side of Eq. (7.13) requires linearization. As only material nonlinearity is considered, the stress is linearized using a first-order Taylor series expansion, that is,

$$^{n+1}\sigma^{k+1} \approx {}^{n+1}\sigma^{k} + \frac{\partial \sigma}{\partial \varepsilon}\delta\varepsilon = {}^{n+1}\sigma^{k} + D^{\text{ep}}\delta\varepsilon, \tag{7.14}$$

Here, D^{ep} denotes the elastoplastic tangent modulus. According to the stress–strain curve in Fig. 7.1, D^{ep} can be determined by the following equation:

$$D^{\text{ep}} = \begin{cases} E & \textit{Elasticity} \\ E_t & \textit{Plasticity} \end{cases} \tag{7.15}$$

By substituting Eq. (7.14) into Eq. (7.13) and moving the known terms to the right-hand side, the following linearized variational equation is obtained:

$$\bar{\mathbf{d}}^{\mathrm{T}}\left[\int_0^L \mathbf{B}^{\mathrm{T}} D^{\text{ep}} \mathbf{B} A \mathrm{d}x\right]\delta\mathbf{d} = \bar{\mathbf{d}}^{\mathrm{T}}\mathbf{F} - \bar{\mathbf{d}}^{\mathrm{T}}\int_0^L \mathbf{B}^{\mathrm{T}}\,{}^{n+1}\sigma^k A\mathrm{d}x, \tag{7.16}$$

Here, the terms in parentheses are referred to as the tangential stiffness matrix of the bar element, which can be expressed as

$$\mathbf{k}_{\mathrm{T}} = \frac{AD^{\text{ep}}}{L}\begin{bmatrix} 1 & -1 \\ -1 & 1 \end{bmatrix}$$

The right-hand side of Eq. (7.16) is referred to as the residual $\boldsymbol{R}$, defined as

$$^{n+1}\mathbf{R}^k = {}^{n+1}\mathbf{F} - \int_0^L \mathbf{B}^{\mathrm{T}}\,{}^{n+1}\sigma^k A\mathrm{d}x = \begin{Bmatrix} ^{n+1}F_1 + {}^{n+1}\sigma^k A \\ ^{n+1}F_2 - {}^{n+1}\sigma^k A \end{Bmatrix}. \tag{7.17}$$

Note: If the residual in Eq. (7.17) becomes zero, it indicates that the original nonlinear Eq. (7.13) has been satisfied. At this point, the Newton–Raphson iteration stops and proceeds to the next load increment. To compute the residual, it is necessary to calculate the stress $^{n+1}\sigma^k$. Due to the nonlinearity of the stress–strain relationship and its dependence on the previous stress history, the stress calculation becomes more complex.

For elastoplastic materials, stress is a function of the previous state, plastic variables, and strain increment, and can be expressed as

$$^{n+1}\sigma^k = f\left({}^n\sigma, {}^n\varepsilon_{\mathrm{p}}, \Delta\varepsilon^k, \cdots\right). \tag{7.18}$$

In stress calculations, it is essential to consider whether the material is in the elastic or strain hardening region of the curve, whether it is being loaded or unloaded, and the changes in yield stress according to the hardening rule applied. This process is known as state determination and will be discussed in detail in the next section.

If the residual has not vanished, another iteration must be performed according to Eq. (7.16). Since Eq. (7.16) must be satisfied for any $\overline{\boldsymbol{d}} \in \mathbb{R}^2$, which is equivalent to solving the following incremental equations

$$\mathbf{k}_{\mathrm{T}} \cdot \delta \mathbf{d}^k =^{n+1} \mathbf{R}^k. \tag{7.19}$$

Therefore, the Newton–Raphson iteration uses the residual from the most recent iteration to solve for the displacement increment of the previous iteration. When the system consists of multiple bar elements, each individual element is assembled in the usual manner and solved after applying the boundary conditions. After obtaining the nodal displacement increment, the displacement increment $\Delta \boldsymbol{d}$ is given by $\Delta \boldsymbol{d}^{k+1} = \Delta \boldsymbol{d}^k + \delta \mathbf{d}^k$.

7.1.3 *Determination of the Stress State*

As described in the previous section, the relationship between stress and strain becomes nonlinear during plastic deformation. In this section, we will discuss how to determine the stress when the strain and its increment are known. For clarity, we will introduce the isotropic hardening model, the kinematic hardening model, and their combined model separately.

7.1.3.1 Isotropic Hardening Model

In the Newton–Raphson iterative method, it is assumed that the following are given: the strain increment ($\Delta\varepsilon$) from the last load increment to the previous iteration, the plastic strain (${}^n\varepsilon_{\mathrm{p}}$), and the stress (${}^n\sigma$) at the load increment t_n. The basic steps for calculating the stress are as follows:

1. **Calculate the current yield stress**

The current yield stress depends on the accumulated plastic strain ${}^n\varepsilon_{\mathrm{p}}$ and the plastic modulus H, that is,

$${}^n\sigma_{\mathrm{Y}} = {}^0\sigma_{\mathrm{Y}} + H\,{}^n\varepsilon_{\mathrm{p}}, \tag{7.20}$$

Here, ${}^0\sigma_{\mathrm{Y}}$ denotes the initial yield stress. Due to strain hardening, the yield stress increases with increasing plastic strain.

Note: The yield stress in tension and compression is identical and increases with increasing plastic strain.

2. **Elastic Prediction**

Assuming elastic behavior occurs during the strain increment, the stress increment and experimental stress can be expressed as follows:

$$\Delta\,{}^{tr}\sigma = E\Delta\varepsilon, \tag{7.21}$$

$${}^{tr}\sigma = {}^{n}\sigma + \Delta\,{}^{tr}\sigma. \tag{7.22}$$

3. **Verification of Yield Condition**

Check whether the test stress satisfies the yield criterion, that is, the stress must be below the yield stress. For this purpose, the trial yield function is defined as follows:

$${}^{tr}f = \left|{}^{tr}\sigma\right| - {}^{n}\sigma_{\mathrm{Y}}. \tag{7.23}$$

If ${}^{tr}f \le 0$, the material remains elastic. As shown in Fig. 7.4, the material is either on the initial loading curve below the yield stress or on the unloading/reloading curve. In either case, the stress calculated based on the elastic assumption is valid. Let ${}^{n+1}\sigma = {}^{tr}\sigma$ after ${}^{n+1}\varepsilon_{\mathrm{e}} = {}^{n}\varepsilon_{\mathrm{e}} + \Delta\varepsilon$ and ${}^{n+1}\varepsilon_{\mathrm{p}} = {}^{n}\varepsilon_{\mathrm{p}}$. In this case, the strain increment is purely elastic, and the plastic strain remains unchanged.

4. **Plastic Correction**

If ${}^{tr}f > 0$, the material yields during this increment. As shown in Fig. 7.5, at load increment t_n, assume the material is at point a (elastic), and this elastic state is reached by unloading from a plastic state (for example, point d). Therefore, the current yield stress is ${}^{n}\sigma_{\mathrm{Y}}$. At load increment t_{n+1}, the strain increment $\Delta\varepsilon$ is known, and the corresponding updated stress state is sought. If the material remains elastic, the updated stress will be at point c (trial stress ${}^{tr}\sigma$). However, the stress state cannot exceed the stress–strain curve shown in Fig. 7.5. Therefore, a transition from the elastic state to the plastic state occurs in this step, and the material behavior shifts

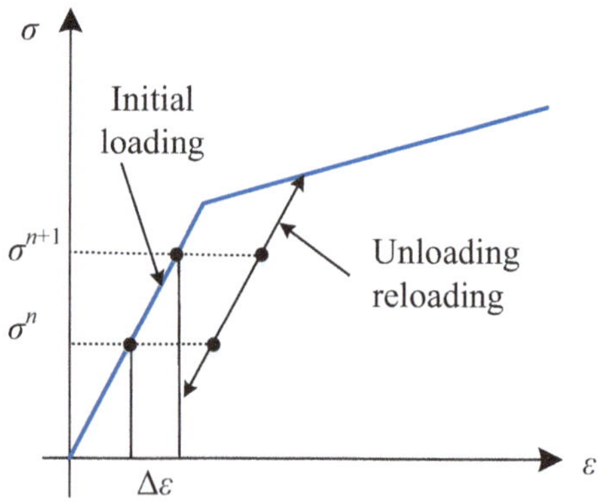

Fig. 7.4 Stress state in the elastic region

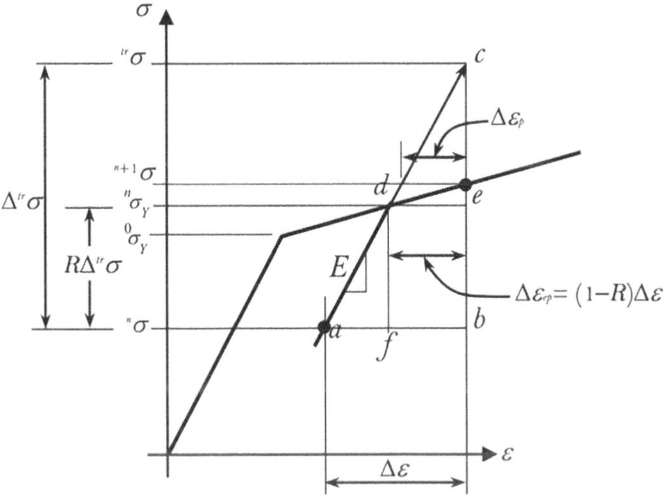

Fig. 7.5 Transition from elasticity to plasticity

upward to point *e* (plastic). Considering that the plastic strain increment does not contribute to the stress increment, the trial stress is updated by subtracting the portion corresponding to the plastic strain increment, that is,

$$ {}^{n+1}\sigma = {}^{tr}\sigma - \mathrm{sgn}\left({}^{tr}\sigma\right) E \Delta\varepsilon_{\mathrm{p}}, \tag{7.24} $$

Here, sgn() denotes the sign function, which takes the value "+ 1" when its argument is positive and "−1 " when negative (the inclusion of this sign function accounts for the fact that the material can also yield under compression). As Eq. (7.24) corrects the trial stress using the plastic strain increment, this process is commonly referred to as plastic correction. Additionally, because it returns the trial stress to the yield stress, it is also known as return mapping. As shown in Fig. 7.5, the challenge in this calculation lies in the fact that, as the trial stress returns, the yield stress also increases (from point *d* to *e*). The return mapping for multidimensional stress states will be discussed in Sect. 7.2.

5. **Plastic Consistency Condition**

In the plastic correction of Eq. (7.24), the plastic strain increment remains unknown. To compute it, the plastic consistency condition is employed here, which requires that the corrected stress during loading must remain on the yield surface. This condition can be expressed as

$$ {}^{n+1}f = \left|{}^{n+1}\sigma\right| - {}^{n+1}\sigma_{\mathrm{Y}} = 0. \tag{7.25} $$

Therefore, the objective is to find the plastic strain increment that satisfies the above conditions.

Note: The plastic strain increment not only reduces the corrected stress ${}^{n+1}\sigma$, but also increases the yield stress ${}^{n+1}\sigma_Y$.

To calculate the plastic strain increment, Eq. (7.25) can be extended as follows:

$$\begin{aligned} &\left|{}^{tr}\sigma - \mathrm{sgn}\left({}^{tr}\sigma\right)E\Delta\varepsilon_\mathrm{p}\right| - \left({}^{n}\sigma_\mathrm{Y} + H\Delta\varepsilon_\mathrm{p}\right) = 0 \\ &\Rightarrow \left|{}^{tr}\sigma\right| - {}^{n}\sigma_\mathrm{Y} - (E+H)\Delta\varepsilon_\mathrm{p} = 0. \end{aligned}$$

Note: The above formula applies to both positive and negative test stresses.

Therefore, the plastic strain increment can be expressed as

$$\Delta\varepsilon_\mathrm{p} = \frac{\left|{}^{tr}\sigma\right| - {}^{n}\sigma_\mathrm{Y}}{E+H} = \frac{{}^{tr}f}{E+H}. \tag{7.26}$$

Since ${}^{tr}f > 0$, the plastic strain increment is always positive. With this plastic strain increment, the stress in Eq. (7.24) can be updated, thus completing the state determination. The plastic strain increment reduces the trial stress and increases the yield stress, thereby satisfying the plastic consistency condition.

Since a linear hardening model with a constant plastic modulus H is used, the plastic consistency condition in Eq. (7.25) can be explicitly solved based on the plastic strain increment. However, if a nonlinear hardening model is employed, the plastic consistency condition must be solved iteratively using methods such as the Newton–Raphson method.

To express the plastic strain increment (or the purely elastic component of the strain increment) in terms of the total strain increment, that is, the portion denoted by R in Fig. 7.5, it can be calculated using the similar triangles *abc* and *dec* as follows:

$$\frac{|\Delta\varepsilon|}{|\Delta\,{}^{tr}\sigma|} = \frac{(1-R)|\Delta\varepsilon|}{|{}^{tr}\sigma| - {}^{n}\sigma_\mathrm{Y}} \Rightarrow R = 1 - \frac{{}^{tr}f}{|\Delta\,{}^{tr}\sigma|}. \tag{7.27}$$

Here, R is the interpolation factor between the elastic modulus and the tangent modulus. When $R = 1$, the material is purely elastic; when $R = 0$, the initial state of the material is plastic, and the strain increment consists of both elastic and plastic components. According to the relationship ${}^{tr}f = (1-R)E|\Delta\varepsilon|$, the plastic strain increment in Eq. (7.26) can be expressed as:

$$\Delta\varepsilon_\mathrm{p} = \frac{(1-R)E}{E+H}|\Delta\varepsilon|. \tag{7.28}$$

Note: When $R = 0$, Eq. (7.28) becomes identical to Eq. (7.9).

Similarly, the updated stress in Eq. (7.24) can be expressed as:

$${}^{n+1}\sigma = {}^{tr}\sigma - \mathrm{sgn}\left({}^{tr}\sigma\right)\frac{(1-R)E^2}{E+H}|\Delta\varepsilon|. \tag{7.29}$$

Here, the first term on the right-hand side of the equation arises from the elastic predictor, while the second term results from the plastic corrector. Therefore, the algorithm for determining the stress state in elastoplasticity is divided into an elastic prediction and a plastic correction.

6. **Algorithmic Tangent Stiffness**

In Eq. (7.15), the elastoplastic tangent modulus D^{ep} represents the relationship between stress increment and strain increment (hypoelasticity), corresponding to the slope of the stress–strain curve in Fig. 7.2, and is referred to as the algorithmic tangent modulus. The tangent modulus determined by the state algorithm given in Eq. (7.24) is called the consistent tangent modulus. When the material is elastic or in the elastic state during unloading/reloading, the plastic strain increment becomes zero, and the trial stress becomes the updated stress. Therefore, the algorithmic tangent modulus becomes identical to the elastic modulus. When the material is in a plastic state, the algorithmic tangent modulus can be obtained by differentiating the updated stress increment with respect to the strain increment, that is,

$$D^{\mathrm{alg}} = \frac{\partial \Delta\sigma}{\partial \Delta\varepsilon} = \frac{\partial\, {}^{tr}\sigma}{\partial \Delta\varepsilon} - \mathrm{sgn}\left({}^{tr}\sigma\right) E \frac{\partial \Delta\varepsilon_{\mathrm{p}}}{\partial \Delta\varepsilon}. \tag{7.30}$$

Since the predicted stress is elastic, it is clear that the first term on the right side of Eq. (7.30) is the elastic modulus. The derivative of the plastic strain increment can be obtained by differentiating Eq. (7.26) with respect to the strain increment, that is,

$$\frac{\partial \Delta\varepsilon_{\mathrm{p}}}{\partial \Delta\varepsilon} = \frac{1}{E+H}\frac{\partial\, {}^{tr}f}{\partial \Delta\varepsilon} = \mathrm{sgn}\left({}^{tr}\sigma\right)\frac{E}{E+H}.$$

Substituting the above equation into Eq. (7.30) yields the algorithmic tangent modulus.

$$D^{\mathrm{alg}} = \begin{cases} E & \text{Elasticity} \\ E_t & \text{Plasticity} \end{cases} \tag{7.31}$$

It should be noted that the algorithmic (elastoplastic) tangent modulus D^{ep} in Eq. (7.15) is identical to the corresponding continuous tangent modulus in Eq. (7.31), meaning that the state determination algorithm is consistent with the stress–strain curve. However, in the multidimensional elastoplasticity discussed in Sect. 7.2, it will be demonstrated that the two tangent moduli are different and exhibit completely distinct convergence behaviors during Newton–Raphson iteration. Furthermore, when a nonlinear hardening model is used, the two tangent moduli will also differ.

7.1.3.2 Dynamic Hardening Model

The main difference between the kinematic hardening model and the isotropic hardening model lies in the evolution of the yield surface. In the isotropic hardening model, the elastic range (twice the yield stress) increases, whereas in the kinematic hardening model, it remains unchanged. As plastic strain increases, the center of the elastic range shifts parallel to the hardening curve. To simulate this effect, the following back stress is defined:

$$\eta = \sigma - \alpha, \tag{7.32}$$

Here, α is referred to as the back stress, representing the center of the elastic domain. In kinematic hardening models, the material state is determined using the shifted stress η instead of σ. Therefore, the back stress is regarded as a plastic variable that must be stored and updated at each load increment step. The loading history information is stored in the back stress.

In the incremental formulation, it is assumed that the following are given: the strain increment ($\Delta\varepsilon$) at the load increment t_n, the stress (${}^{n}\sigma$), and the back stress (${}^{n}\alpha$). The basic steps for calculating the stress are as follows:

1. **Elastic Prediction**

Assuming the material remains in an elastic state during the strain increment, the stress increment and experimental stress can be expressed as follows:

$$\Delta\, {}^{tr}\sigma = E\Delta\varepsilon. \tag{7.33}$$

$${}^{tr}\sigma = {}^{n}\sigma + \Delta\, {}^{tr}\sigma. \tag{7.34}$$

During the elastic prediction phase, the plastic variables remain unchanged, thus

$${}^{tr}\alpha = {}^{n}\alpha. \tag{7.35}$$

Therefore, the elastic prediction of the deviatoric stress is

$${}^{tr}\eta = {}^{tr}\sigma - {}^{tr}\alpha. \tag{7.36}$$

2. **Check Yield Condition**

Check whether the test stress satisfies the yield condition; that is, the stress must be below the yield stress. To this end, the trial yield function is defined as follows:

$${}^{tr}f = \left|{}^{tr}\eta\right| - {}^{0}\sigma_{\mathrm{Y}}. \tag{7.37}$$

Note: In the kinematic hardening model, the initial yield stress remains unchanged, meaning the elastic range does not increase. If ${}^{tr}f \le 0$, then the material remains elastic. Set ${}^{n+1}\sigma = {}^{tr}\sigma$ and proceed to the next load increment. In this case, the strain increment is purely elastic; the back stress is noy changed, namely, ${}^{n+1}\alpha = {}^{n}\alpha$; and all strain increments are elastic, i.e., ${}^{n+1}\varepsilon_\mathrm{e} = {}^{n}\varepsilon_\mathrm{e} + \Delta\varepsilon$.

3. **Plastic Correction**

If ${}^{tr}f > 0$, the material yields during this increment. When the material undergoes plastic deformation, the trial stress and back stress are updated as follows:

$$ {}^{n+1}\sigma = {}^{tr}\sigma - \mathrm{sgn}\left({}^{tr}\eta\right) E \Delta\varepsilon_\mathrm{p}. \tag{7.38} $$

$$ {}^{n+1}\alpha = {}^{tr}\alpha + \mathrm{sgn}\left({}^{tr}\eta\right) H \Delta\varepsilon_\mathrm{p}. \tag{7.39} $$

Note: The trial stress decreases with increasing plastic strain proportional to the elastic modulus, while the back stress increases with increasing plastic strain proportional to the plastic modulus. The stress update formula is almost identical to that in the isotropic hardening model, except that the offset stress is used in the sgn() function. The back stress is updated in a manner similar to the yield stress in the isotropic hardening model of Eq. (7.20), since the stress–strain curves under proportional loading are the same for both hardening models, and thus the same plastic modulus is used. Even if the plastic strain increment is always positive, the back stress may become negative, depending on the stress history; that is, if yielding occurs in tension, the back stress increases, while it decreases in compression.

4. **Plastic Consistency Condition**

In the plastic correction of Eqs. (7.38) and (7.39), the unknown plastic strain increment is expressed by the plastic consistency condition as follows:

$$ {}^{n+1}f = \left|{}^{n+1}\eta\right| - {}^{0}\sigma_\mathrm{Y} = 0. \tag{7.40} $$

To calculate the plastic strain increment, Eq. (7.40) can be expanded as follows:

$$ \begin{aligned} &\left|{}^{tr}\sigma - \mathrm{sgn}\left({}^{tr}\eta\right) E \Delta\varepsilon_\mathrm{p} - {}^{tr}\alpha - \mathrm{sgn}\left({}^{tr}\eta\right) H \Delta\varepsilon_\mathrm{p}\right| - {}^{0}\sigma_\mathrm{Y} = 0 \\ \Rightarrow &\left|{}^{tr}\sigma - {}^{tr}\alpha\right| - {}^{0}\sigma_\mathrm{Y} - (E + H)\Delta\varepsilon_\mathrm{p} = 0. \end{aligned} $$

Note: The above formula applies to both positive and negative test stresses. Therefore, the plastic strain increment can be expressed as

$$ \Delta\varepsilon_\mathrm{p} = \frac{{}^{tr}f}{E + H}. \tag{7.41} $$

It can be seen that the plastic strain increment Eq. (7.41) is identical to the plastic strain increment Eq. (7.26) in the isotropic hardening model. The force and back stress are updated based on the plastic strain increment and Eq. (7.38) and (7.39), thereby obtaining the final stress state.

In the kinematic hardening model, it is not necessary to store the plastic strain increment, as the load history information is contained within the back stress. Since the plastic strain increment and stress update formulas are the same for both the isotropic and kinematic hardening models, the same algorithmic tangent stiffness can be used for both models.

7.1.3.3 Combination of Isotropic and Kinematic Hardening Models

Many real-world materials exhibit a combined effect of isotropic hardening and kinematic hardening, especially in polycrystalline metals. In such cases, the yield stress initially increases due to plastic hardening, but decreases as the direction of strain changes. This phenomenon is related to the dislocation structures in cold-worked metals. As deformation occurs, dislocations accumulate at barriers, leading to dislocation pile-up and entanglement. This effect is commonly referred to as the Bauschinger effect.

To simulate the combined effects of kinematic hardening and isotropic hardening, a new parameter β is introduced, which varies between 0 and 1. Considering that isotropic hardening alters the yield stress while kinematic hardening modifies the back stress, the parameter β interpolates between the two hardening models as follows:

$$ {}^{n+1}\sigma_{\mathrm{Y}} = {}^{tr}\sigma_{\mathrm{Y}} + (1-\beta)H\Delta\varepsilon_{\mathrm{p}}, \tag{7.42} $$

$$ {}^{n+1}\alpha = {}^{n}\alpha + \mathrm{sgn}\left({}^{tr}\eta\right)\beta H\Delta\varepsilon_{\mathrm{p}}. \tag{7.43} $$

When $\beta = 0$, it corresponds to isotropic hardening, and when $\beta = 1$, it corresponds to kinematic hardening. In the combined hardening model, both plastic strain and back stress are plastic variables, and both need to be updated and stored at each load increment. Readers may attempt to prove that the plastic strain increment obtained from the plastic consistency condition is exactly the same as in the isotropic hardening model of Eq. (7.26) and the kinematic hardening model of Eq. (7.41).

7.2 Multidimensional Elastoplastic Problems

The fundamental concepts of one-dimensional systems introduced in the previous section can be extended to multidimensional systems. However, in the one-dimensional case, obtaining the required stress–strain relationship is relatively straightforward, as experiments are typically conducted using uniaxial tensile tests. This relationship is expressed in terms of axial stress and strain. In addition, strain hardening models are also relatively simple, since either the magnitude of the yield stress increases (isotropic hardening model), or the range remains the same between tension and compression (kinematic hardening model). These strain hardening behaviors can also be determined through uniaxial tension/compression tests. However, one-dimensional elastoplastic theory has very limited applications, such as in rods and trusses.

When the structural system is two- or three-dimensional, it becomes more difficult to apply the elastoplastic theory discussed in the previous section, because in this case, stress is not a scalar but a tensor with up to six components. To use a process similar to that of the previous section, material testing would have to be conducted using different combinations of stress components. In practice, this is impossible due to the infinite number of possible combinations. For example, biaxial loading in a plane stress structure is one of the simplest stress states beyond uniaxial loading, with the only nonzero stress components being σ_{11} and σ_{22}. One possible combination is to fix σ_{11} at 100 MPa and gradually increase σ_{22} beyond the yield strength of the material, then fix σ_{11} at different values and again gradually increase σ_{22}, thereby achieving various combinations. To obtain the stress–strain relationship under biaxial loading, all these possible combinations of stress components must be tested. Therefore, except for very limited cases, it is practically impossible to obtain the stress–strain relationship for multidimensional systems.

A key concept in multidimensional elastoplastic theory is the use of physical models that can represent all possible scenarios, rather than establishing stress–strain relationships for every possible combination. As illustrated in the previous example, even when only two stress components are involved, the number of possible combinations is infinite. Therefore, from a practical standpoint, it is preferable to use scalar measures of stress and strain that can represent multidimensional stress states. Moreover, these measures should be independent of the coordinate system for isotropic materials, meaning they should remain unchanged under coordinate transformations, since the material exhibits identical properties in all directions. To address this, the concepts of equivalent stress and effective strain will be introduced below. Since multidimensional models must also satisfy the one-dimensional case, the stress–strain relationship can be obtained from uniaxial tension/compression tests. Therefore, the key lies in selecting an appropriate yield criterion for a given hardening rule, as well as establishing the elastoplastic stress–strain relationship that links incremental stress and strain in the plastic region. Among the various yield criteria, the von Mises yield criterion is widely used for isotropic metal plasticity, which will be discussed in Sect. 7.2.2. As in Sect. 7.1, the following will discuss kinematic and isotropic hardening models for multidimensional systems, introduce the general formulation of incremental elastoplastic stress–strain relationships, and present some practical computational techniques commonly used in engineering practice.

7.2.1 Yield Function and Yield Criterion

Material yielding occurs due to the relative sliding of molecules within the crystal lattice structure, which is analogous to shear deformation. After the applied load is removed, the material does not return to its original shape. As shown in Fig. 7.6, this type of sliding deformation preserves the volume of the material. If the intermolecular distances change, the volume will also change. However, it is very difficult to cause permanent deformation in a material by altering the intermolecular distances. Therefore, it is generally believed that material failure is related to shear deformation.

7.2.1.1 Maximum Shear Stress Criterion

One of the simplest failure criteria for ductile materials, proposed by Tresca in 1864, is the maximum shear stress criterion. This criterion uses the maximum shear stress $\tau_{\max}$ as the equivalent stress. The maximum shear stress corresponds to the radius of the largest Mohr's circle. Let σ_1, σ_2, and σ_3 be the three principal stresses, ordered such that $\sigma_1 \geq \sigma_2 \geq \sigma_3$, then the maximum shear stress is

$$\tau_{\max} = \frac{\sigma_1 - \sigma_3}{2}. \tag{7.44}$$

Note: $\tau_{\max}$ is independent of the chosen coordinate system, as principal stresses are also independent of the coordinate system used.

This criterion assumes that failure occurs when the maximum shear stress $\tau_{\max}$ equals the shear stress at yield τ_Y in a tensile specimen. In one-dimensional yield in a uniaxial tensile test, $\sigma_1 = \sigma_Y$ and $\sigma_2 = \sigma 3 = 0$. Therefore, it is easy to see from the Mohr's circle that σ_Y is half of τ_Y. When $\tau_{\max}$ is less than τ_Y, the material is elastic. $\tau_{\max}$ cannot be greater than τ_Y. In this criterion, the elastic range of the material is defined as

$$\tau_{\max} \leq \tau_Y \left(= \frac{1}{2}\sigma_Y \right). \tag{7.45}$$

The yield criterion defines the boundary of the elastic region, that is,

$$f(\boldsymbol{\sigma}) = \tau_{\max} - \tau_Y = 0, \tag{7.46}$$

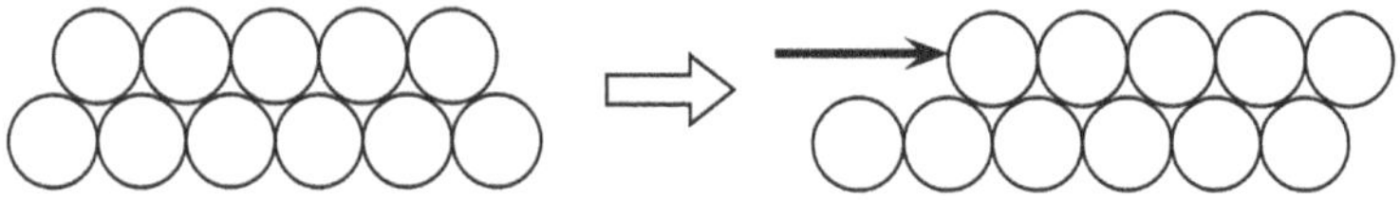

Fig. 7.6 Material failure caused by relative sliding

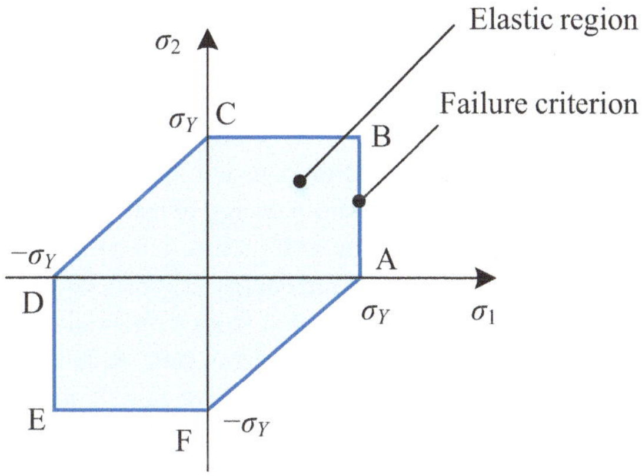

Fig. 7.7 Maximum shear stress criterion

Here, f $(\boldsymbol{\sigma})$ is the yield function, and $f(\boldsymbol{\sigma}) = 0$ is the yield criterion. Although the above equation has a simple form, the actual combination of multidimensional stress states can be complex. Figure 7.7 shows the hexagonal failure envelope of the two-dimensional maximum shear stress theory, i.e., $\sigma_3 = 0$. Based on the uniaxial tensile test, the material fails at $\sigma_1 = \sigma_Y$ and $\sigma_2 = 0$ (point A). If σ_2 increases from point A, it does not affect the yield criterion until it reaches σ_Y (point B), as τmax is determined by σ_1 and σ_3. Along line BC, τ_{max} is determined by σ_2 and σ_3. Along line CD, τ_{max} is determined by σ_1 and σ_2, and as σ_1 becomes more negative, σ_2 must decrease. The elastic norm defined by Eq. (7.45) corresponds to the interior of the hexagon, in which the material is elastic. The yield criterion defined by Eq. (7.46) corresponds to the boundary and is commonly called the yield surface.

7.2.1.2 Distortion Energy Criterion

The maximum shear stress criterion is based solely on shear stress. In general, deformation can be divided into a volumetric (dilatational) component and a volume-preserving (distortional) component. For example, for a given strain tensor $\boldsymbol{\varepsilon}$, the volumetric strain in small deformations is defined as

$$\varepsilon_V = \text{tr}(\boldsymbol{\varepsilon}) = \varepsilon_{11} + \varepsilon_{22} + \varepsilon_{33}. \tag{7.47}$$

In more general terms, plastic deformation is related to the deviatoric part of the strain, which is often referred to as the deviatoric strain. For example, in the case of pure shear deformation, $\varepsilon_{12} = \varepsilon_{21}$ is the only nonzero strain component. Therefore, the volumetric strain becomes zero, and the shear deformation corresponds to the deviatoric strain. It depends on the coordinate system used. Therefore, in order to use deviatoric strain or stress, a scalar defined according to deviatoric strain or stress is required.

As stress gradually increases, the strain energy density of the material also increases. Since these two quantities are related, strain energy density can be used as a failure criterion. An advantage of using strain energy density is that it remains a scalar quantity even when all six stress components are nonzero. As previously mentioned, since volumetric deformation does not lead to material failure, this component must be excluded before using strain energy density as a failure criterion. The strain energy density after removing the volumetric deformation is referred to as the distortion energy density. The distortion energy criterion compares the distortion energy under a multiaxial stress state with that from a uniaxial tensile yield test: when the distortion energy under a multiaxial stress state equals that of the uniaxial tensile yield test, the material is considered to have failed; when the distortion energy under a multiaxial stress state is less than that of the uniaxial tensile yield test, the material is considered to remain elastic.

Distortion energy is defined as the difference between the strain energy and its volumetric component. To calculate the distortion energy, first define the volumetric components of stress (hydrostatic pressure) and strain (mean strain) as follows:

$$\sigma_{\mathrm{m}} = \frac{1}{3}\mathrm{tr}(\boldsymbol{\sigma}) = \frac{1}{3}(\sigma_{11} + \sigma_{22} + \sigma_{33}) \tag{7.48}$$

$$\varepsilon_{\mathrm{m}} = \frac{1}{3}\mathrm{tr}(\boldsymbol{\varepsilon}) = \frac{1}{3}(\varepsilon_{11} + \varepsilon_{22} + \varepsilon_{33}), \tag{7.49}$$

where tr(□) is the trace operator, such that $\mathrm{tr}(\boldsymbol{\sigma}) = \sigma_{kk}$.

Note: The average strain is three times the volume strain.

Next, the deviatoric stress and deviatoric strain tensors are defined as

$$\mathbf{s} \equiv \boldsymbol{\sigma} - \sigma_m \mathbf{1} = \mathbf{I}_{\mathrm{dev}} : \boldsymbol{\sigma} \tag{7.50}$$

$$\mathbf{e} \equiv \boldsymbol{\varepsilon} - \varepsilon_m \mathbf{1} = \mathbf{I}_{\mathrm{dev}} : \boldsymbol{\varepsilon}, \tag{7.51}$$

Here, $\mathbf{1} = [\delta_{ij}]$ denotes the second-order identity tensor; $\mathbf{I}$ represents the fourth-order symmetric identity tensor, defined as $\mathbf{I}_{ijkl} = (\delta_{ik}\delta_{jl} + \delta_{il}\delta_{jk})/2$, $\mathbf{I}_{dev} = \mathbf{I} - {}^1/_3 \mathbf{1} \otimes \mathbf{1}$ denotes the fourth-order identity tensor, ⊗ represents the tensor product, and ":" denotes the double contraction operator for tensors.

Note: $\mathrm{tr}(\boldsymbol{s}) = 0$ and $\mathrm{tr}(\boldsymbol{e}) = 0$.

For isotropic linearly elastic materials, the constitutive relation between stress and strain can be expressed as

$$\boldsymbol{\sigma} = [\lambda \mathbf{1} \otimes \mathbf{1} + 2\mu \mathbf{I}] : \boldsymbol{\varepsilon} \equiv \mathbf{D} : \boldsymbol{\varepsilon}, \tag{7.52}$$

where, λ and μ are the Lamé constants, and D is the fourth-order constitutive tensor. Using Property **1**: $\boldsymbol{\varepsilon} = \mathrm{tr}(\boldsymbol{\varepsilon}) = 3\varepsilon_{\mathrm{m}}$, Eq. (7.52) can be rewritten as follows

$$\boldsymbol{\sigma} = \lambda(3\varepsilon_{\mathrm{m}})\mathbf{1} + 2\mu(\mathbf{e} + \varepsilon_m \mathbf{1}) = (3\lambda + 2\mu)\varepsilon_m \mathbf{1} + 2\mu \mathbf{e}. \tag{7.53}$$

Comparing the above equation with Eq. (7.50), the first term on the right-hand side represents the volumetric component of the stress, while the second term represents the deviatoric component. Therefore, the following decomposed constitutive relation can be obtained

$$\mathbf{s} = 2\mu\mathbf{e} \tag{7.54}$$

$$\sigma_m = (3\lambda + 2\mu)\varepsilon_{\mathrm{m}}. \tag{7.55}$$

It is easy to show that the bulk modulus (the constant relating σ_m and ε_V) is defined as $K = (3\lambda + 2\mu)/3$. Interestingly, only the shear modulus μ appears in the relationship between deviatoric stress and deviatoric strain.

According to Eqs. (7.50) and (7.51), the distortion energy density can be defined as

$$U_{\mathrm{d}} = \frac{1}{2}\boldsymbol{s} : \boldsymbol{e}. \tag{7.56}$$

Using the relationship $\boldsymbol{s} = 2\mu\mathbf{e}$, the deviatoric strain energy density can be expressed in terms of the deviatoric stress as

$$U_{\mathrm{d}} = \frac{1}{4\mu}\mathbf{s} : \mathbf{s}. \tag{7.57}$$

In the case of a uniaxial tensile test, when $\sigma_{11} = \sigma_{\mathrm{Y}}$ and all other stress components are zero, the material fails, and the deviatoric stress is

$$\mathbf{s} = \begin{bmatrix} \frac{2}{3}\sigma_{\mathrm{Y}} & 0 & 0 \\ 0 & -\frac{1}{3}\sigma_{\mathrm{Y}} & 0 \\ 0 & 0 & -\frac{1}{3}\sigma_{\mathrm{Y}} \end{bmatrix}. \tag{7.58}$$

Therefore, in the uniaxial tensile test, the distortion energy density at material failure becomes

$$U_{\mathrm{d}}|_{1\mathrm{D}} = \frac{1}{6\mu}\sigma_{\mathrm{Y}}^2. \tag{7.59}$$

When the distortion energy density equals the energy density at the yield point in a tensile test, the material under a multiaxial stress state yields. By setting Eq. (7.57) equal to Eq. (7.59), it can be concluded that the stress measure at yielding under a multiaxial stress state is equal to the uniaxial yield stress.

$$\sigma_{\mathrm{e}} \equiv \sqrt{\frac{3}{2}\mathbf{s} : \mathbf{s}} = \sigma_{\mathrm{Y}}. \tag{7.60}$$

Here, σ_e is referred to as the equivalent stress or Von Mises stress. When the equivalent stress is less than the uniaxial yield stress, the material is considered to be elastic. Therefore, the use of equivalent stress allows the method for determining uniaxial stress states to be applied to multidimensional cases.

Note: Although equivalent stress is used to determine material failure, this approach is also based on the distortion energy criterion.

The corresponding quantity to equivalent stress is effective strain, and the two are defined through the distortion energy density, that is,

$$U_{\mathrm{d}} = \frac{1}{2}\mathbf{s} : \mathbf{e} = \frac{1}{2}\sigma_{\mathrm{e}} e_{\mathrm{e}}, \tag{7.61}$$

Here, e_{e} denotes the effective strain. According to Eqs. (7.57) and (7.60), the distortion energy density can be expressed in terms of the equivalent stress as follows:

$$U_{\mathrm{d}} = \frac{1}{4\mu}\left(\frac{2}{3}\sigma_{\mathrm{e}}^2\right) = \frac{1}{6\mu}\sigma_{\mathrm{e}}^2. \tag{7.62}$$

By comparing Eqs. (7.61) and (7.62), the effective strain can be expressed in terms of the equivalent stress as follows:

$$e_{\mathrm{e}} = \frac{1}{3\mu}\sigma_{\mathrm{e}} = \frac{1}{3\mu}\sqrt{\frac{3}{2}\mathbf{s} : \mathbf{s}}. \tag{7.63}$$

Using Eq. (7.54), the effective strain can be expressed in terms of the deviatoric strain as

$$e_{\mathrm{e}} = \sqrt{\frac{2}{3}\mathbf{e} : \mathbf{e}}. \tag{7.64}$$

Note: The equivalent stress in Eq. (7.60) and the effective strain in Eq. (7.64), aside from their coefficients, are defined similarly to deviatoric stress and strain, respectively. The use of effective strain is related to plastic deformation. The scalar plastic strain in one-dimensional plasticity becomes a tensor in multidimensional plasticity.

7.2.2 Von Mises Yield Criterion

The distortion energy theory discussed in the previous section is also known as the Von Mises yield criterion. It states that yielding occurs when the equivalent stress reaches the yield stress of the material under uniaxial tension. The equivalent stress σ_{e} in Eq. (7.60) can be expressed as

$$\sigma_e = \sqrt{\frac{3}{2}\mathbf{s} : \mathbf{s}} \equiv \sqrt{3J_2}, \tag{7.65}$$

Here, J_2 is the second invariant of the deviatoric stress. It can be expressed in the following alternative forms: in terms of stress components,

$$J_2 = \frac{1}{6}\left[\left(\sigma_x - \sigma_y\right)^2 + \left(\sigma_y - \sigma_z\right)^2 + (\sigma_z - \sigma_x)^2\right] + \tau_{xy}^2 + \tau_{yz}^2 + \tau_{zx}^2. \tag{7.66}$$

or in terms of principal stress

$$J_2 = \frac{1}{6}\left[(\sigma_1 - \sigma_2)^2 + (\sigma_2 - \sigma_3)^2 + (\sigma_3 - \sigma_1)^2\right]. \tag{7.67}$$

Using the equivalent stress, the yield function and the corresponding yield criterion can be defined as follows

$$f'(\boldsymbol{\sigma}) \equiv \sigma_e^2 - \sigma_Y^2 = 3J_2 - \sigma_Y^2 = 0, \tag{7.68}$$

Here, σ_Y is the yield strength obtained from tensile test. Therefore, even when the stress state is multiaxial, by treating the equivalent stress as a tensile stress, one-dimensional experimental data can still be used.

According to the definition of J_2, it is evident that the von Mises yield function corresponds to a two-dimensional ellipse. As shown in Fig. 7.8, any point inside the ellipse $\left(f' < 0\right)$ represents an elastic stress state. The interior of the ellipse is referred to as the elastic domain of the material. Points on the yield surface $\left(f' = 0\right)$ correspond to stress states that cause the material to yield. Stress states cannot exist outside the yield surface.

Figure 7.8 also illustrates the maximum shear stress criterion.

Note: The two criteria coincide at six vertices, but the maximum shear stress criterion lies inside the von Mises criterion, meaning the former is more conservative than the latter. For instance, in the case of uniaxial tension, corresponding to a stress state along the σ_1-axis, both criteria predict the same yield point at $\sigma_1 = \sigma_Y$. However, in the case of pure shear stress, along line OA in Fig. 7.8, the maximum shear stress criterion predicts material failure earlier than the von Mises criterion. It should also be noted that the von Mises yield surface is smooth, whereas the maximum shear stress criterion has six vertices. The smoothness of the yield surface aids in numerically locating the yield point.

Plastic deformation can be physically explained by atomic dislocations. Elastic deformation corresponds to changes in intermolecular distances without causing atomic dislocations, while plastic deformation implies relative sliding of atomic layers and permanent deformation, though the volume of the structure remains unchanged. Therefore, plastic behavior can be effectively described by the deviatoric part of the tensor, which preserves volume. As previously discussed in the equations, J_2 is the second invariant of the deviatoric stress. Specifically, for the

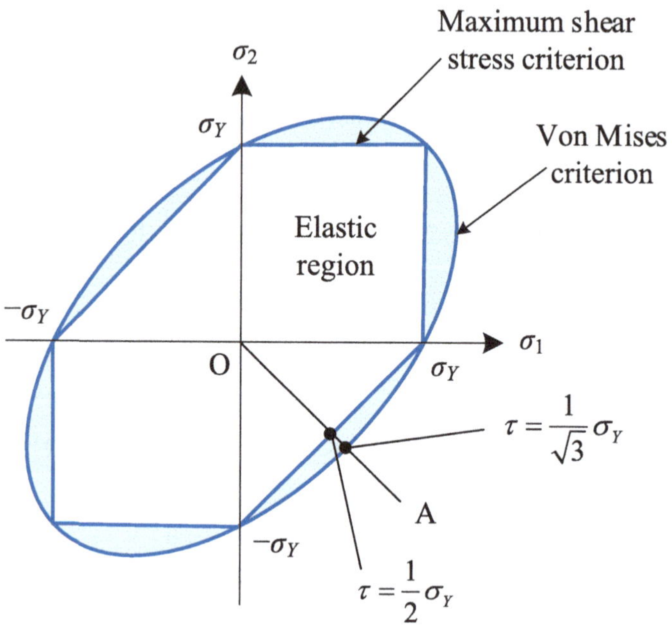

Fig. 7.8 Von mises yield criterion

stress tensor $\boldsymbol{\sigma} = [\sigma_{ij}]$, $(i, j = 1,\ 2,\ 3)$, the second invariant of the deviatoric stress is given by:

$$J_2 = \frac{1}{2}\left[\mathbf{s} : \mathbf{s} - \mathrm{tr}(\mathbf{s})^2\right] = \frac{1}{2}\mathbf{s} : \mathbf{s}. \tag{7.69}$$

Note: $\mathrm{tr}(s) = 0$, because the trace component of the stress tensor has been transferred to σ_m.

The yield function defined in Eq. (7.68) is expressed in terms of the square of the effective stress. Since both the effective stress and the yield stress are always positive, it is unnecessary to use the squared term to define the yield criterion. Defining the yield criterion without the square is also more convenient, as it ensures the units are consistent with those of stress. Therefore, the Von Mises yield criterion can be rewritten as

$$f(\sigma) = \|\mathbf{s}\| - \sqrt{\frac{2}{3}}\sigma_Y = 0, \tag{7.70}$$

Here, $\|s\| = (s{:}s)^{1/2}$ is the norm of the deviatoric stress. If the ellipse in Fig. 7.8 is plotted in terms of the principal deviatoric stresses, it becomes $\sqrt{2/3}\sigma_Y$, which can be regarded as the radius of the yield circle in the deviatoric stress space.

7.2.3 Hardening Models

The yield criteria discussed in the previous section determine whether a material yields based on a given yield stress. In Eq. (7.70), it is assumed that the yield stress of the material remains constant. However, as discussed in Sect. 7.1, the yield stress itself can change with plastic deformation, a phenomenon known as strain hardening. In general, materials whose yield stress varies with plastic deformation can be classified into three categories (see Fig. 7.9): yield stress increases proportionally with plastic deformation (strain hardening); yield stress remains constant (ideal plasticity); and yield stress decreases with increasing plastic deformation (strain softening). Typically, metals are strain-hardening materials, while geomaterials may exhibit strain softening under certain conditions. Strain-hardening materials are considered stable and will be discussed in this section.

As shown in Fig. 7.10, two different strain hardening models are discussed: the isotropic hardening model and the kinematic hardening model. In the isotropic hardening model, the elastic domain expands continuously due to plastic deformation, whereas in the kinematic hardening model, the elastic domain remains unchanged but shifts in parallel as hardening progresses. The definitions of these hardening models can be extended to multidimensional plasticity. The yield criterion expressed by Eq. (7.70) can be regarded as the equation of a circle, with its center at the origin and a radius of $\sqrt{2/3}\sigma_{\mathrm{Y}}$. In the isotropic hardening model, the center remains fixed while the radius (i.e., σ_{Y}) increases uniformly; in the kinematic hardening model, the radius remains constant while the center moves in stress space, that is, the norm of the deviatoric stress becomes $\|\boldsymbol{s}-\boldsymbol{\alpha}\|$, where $\boldsymbol{\alpha}$ denotes the position of the yield surface center. Figure 7.10 illustrates these two hardening models. Since tensile tests are used to characterize material behavior beyond yielding, the same hardening parameters used in one-dimensional plasticity can be applied to multidimensional hardening.

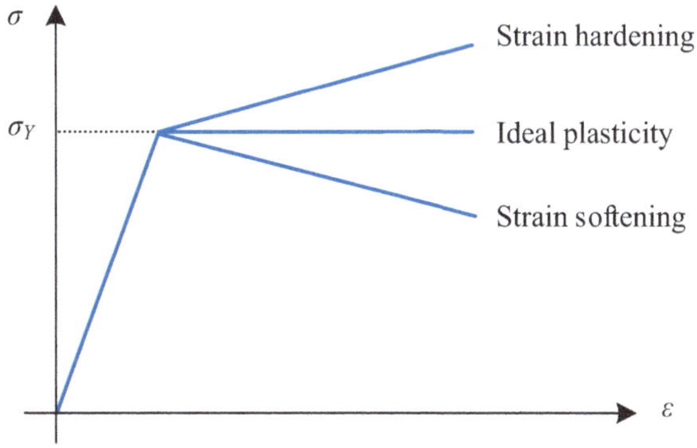

Fig. 7.9 The behavior of materials after plastic deformation

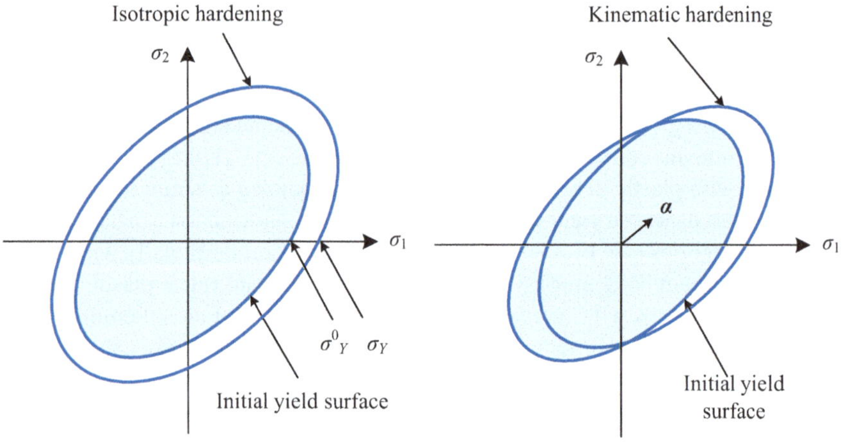

(a) Isotropic hardening model (b) Kinematic hardening model

Fig. 7.10 Two-dimensional hardening models

7.2.3.1 Isotropic Hardening Model

In the isotropic hardening model, the post-yield behavior depends on the accumulated effective plastic strain, e_p. For the linear isotropic hardening model, the yield stress increases with the effective plastic strain, that is,

$$\sigma_Y = \sigma_Y^0 + He_p, \tag{7.71}$$

Here, the plastic modulus H can be obtained from the uniaxial stress–strain relationship.

$$H = \frac{\Delta\sigma}{\Delta e_p}. \tag{7.72}$$

The radius of the yield surface increases in proportion to the effective plastic strain and the plastic modulus. The definition of the plastic modulus above can be applied to general nonlinear hardening, as it is defined in rate form. In this case, H is the slope of the stress-plastic strain curve at a given total plastic strain. For linear hardening, H is a constant.

When $H = 0$, the material is referred to as an ideal elastic–plastic material. Ideal plastic material is an idealized model intended to simplify the constitutive equations. This idealization is reasonable for materials that do not exhibit significant strain hardening. The adequacy of this idealization depends on the purpose and requirements of the specific application. If only monotonic loading is of interest and highly detailed results are not required, this idealization may yield satisfactory outcomes. However, due to industrial advancements, structures are often subjected to complex loading conditions and stringent requirements, making this idealization inadequate for many applications. Strain hardening should therefore be taken into consideration.

7.2.3.2 Kinematic Hardening Model

In the kinematic hardening model, the yield surface translates in stress space (see Fig. 7.10). Therefore, the equation of the yield surface can be obtained by introducing a back stress to the initial yield surface, and this back stress at the center of the yield surface is denoted by $\boldsymbol{\alpha}$. The distance from the center of the yield surface to the yield surface itself can be measured by the difference between $\mathbf{s}$ and $\boldsymbol{\alpha}$, so the back stress of the yield surface can be expressed as

$$\boldsymbol{\eta} = \mathbf{s} - \boldsymbol{\alpha}. \tag{7.73}$$

It should be noted that both the back stress $\boldsymbol{\alpha}$ and the shift stress $\boldsymbol{\eta}$ are deviatoric stresses. Therefore, the yield surface equation can be defined as follows:

$$\|\boldsymbol{\eta}\| - \sqrt{\frac{2}{3}}\sigma_Y = 0. \tag{7.74}$$

In the kinematic hardening model, σ_Y is the initial yield stress and remains constant, while the back stress depends on the current stress and the accumulated effective plastic strain e_p. According to Ziegler's rule, the increment of back stress in the linear kinematic hardening model can be expressed as

$$\Delta\boldsymbol{\alpha} = \sqrt{\frac{2}{3}}H\Delta e_p\frac{\boldsymbol{\eta}}{\|\boldsymbol{\eta}\|}. \tag{7.75}$$

As with the isotropic hardening model, the effective plastic strain plays a crucial role in the computation of back stress evolution. Additionally, the back stress increases in the direction parallel to the deviatoric stress. Since $\boldsymbol{\eta} = \boldsymbol{s} - \boldsymbol{\alpha}$ is the radial direction of the yield surface, the increment is always radial.

7.2.3.3 Combined Hardening Model

The distinction between isotropic hardening models and kinematic hardening models is clear: the former increases the radius of the yield surface, while the latter shifts the center of the yield surface. Many materials exhibit a combination of these two behaviors: the yield stress increases with plastic deformation, but the material yields earlier when loaded in the reverse direction. This phenomenon is caused by dislocation pile-up and entanglement (back stress), which facilitate dislocation movement when the direction of strain changes. For the combined linear isotropic/kinematic hardening model, a parameter $\beta \in [0, 1]$ is used to account for this combined effect, known as the Bauschinger effect. In this model, the equation of the yield surface is defined as follows:

$$\|\boldsymbol{\eta}\| - \sqrt{\frac{2}{3}}\left[\sigma_{\mathrm{Y}}^{0} + (1-\beta)He_{\mathrm{p}}\right] = 0 \tag{7.76}$$

The increment of back stress is defined as

$$\Delta\boldsymbol{\alpha} = \sqrt{\frac{2}{3}}\beta H\,\Delta e_{\mathrm{p}}\frac{\boldsymbol{\eta}}{\|\boldsymbol{\eta}\|}. \tag{7.77}$$

This model is sufficiently general to represent both isotropic hardening and kinematic hardening models as special cases: for the kinematic hardening model $\beta = 1$; for the isotropic hardening model $\beta = 0$.

7.2.4 *Classical Elastoplastic Models*

In the previous subsection, the yield function and strain hardening model for multi-dimensional plasticity were discussed. This section will address how to apply these models to determine the current stress state and the evolution of plastic variables. Based on the Von Mises yield criterion and isotropic/kinematic hardening models, the plastic variables are the effective plastic strain e_{p} and the back stress $\boldsymbol{\alpha}$. The former is a scalar, while the latter is a second-order tensor. All relationships in this section are presented in rate form and expressed in incremental form to facilitate numerical integration in the next section. For example, the relationship between the strain rate and the strain increment $\Delta\varepsilon$ is given by $\Delta\varepsilon = \Delta t\,\dot{\varepsilon}$, where Δt is the time increment.

7.2.4.1 Additive Decomposition

The fundamental assumption of small deformation elastoplasticity is that the elastic and plastic components can be additively decomposed. This assumption represents a fundamental distinction from finite deformation elastoplasticity. Assuming a small elastoplastic strain, the strain and its rate of change can be additively decomposed into elastic and plastic parts, that is,

$$\boldsymbol{\varepsilon} = \boldsymbol{\varepsilon}^{\mathrm{e}} + \boldsymbol{\varepsilon}^{\mathrm{p}}, \quad \dot{\varepsilon} = \dot{\varepsilon}^{\mathrm{e}} + \dot{\varepsilon}^{\mathrm{p}}, \tag{7.78}$$

where the superscripts e and p denote the elastic and plastic parts, respectively. The dot over the superscript indicates the rate of change. It is important to note that in static problems, the rate of change corresponds to load increments. It is assumed that plastic deformation occurs only in the deviatoric space, meaning the plastic strain

ε_p and its rate are deviatoric tensors, i.e., tr(ε_p) = 0. Similar to one-dimensional plasticity, the total strain and its rate of change are given. However, it is not yet clear how much of them is elastic or plastic. Therefore, given the total strain and its rate of change, the corresponding elastic or plastic strains need to be determined. Elastic strain generates stress in the material, while plastic strain is independent of stress. However, plastic strain influences the yield stress in strain hardening models.

7.2.4.2 Strain Energy Density

It is generally assumed that the elastic component possesses a strain energy density, so the stress can be determined by differentiating the strain energy density with respect to the elastic strain. The elastic part of an elastoplastic model is identical to that of a linear elastic material. Since the relationship between stress and elastic strain is linear, the strain energy density function adopts a quadratic form, which can be expressed as follows:

$$W(\boldsymbol{\varepsilon}^{\mathrm{e}}) = \frac{1}{2}\boldsymbol{\varepsilon}^{\mathrm{e}} : \mathbf{D} : \boldsymbol{\varepsilon}^{\mathrm{e}} = \frac{1}{2}(\boldsymbol{\varepsilon} - \boldsymbol{\varepsilon}^{\mathrm{p}}) : \mathbf{D} : (\boldsymbol{\varepsilon} - \boldsymbol{\varepsilon}^{\mathrm{p}}), \tag{7.79}$$

Here, $\boldsymbol{D}$ denotes the fourth-order constitutive tensor. Before determining the plastic behavior of the material, the elastic component of the strain is typically unknown. By distinguishing the definitions above, the relationship between stress and elastic strain is given by

$$\boldsymbol{\sigma} = \frac{\partial W(\boldsymbol{\varepsilon}^{\mathrm{e}})}{\partial \boldsymbol{\varepsilon}^{\mathrm{e}}} = \boldsymbol{D} : \boldsymbol{\varepsilon}^{\mathrm{e}} = \boldsymbol{D} : (\boldsymbol{\varepsilon} - \boldsymbol{\varepsilon}^{\mathrm{p}}) \tag{7.80}$$

Clearly, when the strain is considered elastic, $\boldsymbol{D}$ is the constitutive tensor introduced in Eq. (7.52). Based on the assumption in Eq. (7.78), the rate form of Eq. (7.80) can be expressed as

$$\dot{\sigma} = \mathbf{D} : (\dot{\varepsilon} - \dot{\varepsilon}^{\mathrm{p}}), \tag{7.81}$$

Here, $\mathbf{D} = (\lambda + (2/3)\mu)\mathbf{1} \otimes \mathbf{1} + 2\mu I_{dev}$ is the fourth-order isotropic constitutive tensor. According to the decompositions in Eqs. (7.54) and (7.55), the relation in Eq. (7.81) can be further separated into volumetric and deviatoric components, namely,

$$\dot{\boldsymbol{\sigma}}_{\mathrm{m}} = \frac{1}{3}\,\mathrm{tr}(\dot{\boldsymbol{\sigma}}) = \left(\lambda + \frac{2}{3}\mu\right)\mathrm{tr}(\dot{\boldsymbol{\varepsilon}}) = (3\lambda + 2\mu)\dot{\boldsymbol{\varepsilon}}_{\mathrm{m}} \tag{7.82}$$

$$\dot{\mathbf{s}} = 2\mu(\dot{\mathbf{e}} - \dot{\mathbf{e}}^{\mathrm{p}}), \tag{7.83}$$

Equation (7.82) utilizes the property $tr(\varepsilon^p) = 0$. From Eq. (7.82), it can be observed that the volumetric stress (hydrostatic pressure) is unrelated to plastic deformation, which is consistent with the fact that the yield function is defined solely in terms of deviatoric stress.

7.2.4.3 Yield Function

For the plastic behavior of metallic materials, the von Mises yield criterion with an associated flow rule is commonly employed to describe material response beyond the elastic limit. Hence, the yield function and yield criterion can be stated as:

$$f\left(\boldsymbol{\eta}, e_{\mathrm{p}}\right) \equiv \|\boldsymbol{\eta}\| - \sqrt{\frac{2}{3}}\kappa\left(e_{\mathrm{p}}\right) \leq 0, \tag{7.84}$$

where $\boldsymbol{\eta} = \boldsymbol{s} - \boldsymbol{\alpha}$ is the deviatoric stress; $\boldsymbol{\alpha}$ is the back stress, which is the center of the yield surface (elastic domain) determined by a kinematic hardening model; $\kappa(e_{\mathrm{p}})$ is the radius of the elastic domain determined by an isotropic hardening model; e_{p} is the effective plastic strain. Equation (7.84) employs a combined isotropic/kinematic hardening model. The elastic domain generated by the yield function in Eq. (7.84) forms a convex set, given by

$$E = \left\{\left(\boldsymbol{\eta}, e_{\mathrm{p}}\right) \middle| f\left(\boldsymbol{\eta}, e_{\mathrm{p}}\right) \leq 0\right\}. \tag{7.85}$$

In general, the yield surface defined by f in the above equation is smooth and convex. Mathematically, plasticity can be regarded as the projection of stress onto the yield surface. If the material is assumed to be purely elastic, then for a given strain, the resulting stress would be much higher than that of an elastoplastic material. Since stress cannot lie outside the elastic domain, elastoplasticity projects this stress back onto the yield surface. Because the yield surface is convex, this projection becomes a contraction mapping, thereby ensuring the existence of a unique projection. The same concept was applied in the one-dimensional elastoplasticity discussed in Sect. 7.1. Initially, the stress increment is assumed to be elastic. If the predicted stress exceeds the yield stress, it is returned to the yield stress. If the yield stress is fixed, this process is relatively straightforward. However, when the stress is brought back to the yield surface, the yield surface itself changes according to the hardening model. Therefore, it is necessary to determine how plastic deformation alters the yield function.

7.2.4.4 Associated Flow Rule

The flow rule governs the evolution of plastic strain $\boldsymbol{\varepsilon}^{\mathrm{p}}$. In one-dimensional plasticity, the plastic strain is a scalar and can only increase in value. In multidimensional plasticity, since $\boldsymbol{\varepsilon}^{\mathrm{p}}$ is a tensor, both its magnitude and direction must be determined. Therefore, the general form of the flow rule can be expressed as follows:

$$\dot{\varepsilon}^{\mathrm{p}} = \gamma \mathbf{r}(\sigma, \boldsymbol{\xi}), \tag{7.86}$$

Here, $\xi = (\boldsymbol{\alpha}, \boldsymbol{e}_{\mathrm{p}})$ denotes the plastic variables, and γ is referred to as the plastic consistency parameter. In general, $\gamma \geq 0$, which means that there is no plasticity at 0.

Advanced elastoplastic mechanics demonstrates that flow rules can be derived from constrained optimization theory, in which Eq. (7.84) represents an inequality constraint. According to the principle of minimum potential energy, the equilibrium of elastic structures can be obtained by minimizing the potential energy, which is the sum of the strain energy and the potential energy of external forces. Therefore, the equilibrium equations are derived from optimality conditions, namely, the first derivative of the potential energy should become zero. In elastoplasticity, this optimization problem is modified so that the stress must remain within the elastic domain defined by Eq. (7.84). This condition can be regarded as a constraint in the optimization problem. If the stress computed by minimizing the potential energy remains within the elastic domain, no constraint is needed. However, if the computed stress lies outside the elastic domain, it must be projected back onto the boundary of the elastic domain. In this case, the plastic consistency parameter becomes a Lagrange multiplier to enforce the constraint, and it is always non-negative.

The expression for $\boldsymbol{r}(\boldsymbol{\sigma}, \xi)$ depends on the plasticity model. It is generally assumed that there exists a flow potential (or plastic potential) g, such that the plastic strain evolves in the direction normal to the flow potential, that is,

$$\dot{\varepsilon}^{\mathrm{p}} = \gamma \frac{\partial g(\sigma, \boldsymbol{\xi})}{\partial \sigma}, \tag{7.87}$$

Here, $g(\boldsymbol{\sigma}, \xi)$ is the flow potential function. When the flow potential function is identical to the yield function, the plasticity model is referred to as an associated flow model. Therefore,

$$\dot{\varepsilon}^{\mathrm{p}} = \gamma \frac{\partial f(\boldsymbol{\eta}, e_{\mathrm{p}})}{\partial \boldsymbol{\eta}} = \gamma \frac{\boldsymbol{\eta}}{\|\boldsymbol{\eta}\|} = \gamma \mathbf{N}, \tag{7.88}$$

Here, $\boldsymbol{N}$ is the unit deviatoric tensor normal to the yield surface; γ is the plastic consistency parameter, which is non-negative. If the material is in an elastic state, γ must be zero; if the material is in a plastic state, γ must be positive. Therefore, the plastic strain increases in the direction normal to the yield surface, with a magnitude determined by the plastic consistency parameter γ.

As the material undergoes plastic deformation, plastic variables (back stress and effective plastic strain) also change according to the hardening model. The general form of the hardening law can be expressed as

$$\dot{\xi} = \gamma \mathbf{h}(\sigma, \boldsymbol{\xi}). \tag{7.89}$$

It should be noted that the evolution of plastic variables is also proportional to the plastic consistency parameter. In particular, the back stress rate can be determined by the kinematic hardening model as follows:

$$\dot{\alpha} = H_\alpha(e_\mathrm{p})\gamma \frac{\partial f(\boldsymbol{\eta}, e_\mathrm{p})}{\partial \boldsymbol{\eta}} = H_\alpha(e_\mathrm{p})\gamma \mathbf{N}, \tag{7.90}$$

Here, $H_\alpha(e_\mathrm{p})$ represents the nonlinear form of the kinematic hardening plastic modulus. In the case of linear hardening, it becomes a constant, i.e., $H_\alpha(e_\mathrm{p}) = H$. The effective strain rate can be expressed as

$$\dot{e}_\mathrm{p} = \sqrt{\frac{2}{3}} \|\dot{\mathbf{e}}^\mathrm{p}(t)\| = \sqrt{\frac{2}{3}}\gamma, \tag{7.91}$$

Here, $\dot{\boldsymbol{e}}^\mathrm{p}$ denotes the deviatoric strain rate.

In the case of nonlinear kinematic hardening, the back stress rate can be defined as

$$\dot{\alpha} = H(e_\mathrm{p})\dot{\mathbf{e}}^\mathrm{p}, \quad H(e_\mathrm{p}) = H_0 \exp\left(-\frac{e_\mathrm{p}}{e_\mathrm{p}^\infty}\right), \tag{7.92}$$

Here, e_p^∞ represents the asymptotic limit of plastic strain, and H_0 denotes the initial hardening modulus. This is also referred to as the saturation hardening model. In the case of nonlinear isotropic hardening, the yield surface radius can be defined as

$$\kappa(e_\mathrm{p}) = \sigma_\mathrm{Y}^0 + (\sigma_\mathrm{Y}^\infty - \sigma_\mathrm{Y}^0)\left[1 - \exp\left(-\frac{e_\mathrm{p}}{e_\mathrm{p}^\infty}\right)\right], \tag{7.93}$$

Here, σ_Y^∞ represents the asymptotic limit of the yield stress.

7.2.4.5 Plastic Consistency Parameter

As previously mentioned, when the material is in an elastic state, the plastic consistency parameter γ is 0 ($f < 0$); when the material is in a plastic state, the consistency parameter γ is positive ($f < 0$). In optimization, this is referred to as the Kuhn-Tucker condition and can be written as

$$\gamma \geq 0, \quad f \leq 0, \quad \gamma f = 0. \tag{7.94}$$

The non-positivity of the yield function f can be regarded as a constraint, while the plastic consistency parameter γ can be considered as the Lagrange multiplier corresponding to the inequality constraint. The above Kuhn-Tucker conditions encompass all possible states of the material. For example, when the material is in the elastic state, that is, when the stress lies within the elastic domain, these conditions reduce to

$$f < 0, \quad \gamma = 0 \Rightarrow \gamma f = 0.$$

When the stress lies on the yield surface, that is, when the material is in a plastic state, the condition of Eq. (7.94) is satisfied because $f = 0$. However, when the state changes, three different situations may occur:

- Elastic unloading $\dot{f} < 0, \quad \gamma = 0 \Rightarrow \gamma\dot{f} = 0$.
- Neutral loading $\dot{f} = 0, \quad \gamma = 0 \Rightarrow \gamma\dot{f} = 0$.
- Plastic loading $\dot{f} = 0, \quad \gamma > 0 \Rightarrow \gamma\dot{f} = 0$.

Therefore, when the stress lies on the yield surface, $\gamma f = 0$ in Eq. (7.94) is equivalent to $\gamma\dot{f} = 0$. Thus, the rate form of the Kuhn-Tucker condition can be used to calculate the plastic consistency parameters. Among the above three possible scenarios, we only focus on scenario □, namely, plastic loading, since the other cases can be identified by $\gamma = 0$, which does not cause any change in the plastic variables. Therefore, when the plastic loading state continues, we have

$$\gamma > 0 \quad \dot{f}(\boldsymbol{\sigma}, \boldsymbol{\xi}) = 0,$$

This means that the yield function remains unchanged during plastic loading, i.e.,

$$\dot{f}(\boldsymbol{\sigma}, \boldsymbol{\xi}) = \frac{\partial f}{\partial \boldsymbol{\sigma}} : \dot{\sigma} + \frac{\partial f}{\partial \boldsymbol{\xi}} : \dot{\xi} = 0. \tag{7.95}$$

Substituting the stress rate and plastic variables into Eq. (7.95), it can be rewritten as

$$\frac{\partial f}{\partial \boldsymbol{\sigma}} : \mathbf{D} : (\dot{\varepsilon} - \dot{\varepsilon}^{\mathrm{p}}) + \frac{\partial f}{\partial \boldsymbol{\xi}} \cdot \gamma \mathbf{h} = 0. \tag{7.96}$$

Since the plastic strain rate can also be expressed in terms of the plastic consistency parameter, Eq. (7.96) can be rewritten as follows:

$$\frac{\partial f}{\partial \boldsymbol{\sigma}} : \mathbf{D} : \dot{\varepsilon} - \frac{\partial f}{\partial \boldsymbol{\sigma}} : \mathbf{D} : \gamma \mathbf{r} + \frac{\partial f}{\partial \boldsymbol{\xi}} \cdot \gamma \mathbf{h} = 0.$$

Therefore, the plastic consistency parameter can be obtained as

$$\gamma = \frac{\left\langle \frac{\partial f}{\partial \sigma} : \mathbf{D} : \dot{\varepsilon} \right\rangle}{\frac{\partial f}{\partial \sigma} : \mathbf{D} : \mathbf{r} - \frac{\partial f}{\partial \boldsymbol{\xi}} \cdot \mathbf{h}}. \tag{7.97}$$

Here, due to the requirement that $\gamma > 0$, the numerator in Eq. (7.97) must be non-negative. The physical meaning of this condition is that when the material is under plastic loading, the angle between the normal direction of the yield surface and the rate of stress increment must be an acute angle (see Fig. 7.11), that is

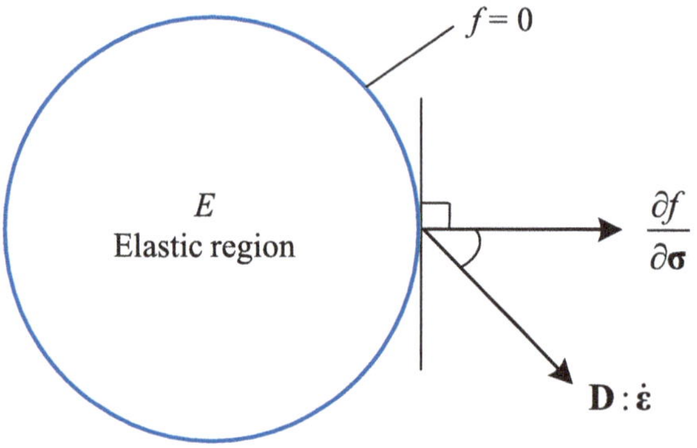

Fig. 7.11 Angle between the elastic trial stress and the normal of the yield surface

$$\cos\theta = \frac{\frac{\partial f}{\partial \sigma} : \mathbf{D} : \dot{\varepsilon}}{\left\| \frac{\partial f}{\partial \sigma} \right\| \|\mathbf{D} : \dot{\varepsilon}\|}. \tag{7.98}$$

If $\theta < 90°$, the material is in a plastic loading state; if $\theta = 90°$, the material is in a neutral loading state; if $\theta > 90°$, the material is in a state of elastic unloading.

7.2.4.6 Elastoplastic Tangent Stiffness

In one-dimensional systems, the elastoplastic modulus D^{ep} is calculated based on the elastic modulus and the plastic modulus. In multidimensional systems, the corresponding modulus is referred to as the consistent elastoplastic tangent stiffness. It characterizes the relationship between stress and strain rate. By substituting the plastic consistency parameter Eq. (7.97) into Eq. (7.81), we obtain

$$\dot{\sigma} = \mathbf{D} : \dot{\varepsilon} - \mathbf{D} : \gamma \mathbf{r} = \mathbf{D} : \dot{\varepsilon} - \mathbf{D} : \mathbf{r} \frac{\left\langle \frac{\partial f}{\partial \sigma} : \mathbf{D} : \dot{\varepsilon} \right\rangle}{\frac{\partial f}{\partial \sigma} : \mathbf{D} : \mathbf{r} - \frac{\partial f}{\partial \xi} \cdot \mathbf{h}}. \tag{7.99}$$

The above equation can be expressed in terms of the rate of change of stress and strain as follows:

$$\dot{\sigma} = \left[\mathbf{D} - \frac{\left\langle \mathbf{D} : \mathbf{r} \otimes \frac{\partial f}{\partial \sigma} : \mathbf{D} \right\rangle}{\frac{\partial f}{\partial \sigma} : \mathbf{D} : \mathbf{r} - \frac{\partial f}{\partial \xi} \cdot \mathbf{h}} \right] : \dot{\varepsilon} = \mathbf{D}^{\mathrm{ep}} : \dot{\varepsilon}, \tag{7.100}$$

Here, $\boldsymbol{D}^{\mathrm{ep}}$ denotes the continuous elastoplastic tangent stiffness, which is generally not symmetric. However, when the associated flow rule is used, i.e., $\boldsymbol{r} = \partial f / \partial \boldsymbol{\sigma}$, it becomes symmetric. When the flow rule and hardening model are specified, an explicit expression for $\boldsymbol{D}^{\mathrm{ep}}$ can be obtained. Since part of the strain rate is plastic and does not contribute to stress increase, the elastic stiffness $\boldsymbol{D}$ is reduced by the plastic consistency parameter. For the one-dimensional case in Eq. (7.8), a similar conclusion can be drawn.

For a stable material, the power generated by the stress rate must be positive, i.e. $\dot{\sigma} : \dot{\varepsilon} > 0$. Equation (7.100) implies that the elastoplastic tangent stiffness $\boldsymbol{D}^{\text{ep}}$ must be positive definite for a stable material. In addition, to ensure stable hardening behavior, the power during the plastic deformation process must also be positive, i.e., $\dot{\sigma} : \dot{\varepsilon}^{\text{p}} > 0$. These two conditions are known as the Drucker hypothesis.

7.2.5 Numerical Integration

Since the constitutive relations and evolution of plastic variables in elastoplastic models are expressed in rate form, they must be integrated over increments of time (or load). In static problems, the time increment should be interpreted as a load increment. The total load is first divided into N increments, and the load increment method is used to find structural equilibrium at each increment. It is assumed that the solution and material state at time t_n, including stress and plastic variables, are known. Then, at time t_{n+1}, the Newton–Raphson method is employed to solve for the incremental displacement during the convergence iterations. Therefore, the objective is to update the stress and plastic variables from time t_n to t_{n+1} using a given displacement increment, or equivalently, a given strain increment.

Note: Structural equilibrium is satisfied only at discrete time increments. Therefore, temporal discretization errors may occur, especially when the material state changes within a time increment. Using smaller time increments can reduce these errors. In the case of nonlinear elastic systems, selecting an appropriate time increment size facilitates convergence. For elastoplastic systems, the size of the time increment may affect the accuracy of the solution.

Although there are many integration methods for solving differential equations, it is crucial that the chosen method provides accurate and robust results. In the following derivation, the backward Euler time integration method is employed, which is widely favored for its simplicity and unconditional stability. Since the return mapping algorithm (with the radial return method as a special case) is an efficient and robust approach for plasticity computations, this section discusses the time integration of elastoplastic models using the return mapping algorithm. The return mapping algorithm commonly utilizes a two-step procedure: elastic predictor and plastic corrector. First, the elastic predictor state is computed, in which all strain increments are assumed to be elastic. If the predicted stress lies outside the elastic domain, it is projected back onto the yield surface, which is a convex set. This step is referred to as the return mapping to the yield surface. During the return mapping step, the yield surface itself evolves due to the evolution of plastic variables (strain hardening). Therefore, as the radius and center of the yield surface continuously change, it becomes challenging to locate the return mapping point on the yield surface.

7.2.5.1 Return-Mapping Algorithm

For associated plasticity, the backward Euler time integration method yields the closest point projection. Since the displacement increment at time t_{n+1} is known, the strain increment at time t_{n+1} can be calculated using the definition of strain. The first step, called the elastic prediction, uses this strain increment. The elastic predictions for the stress and hardening parameters are given by

$$^{tr}\mathbf{s} = {}^{n}\mathbf{s} + 2\mu\Delta\mathbf{e}, \quad {}^{tr}\boldsymbol{\alpha} = {}^{n}\boldsymbol{\alpha}, \quad {}^{tr}e_{\mathrm{p}} = {}^{n}e_{\mathrm{p}}, \tag{7.101}$$

$$^{tr}\boldsymbol{\eta} = {}^{tr}\mathbf{s} - {}^{tr}\boldsymbol{\alpha}, \tag{7.102}$$

Here, the superscript n on the variables denotes the time t_n, and tr represents the predicted state. In the elastic prediction step, all strain increments are considered elastic, and all plastic variables are held constant. Therefore, the plastic variables do not change. Although both the volumetric and deviatoric components of the stress may vary, Eq. (7.102) only considers the change in deviatoric stress, as hydrostatic stress does not affect plasticity.

If the predicted stress $^{tr}\boldsymbol{\eta}$ is within the elastic domain, i.e., $f({}^{tr}\boldsymbol{\eta}, {}^{tr}e_{\mathrm{p}}) < 0$, the material is in an elastic state, and the stress and plastic variables are updated using the trial predicted variables as follows

$$^{n+1}\mathbf{s} = {}^{tr}\mathbf{s}, \quad {}^{n+1}\boldsymbol{\alpha} = {}^{tr}\boldsymbol{\alpha}, \quad {}^{n+1}e_{\mathrm{p}} = {}^{tr}e_{\mathrm{p}}. \tag{7.103}$$

When the material is in an elastic state, it is considered as the end of the time integration.

If the predicted stress $^{tr}\boldsymbol{\eta}$ lies outside the elastic domain, that is, if $f({}^{tr}\boldsymbol{\eta}, {}^{tr}e_{\mathrm{p}}) > 0$, the material enters a plastic state and a plastic correction step is required to determine the material's plastic state. The stress and plastic variables are then updated by accounting for plastic deformation. As shown in Fig. 7.12, this illustrates the process of elastic prediction and plastic correction steps. Since plastic strain does not contribute to the stress, the trial stress decreases in proportion to the increment of plastic strain, that is,

$$^{n+1}\mathbf{s} = {}^{tr}\mathbf{s} - 2\mu\Delta\boldsymbol{\varepsilon}^{\mathrm{p}} = {}^{tr}\mathbf{s} - 2\mu\Delta\hat{\gamma}\mathbf{N}. \tag{7.104}$$

According to the flow rule, the plastic variables and stress are updated simultaneously as follows

$$^{n+1}\boldsymbol{\alpha} = {}^{tr}\boldsymbol{\alpha} + H_{\alpha}\hat{\gamma}\mathbf{N}, \tag{7.105}$$

$$^{n+1}e_{\mathrm{p}} = {}^{n}e_{\mathrm{p}} + \sqrt{\frac{2}{3}}\hat{\gamma}, \tag{7.106}$$

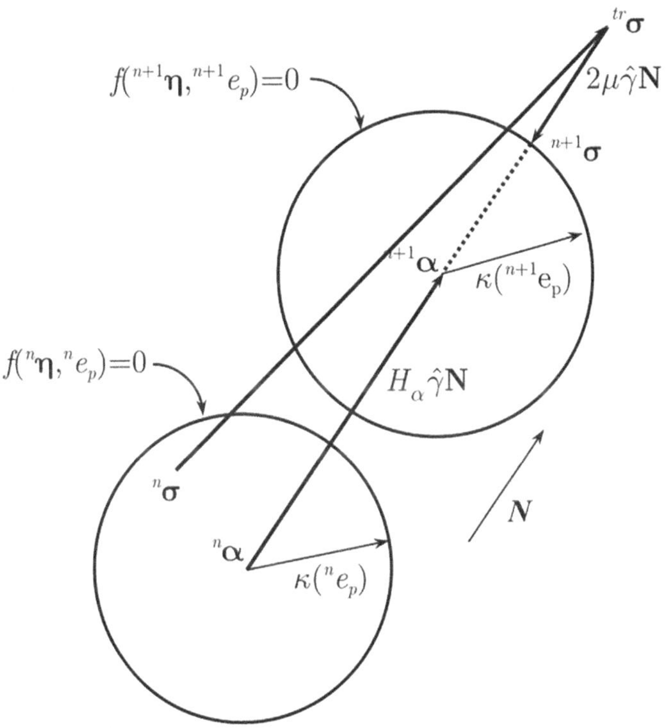

Fig. 7.12 Isotropic elastoplastic return mapping algorithm

Here, $\hat{\gamma} = \gamma\Delta t$ is the plastic consistency parameter, $\boldsymbol{N} = {}^{n+1}\boldsymbol{\eta}/\|{}^{n+1}\boldsymbol{\eta}\|$ is the unit deviatoric tensor, which is perpendicular to the yield surface at time t_{n+1}.

Note: Stress and back stress are corrected in the direction parallel to $\boldsymbol{N}$; the trial stress decreases, while the back stress increases. This facilitates locating the updated stress on the updated yield surface.

Furthermore, the plastic strain increment $\Delta\varepsilon^{\mathrm{p}}$ or its equivalent value $\hat{\gamma}\boldsymbol{N}$ is not yet clearly defined. To simplify subsequent calculations, the deviatoric stress at time t_{n+1} can be expressed as

$$ {}^{n+1}\boldsymbol{\eta} = {}^{n+1}\mathbf{s} - {}^{n+1}\boldsymbol{\alpha} = {}^{tr}\boldsymbol{\eta} - (2\mu + H_\alpha)\hat{\gamma}\mathbf{N}. \tag{7.107} $$

Since ${}^{n+1}\boldsymbol{\eta}$ is parallel to $\boldsymbol{N}$, ${}^{tr}\boldsymbol{\eta}$ must also be parallel to $\boldsymbol{N}$, which means that the direction of the final updated stress coincides with the direction of the trial stress increment. Therefore, the unit normal vector to the yield surface can be expressed in terms of the trial stress as follows:

$$ \mathbf{N} = \frac{{}^{tr}\boldsymbol{\eta}}{\|{}^{tr}\boldsymbol{\eta}\|}, \tag{7.108} $$

It can be obtained from the elastic predictor step. Therefore, the plastic correction step is reduced to determining the plastic consistency parameter $\hat{\gamma}$, from which the plastic strain increment can be derived. The fundamental idea is to ensure that the yield function satisfies the yield condition in the updated state. Thus, at the return mapping point, the following yield condition must be satisfied:

$$\begin{gathered} f\left({}^{n+1}\boldsymbol{\eta}, {}^{n+1}e_{\mathrm{p}}\right) \equiv \left\|{}^{n+1}\boldsymbol{\eta}\right\| - \sqrt{\frac{2}{3}}\kappa\left({}^{n+1}e_{\mathrm{p}}\right) = \\ \left\|{}^{tr}\boldsymbol{\eta}\right\| - \left(2\mu + H_{\alpha}\left({}^{n+1}e_{\mathrm{p}}\right)\right)\hat{\gamma} - \sqrt{\frac{2}{3}}\kappa\left({}^{n+1}e_{\mathrm{p}}\right) = 0, \end{gathered} \tag{7.109}$$

This is a nonlinear equation in $\hat{\gamma}$. The value of $\hat{\gamma}$ can be obtained by solving Eq. (7.109) using the local Newton–Raphson method. It should be noted that this is different from the Newton–Raphson method used for convergent iteration, which is applied to find the equilibrium solution between internal and external forces (see Eq. (7.19)). Here, the local iteration is performed only to locate the stress point on the yield surface. Therefore, the solution process for elastoplastic systems involves a double iteration loop. The inner local iteration loop typically converges rapidly within 5 to 6 iterations, but must be performed at each plastic integration point. The following are the steps of the local Newton–Raphson iteration method for solving the stress on the yield surface:

- Initialize variables:

$$k = 0, \quad e_{\mathrm{p}}^{k} = {}^{n}e_{\mathrm{p}}, \quad \gamma^{k} = 0, \quad f_{\mathrm{TOL}} = \sigma_{\mathrm{Y}}^{0} \times 10^{-7}, \quad k_{\mathrm{MAX}} = 20.$$

- Calculate the yield function:

$$f^{k} = \left\|{}^{tr}\boldsymbol{\eta}\right\| - \left(2\mu + H_{\alpha}\left(e_{\mathrm{p}}^{k}\right)\right)\gamma^{k} - \sqrt{\frac{2}{3}}\kappa\left(e_{\mathrm{p}}^{k}\right).$$

- Calculate the Jacobian relation:

$$\frac{\partial f}{\partial \gamma} = 2\mu + H_{\alpha} + \sqrt{\frac{2}{3}}H_{\alpha,e_{\mathrm{p}}}\gamma^{k} + \frac{2}{3}\kappa_{,e_{\mathrm{p}}}.$$

- Update the plastic consistency parameter and effective plastic strain:

$$\gamma^{k+1} = \gamma^{k} + \frac{f^{k}}{\partial f/\partial \gamma}, \quad e_{\mathrm{p}}^{k+1} = {}^{n}e_{\mathrm{p}} + \sqrt{\frac{2}{3}}\gamma^{k+1}.$$

- Check for convergence:

If $|f^{k}| > f_{TOL}$, then set $k = k + 1$ and return to step □.
If $k > k_{MAX}$, stop and display an error message.
If isotropic/kinematic hardening is a linear function of $\hat{\gamma}$ or the effective plastic strain, the equation becomes linear, and the return mapping point can be computed with a single iteration.

7.2.5.2 Update of Stress and Plastic Variables

After obtaining the plastic consistency parameter $\hat{\gamma}$, the deviatoric stress at time t_{n+1} can be updated as follows:

$$
{}^{n+1}\mathbf{s} = {}^{n}\mathbf{s} + 2\mu\Delta\mathbf{e} - 2\mu\Delta\hat{\gamma}\mathbf{N}. \tag{7.110}
$$

However, once the return mapping point is found, the deviatoric stress is no longer needed. The stress can be updated as follows:

$$
{}^{n+1}\boldsymbol{\sigma} = {}^{n}\boldsymbol{\sigma} + \Delta\boldsymbol{\sigma}, \tag{7.111}
$$

in which the stress increment

$$
\Delta\boldsymbol{\sigma} = \mathbf{D} : \Delta\boldsymbol{\varepsilon} - 2\mu\hat{\gamma}\mathbf{N}. \tag{7.112}
$$

Additionally, the back stress and effective plastic strain can be updated as follows:

$$
{}^{n+1}\boldsymbol{\alpha} = {}^{n}\boldsymbol{\alpha} + H_{\alpha}\hat{\gamma}\mathbf{N}, \tag{7.113}
$$

$$
{}^{n+1}e_{\mathrm{p}} = {}^{n}e_{\mathrm{p}} + \sqrt{\frac{2}{3}}\hat{\gamma}. \tag{7.114}
$$

Note: For Eqs. (7.111) and (7.113), the directions of the stress increment and back stress increment corresponding to the plastic correction component are both in the direction of $\boldsymbol{N}$, as shown in Fig. 7.12.

7.2.5.3 Consistent Tangent Stiffness

As discussed in Sect. 2.4 on solving nonlinear systems of algebraic equations, if the Jacobian matrix (or tangent stiffness matrix) is exact, the Newton–Raphson method exhibits quadratic convergence. In structural analysis, since the residual force depends on the stress, computing the Jacobian matrix requires the derivative of stress with respect to strain, which is referred to as the tangent stiffness. The continuous elastoplastic tangent stiffness $\boldsymbol{D}^{\mathrm{ep}}$ in Eq. (7.100) can be used for this purpose; however, numerical tests have shown that when $\boldsymbol{D}^{\mathrm{ep}}$ is used, the Newton–Raphson iteration does not achieve quadratic convergence. Simo and Taylor demonstrated that this is because $\boldsymbol{D}^{\mathrm{ep}}$ is inconsistent with the time integration algorithm: $\boldsymbol{D}^{\mathrm{ep}}$ represents the tangent stiffness between stress and strain rate, whereas the time integration algorithm operates with finite time increments. To ensure that the tangent stiffness $\boldsymbol{D}^{\mathrm{ep}}$ is consistent with the time integration algorithm and to achieve quadratic convergence in the Newton–Raphson iteration, this section derives a tangent stiffness that is consistent with the time integration algorithm. The stress increment in Eq. (7.111) is the differentiation with respect to the strain increment, which, together with the return mapping algorithm, yields a consistent constitutive relation.

$$\mathbf{D}^{\mathrm{alg}} = \frac{\partial \Delta \boldsymbol{\sigma}}{\partial \Delta \boldsymbol{\varepsilon}} = \mathbf{D} - 2\mu \mathbf{N} \otimes \frac{\partial \hat{\gamma}}{\partial \Delta \boldsymbol{\varepsilon}} - 2\mu \hat{\gamma} \frac{\partial \mathbf{N}}{\partial \Delta \boldsymbol{\varepsilon}}, \tag{7.115}$$

Here, $\boldsymbol{D}^{\mathrm{alg}}$ denotes the consistent (or algorithmic) tangent stiffness. The above equation requires the derivatives of the plastic consistency parameter and the unit deviatoric tensor.

Since the consistency condition in Eq. (7.109) must satisfy all strain states between the load steps t_n and t_{n+1}, the differential of f with respect to $\Delta\varepsilon$ must be zero, and the relationship between $\hat{\gamma}$ and $\Delta\varepsilon$ can be obtained. To differentiate Eq. (7.109), the following relation can be used.

$$\frac{\partial \left\| {}^{tr}\boldsymbol{\eta} \right\|}{\partial \Delta \boldsymbol{\varepsilon}} = 2\mu \frac{{}^{tr}\boldsymbol{\eta}}{\left\| {}^{tr}\boldsymbol{\eta} \right\|} = 2\mu \mathbf{N}, \tag{7.116}$$

$$\frac{\partial H_\alpha \left({}^{n+1}e_{\mathrm{p}} \right)}{\partial \Delta \boldsymbol{\varepsilon}} = \sqrt{\frac{2}{3}} \frac{\partial H_\alpha}{\partial e_{\mathrm{p}}} \frac{\partial \hat{\gamma}}{\partial \Delta \boldsymbol{\varepsilon}}. \tag{7.117}$$

In deriving the above expression, the deviatoric nature of $\boldsymbol{N}$ and the relation $\Delta e_p = \sqrt{2/3}\hat{\gamma}$ have been used. The yield function in Eq. (7.109) can be obtained through differentiating with respect to the strain variable,

$$\frac{\partial f}{\partial \Delta \varepsilon} = 2\mu \mathbf{N} - \left(2\mu + H_\alpha + \sqrt{\frac{2}{3}} H_{\alpha, e_p} \hat{\gamma} + \frac{2}{3} \kappa_{,e_p} \right) \frac{\partial \hat{\gamma}}{\partial \Delta \varepsilon} = 0,$$

where $H_{\alpha,ep} = \partial H_\alpha = \partial H_\alpha / \partial e_p$, $\kappa_{,ep} = \partial \kappa / \partial e_p$. Therefore, the derivative of the plastic consistency parameter with respect to the strain increment can be expressed as

$$\frac{\partial \hat{\gamma}}{\partial \Delta \boldsymbol{\varepsilon}} = \frac{2\mu \mathbf{N}}{2\mu + H_\alpha + \sqrt{(2/3)} H_{\alpha, e_{\mathrm{p}}} \hat{\gamma} + (2/3) \kappa_{,e_{\mathrm{p}}}}.$$

The increment of the unit normal tensor with respect to the yield function can be expressed as

$$\frac{\partial \mathbf{N}}{\partial \Delta \boldsymbol{\varepsilon}} = \frac{\partial \mathbf{N}}{\partial^{tr}\boldsymbol{\eta}} : \frac{\partial \|^{tr}\boldsymbol{\eta}\|}{\partial \Delta \boldsymbol{\varepsilon}} = \left[\frac{\mathbf{I}}{\|^{tr}\boldsymbol{\eta}\|} - \frac{^{tr}\boldsymbol{\eta} \otimes^{tr} \boldsymbol{\eta}}{\|^{tr}\boldsymbol{\eta}\|^3} \right] : 2\mu \mathbf{I}_{\text{dev}} = \frac{2\mu}{\|^{tr}\boldsymbol{\eta}\|}[\mathbf{I}_{\text{dev}} - \mathbf{N} \otimes \mathbf{N}]. \tag{7.118}$$

Therefore, using Eq. (7.112), the consistent (or algorithmic) tangent stiffness can be expressed as

$$\begin{aligned} \mathbf{D}^{\text{alg}} = \frac{\partial \Delta \boldsymbol{\sigma}}{\partial \Delta \boldsymbol{\varepsilon}} &= \mathbf{D} - \frac{4\mu^2 \mathbf{N} \otimes \mathbf{N}}{2\mu + H_\alpha + \sqrt{(2/3)} H_{\alpha, e_p} \hat{\gamma} + (2/3)\kappa_{,e_p}} \\ &- \frac{4\mu^2 \hat{\gamma}}{\|^{tr}\boldsymbol{\eta}\|}[\mathbf{I}_{\text{dev}} - \mathbf{N} \otimes \mathbf{N}]. \end{aligned} \tag{7.119}$$

It is necessary to compare the tangent stiffness described above with the continuous elastoplastic tangent stiffness in Eq. (7.100). Since Eq. (7.100) applies to general hardening models, it can be simplified for the cases of isotropic hardening and kinematic hardening models as follows:

$$\mathbf{D}^{\text{ep}} = \mathbf{D} - \frac{4\mu^2 \mathbf{N} \otimes \mathbf{N}}{2\mu + H_\alpha + (2/3)\kappa_{,e_p}}. \tag{7.120}$$

By comparing Eq. (7.119) and (7.120), it is evident that $\boldsymbol{D}^{\text{ep}}$ lacks the third term presented in $\boldsymbol{D}^{\text{alg}}$, which accounts for the effect of changes in $\boldsymbol{N}$ induced by strain increments. This is because the rate form only considers infinitesimal strain increments (strain rates) and does not account for changes in their direction. However, when the strain increment is not infinitesimal, it may alter the direction of the deviatoric stress, thereby changing $\boldsymbol{N}$. This effect does not appear in the one-dimensional elastoplastic model discussed in Sect. 7.1, as it employs scalar stress and a constant $\boldsymbol{N}$. Another distinction lies in the denominator of the second term on the right-hand side of the equations. While they are similar, $\boldsymbol{D}^{\text{ep}}$ does not include the effects of nonlinear hardening. Therefore, when hardening is linear, the two become equivalent. This is because, in a certain sense, the rate form differentiates the hardening model first and then takes the increment, whereas the incremental form differentiates after taking the increment.

Note: When $\hat{\gamma} = 0$ the two tangential stiffnesses are identical.

7.2.5.4 Elastoplastic Incremental Equations

For convenience, the energy functional and its linearization are defined as follows:

$$a\left(^{n}\boldsymbol{\xi};\ ^{n+1}\mathbf{u}\bar{\mathbf{u}}\right) \equiv \iint_\Omega \boldsymbol{\varepsilon}(\bar{\mathbf{u}}) : {}^{n+1}\boldsymbol{\sigma}\, d\Omega, \tag{7.121}$$

$$a^*\left({}^{n}\boldsymbol{\xi},\ {}^{n+1}\mathbf{u};\ \delta\mathbf{u},\overline{\mathbf{u}}\right) \equiv \iint_{\Omega} \boldsymbol{\varepsilon}(\overline{\mathbf{u}}) : \mathbf{D}^{\text{alg}} : \boldsymbol{\varepsilon}(\delta\mathbf{u})\mathrm{d}\Omega. \tag{7.122}$$

The notation $a*(\boldsymbol{\xi}, \boldsymbol{u}; \delta\boldsymbol{u}, \overline{\boldsymbol{u}})$ is used so that the functional depends implicitly on the plastic variable ξ and the total displacement $\boldsymbol{u}$, and is bilinear with respect to δu and $\overline{\boldsymbol{u}}$. The energy form also depends implicitly on the plastic variable.

Note: Unlike geometric nonlinearity, since only infinitesimal deformations are considered, the initial stiffness term does not appear. For small deformation problems, this is entirely equivalent to the updated Lagrangian formulation.

Since only material nonlinearity is considered, the weak form of structural equilibrium at load step t_n can be expressed as

$$a\left({}^{n}\boldsymbol{\xi};\ {}^{n+1}\mathbf{u},\overline{\mathbf{u}}\right) = l(\overline{\mathbf{u}}),\quad \forall\overline{\mathbf{u}} \in \mathbb{Z}. \tag{7.123}$$

Let the current load step be t_{n+1} and the current iteration count be k. Assuming that the applied load is independent of displacement, the linearized incremental equation is given by

$$a^*\left({}^{n}\boldsymbol{\xi},\ {}^{n+1}\mathbf{u};\ \delta\mathbf{u},\overline{\mathbf{u}}\right) = l(\overline{\mathbf{u}}) - a\left({}^{n}\boldsymbol{\xi};\ {}^{n+1}\mathbf{u},\overline{\mathbf{u}}\right),\quad \forall\overline{\mathbf{u}} \in \mathbb{Z}, \tag{7.124}$$

and the total displacement is updated as

$${}^{n+1}\mathbf{u}^{k+1} = {}^{n+1}\mathbf{u}^{k} + \delta\mathbf{u}^{k}. \tag{7.125}$$

Note: The linearized incremental Eq. (7.124), after applying finite element discretization, takes the form $\left[{}^{n+1}\boldsymbol{K}^k\right]\cdot\{\delta\mathbf{u}^k\} = \left\{{}^{n+1}\boldsymbol{R}^k\right\}$. Equation (7.124) is solved iteratively until the residual vanishes, indicating that the original nonlinear Eq. (7.123) is satisfied. It is important to emphasize the linearized incremental Eq. (7.123) here, which is used to solve for the displacement increment $\delta\boldsymbol{u}^k =^{n+1} \boldsymbol{u}^{k+1} -^{n+1} \boldsymbol{u}^k$ between two consecutive iterations. But the strain increment must be computed using the total displacement increment $\delta\mathbf{u}^k =^{n+1} \boldsymbol{u}^{k+1} +^{n+1} \boldsymbol{u}^k$, as given in Eq. (7.10). *This is because stresses and all history-dependent variables are updated from the previously converged load increment, not from the previous iteration.*

Unlike nonlinear elastic systems, after the convergence of Eq. (7.123) at load step t_{n+1}, an additional computation step is required for elastoplastic systems. Since the stress and plastic variables will be used in the next load step, they must be updated at the end of the current load step. This procedure is identical to the update process for stress and plastic variables described in Eqs. (7.111) to (7.114). During the iteration process, the values of these variables are computed but not stored, as they are not converged values. Once the nonlinear equations have converged, the values of these variables are updated and stored.

7.2.6 Computational Implementation of Elastoplasticity

This section introduces the computational implementation of elastoplasticity using the Von Mises yield criterion and a combined isotropic/kinematic hardening model. Although finite element formulations can be developed for various element types, here we use an 8-node hexahedral solid element as an example. Since only material nonlinearity is considered, both strains and rigid body rotations are assumed to be small.

In the computer implementation of finite element programs, matrix–vector representations are more convenient than tensor representations. In matrix–vector notation, second-order symmetric tensors are represented as vectors, while fourth-order symmetric tensors are represented as matrices. For example, the Cauchy stress and incremental strain vectors are defined as follows:

$$\{\boldsymbol{\sigma}\} = \left\{ \sigma_{11}\ \sigma_{22}\ \sigma_{33}\ \sigma_{12}\ \sigma_{23}\ \sigma_{13} \right\}^{\mathrm{T}}$$

$$\{\Delta\boldsymbol{\varepsilon}\} = \left\{ \Delta\varepsilon_{11}\ \Delta\varepsilon_{22}\ \Delta\varepsilon_{33}\ 2\Delta\varepsilon_{12}\ 2\Delta\varepsilon_{23}\ 2\Delta\varepsilon_{13} \right\}^{\mathrm{T}},$$

The above definition makes use of the symmetry property of tensors.

Assume that each node of the element is assigned a displacement vector increment $\Delta\mathbf{d}_I = \{\Delta d_{I1}, \Delta d_{I2}, \Delta d_{I3}\}^T$, where the subscript I denotes an element node. In each element, the element node numbers are locally defined. Therefore, for a hexahedral element, $I = 1, 2, \ldots\ldots, 8$, (see Fig. 7.13). The following interpolation scheme can be used to compute values within the element.

$$\Delta\mathbf{u} = \sum_{I=1}^{8} N_I(\boldsymbol{\xi})\Delta\mathbf{d}_I, \tag{7.126}$$

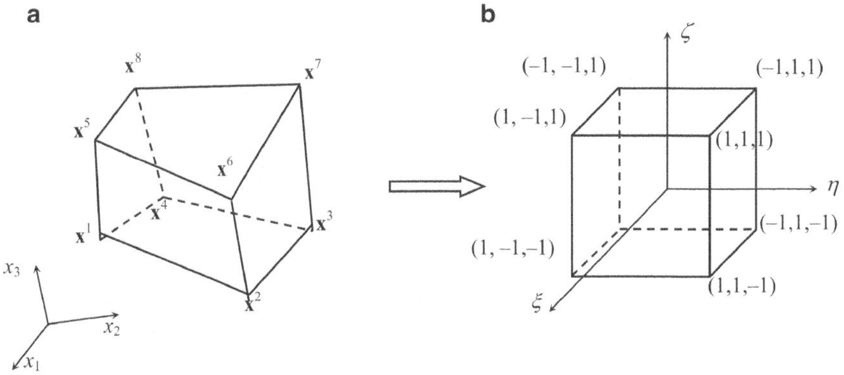

Fig. 7.13 An 8-node three-dimensional isoparametric solid element

where $\xi = \{\xi, \eta, \zeta\}^{\mathrm{T}}$ is the vector of natural coordinates at a unit reference element, $N_I(\boldsymbol{\xi})$ is the interpolation function or shape function, and $\Delta\mathbf{d}_I$ is the nodal displacement increment vector. Since stresses and plastic variables are computed at the integration points, the natural coordinate values are selected at those integration points.

For a given displacement increment, the calculation of strain increment is similar to that for linear elastic materials, as the deformation is assumed to be infinitesimal. Therefore, the strain increment can be expressed as follows:

$$\{\Delta\boldsymbol{\varepsilon}\} = \sum_{I=1}^{8} \mathbf{B}_I \Delta\mathbf{d}_I, \tag{7.127}$$

where,

$$\mathbf{B}_I = \begin{bmatrix} N_{I,1} & 0 & 0 \\ 0 & N_{I,2} & 0 \\ 0 & 0 & N_{I,3} \\ N_{I,2} & N_{I,1} & 0 \\ 0 & N_{I,3} & N_{I,2} \\ N_{I,3} & 0 & N_{I,1} \end{bmatrix} \tag{7.128}$$

is the discrete displacement–strain matrix of the solid element, and $N_{I,j} = \partial N_I / \partial x_j$ is the derivative of the shape function N_I with respect to the physical coordinates.

Note: Since shape functions are defined in natural coordinates, it is necessary to use the Jacobian relationship to convert between physical and natural coordinates.

Under the small deformation assumption, the total displacement and total strain can be obtained by adding the incremental displacement and strain, respectively. Therefore, the total displacement and total strain at the load increment t_{n+1} can be expressed as follows:

$$\begin{aligned} {}^{n+1}\mathbf{u} &= {}^{n}\mathbf{u} + \Delta\mathbf{u} \\ \left\{{}^{n+1}\boldsymbol{\varepsilon}\right\} &= \left\{{}^{n}\boldsymbol{\varepsilon}\right\} + \{\Delta\boldsymbol{\varepsilon}\}. \end{aligned}$$

In addition to the strain increment, the stress and plastic variables at the previous load increment t_n are also required. In the case of linear isotropic/kinematic hardening combined model, the variables ${}^{n}e_{\mathrm{p}}$ and the following vectors are needed

$$\begin{aligned} \left\{{}^{n}\boldsymbol{\sigma}\right\} &= \left\{{}^{n}\sigma_{11} \; {}^{n}\sigma_{22} \; {}^{n}\sigma_{33} \; {}^{n}\sigma_{12} \; {}^{n}\sigma_{23} \; {}^{n}\sigma_{13}\right\}^{\mathrm{T}}, \\ \left\{{}^{n}\boldsymbol{\alpha}\right\} &= \left\{{}^{n}\alpha_{11} \; {}^{n}\alpha_{22} \; {}^{n}\alpha_{33} \; {}^{n}\alpha_{12} \; {}^{n}\alpha_{23} \; {}^{n}\alpha_{13}\right\}^{\mathrm{T}}. \end{aligned}$$

Based on the above stress, plastic variables, and material parameters $(\lambda, \mu, \beta, H, \sigma_{\mathrm{Y}}^{0})$, the updated values can be computed using the return mapping algorithm stress and plastic variables.

7.2.6.1 Return Map Algorithm

At each integration point within the element, stresses and plastic variables are determined using the return mapping algorithm. In the following computational steps, vector notation is omitted for simplicity:

① Define the unit tensor

$$\mathbf{1} = [1 1 1 0 0 0]^{\mathrm{T}}$$

② Calculate the predicted stress

$$^{tr}\boldsymbol{\sigma} =^{n} \boldsymbol{\sigma} + \mathbf{D} \cdot \Delta\boldsymbol{\varepsilon}$$

③ Calculate the trace of the stress tensor

$$tr(^{tr}\boldsymbol{\sigma}) =^{tr} \sigma_{11} +^{tr} \sigma_{22} +^{tr} \sigma_{33}$$

④ Calculate the deviatoric stress

$$^{tr}\boldsymbol{\eta} =^{tr} \boldsymbol{\sigma} -^{n} \boldsymbol{\alpha} - (1/3)tr(^{tr}\boldsymbol{\sigma})1$$

⑤ Calculate the norm

$$\left\|^{tr}\boldsymbol{\eta}\right\| = \sqrt{(^{tr}\eta_{11})^2 + (^{tr}\eta_{22})^2 + (^{tr}\eta_{33})^2 + 2\left[(^{tr}\eta_{12})^2 + (^{tr}\eta_{23})^2 + (^{tr}\eta_{13})^2\right]}.$$

⑥ Calculate the yield function

$$f = \left\|^{tr}\eta\right\| - \sqrt{\frac{2}{3}}\left[\sigma_{\mathrm{Y}}^{0} + (1-\beta)H^{n}e_{\mathrm{p}}\right]$$

⑦ Check the yield condition. If f < 0, the material is elastic, and thus

$$^{n+1}\boldsymbol{\sigma} = {}^{tr}\boldsymbol{\sigma},$$

$$\mathbf{D}^{\mathrm{alg}} = \mathbf{D} = \begin{bmatrix} \lambda+2\mu & \lambda & \lambda & 0 & 0 & 0 \\ \lambda & \lambda+2\mu & \lambda & 0 & 0 & 0 \\ \lambda & \lambda & \lambda+2\mu & 0 & 0 & 0 \\ 0 & 0 & 0 & \mu & 0 & 0 \\ 0 & 0 & 0 & 0 & \mu & 0 \\ 0 & 0 & 0 & 0 & 0 & \mu \end{bmatrix}.$$

⑧ Calculate the consistency index

$$\hat{\gamma} = \frac{f}{2\mu + (2/3)H}$$

⑨ Calculate the unit deviation vector

$$\mathbf{N} = \frac{{}^{tr}\boldsymbol{\eta}}{\|{}^{tr}\boldsymbol{\eta}\|}$$

⑩ Update Stress

$$^{n+1}\boldsymbol{\sigma} = ^{tr}\boldsymbol{\sigma} - 2\mu\hat{\gamma}\mathbf{N}$$

⑪ Update Back Stress

$$^{n+1}\boldsymbol{\alpha} = ^{n}\boldsymbol{\alpha} + (2/3)\beta H\hat{\gamma}\mathbf{N}$$

⑫ Update of Equivalent Plastic Strain

$$^{n+1}e_p = ^{n}e_p + \sqrt{(2/3)}\hat{\gamma}$$

⑬ Calculation of Consistent Tangent Stiffness

$$c_1 = \frac{4\mu^2}{2\mu + (2/3)H}, \quad c_2 = \frac{4\mu^2\hat{\gamma}}{\|{}^{tr}\boldsymbol{\eta}\|},$$

$$\mathbf{I}^{\text{dev}} = \begin{bmatrix} 2/3 & -1/3 & -1/3 & 0 & 0 & 0 \\ -1/3 & 2/3 & -1/3 & 0 & 0 & 0 \\ -1/3 & -1/3 & 2/3 & 0 & 0 & 0 \\ 0 & 0 & 0 & 1/2 & 0 & 0 \\ 0 & 0 & 0 & 0 & 1/2 & 0 \\ 0 & 0 & 0 & 0 & 0 & 1/2 \end{bmatrix},$$

Note: The above calculations must be performed for each integration point of the element. Therefore, the stresses and plastic variables at each integration point must be stored and updated. There is no need to store stresses and plastic variables during non-convergent iterations. Once the Newton–Raphson method converges at the load increment t_{n+1}, store them, and then proceed to the next load increment.

7.2.6.2 The Elastic–plastic Solution for Plane Stress Problems

1. **Plane Stress Elastoplastic J_2 Flow Model**

Introducing the vector representation of stress and the deviatoric stress tensor, namely

$$\begin{aligned} \boldsymbol{\sigma} &:= [\sigma_{11}\ \sigma_{22}\ \sigma_{12}]^{\mathrm{T}} \\ \mathbf{s} &:= [s_{11}\ s_{22}\ s_{12}]^{\mathrm{T}} \end{aligned} \tag{7.129}$$

Introducing a mapping $\boldsymbol{P}$, so that the relationship between the stress tensor and the deviatoric stress tensor is established as follows:

$$\begin{aligned} \mathbf{s} &:= \mathrm{dev}[\boldsymbol{\sigma}] = \overline{\mathbf{P}}\boldsymbol{\sigma} \\ \overline{\mathbf{P}} &:= \frac{1}{3}\begin{bmatrix} 2 & -1 & 0 \\ -1 & 2 & 0 \\ 0 & 0 & 3 \end{bmatrix} \end{aligned} \tag{7.130}$$

We use the plane stress conditions to establish equations, and define the following relationships

$$[\alpha_{11}\ \alpha_{22}\ \alpha_{12}]^{\mathrm{T}} =: \overline{\mathbf{P}}\alpha, \quad \tilde{\alpha} := [\tilde{\alpha}_{11}\ \tilde{\alpha}_{22}\ \tilde{\alpha}_{12}]^{\mathrm{T}} \tag{7.131}$$

where $\boldsymbol{\alpha}$ denotes the back stress. The strain and plastic strain tensors are defined as follows:

$$\begin{aligned} \boldsymbol{\varepsilon} &:= [\varepsilon_{11}\ \varepsilon_{22}\ \varepsilon_{12}]^{\mathrm{T}} \\ \boldsymbol{\varepsilon}^p &:= \left[\varepsilon^p_{11}\ \varepsilon^p_{22}\ \varepsilon^p_{12}\right]^{\mathrm{T}} \end{aligned} \tag{7.132}$$

To facilitate the expression of strain deviator in terms of stress deviator, and to account for the two factors in the shear strain components, the mapping $\overline{\boldsymbol{P}}$ is modified as follows:

$$\mathbf{P} := \frac{1}{3}\begin{bmatrix} 2 & -1 & 0 \\ -1 & 2 & 0 \\ 0 & 0 & 6 \end{bmatrix} \tag{7.133}$$

The plane stress form of the fundamental equations of three-dimensional $\boldsymbol{J}_2$ flow theory can be expressed as

$$\begin{aligned} &\boldsymbol{\varepsilon} = \boldsymbol{\varepsilon}^e + \boldsymbol{\varepsilon}^p \\ &\boldsymbol{\eta} := \boldsymbol{\sigma} - \tilde{\alpha} \\ &\boldsymbol{\sigma} = \mathbf{D}\varepsilon^e \\ &\dot{\varepsilon}^p = \gamma \mathbf{P}\eta \\ &\dot{\tilde{\alpha}} = \gamma \frac{2}{3} H_\alpha \boldsymbol{\eta} \\ &f := \sqrt{\boldsymbol{\eta}^T \mathbf{P}\eta} - \sqrt{\frac{2}{3}}\kappa(e_p) \le 0 \end{aligned} \tag{7.134}$$

where $\boldsymbol{D}$ is the elastic constitutive matrix for plane stress.

2. **Return Mapping Algorithm for Plane Stress Elastoplastic Problems**

The approach of solving plane stress elastoplastic problems is to apply the aforementioned equations directly to the return mapping algorithm within the constrained plane stress subspace. Under kinematic/isotropic hardening conditions, for plane stress problems, the $\boldsymbol{J}_2$ flow theory solution reduces to the iterative solution of a nonlinear equation.

Under plane stress conditions, the procedural steps of the return mapping algorithm are as follows:

① Update the strain tensor and calculate the elastic trial stress:

$$\begin{aligned} \boldsymbol{\varepsilon}^{n+1} &= \boldsymbol{\varepsilon}^n + \Delta\boldsymbol{\varepsilon} \\ \boldsymbol{\sigma}^{tr} &= \boldsymbol{\sigma}^n + \mathbf{D}\Delta\boldsymbol{\varepsilon} \\ \boldsymbol{\eta}^{tr} &= \boldsymbol{\sigma}^{tr} - \tilde{\alpha}^n \end{aligned}$$

② If $f(\boldsymbol{\eta}^{tr}, e_p^{tr}) < 0$, exit the process; otherwise, solve $f(\hat{\gamma}) = 0$:

$$\begin{aligned} f^2(\hat{\gamma}) &:= \frac{1}{2}\bar{f}^2(\hat{\gamma}) - R^2(\hat{\gamma}) \equiv 0 \\ \bar{f}^2(\hat{\gamma}) &:= \frac{1}{2}\frac{\frac{1}{3}(\eta_{11}^{tr} + \eta_{22}^{tr})^2}{\left\{1 + \left(\frac{E}{3(1-\nu)} + \frac{2}{3}H_\alpha\right)\hat{\gamma}\right\}^2} + \frac{\frac{1}{2}(\eta_{11}^{tr} - \eta_{22}^{tr})^2 + 2(\eta_{12}^{tr})^2}{\left\{1 + (2\mu + \frac{2}{3}H_\alpha)\hat{\gamma}\right\}^2} \\ R^2(\Delta\gamma) &:= \frac{1}{3}\kappa^2\left[e_p^n + \sqrt{\frac{2}{3}}\Delta\gamma\bar{f}(\hat{\gamma})\right] \end{aligned}$$

Here, $\bar{f} = \sqrt{\boldsymbol{\eta}^{\mathrm{T}}\mathbf{P}\eta}$, μ denotes the shear modulus of elasticity.

③ Calculate the corrected (algorithmic) tangent stiffness:

$$\boldsymbol{\Xi} := \left[\mathbf{D}^{-1} + \frac{\hat{\gamma}}{1 + \frac{2}{3}\hat{\gamma}H_\alpha}\mathbf{P}\right]^{-1}$$

④ Update relevant variables:

$$\begin{aligned}
\boldsymbol{\eta}^{n+1} &= \frac{1}{1+(2/3)\hat{\gamma}H_\alpha}\Xi(\hat{\gamma})(\mathbf{D}^{-1})\,\boldsymbol{\eta}^{tr} \\
\tilde{\alpha}^{n+1} &= \tilde{\alpha}^n + \hat{\gamma}\frac{2}{3}H_\alpha\boldsymbol{\eta}^{n+1} \\
\boldsymbol{\sigma}^{n+1} &= \boldsymbol{\eta}^{n+1} + \tilde{\alpha}^{n+1} \\
e_p^{n+1} &= e_p^n + \sqrt{\frac{2}{3}}\hat{\gamma}\bar{f}(\hat{\gamma})
\end{aligned}$$

⑤ Calculate the consistent elastoplastic tangent stiffness:

$$\begin{aligned}
\left.\frac{\mathrm{d}\boldsymbol{\sigma}}{\mathrm{d}\boldsymbol{\varepsilon}}\right|_{n+1} &= \Xi - \frac{[\Xi\mathbf{P}\eta^{n+1}][\Xi\mathbf{P}\eta^{n+1}]^T}{(\boldsymbol{\eta}^{n+1})^T\mathbf{P}\Xi\mathbf{P}(\boldsymbol{\eta}^{n+1}) + \overline{\beta}^{n+1}} \\
\overline{\beta}^{n+1} &:= \frac{2}{3}\frac{\theta_1}{\theta_2}(\kappa'_{n+1}\theta_1 + H_\alpha\theta_2)\boldsymbol{\eta}^{n+1}\mathbf{P}\boldsymbol{\eta}^{n+1} \\
&\theta_1 := 1 + \frac{2}{3}H_\alpha\hat{\gamma},\ \theta_2 := 1 - \frac{2}{3}\kappa'_{n+1}\hat{\gamma}
\end{aligned}$$

7.2.6.3 Elastoplastic Finite Element Analysis Procedure

Once the stress and plastic variables have been determined, they can be used to solve the nonlinear equilibrium equations. First, the same strain–displacement matrix $\mathbf{B}_I$ can be used to interpolate the variation of strain.

$$\{\boldsymbol{\varepsilon}(\bar{\mathbf{u}})\} = \sum_{I=1}^{8}\mathbf{B}_I\bar{\mathbf{d}}_I = \mathbf{B}\{\bar{\mathbf{d}}\}, \tag{7.135}$$

Here, $\mathbf{d}_I = \{d_{I1}, d_{I2}, d_{I3}\}^T$ represents the displacement variation at node I, while $\bar{\mathbf{d}} = \{d_1, d_2, {}^1\!/_4, d_8\}^{\mathrm{T}}$ denotes those within the element $[\mathbf{B}] = [\mathbf{B}_1, \mathbf{B}_2, \ldots, \mathbf{B}_8]$.

Using the updated stress from Eqs. (7.135) and (7.111), the discrete form of the energy equation can be expressed as:

$$a({}^n\boldsymbol{\xi};\ {}^{n+1}\mathbf{u}, \bar{\mathbf{u}}) \equiv \iint_\Omega \boldsymbol{\varepsilon}(\bar{\mathbf{u}})^{\mathrm{T}}\{{}^{n+1}\boldsymbol{\sigma}\}\,\mathrm{d}\Omega = \{\bar{\mathbf{d}}\}^{\mathrm{T}}\iint_\Omega [\mathbf{B}]^{\mathrm{T}}\{{}^{n+1}\boldsymbol{\sigma}\}\,\mathrm{d}\Omega \equiv \{\bar{\mathbf{d}}\}^{\mathrm{T}}\{\mathbf{f}^{\mathrm{int}}\}, \tag{7.136}$$

Here, $\mathbf{f}^{\mathrm{int}}$ denotes the discrete internal force vector. When numerical integration is employed, $\mathbf{f}^{\mathrm{int}}$ can be expressed as

$$\{\mathbf{f}^{\text{int}}\} = \sum_{K=1}^{NG} \left([\mathbf{B}]^{\text{T}}\{{}^{n+1}\boldsymbol{\sigma}\}|\mathbf{J}|\right)_K \omega_K,$$

Here, NG is the number of integration points, $|J|$ is the determinant of the Jacobian matrix between the physical element and the reference element, and ω_K is the integration weight coefficient.

In addition, the discrete external force vector can be derived from the definition of the load form, namely

$$\begin{gathered} l(\bar{\mathbf{u}}) \equiv \iint_{\Omega} \bar{\mathbf{u}}^{\text{T}}\mathbf{f}^{\text{B}}\text{d}\Omega + \int_{\Gamma^{\text{S}}} \bar{\mathbf{u}}^{\text{T}}\mathbf{f}^{\text{S}}\text{d}\Gamma = \\ \sum_{I=1}^{4} \bar{\mathbf{d}}_I^{\text{T}} \left\{ \iint_{\Omega} N_I(\boldsymbol{\xi})\mathbf{f}^{\text{B}}\text{d}\Omega + \int_{\Gamma^{\text{S}}} N_I(\boldsymbol{\xi})\mathbf{f}^{\text{S}}\text{d}\Gamma \right\} \equiv \bar{\mathbf{d}}^{\text{T}}\{\mathbf{f}^{\text{ext}}\}. \end{gathered} \tag{7.137}$$

When concentrated nodal forces are applied, they can be directly added to the corresponding positions in $\mathbf{f}^{\text{ext}}$. Since the applied loads are assumed to be independent of deformation, the external force $\mathbf{f}^{\text{ext}}$ is a fixed vector. The discrete form of the nonlinear equilibrium equation seeks to find the internal force that matches the value of the external force, that is,

$$\bar{\mathbf{d}}^{\text{T}}\{\mathbf{f}^{\text{int}}(\mathbf{d})\} = \bar{\mathbf{d}}^{\text{T}}\{\mathbf{f}^{\text{ext}}\}, \quad \forall\{\bar{\mathbf{d}}\} \in \mathbb{Z}_{\text{h}}, \tag{7.138}$$

Here, $\mathbb{Z}_{\text{h}}$ is the discrete form of the space $\mathbb{Z}$. Since the displacement variation is zero at nodes with prescribed displacements, Eq. (7.137) holds as $\mathbf{f}^{\text{int}}(\mathbf{d}) = \mathbf{f}^{\text{ext}}$ for all nodes without prescribed displacements.

Since the internal forces are nonlinear functions of deformation, Eq. (7.138) must be solved using an iterative method such as the Newton–Raphson method, which requires the Jacobian matrix or the equivalent tangent stiffness matrix. Employing the consistent tangent stiffness, the linearized energy formulation can be discretized as:

$$a^*\left({}^{n}\boldsymbol{\xi}, {}^{n+1}\mathbf{u}; \delta\mathbf{u}, \bar{\mathbf{u}}\right) = \{\bar{\mathbf{d}}\}^T \left\{ \iint_{\Omega} [\mathbf{B}]^T \left[\mathbf{D}^{a\text{lg}}\right][\mathbf{B}]\,\text{d}\Omega \right\} \{\delta\mathbf{d}\}. \tag{7.139}$$

The integral term in Eq. (7.139) is referred to as the tangent stiffness matrix. After applying numerical integration, it becomes

$$[\mathbf{K}_{\text{T}}] = \sum_{K=1}^{NG} \left([\mathbf{B}]^{\text{T}}\left[\mathbf{D}^{\text{alg}}\right][\mathbf{B}]|\mathbf{J}|\right)_K \omega_K. \tag{7.140}$$

In general, the above integrals and the internal force integrals in Eq. (7.136) are computed using the Gaussian quadrature rule. Typically, a quadrilateral element uses 2×2 integration points.

The discrete form of the incremental equation in Eq. (7.124) can now be expressed as a finite element matrix equation, namely,

$$\{\bar{\mathbf{d}}\}^{\mathrm{T}}[\mathbf{K}_{\mathrm{T}}]\{\delta\mathbf{d}\} = \bar{\mathbf{d}}^{\mathrm{T}}\{\mathbf{f}^{\mathrm{int}} - \mathbf{f}^{\mathrm{ext}}\}, \quad \forall\{\bar{\mathbf{d}}\} \in \mathbb{Z}_{\mathrm{h}}. \tag{7.141}$$

The above system of linear equations must be solved iteratively until the residual force (right-hand side) vanishes. Various methods for solving nonlinear systems of equations can be employed. For example, in the modified Newton–Raphson method, the tangent stiffness matrix $[\boldsymbol{K}_{\mathrm{T}}]$ from the first iteration is reused throughout. In the load increment method, the external force vector $\boldsymbol{f}^{\mathrm{ext}}$ is divided into the required number of increments, and the Newton–Raphson method is applied at each load increment.

Note: The above algorithm is only valid under the assumptions of small deformations and small rotations. When the deformation becomes large, the additive decomposition assumption, i.e., $\Delta\boldsymbol{\varepsilon} = \Delta\boldsymbol{\varepsilon}_e + \Delta\boldsymbol{\varepsilon}_p$, no longer holds.

7.2.6.4 Discussion of the Elastoplastic Calculation Procedure

The overall process for the numerical solution of elastoplastic problems can be divided into two stages: preliminary preparation and incremental iteration. The specific steps of the preliminary preparation are as follows:

① Initialize the relevant material parameters (such as elastic modulus, Poisson's ratio, etc.) based on the specific problem to be solved, as well as the order and coordinate information of the hierarchical quadrature elements used. According to the element order, coordinate information, and the construction method of the hierarchical quadrature elements, the element stiffness matrix in the elastic state can be initialized and assembled into the global stiffness matrix.

② Apply boundary conditions and determine the number of incremental loading steps.

③ Initialize the external force vector, internal force vector, and total displacement to zero, then proceed to the incremental iteration stage.

For the incremental iteration stage, the specific steps are as follows:

① After adding an incremental step to the external force vector, initialize the total displacement increment (used to record the final displacement increment of the current loading step) to zero. Calculate the difference between the external and internal force vectors (the residual) and compute the displacement increment.

② Add the displacement increment to the total displacement increment. At each integration point, use this updated displacement increment in combination with the return-mapping algorithm to calculate the true stress at that integration point, derive its elastoplastic matrix, and update the relevant elastoplastic parameters.

③ Update the internal force vector based on the true stress at the integration points, and update the element stiffness matrix through numerical integration using the elastoplastic matrix at the integration points, thereby updating the global stiffness matrix.

Repeat the above process. If the calculated difference between the external and internal force vectors is less than the tolerance, the iteration for the current incremental loading step ends. Add the total displacement increment to the total displacement. If the number of incremental loading steps has not yet reached the specified number, proceed to the next round of incremental iteration until all incremental iterations are completed. If the difference between the external and internal force vectors exceeds the tolerance and the iteration count has reached the maximum allowable number, it indicates that the current incremental iteration process did not achieve a converged result.

Figure 7.14 provides a more intuitive illustration of the overall process for solving elastoplastic problems in this study. It is evident that the key to the implementation lies in the application of the return-mapping algorithm at each integration point. As elaborated earlier, the return-mapping algorithm primarily consists of two parts: elastic prediction and plastic correction. Figure 7.15 details the specific operational flow of the return-mapping algorithm for each integration point under nonlinear isotropic hardening conditions. In practice, the operation of the return-mapping algorithm can be intuitively understood as: after multiple return-mapping operations, the trial stress is projected onto the new yield surface, as shown in Fig. 7.16.

7.2.7 Hierarchical Quadrature Element Method

Chapter 4 introduced the hierarchical quadrature element formulations for two-dimensional and three-dimensional solids. Taking the hexahedral element as an example, its displacement field can be expressed as follows:

$$\mathbf{u} = \tilde{\mathbf{S}}^{\mathrm{T}}\overline{\mathbf{u}} \tag{7.142}$$

The displacement field at the integration point can be expressed as

$$\hat{\mathbf{u}} = \mathbf{G}\overline{\mathbf{u}} \tag{7.143}$$

where

$$\begin{aligned} \mathbf{G}^{\mathrm{T}} &= [S(\xi_1, \eta_1, \zeta_1)\, S(\xi_2, \eta_2, \zeta_2) \ldots S(\xi_K, \eta_K, \zeta_K)] \\ \overline{\mathbf{u}} &= [u(\xi_1, \eta_1, \zeta_1)\, u(\xi_2, \eta_2, \zeta_2) \ldots u(\xi_K, \eta_K, \zeta_K)]^{\mathrm{T}} \end{aligned} \tag{7.144}$$

Using the chain rule of differentiations, the derivatives of the displacement field $\boldsymbol{u}$ with respect to the global coordinates x, y, and z can be obtained as

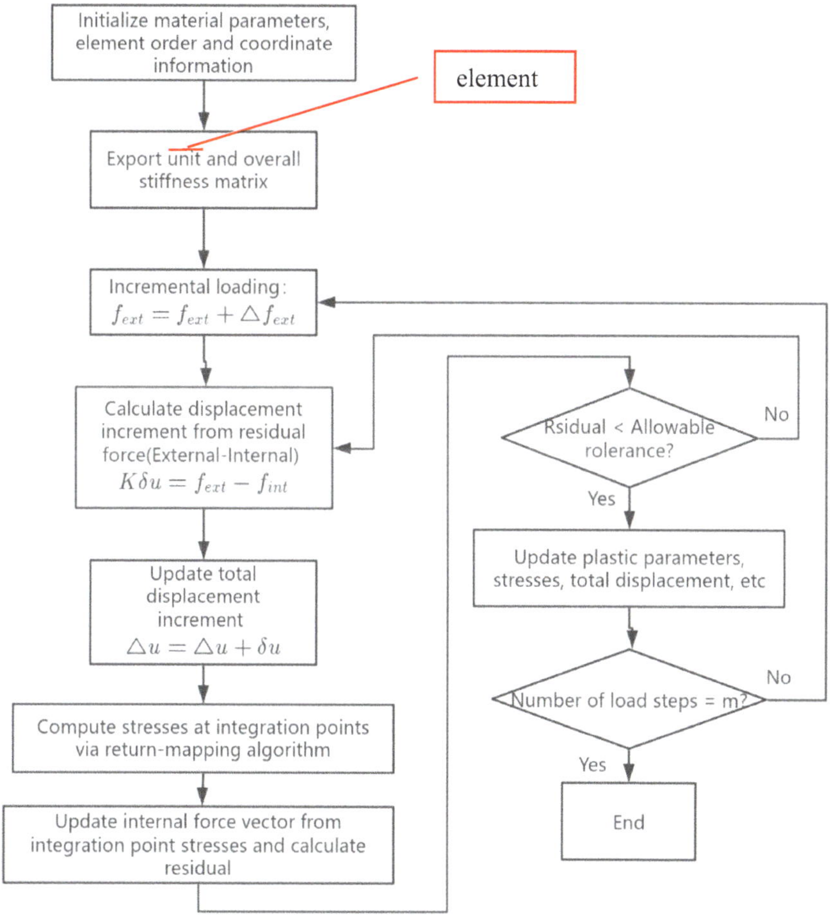

Fig. 7.14 Elastoplastic global solution procedure

$$\begin{bmatrix} \dfrac{\partial u}{\partial x} \\ \dfrac{\partial u}{\partial y} \\ \dfrac{\partial u}{\partial z} \end{bmatrix} = \mathbf{J}^{-1} \begin{bmatrix} \dfrac{\partial u}{\partial \xi} \\ \dfrac{\partial u}{\partial \eta} \\ \dfrac{\partial u}{\partial \zeta} \end{bmatrix} \tag{7.145}$$

where $\boldsymbol{J}$ is the Jacobian determinant

$$\mathbf{J} = \begin{vmatrix} \frac{\partial x}{\partial \xi} & \frac{\partial y}{\partial \xi} & \frac{\partial z}{\partial \xi} \\ \frac{\partial x}{\partial \eta} & \frac{\partial y}{\partial \eta} & \frac{\partial z}{\partial \eta} \\ \frac{\partial x}{\partial \zeta} & \frac{\partial y}{\partial \zeta} & \frac{\partial z}{\partial \zeta} \end{vmatrix} \tag{7.146}$$

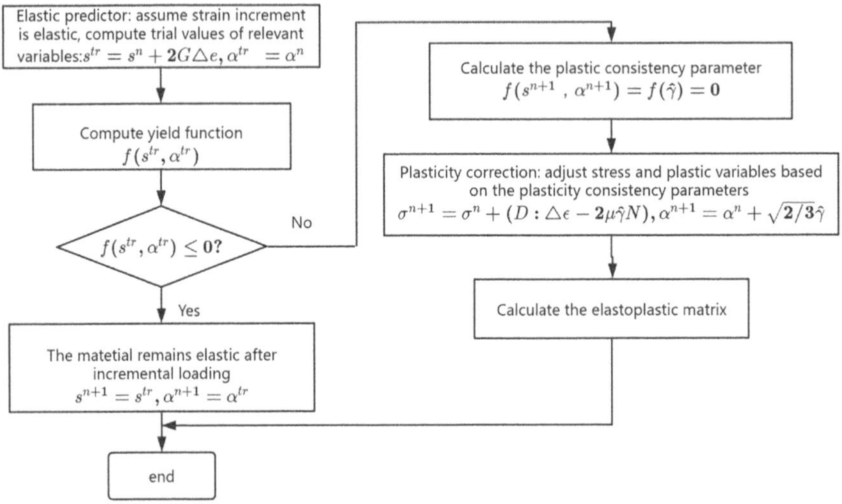

Fig. 7.15 Return mapping algorithm flowchart

The derivatives at the integration point $x_{ijk} = x(\xi_i, \eta_j, \zeta_k), y_{ijk} = y(\xi_i, \eta_j, \zeta_k), z_{ijk} = z(\xi_i, \eta_j, \zeta_k)$ can be expressed as

$$\begin{aligned}
\left(\frac{\partial u}{\partial x}\right)_{ijk} &= \frac{1}{|\mathbf{J}|_{ijk}} \begin{bmatrix} \left(\frac{\partial y}{\partial \eta}\frac{\partial z}{\partial \zeta} - \frac{\partial y}{\partial \zeta}\frac{\partial z}{\partial \eta}\right)_{ijk} \\ -\left(\frac{\partial y}{\partial \xi}\frac{\partial z}{\partial \zeta} - \frac{\partial y}{\partial \zeta}\frac{\partial z}{\partial \xi}\right)_{ijk} \\ \left(\frac{\partial y}{\partial \xi}\frac{\partial z}{\partial \eta} - \frac{\partial y}{\partial \eta}\frac{\partial z}{\partial \xi}\right)_{ijk} \end{bmatrix} \begin{bmatrix} \mathbf{S}_\xi^{\mathrm{T}}(\xi_i, \eta_j, \zeta_k) \\ \mathbf{S}_\eta^{\mathrm{T}}(\xi_i, \eta_j, \zeta_k) \\ \mathbf{S}_\zeta^{\mathrm{T}}(\xi_i, \eta_j, \zeta_k) \end{bmatrix} \mathbf{u} \\
\left(\frac{\partial u}{\partial y}\right)_{ijk} &= \frac{1}{|\mathbf{J}|_{ijk}} \begin{bmatrix} -\left(\frac{\partial x}{\partial \eta}\frac{\partial z}{\partial \zeta} - \frac{\partial x}{\partial \zeta}\frac{\partial z}{\partial \eta}\right)_{ijk} \\ \left(\frac{\partial x}{\partial \xi}\frac{\partial z}{\partial \zeta} - \frac{\partial x}{\partial \zeta}\frac{\partial z}{\partial \xi}\right)_{ijk} \\ -\left(\frac{\partial x}{\partial \xi}\frac{\partial z}{\partial \eta} - \frac{\partial x}{\partial \eta}\frac{\partial z}{\partial \xi}\right)_{ijk} \end{bmatrix} \begin{bmatrix} \mathbf{S}_\xi^{\mathrm{T}}(\xi_i, \eta_j, \zeta_k) \\ \mathbf{S}_\eta^{\mathrm{T}}(\xi_i, \eta_j, \zeta_k) \\ \mathbf{S}_\zeta^{\mathrm{T}}(\xi_i, \eta_j, \zeta_k) \end{bmatrix} \mathbf{u} \\
\left(\frac{\partial u}{\partial z}\right)_{ijk} &= \frac{1}{|\mathbf{J}|_{ijk}} \begin{bmatrix} -\left(\frac{\partial x}{\partial \eta}\frac{\partial y}{\partial \zeta} - \frac{\partial x}{\partial \zeta}\frac{\partial y}{\partial \eta}\right)_{ijk} \\ \left(\frac{\partial x}{\partial \xi}\frac{\partial y}{\partial \zeta} - \frac{\partial x}{\partial \zeta}\frac{\partial y}{\partial \xi}\right)_{ijk} \\ -\left(\frac{\partial x}{\partial \xi}\frac{\partial y}{\partial \eta} - \frac{\partial x}{\partial \eta}\frac{\partial y}{\partial \xi}\right)_{ijk} \end{bmatrix} \begin{bmatrix} \mathbf{S}_\xi^{\mathrm{T}}(\xi_i, \eta_j, \zeta_k) \\ \mathbf{S}_\eta^{\mathrm{T}}(\xi_i, \eta_j, \zeta_k) \\ \mathbf{S}_\zeta^{\mathrm{T}}(\xi_i, \eta_j, \zeta_k) \end{bmatrix} \mathbf{u}
\end{aligned} \tag{7.147}$$

For clarity, we denote the partial derivatives of the bases vector $\boldsymbol{S}$ and the displacement field $\boldsymbol{u}$ as follows:

$$\begin{aligned}
\mathbf{S}_\xi &= \frac{\partial \mathbf{S}}{\partial \xi}, \quad \mathbf{S}_\eta = \frac{\partial \mathbf{S}}{\partial \eta}, \quad \mathbf{S}_\zeta = \frac{\partial \mathbf{S}}{\partial \zeta} \\
\mathbf{u}_x &= \frac{\partial \mathbf{u}}{\partial x}, \quad \mathbf{u}_y = \frac{\partial \mathbf{u}}{\partial y}, \quad \mathbf{u}_z = \frac{\partial \mathbf{u}}{\partial z}
\end{aligned} \tag{7.148}$$

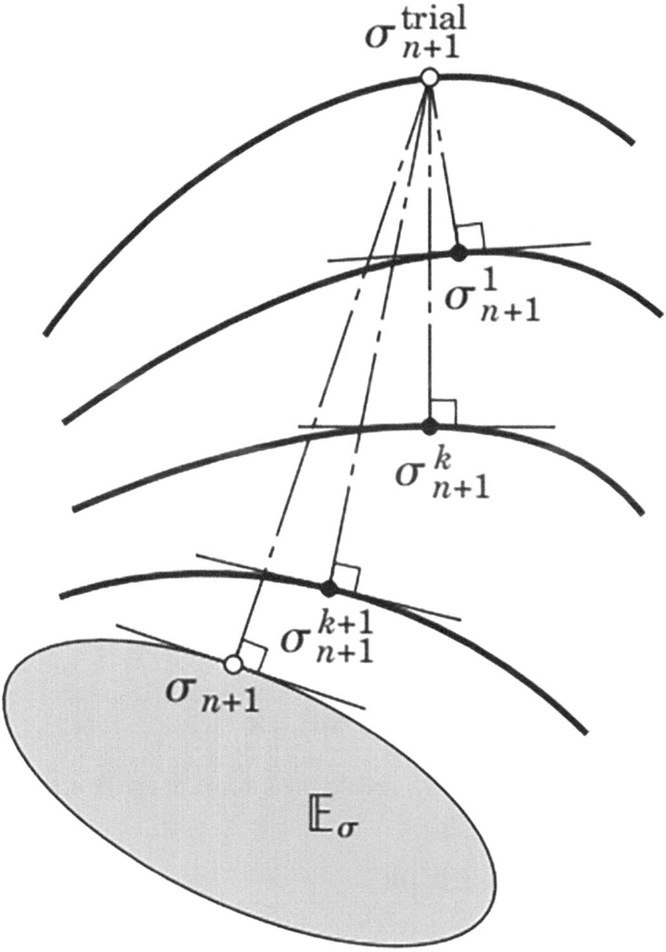

Fig. 7.16 Intuitive understanding of the return mapping algorithm

Thus, the differentiations in Eq. (7.147) can be simplified as

$$\bar{\mathbf{u}}_x = \mathrm{G}_x\mathbf{u},\ \bar{\mathbf{u}}_y = \mathrm{G}_y\mathbf{u},\ \bar{\mathbf{u}}_z = \mathbf{G}_z\mathbf{u} \tag{7.149}$$

Similarly, the derivatives with respect to v and w can be expressed as

$$\begin{aligned} \bar{\mathbf{v}}_x = \mathbf{G}_x\mathbf{v},\ \bar{\mathbf{v}}_y = \mathbf{G}_y\mathbf{v},\ \bar{\mathbf{v}}_z = \mathbf{G}_z\mathbf{v} \\ \bar{\mathbf{w}}_x = \mathbf{G}_x\mathbf{w},\ \bar{\mathbf{w}}_y = \mathbf{G}_y\mathbf{w},\ \bar{\mathbf{w}}_z = \mathbf{G}_z\mathbf{w} \end{aligned} \tag{7.150}$$

Consider the discrete geometric equation $\boldsymbol{\varepsilon} = \boldsymbol{Bu}$, or

$$\begin{Bmatrix} \varepsilon_{xx} \\ \varepsilon_{yy} \\ \varepsilon_{zz} \\ \gamma_{yz} \\ \gamma_{xz} \\ \gamma_{xy} \end{Bmatrix} = \begin{bmatrix} \mathbf{G}_x & \mathbf{0} & \mathbf{0} \\ \mathbf{0} & \mathbf{G}_y & \mathbf{0} \\ \mathbf{0} & \mathbf{0} & \mathbf{G}_z \\ \mathbf{0} & \mathbf{G}_z & \mathbf{G}_y \\ \mathbf{G}_z & \mathbf{0} & \mathbf{G}_x \\ \mathbf{G}_y & \mathbf{G}_x & \mathbf{0} \end{bmatrix} \begin{bmatrix} \mathbf{u} \\ \mathbf{v} \\ \mathbf{w} \end{bmatrix} \tag{7.151}$$

Using the following weights for integration on a hexahedral domain

$$\begin{aligned} \mathbf{C} = \mathrm{diag}\Big\{ & |\boldsymbol{J}|_{111} C_1^{\xi} C_1^{\eta} C_1^{\zeta}, \ldots, |\boldsymbol{J}|_{N_\xi 11} C_{N_\xi}^{\xi} C_1^{\eta} C_1^{\zeta}, \ldots, \\ & |\boldsymbol{J}|_{N_\xi N_\eta 1} C_{N_\xi}^{\xi} C_{N_\eta}^{\eta} C_1^{\zeta}, \ldots, |\boldsymbol{J}|_{N_\xi N_\eta N_\zeta} C_{N_\xi}^{\xi} C_{N_\eta}^{\eta} C_{N_\zeta}^{\zeta} \Big\} \end{aligned} \tag{7.152}$$

The following energy functional can be discretized.

$$\delta\Pi = \int_{\Omega} (\sigma_{xx}\delta\varepsilon_{xx} + \sigma_{yy}\delta\varepsilon_{yy} + \sigma_{zz}\delta\varepsilon_{zz} + \tau_{yz}\delta\gamma_{yz} + \tau_{zx}\delta\gamma_{zx} + \tau_{xy}\delta\gamma_{xy})\mathrm{d}\Omega \tag{7.153}$$

Setting it equal to zero allows the energy functional to attain a stationary value, yielding

$$\mathbf{K}_L\mathbf{U} + \mathbf{M}\ddot{\mathbf{U}} = \mathbf{R} \tag{7.154}$$

where the linear stiffness matrix $\boldsymbol{K}_L$ and the mass matrix $\boldsymbol{M}$ can be derived using the variational principle, that is,

$$\begin{aligned} \mathbf{K}_L &= \mathbf{B}^T\mathbf{D}\mathbf{B} \\ \mathbf{M} &= \rho \begin{bmatrix} \mathbf{G}^T\mathbf{C}\mathbf{G} & & \\ & \mathbf{G}^T\mathbf{C}\mathbf{G} & \\ & & \mathbf{G}^T\mathbf{C}\mathbf{G} \end{bmatrix} \end{aligned} \tag{7.155}$$

When the element is in the elastic state, the $\boldsymbol{D}$ matrix is the elastic matrix $\boldsymbol{D}_e$, which is given by

$$\mathbf{D}_e = \frac{E(1-\upsilon)}{(1+\upsilon)(1-2\upsilon)} \times \begin{bmatrix} \mathbf{C} & \frac{\upsilon}{1-\upsilon}\mathbf{C} & \frac{\upsilon}{1-\upsilon}\mathbf{C} & \mathbf{0} & \mathbf{0} & \mathbf{0} \\ \frac{\upsilon}{1-\upsilon}\mathbf{C} & \mathbf{C} & \frac{\upsilon}{1-\upsilon}\mathbf{C} & \mathbf{0} & \mathbf{0} & \mathbf{0} \\ \frac{\upsilon}{1-\upsilon}\mathbf{C} & \frac{\upsilon}{1-\upsilon}\mathbf{C} & \mathbf{C} & \mathbf{0} & \mathbf{0} & \mathbf{0} \\ \mathbf{0} & \mathbf{0} & \mathbf{0} & \frac{1-2\upsilon}{2(1-\upsilon)}\mathbf{C} & \mathbf{0} & \mathbf{0} \\ \mathbf{0} & \mathbf{0} & \mathbf{0} & \mathbf{0} & \frac{1-2\upsilon}{2(1-\upsilon)}\mathbf{C} & \mathbf{0} \\ \mathbf{0} & \mathbf{0} & \mathbf{0} & \mathbf{0} & \mathbf{0} & \frac{1-2\upsilon}{2(1-\upsilon)}\mathbf{C} \end{bmatrix} \tag{7.156}$$

When an element is in an elastoplastic state, the $\boldsymbol{D}$ matrix becomes the elastoplastic matrix $\boldsymbol{D}_{\text{ep}}$, which is composed of the tangent stiffness matrices at the integration points of the element and the integration weighting coefficients, and can be expressed as:

$$\mathbf{D}_{ep} = \begin{bmatrix} \boldsymbol{D}_{11} & \boldsymbol{D}_{12} & \boldsymbol{D}_{13} & \boldsymbol{D}_{14} & \boldsymbol{D}_{15} & \boldsymbol{D}_{16} \\ \boldsymbol{D}_{21} & \boldsymbol{D}_{22} & \boldsymbol{D}_{23} & \boldsymbol{D}_{24} & \boldsymbol{D}_{25} & \boldsymbol{D}_{23} \\ \boldsymbol{D}_{31} & \boldsymbol{D}_{32} & \boldsymbol{D}_{33} & \boldsymbol{D}_{34} & \boldsymbol{D}_{35} & \boldsymbol{D}_{36} \\ \boldsymbol{D}_{41} & \boldsymbol{D}_{42} & \boldsymbol{D}_{43} & \boldsymbol{D}_{44} & \boldsymbol{D}_{45} & \boldsymbol{D}_{46} \\ \boldsymbol{D}_{51} & \boldsymbol{D}_{52} & \boldsymbol{D}_{53} & \boldsymbol{D}_{54} & \boldsymbol{D}_{55} & \boldsymbol{D}_{56} \\ \boldsymbol{D}_{61} & \boldsymbol{D}_{62} & \boldsymbol{D}_{63} & \boldsymbol{D}_{64} & \boldsymbol{D}_{65} & \boldsymbol{D}_{66} \end{bmatrix} \tag{7.157}$$

where

$$\boldsymbol{D}_{ij} = \begin{bmatrix} C_{111}d_{ij}^{(111)} & 0 & 0 \cdots & 0 \\ 0 & C_{211}d_{ij}^{(211)} & 0 \cdots & 0 \\ \vdots & \vdots & \vdots \;\; \vdots & \vdots \\ 0 & 0 & 0 \cdots & C_{N_\xi N_\eta N_\zeta} d_{ij}^{(N_\xi N_\eta N_\zeta)} \end{bmatrix}$$

$$\boldsymbol{D}_{ep}^{(klm)} = \begin{bmatrix} d_{11}^{(klm)} & d_{12}^{(klm)} & d_{13}^{(klm)} & d_{14}^{(klm)} & d_{15}^{(klm)} & d_{16}^{(klm)} \\ d_{21}^{(klm)} & d_{22}^{(klm)} & d_{23}^{(klm)} & d_{24}^{(klm)} & d_{25}^{(klm)} & d_{26}^{(klm)} \\ d_{31}^{(klm)} & d_{32}^{(klm)} & d_{33}^{(klm)} & d_{34}^{(klm)} & d_{35}^{(klm)} & d_{36}^{(klm)} \\ d_{41}^{(klm)} & d_{42}^{(klm)} & d_{43}^{(klm)} & d_{44}^{(klm)} & d_{45}^{(klm)} & d_{46}^{(klm)} \\ d_{51}^{(klm)} & d_{52}^{(klm)} & d_{53}^{(klm)} & d_{54}^{(klm)} & d_{55}^{(klm)} & d_{56}^{(klm)} \\ d_{61}^{(klm)} & d_{62}^{(klm)} & d_{63}^{(klm)} & d_{64}^{(klm)} & d_{65}^{(klm)} & d_{66}^{(klm)} \end{bmatrix} \tag{7.158}$$

$$klm = 111, 211, \cdots N_\xi 11, \cdots N_\xi N_\eta N_\zeta$$

The consistent tangent stiffness matrix at the integration point is derived from the return mapping algorithm presented in the previous section.

In the preceding discussion before this subsection, the fundamental theory for two-dimensional and three-dimensional elastoplastic problems under isotropic hardening conditions was clarified, and the corresponding numerical solution algorithms were derived. Based on the hierarchical quadrature element method introduced in Chap. 4, the numerical solution of two-dimensional and three-dimensional elastoplastic problems can be achieved. This subsection integrates the aforementioned two aspects, presenting the overall solution process for two-dimensional and three-dimensional elastoplastic problems under isotropic hardening conditions using the hierarchical quadrature element method in combination with the return-mapping algorithm.

7.2.8 Summary of This Section

In contrast to the history-independent behavior of elastic materials, the stress–strain state of elastoplastic materials is not uniquely determined, but rather depends on the specific loading history. This section first introduces the fundamental concepts related to elastoplastic problems, presents the explicit forms of the initial and subsequent yield criteria, and clarifies how to determine when a material yields—i.e., enters the plastic state—as well as how the material evolves after yielding. Next, under the assumptions of small deformations and small strains, the basic theory of elastoplasticity is discussed in detail for nonlinear isotropic hardening conditions, and the rate form expressions of elastoplastic models in both two and three dimensions are provided. The rate form of the elastoplastic model is then converted to an incremental form using the backward implicit Euler time integration method, and the numerical solution algorithm for elastoplastic problems—the return mapping algorithm is derived. This algorithm consists of two steps: elastic prediction and plastic correction. Through multiple iterations, the trial stress is projected onto the new yield surface. In practical elastoplastic problem solving, elastoplastic analysis is performed at each integration point to derive the elastoplastic matrix, which is then used to obtain the element stiffness matrix via numerical integration, and subsequently assembled into the global stiffness matrix for solution. For each integration point, the specific operational procedure of the return mapping algorithm is provided. Finally, the overall procedure for solving elastoplastic problems using the hierarchical quadrature element method is presented.

7.3 Numerical Examples for Elastoplastic Problems

7.3.1 Overview

The previous section introduced the fundamental theory and algorithms for solving elastoplastic problems, and developed two-dimensional and three-dimensional hierarchical quadrature elements. Explicit numerical solution algorithms for elastoplastic problems in both two and three dimensions were presented. In this section, based on the theory and algorithms from the previous section, the hierarchical quadrature element method will be applied to two two-dimensional elastoplastic examples (including a plane strain problem and a plane stress problem) and one three-dimensional elastoplastic example, in order to demonstrate the high accuracy and efficiency of the hierarchical quadrature element method in solving elastoplastic problems.

This section considers two sets of material parameters (see Table 7.1): the first set is applied to the ideal elastic–plastic model, while the second set is used for the nonlinear isotropic hardening model. Section 7.3.2 addresses a thick-walled cylinder under internal pressure in plane strain conditions, using the ideal elastic–plastic

Table 7.1 Material parameters

Material parameters	Type 1	Type 2
Bulk modulus	166.7 (MPa)	164,206.0 (MPa)
Shear modulus	3.36 (MPa)	80,193.8 (MPa)
Initial yield stress	58.0 (KPa)	450.0 (MPa)
Ultimate stress	58.0 (KPa)	715.0 (MPa)
Linear hardening parameter	0.0 (KPa)	129.24 (MPa)
Hardening exponent	0.0	16.93

model. Section 7.3.3 discusses a thin-walled plate with a circular hole under plane stress conditions. Section 7.3.4 covers a thick-walled plate with a circular hole, which is a three-dimensional elastic–plastic problem. The examples in Sects. 7.3.3 and 7.3.4 employ the nonlinear isotropic hardening model.

In each example, the computational results of the hierarchical quadrature element method are compared with the analytical solution (if available) and the results obtained by other finite element methods (primarily the h-version finite element results calculated using ABAQUS). For convenience, in the comparative figures presented in this section, the hierarchical quadrature element method is uniformly denoted as HQEM.

7.3.2 *Thick-Walled Cylinders Subjected to Internal Pressure*

7.3.2.1 Derivation of the Solution for a Thick-Walled Cylinder Subjected to Internal Pressure with Closed Ends

Consider a long cylindrical tube subjected to a uniform internal pressure p. According to the axisymmetric condition, in the cylindrical coordinate system (r, θ, z), the following simplifications can be applied:

$$\begin{aligned}
&\tau_{rz} = \tau_{z\theta} = \tau_{\theta r} = 0,\ u_\theta = 0 \\
&\frac{\partial \sigma_r}{\partial \theta} = \frac{\partial \sigma_\theta}{\partial \theta} = \frac{\partial u_r}{\partial \theta} = \frac{\partial u_z}{\partial \theta} = 0 \\
&\frac{\partial \sigma_r}{\partial z} = \frac{\partial \sigma_\theta}{\partial z} = \frac{\partial \sigma_r}{\partial \theta} = \frac{\partial u_r}{\partial z} = 0,\ \frac{\partial u_z}{\partial r} = 0
\end{aligned} \tag{7.159}$$

Under small deformation conditions, whether in the elastic or plastic state, the equilibrium equations, geometric equations, and boundary conditions remain the same. For axisymmetric cases, they take the following simplified forms:

Equilibrium Equation:

$$r\frac{\mathrm{d}\sigma_r}{\mathrm{d}r} = \sigma_\theta - \sigma_r \tag{7.160}$$

Geometric Equation:

$$\varepsilon_r = \frac{du}{dr}, \varepsilon_\theta = \frac{u}{r}, \varepsilon_z = \varepsilon_0 = const \tag{7.161}$$

Boundary Conditions:

$$\begin{aligned} r = a, \sigma_r = p \\ r = b, \sigma_r = 0 \end{aligned} \tag{7.162}$$

Under elastic conditions, the constitutive equation is

$$\begin{aligned} \varepsilon_r &= \frac{1}{E}[\sigma_r - \mu(\sigma_\theta + \sigma_z)] \\ \varepsilon_\theta &= \frac{1}{E}[\sigma_\theta - \mu(\sigma_z + \sigma_r)] \\ \varepsilon_z &= \frac{1}{E}[\sigma_z - \mu(\sigma_r + \sigma_\theta)] \end{aligned} \tag{7.163}$$

From the above conditions, the elastic solution can be obtained as follows:

$$\begin{aligned} \sigma_r &= -p\left(\frac{b^2}{r^2} - 1\right)\left(\frac{b^2}{a^2} - 1\right) \\ \sigma_\theta &= p\left(\frac{b^2}{r^2} + 1\right)\left(\frac{b^2}{a^2} - 1\right) \\ \sigma_z &= E\varepsilon_0 + 2\mu p \Big/ \left(\frac{b^2}{a^2} - 1\right) \\ \varepsilon_z &= \varepsilon_0 = -2\mu a^2 p / E(b^2 - a^2) \\ u_r &= \frac{(1+\mu)pa^2}{E(b^2 - a^2)}\left[(1 - 2\mu)r + \frac{b^2}{r}\right] - \mu\varepsilon_0 r \end{aligned} \tag{7.164}$$

In the case of plane strain, $\varepsilon_z = 0$. Therefore,

$$\begin{aligned} \sigma_z &= \frac{2\mu pa^2}{b^2 - a^2} \\ u_r &= \frac{(1+\mu)pa^2}{E(b^2 - a^2)}\left[(1 - 2\mu)r + \frac{b^2}{r}\right] \end{aligned} \tag{7.165}$$

If the material follows a linear hardening law, then

$$\left.\begin{aligned}&\sigma_e = E\varepsilon_e,\ \varepsilon_e \le \varepsilon_s\\&\sigma_e = \sigma_s\left(1-\frac{E'}{E}\right)+E'\varepsilon_e,\ \varepsilon_e > \varepsilon_s\end{aligned}\right\} \tag{7.166}$$

In the elastic region, the expression for stress is

$$\left.\begin{aligned}\sigma_r &= -\frac{\sigma_s}{\sqrt{3}}\left[\left(1-\frac{E\prime}{E}\right)\left(1+2\ln\frac{r_e}{r}\right)-\frac{r_e^2}{b^2}\left(1-\frac{E\prime b^2}{Er^2}\right)\right]\\\sigma_\theta &= \frac{\sigma_s}{\sqrt{3}}\frac{r_e^2}{b^2}\left[\left(1-\frac{E\prime}{E}\right)\left(1-2\ln\frac{r_e}{r}\right)+\frac{r_e^2}{b^2}\left(1+\frac{E\prime b^2}{Er^2}\right)\right]\\\sigma_z &= \frac{\sigma_s}{\sqrt{3}}\left[\frac{r_e^2}{b^2}-2\left(1-\frac{E\prime}{E}\right)\ln\frac{r_e}{r}\right]\end{aligned}\right\}\ (a \le r \le r_e) \tag{7.167}$$

where $a \le r \le r_e$. In the plastic zone, the expression for stress is

$$\left.\begin{aligned}\sigma_r &= -\frac{\sigma_s}{\sqrt{3}}\frac{r_e^2}{b^2}\left(\frac{b^3}{r^2}-1\right)\\\sigma_\theta &= \frac{\sigma_s}{\sqrt{3}}\frac{r_e^2}{b^2}\left(\frac{b^2}{r^2}-1\right)\\\sigma_z &= \frac{1}{\sqrt{3}}\frac{\sigma_s r_e^2}{b^2}\end{aligned}\right\}(r_e \le r \le b) \tag{7.168}$$

where $r_e \le r \le b$. The displacement expression is the same in the elastoplastic region, that is,

$$u = \frac{c}{r} = \frac{\sqrt{3}\sigma_s r_e^2}{2Er} \tag{7.169}$$

Here, the relationship between r_e and p depends on the boundary condition $r = a$, $\sigma_r = -p$, thus

$$p = \frac{\sigma_s}{\sqrt{3}}\left[\left(1-\frac{E\prime}{E}\right)\left(1+2\ln\frac{r_e}{a}\right)-\frac{r_e^2}{b^2}\left(1-\frac{E\prime b^2}{Ea^2}\right)\right] \tag{7.170}$$

7.3.2.2 Solution of the Linear Problem for Thick-Walled Cylinders Under Internal Pressure

In this subsection, the application of the two-dimensional hierarchical quadrature element method to linear problems is investigated using the thick-walled cylinder under internal pressure as an example. Assuming incompressibility, the thick-walled

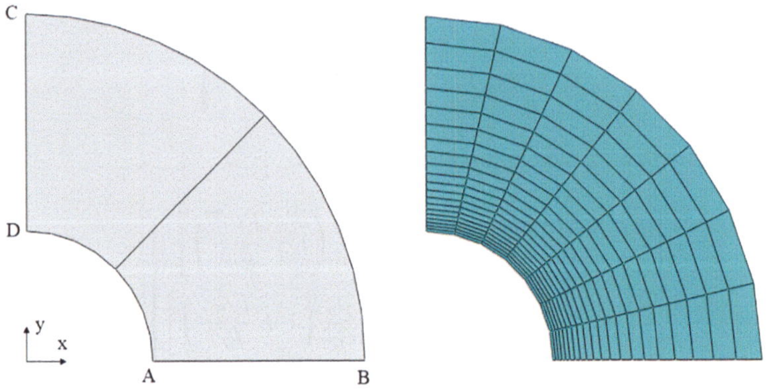

Fig. 7.17 Subdivision of a thick-walled cylinder under internal pressure

cylinder under internal pressure has a closed-form solution. By comparing the results obtained from the hierarchical quadrature element method with the analytical solution and ABAQUS simulation results, the accuracy of the method in solving linear problems is demonstrated. In the process of solving with the hierarchical quadrature element method, the inner radius of the cylinder is set to $a = 1.5$ m, the outer radius to $b = 4$ m, the internal pressure $p = 10{,}000$ Pa, and Young's modulus $E = 2.1 \times 10^8$ Pa, Poisson's ratio $\upsilon = 0.49$. The boundary conditions are: $u_y = 0$ along edge AB, and $u_x = 0$ along edge CD. Due to symmetry, 1/4 is selected for analysis, and the calculation is performed using two HQEM elements, as shown in Fig. 7.17.

The results for the x-direction displacement at point (1.5, 0) calculated by the Hierarchical Quadrature Element Method (HQEM) at different orders are compared with the analytical solution and ABAQUS results, as shown in Table 7.2. It can be seen that the HQEM yields highly accurate results. It is worth noting that, due to the relatively smooth displacement field of a thick-walled cylinder under internal pressure (see Fig. 7.18), the computational efficiency of the HQEM is not fully demonstrated in this elastic analysis. The efficiency advantages of the HQEM are more evident in the next section, where it is applied to the analysis of a rectangular plate with a hole under uniformly distributed load.

7.3.2.3 Solution to the Elastic–Plastic Problem of a Thick-Walled Cylinder Under Internal Pressure

In this subsection, the incremental method is used in conjunction with the hierarchical quadrature element method to solve the elastic–plastic problem of a thick-walled cylinder under internal pressure. The geometric parameters and boundary conditions of the example are the same as those in Sect. 7.3.2.2, and the ideal elastic–plastic material parameters are selected from Table 7.1:

Table 7.2 Comparison of results for the thick-walled cylinder under internal pressure: HQEM, analytical solution, and ABAQUS linear analysis

Method	Node ($m \times n$)	DOFs	Displacement u
Analytical solution			0.000124192
ABAQUS		72	0.000124187
		98	0.000124189
		128	0.00012419
HQEM	4 * 3	76	0.00012320
	6 * 3	104	0.00012414
	7 * 3	118	0.00012419
	8 * 3	132	0.00012419

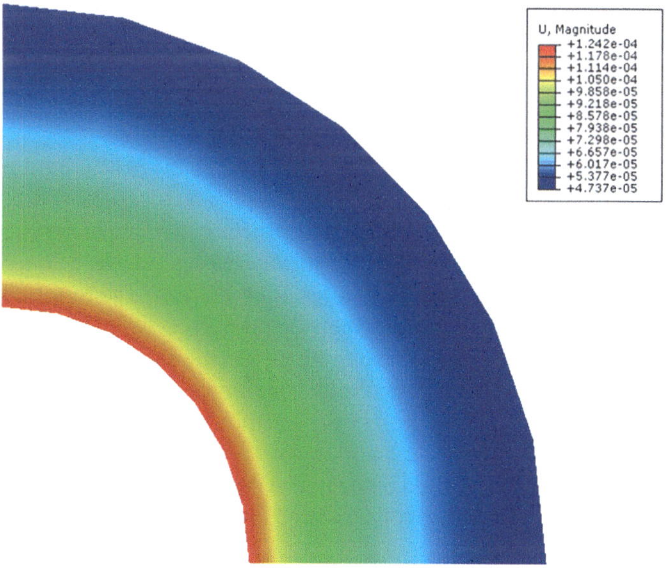

Fig. 7.18 Displacement contour plot of a thick-walled cylinder under internal pressure (elastic problem)

$$\sigma_0 = 58.0 \text{ KPa}, \quad \sigma_\infty = 58.0 \text{ KPa}, \quad h = 0, \quad \omega = 0$$

Take Young's modulus as $E = 1.0 \times 10^7$ Pa, Poisson's ratio $\upsilon = 0.49$, and internal pressure $p = 62$ kPa. Based on the analysis results above, to ensure solution accuracy, the order of hierarchical quadrature element is taken as 8*3. The comparative results obtained by the h-version finite element method use a fine mesh of 140 CPE4R four-node elements, with a total of 336 degrees of freedom (see Fig. 7.17b). In this example, the load is uniformly divided into 25 steps for application.

Under plane strain and incompressible conditions (Poisson's ratio of 0.5), the analytical solution can be calculated using the relevant formulas from Sect. 7.3.2.1.

Although the Poisson's ratio used in the hierarchical quadrature element method calculation in this example is 0.49, it can be demonstrated that the assumption of incompressibility in the elastic region does not significantly affect the validity of the analytical solution under realistic conditions.

Table 7.3 presents a comparison of the results from the hierarchical quadrature element method (HQEM) and ABAQUS under different degrees of freedom. It can be observed that the hierarchical quadrature element method achieves more accurate results with fewer degrees of freedom, and its convergence rate is significantly faster than that of the *h*-version method (achieving two-digit convergence with 104 DOFs).

The radial displacements u_r for the inner surface calculated by the HQEM ($p = 8*3$), ABAQUS (140 elements), and the analytical solution are shown in Fig. 7.19. When the external load exceeds 105 MPa, the radial displacements u_r for the inner surface ceases to increase linearly. Therefore, the elastic limit can be approximately determined to be 27 kPa. It is noteworthy that the elastic limit estimated by the analytical solution is about 28 kPa, indicating that the elastic limit load estimated by the hierarchical quadrature element method is highly consistent with the analytical result.

According to the analytical formula in Sect. 7.3.2.1, the plastic limit of a thick-walled cylinder under internal pressure is calculated to be 65.7 kPa. When the thick-walled tube is in an elastic–plastic state, the structure consists of both elastic and plastic regions. The radius of the elastic–plastic interface, r_e, can be determined using the analytical formula (7.170), yielding a result of 3.101 m. The result obtained by the hierarchical quadrature element method is approximately 3.049 m, as shown in Fig. 7.20, where the dashed line indicates the yield region within the internal area.

Under plane strain and incompressibility conditions, the analytical solutions for each stress component in both the elastic and plastic regions of a thick-walled cylinder, as well as the analytical solution for the radial displacement along the thickness direction, can be obtained based on the previous analysis. A comparison of the radial displacement along the thickness direction, as calculated by the hierarchical

Table 7.3 Comparison of results for thick-walled cylinder under internal pressure: HQEM versus ABAQUS

Method	Nodes ($m \times n$)	DOFs	Displacement u
Analytical solution			0.032 201
HQEM	4×3	76	0.030 819
	6×3	90	0.032028
	8×3	132	0.032 233
	10×3	160	0.032402
ABAQUS		60	0.030955
		176	0.031 718
		336	0.032 211
		682	0.032 324

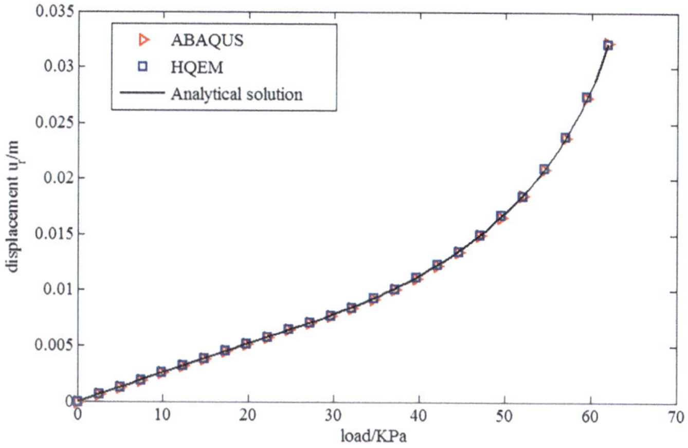

Fig. 7.19 Comparison of inner surface radial displacement calculated by the hierarchical quadrature element method, ABAQUS, and the analytical solution

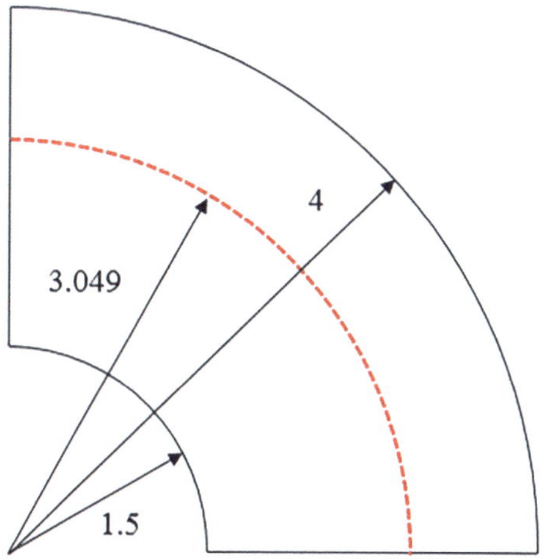

Fig. 7.20 Schematic diagram of the yield zone in a thick-walled cylinder

quadrature element method (HQEM), *h*-version finite element method (ABAQUS), and the analytical solution, is shown in Fig. 7.21. As shown in the figure, the curves obtained from these methods are almost identical, indicating a high degree of consistency among the results. Figure 7.21 demonstrates that the radial displacement gradually decreases across the thickness of the thick-walled cylinder. It is evident that, under internal pressure, the deformation at the inner surface of the thick-walled cylinder is greater than that at the outer surface.

The comparison of radial stress, hoop stress, and axial stress in a thick-walled cylinder under internal pressure in the elastic–plastic regime, as calculated by the

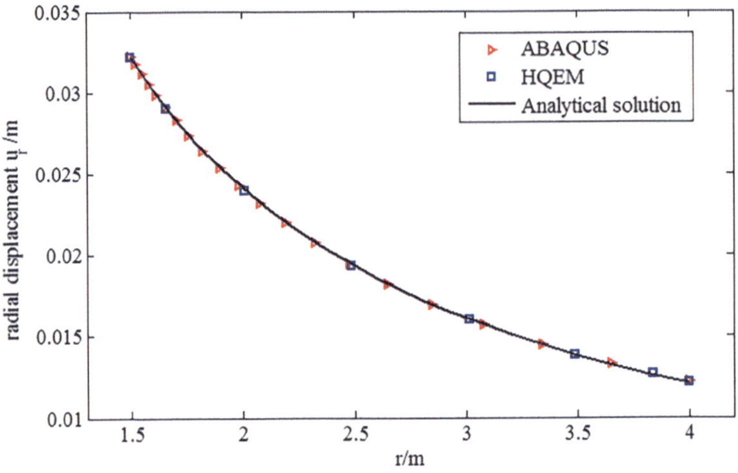

Fig. 7.21 Comparison of radial displacement along the thickness of a thick-walled cylinder obtained by the hierarchical quadrature element method, ABAQUS, and the analytical solution

HQEM, the *h*-version finite element method (ABAQUS), and the analytical solution, is shown in Figs. 7.22, 7.23, and 7.24. These results indicate that the stresses computed by the HQEM closely match the analytical solution and are more accurate than those obtained by the *h*-version finite element method. In addition, the distribution of Von Mises stress in the thick-walled cylinder under internal pressure in the elastic–plastic state was also investigated. The Von Mises stress distributions obtained by the HQEM and the *h*-version finite element method (ABAQUS) are shown in Fig. 7.25, demonstrating good agreement between the two methods.

Fig. 7.22 Comparison of radial stress results for a thick-walled cylinder under internal pressure calculated by the hierarchical quadrature element method, ABAQUS, and the analytical solution

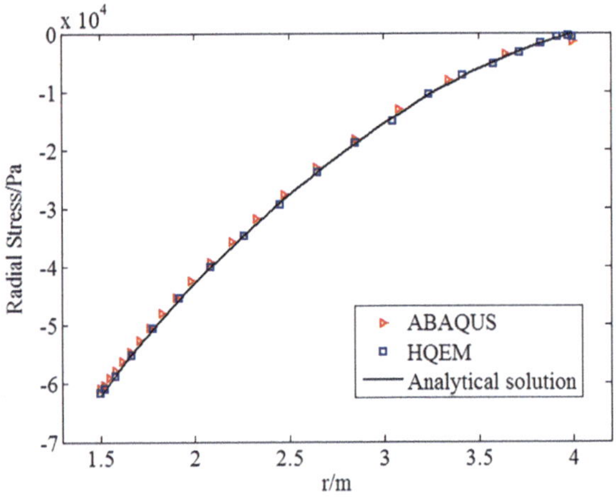

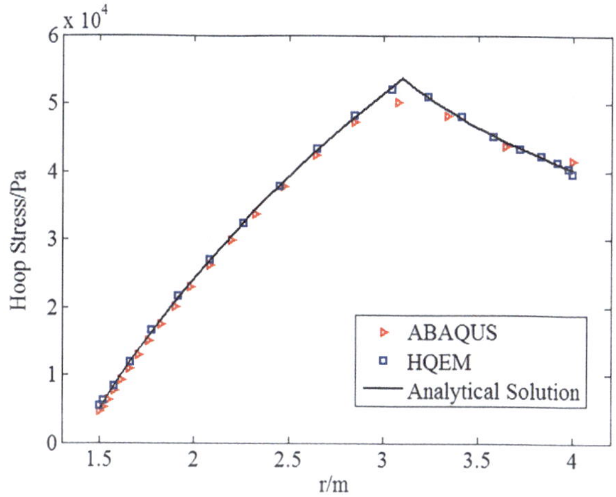

Fig. 7.23 Comparison of circumferential stress results for a thick-walled cylinder under internal pressure calculated by the hierarchical quadrature element method, ABAQUS, and analytical solution

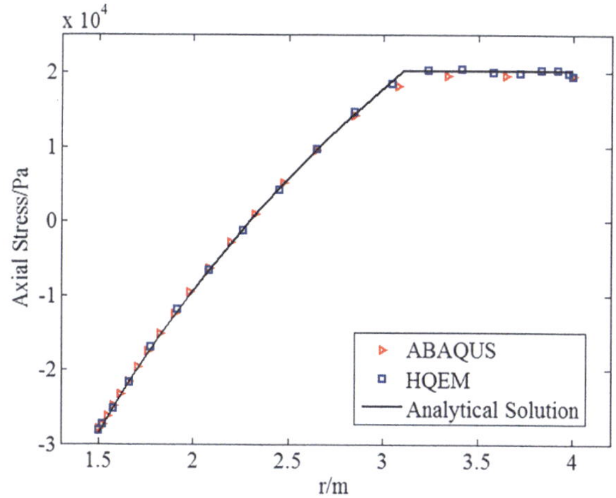

Fig. 7.24 Comparison of axial stress results for a thick-walled cylinder under internal pressure calculated by the hierarchical quadrature element method, ABAQUS, and analytical solution

7.3.3 Thin Plate with a Central Hole Under Uniform Load

This section presents a numerical example for solving the elastoplastic problem of a thin plate with a central hole under plane stress conditions, using the nonlinear isotropic hardening model parameters (see Material Set 2 in Table 7.1), namely,

$$\sigma_0 = 450.0\,\text{MPa}, \quad \sigma_\infty = 715.0\,\text{MPa}, \quad h = 129.24\,\text{MPa}, \quad \omega = 16.93$$

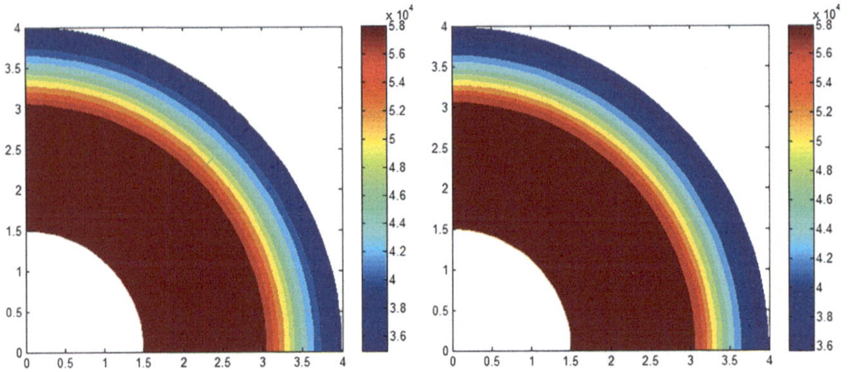

Fig. 7.25 Comparison of von Mises stress results for thick-walled cylinders under internal pressure

Due to the structural symmetry, it is sufficient to perform numerical simulation on 1/4 of the structure. The structural model and geometric dimensions are shown in Fig. 7.26. A uniformly distributed pressure of $P = 4.5 \times 10^8$ Pa is applied along edge 2–3.

For the hierarchical quadrature element method (HQEM), a mesh consisting of two uniform elements is used in numerical computations, with element orders ranging from $p = 4$ to 12 (see Fig. 7.27a). The h-version finite element method (ABAQUS) employs a series of mesh divisions composed of CPS4R four-node elements. Figure 7.27b shows one of the mesh configurations used in the h-version

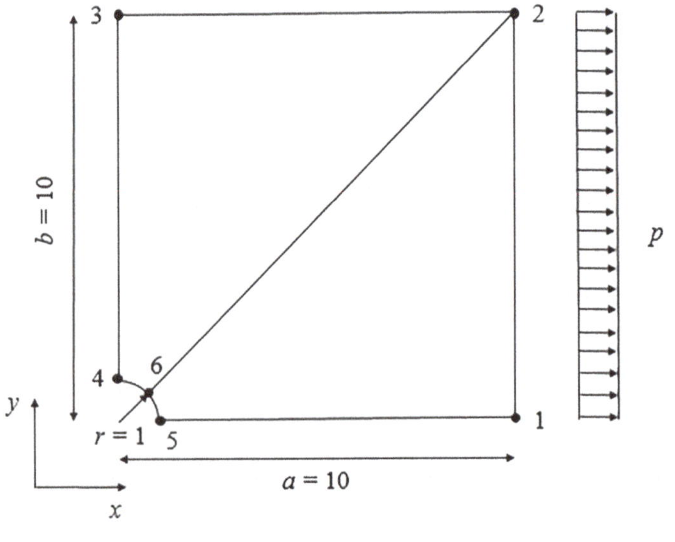

Fig. 7.26 Thin plate with a hole under uniform load 1/4

finite element method (ABAQUS) analysis. In this example, the load is applied in 50 uniform incremental steps.

Table 7.4 presents a comparison of the computational results from HQEM and the *h*-version finite element method (ABAQUS) under different degrees of freedom. The results obtained using the *h*-version finite element method with 72,782 degrees of freedom are taken as the reference solution. The results of both methods under varying degrees of freedom are plotted in the form of relative errors on a double logarithmic scale in Fig. 7.28. It can be observed that the hierarchical quadrature element method yields more accurate results, and its convergence rate is significantly faster than that of the *h*-version finite element method (achieving two-digit convergence with only 622 DOFs).

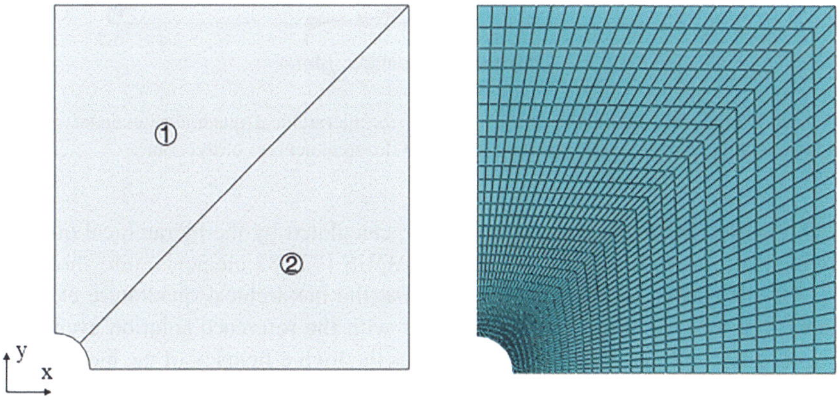

Fig. 7.27 Mesh generation of a thin plate with a central hole under uniform load

Table 7.4 Calculation results for the elastic–plastic problem of a thin plate with a central hole under uniform load using the HQEM and *h*-version finite element method

Method	Node ($m \times n$)	DOFs	u_y at node 4
HQEM	4 * 4	104	−0.009155
	6 * 6	212	−0.008546
	8 * 8	352	−0.007432
	10 * 10	524	−0.006769
	11 * 11	622	−0.006692
	12 * 12	726	−0.006613
ABAQUS		108	−0.007801
		374	−0.007253
		800	−0.006982
		1386	−0.006835
		3038	−0.006703
		5330	−0.006660
		72,782	−0.006621

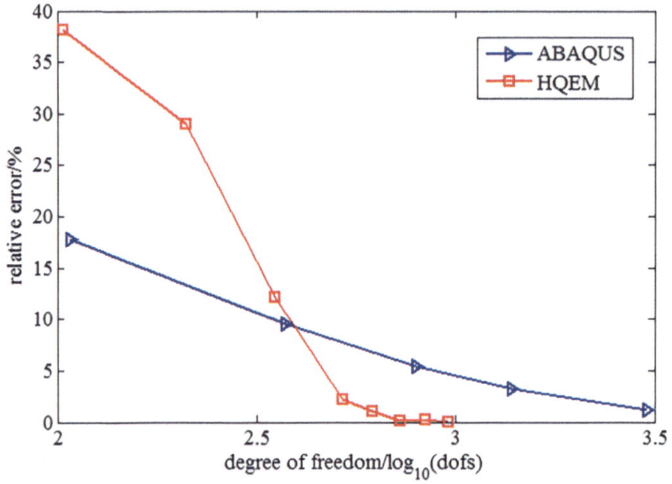

Fig. 7.28 Comparison of relative errors between the hierarchical quadrature element method (HQEM) and the *h*-version finite element method at different degrees of freedom

The *y*-direction displacements at node 4, calculated by the hierarchical quadrature element method ($p = 12*12$) and ABAQUS (72,782 elements), are shown in Fig. 7.29. It can be seen from the figure that the hierarchical quadrature element method (HQEM) achieves good agreement with the reference solution even with fewer degrees of freedom. This demonstrates the high efficiency of the hierarchical quadrature element method in solving elastoplastic problems.

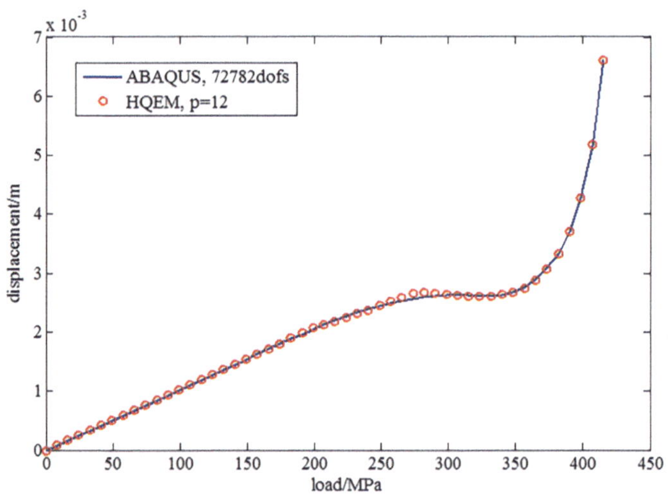

Fig. 7.29 Comparison of the displacement u_y at node 4 calculated by the hierarchical quadrature element method (HQEM) and the *h*-version finite element method

Figure 7.30 presents a comparison between the stress results in the x-direction at node 4, calculated by HQEM, and the reference solution (obtained from ABAQUS with 72,782 degrees of freedom). The results indicate that the stress computed by HQEM aligns well with the reference solution. Additionally, Fig. 7.31 displays the Von Mises stress distribution on a perforated thin plate under uniform load, calculated using both the hierarchical quadrature element method and the h-version finite element method (reference solution), which demonstrates good consistency between the two methods.

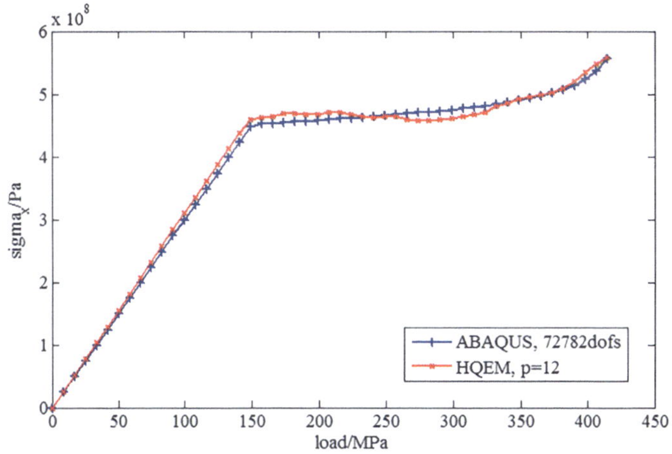

Fig. 7.30 Comparison of x-direction stress at node 4 calculated by the hierarchical quadrature element method (HQEM) and the h-version finite element method

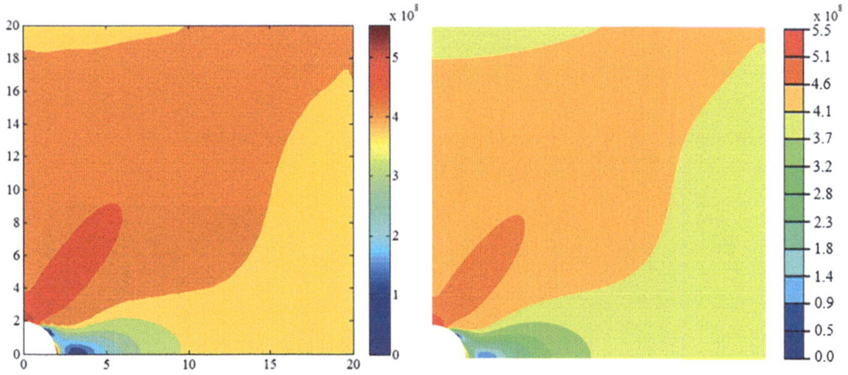

Fig. 7.31 Comparison of Von Mises stress in a thin plate with a central hole under uniform load

7.3.4 Thick Plate with a Central Hole Under Uniform Distributed Load

The final example involves solving an elastoplastic problem of a perforated thick plate under uniform load using a nonlinear isotropic hardening model (Material 2 as listed in Table 7.1). This is a three-dimensional elastoplastic case. Similarly, due to the symmetry of the structure, only one-eighth of the structure is used in the numerical simulation. The specific geometric dimensions are shown in Fig. 7.32. A uniform pressure of $P = 4.5 \times 10^8$ Pa is applied on the surface 2–2'-3–3'.

Similarly, the hierarchical quadrature element method (HQEM) was employed for calculations using a mesh composed of two elements. A series of numerical solutions were performed by increasing the element order from $p = 4$ to $p = 10$ (as shown in Fig. 7.33a). For the *h*-version finite element method, three-dimensional elements in ABAQUS were used for the computation. Figure 7.33b displays one of the meshes used in the ABAQUS analysis. In this example, the load was applied uniformly through 50 incremental steps.

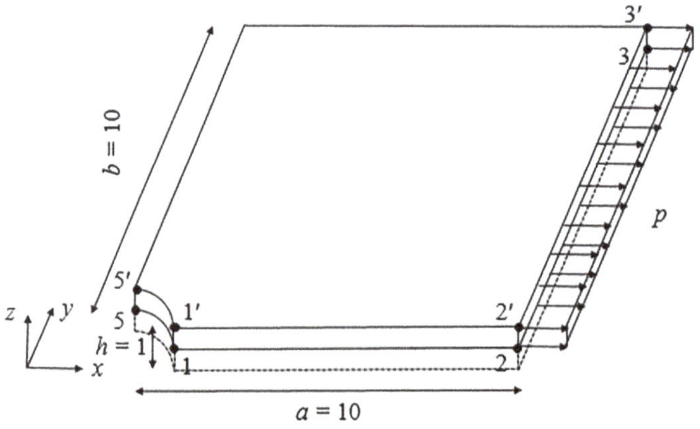

Fig. 7.32 One eighth (1/8) of a thick plate with a central hole under uniform load

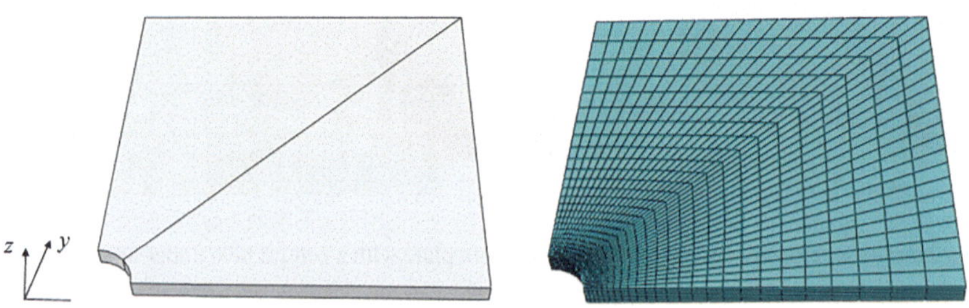

Fig. 7.33 Mesh generation of a thick plate with a central hole under uniform load

Table 7.5 Calculation results for the thick plate with a central hole under elastoplastic conditions using the HQEM and *h*-version finite element method

Method	Node ($l \times m \times n$)	DOFs	u_y at node 5
HQEM	4 * 4 * 4	1188	−0.003048
	5 * 5 * 5	1911	−0.002729
	6 * 6 * 6	2474	−0.002474
	7 * 7 * 7	4131	−0.002289
	8 * 8 * 8	5700	−0.002195
	9 * 9 * 9	7623	−0.002173
	10 * 10 * 8	7008	−0.002180
ABAQUS		1584	−0.002530
		2970	−0.002471
		4788	−0.002369
		6048	−0.002323
		8100	−0.002297
		11,718	−0.002263
		130,174	−0.002167

The computational results for the elastic–plastic problem of a thick plate with a central hole under uniformly distributed load, obtained using the HQEM and the *h*-version finite element method (ABAQUS) at different degrees of freedom, are listed in Table 7.5. Taking the ABAQUS result at 130,174 degrees of freedom as the reference solution, the relative errors of both methods at various degrees of freedom are plotted on a double logarithmic scale in Fig. 7.34. It can be seen that the HQEM yields more accurate results for three-dimensional elastic–plastic problems, and its convergence rate is significantly faster than that of the *h*-version finite element method (achieving two significant digits of convergence at only 5,700 degrees of freedom).

The comparison between the *y*-direction displacement at node 5 of a thick plate with a central hole under uniform load, calculated using the HQEM, and the reference solution (obtained by the *h*-version finite element method with 130 and 174 degrees of freedom) is shown in Fig. 7.35. As can be seen from the figure, although the HQEM uses relatively few degrees of freedom, the displacements computed by HQEM are in good agreement with the reference solution.

7.3.5 Summary of This Section

Building upon the elastoplastic numerical solution methods discussed in the previous section, this section selects three benchmark problems: a thick-walled cylinder under internal pressure in plane strain, a thin plate with a central hole in plane stress, and a thick plate with a central hole. The HQEM is applied to solve two-dimensional and three-dimensional elastoplastic problems. The numerical results are compared with

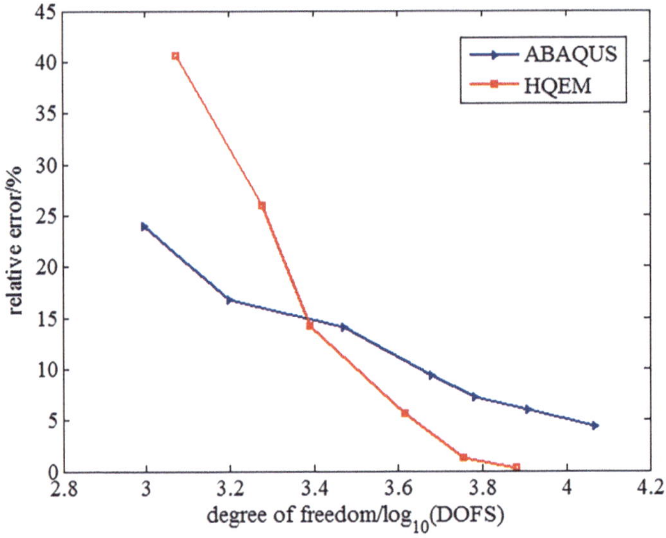

Fig. 7.34 Comparison of convergence rates between the hierarchical quadrature element method (HQEM) and the h-version finite element method in the elastoplastic analysis of a thick plate with a central hole under uniform load

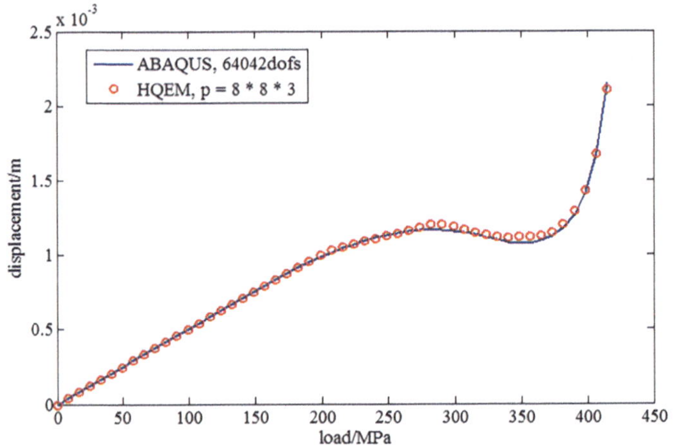

Fig. 7.35 Comparison of the computed displacement u_y at node 5 using the HQEM and the h-version finite element method

analytical solutions (for the thick-walled cylinder under internal pressure) and with solutions obtained using the h-version finite element method (via ABAQUS). In all cases, the hierarchical quadrature element method demonstrates a faster convergence rate, achieving accurate results with significantly fewer degrees of freedom. Since the computational cost of iterative solutions for elastoplastic problems is substantial—primarily determined by the number of degrees of freedom (which dictates the size of the stiffness matrix, among others) and the number of integration points (as nonlinear isotropic hardening requires nonlinear iteration at each integration point to solve for the plastic consistency parameter)—the advantages exhibited by the hierarchical quadrature element method are particularly beneficial for the efficient and accurate solution of elastoplastic problems.

7.4 Summary

This chapter discusses the finite element analysis of elastoplastic problems, which correspond to material nonlinearity. Elastoplastic problems are considered as strong nonlinearity because their response depends on the loading history and the material state may change abruptly. The permanent material dislocations that occur during plastic deformation are represented by the evolution of internal plastic variables. In large deformation problems, both material and geometric nonlinearities are present, making the problem even more challenging to solve.

This chapter first introduces the one-dimensional elastoplastic problem based on the linear hardening model and the small deformation assumption, discussing two different hardening models: isotropic hardening and kinematic hardening. The former increases the size of the elastic domain, while the latter maintains the size of the elastic domain but shifts its center. Under the small strain assumption, the total strain is decomposed into elastic strain and plastic strain, with only the elastic strain being related to stress. Plastic deformation depends on the loading history and is stored in the plastic strain. The determination of the stress state is based on elastic prediction and plastic return mapping. In the elastic trial state, it is assumed that the strain increment is elastic, and the stress increases accordingly. If the predicted stress exceeds the elastic range (i.e., surpasses the current yield stress), it is returned to the yield stress. The plastic strain increment is identified during this return mapping process.

In a multiaxial stress state, it is impractical to test all possible stress combinations. By employing equivalent stress, uniaxial tensile test data can still be used for failure analysis under multiaxial stress conditions. Since failure criteria should be independent of the coordinate system, they can be defined using invariants. The Tresca criterion uses the maximum shear stress, while the Von Mises criterion employs the second invariant (J_2) of the deviatoric stress. As in the uniaxial case, the algorithm consists of an elastic predictor and a plastic return mapping. For the Von Mises criterion, the return mapping occurs in the radial direction of the deviatoric stress. When the stress is restored from the trial state, the yield surface evolves simultaneously. The final return mapping point is determined by the plastic consistency condition. Results indicate that, with linear hardening, this return mapping point can be explicitly determined. Otherwise, the local Newton–Raphson method is required to find the return mapping point. Due to the inconsistency between the consistent tangent stiffness and the finite time step of the integration, the convergence iteration does not exhibit quadratic convergence. To ensure quadratic convergence of the Newton–Raphson iteration, the algorithmic tangent stiffness consistent with the return mapping algorithm is obtained by differentiating the time integration algorithm with respect to the finite time step.

Subsequently, the hierarchical quadrature element method (HQEM) based on the J_2 flow theory with isotropic hardening was applied to analyze two-dimensional and three-dimensional elastoplastic problems. The backward Euler time integration method was employed for the rate form of the classical elastoplastic model, introducing the return mapping algorithm and determining the plastic consistency parameter to address J_2 plasticity. The local return mapping algorithm for pointwise integration of the constitutive equations was embedded within the global Newton–Raphson iteration.

Building on the research of numerical methods for elastoplasticity, this chapter presents three example problems: a plane strain problem, a plane stress problem, and a three-dimensional elastoplastic problem. Numerical solutions are implemented using the hierarchical quadrature element method. By comparing the results with analytical solutions and those obtained from the *h*-version finite element method, the superior convergence rate of the hierarchical quadrature element method is demonstrated. This method achieves high accuracy with very few degrees of freedom, which is particularly advantageous for reducing computational effort in elastoplastic problems that require extensive iterative solutions, thereby enabling efficient and accurate computation. Additionally, the method allows for local order elevation based on specific requirements, further reducing computational cost during the solution process. In summary, the research in this chapter demonstrates that the hierarchical quadrature element method offers unique advantages for solving elastoplastic problems, making it an efficient and highly accurate solution approach.

Chapter 8
Finite Element Analysis of Geometric Nonlinear Problems

In all previous discussions, we have assumed that deformations remain small, allowing the use of linear relationships to describe strain within the material. In reality, however, deformations may become significant during loading, making it necessary to distinguish between the known reference configuration of the body's initial shape and the current (deformed) configuration after loading. It is important to note that the deformed configuration of the body is unknown at the start of the analysis and must therefore be determined as part of the solution process—this procedure is inherently nonlinear. The relationships describing the finite deformation behavior of solids involve equations associated with both the reference and deformed configurations. These relationships can be expressed using index notation or in matrix form (see Chap. 1); however, it is ultimately necessary to return to the matrix form to construct the finite element approximation.

This chapter first introduces the fundamental kinematic relations used to describe finite deformations in solids, then summarizes the various stress and external force descriptions associated with the reference and deformed configurations, the specification of boundary and initial conditions, and provides a brief overview of constitutive models for finite elastic solids. The variational Galerkin formulation for finite elastic materials is presented in the reference configuration. Subsequently, the index or tensor notation of the variational problem is converted into matrix form, and the standard finite element solution procedure is outlined. Up to this point, the process is based on equations associated with the reference configuration. Finally, the relevant equations are transformed into forms associated with the current configuration, demonstrating that, in the current configuration, the finite element formulation still allows for a separation suitable for handling nearly incompressible cases.

This chapter develops a geometric nonlinear analysis method for the HQEM. Within the framework of the total Lagrangian description, three types of nonlinear elements are constructed: two-dimensional, three-dimensional, and Mindlin hierarchical quadrature elements. The virtual work equation is formulated based on the Green–Lagrange strain tensor and the second Piola–Kirchhoff stress tensor in

B. Liu and C. Liu, *Introduction to Linear and Nonlinear Finite Element Analysis*,
https://doi.org/10.1007/978-981-95-9670-6_8

the reference configuration, and discretized in the spatial domain. In the context of two-dimensional elasticity theory, quadrilateral hierarchical quadrature elements are constructed, primarily for the analysis of large in-plane deformations. Under three-dimensional elasticity theory, hexahedral hierarchical quadrature elements are developed, mainly for the geometric nonlinear analysis of complex three-dimensional structures such as spatial curved beams and shells. Based on Mindlin shell theory, Mindlin hierarchical quadrature elements are constructed, which are suitable for the geometric nonlinear analysis of shallow shells. All constructed elements employ non-uniformly distributed Gauss–Lobatto nodes. The various types of elements are analyzed using two nonlinear solution techniques: the Newton–Raphson method and the arc-length method, for geometric nonlinear quasi-static analysis. The results are first compared with those obtained from the commercial software ABAQUS to verify the correctness and accuracy of the proposed method, followed by a convergence analysis, demonstrating the computational efficiency of the method. Some numerical examples are calculated using both single elements and assemblies of multiple elements to verify the easiness of assembly of the proposed method. Compared to lower-order elements, the hierarchical quadrature element method requires significantly fewer degrees of freedom for the analysis of equivalent problems, showing great potential for application in nonlinear finite element analysis where iterative updating of the stiffness matrix is required.

8.1 Measures of Stress and Strain

In this section, we will provide an appropriate description of the statics and kinematics of continuous media under large deformation. The term "large deformation" may seem somewhat imprecise. For example, in the cantilever beam structure shown in Fig. 8.1, by increasing the stiffness EI of the beam, the internal strain of the beam can be made arbitrarily small. However, even if a beam possesses extremely high stiffness EI (and thus extremely small strain), as long as its length is sufficiently large, significant tip displacement and rotation can still occur. Clearly, large strain and large displacement (or large displacement gradient) do not refer to the same concept. Large strain occurs if and only if large displacement phenomena take place, but conversely, the opposite is not true: in many structures, significant displacements can be observed, yet the strain remains very limited (for example, below 2%).

In practice, for most engineering materials, the strain is relatively small. This section adopts the assumption of large deformation with small strain; therefore, no approximations are made to the kinematics and statics of large-deformation continua. Commonly used constitutive relations (such as Hooke's law for linear elastic materials) are still considered valid under this assumption.

In a continuum, the measurement of stress and strain must be defined in a clear and physically meaningful manner. The most straightforward approach is to consider the total motion of a small cubic material element as a combination of translation, rigid-body rotation, and pure deformation. Suppose a small material element occupies the

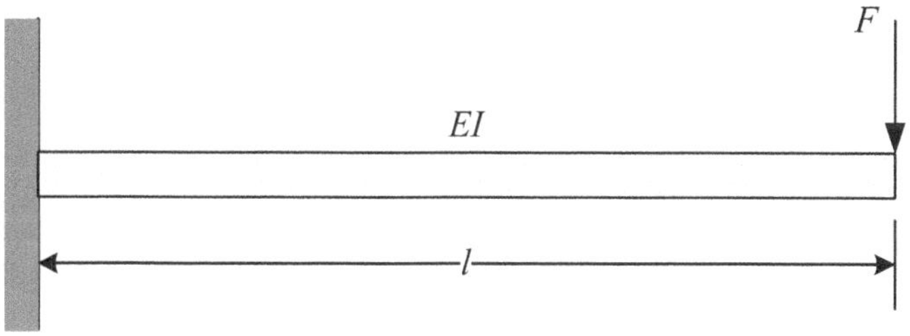

Fig. 8.1 A cantilever beam structure

position ξ in space in the reference configuration (typically the undeformed state), which is referred to as the material coordinates (ξ_1, ξ_2, ξ_3). In Cartesian coordinates, the position vector in the reference configuration can be expressed in terms of its components as follows:

$$\boldsymbol{\xi} = \xi_j \mathbf{E}_j, \quad j = 1,\ 2,\ 3 \tag{8.1}$$

Here, $\boldsymbol{E}_j$ denotes the unit orthogonal basis vectors, and the summation convention applies to repeated indices of the same type (such as j). The position occupied by this infinitesimal element in its current configuration (usually the deformed configuration) is denoted by $\boldsymbol{x}$, that is, the so-called spatial coordinates (x_1, x_2, x_3). In Cartesian coordinates, the position vector in the current configuration can be expressed in terms of its components as follows:

$$\boldsymbol{x} = x_i \boldsymbol{e}_i, \quad i = 1,\ 2,\ 3 \tag{8.2}$$

Here, $\boldsymbol{ei}$ denotes the unit basis vector at the current time t, and the summation convention is applied again. In the context of large deformation problems, the former coordinate system is typically referred to as the Lagrangian coordinate system, while the latter is known as the Eulerian coordinate system. Clearly, the two are not independent: $\boldsymbol{x} = \boldsymbol{x}(\xi)$, or equivalently

$$x_i = \phi_i\left(\xi_j, t\right) \tag{8.3}$$

For any solution method, determining ϕ_i is one of the steps; it is analogous to the displacement vector to be introduced next. When the coordinate systems share a common origin and orientation, the displacement vector can be introduced to represent the transformation between the two coordinate systems.

In differential operations, translational motion can be neglected. Differentiating spatial coordinates with respect to material coordinates yields the deformation gradient tensor.

$$\mathbf{F} = \frac{\partial \mathbf{x}}{\partial \boldsymbol{\xi}} \tag{8.4}$$

or can be expressed in the following form

$$F_{ij} = \frac{\partial x_i}{\partial \xi_j} = \frac{\partial \phi_i}{\partial \xi_j} \tag{8.5}$$

The deformation gradient tensor is a second-order tensor that fully characterizes the motion of an infinitesimal material element. For the quadrilateral element shown in Fig. 8.2, the total deformation is decomposed into a pure deformation $\boldsymbol{U}$ (from configuration A to configuration B) and a rigid-body rotation $\boldsymbol{R}$ (from configuration B to configuration C). In the reference configuration, a line element $\mathrm{d}\boldsymbol{\xi}$ formed by two infinitesimally close material points is first deformed into $\mathrm{d}\boldsymbol{\eta}$ in the intermediate configuration, and finally into $\mathrm{d}\boldsymbol{x}$ in the current configuration. Therefore, the relationship between $\mathrm{d}\boldsymbol{x}$ and $\mathrm{d}\boldsymbol{\eta}$ is given by

$$\mathrm{d}\boldsymbol{x} = \boldsymbol{R} \cdot \mathrm{d}\boldsymbol{\eta} \tag{8.6}$$

The relationship between $\mathrm{d}\boldsymbol{\eta}$ and $\mathrm{d}\boldsymbol{\xi}$ can be expressed as

$$\mathrm{d}\boldsymbol{\eta} = \boldsymbol{U} \cdot \mathrm{d}\boldsymbol{\xi} \tag{8.7}$$

Combining Eqs. (8.6) and (8.7), we obtain

$$\mathrm{d}\boldsymbol{x} = \boldsymbol{R} \cdot \mathrm{d}\boldsymbol{\eta} = \boldsymbol{R} \cdot \boldsymbol{U} \cdot \mathrm{d}\boldsymbol{\xi} \tag{8.8}$$

By comparing with Eq. (8.4), it can be observed that the deformation gradient $\boldsymbol{F}$ can be decomposed into a pure rotation $\boldsymbol{R}$ and a pure deformation $\boldsymbol{U}$, that is,

$$\boldsymbol{F} = \boldsymbol{R} \cdot \boldsymbol{U} \tag{8.9}$$

This decomposition of motion is commonly referred to as the polar decomposition of the deformation gradient, as illustrated in Fig. 8.2.

As a measure of deformation, the right stretch tensor $\boldsymbol{U}$ is not very convenient to use, as its computation involves the resource-intensive operation of taking a square root. A relatively easier deformation measure to compute is the right Cauchy-Green

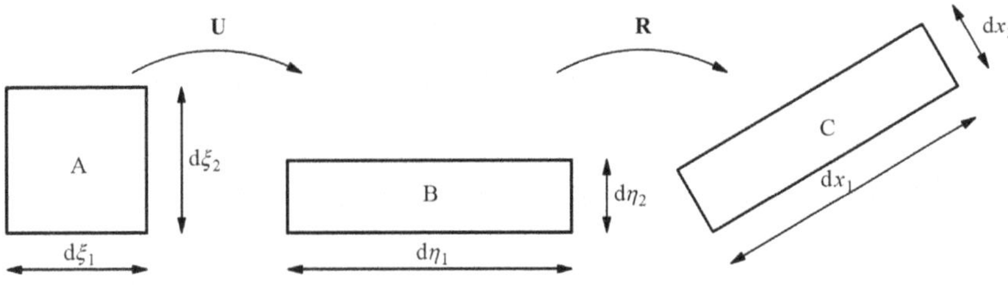

Fig. 8.2 Polar decomposition of the deformation gradient of an infinitesimal element

deformation tensor $\boldsymbol{C}$, which can be obtained by calculating the squared difference between $\mathrm{d}\xi$ and $\mathrm{d}\boldsymbol{x}$, that is,

$$\mathrm{d}\boldsymbol{x} \cdot \mathrm{d}\boldsymbol{x} - \mathrm{d}\boldsymbol{\xi} \cdot \mathrm{d}\boldsymbol{\xi} = (\mathbf{F} \cdot \mathrm{d}\boldsymbol{\xi}) \cdot (\mathbf{F} \cdot \mathrm{d}\boldsymbol{\xi}) - \mathrm{d}\boldsymbol{\xi} \cdot \mathrm{d}\boldsymbol{\xi} = \mathrm{d}\boldsymbol{\xi} \cdot \left(\mathbf{F}^{\mathrm{T}}\mathbf{F} - \mathbf{I}\right) \cdot \mathrm{d}\boldsymbol{\xi} \tag{8.10}$$

Since the right stretch tensor $\boldsymbol{U}$ is a measure of pure deformation, the right Cauchy-Green deformation tensor can likewise describe the complete state of deformation while excluding the effects of rigid body rotation. It can be expressed as

$$\boldsymbol{C} = \boldsymbol{F}^{\mathrm{T}} \cdot \boldsymbol{F} = \boldsymbol{U}^2 \tag{8.11}$$

Thus, the Green–Lagrange strain tensor is defined as

$$\gamma = \frac{1}{2}(\boldsymbol{C} - \boldsymbol{I}) \tag{8.12}$$

Equations (8.4), (8.11), and (8.12) indicate that the Green–Lagrange strain tensor γ is defined with respect to the reference configuration (the undeformed configuration). This measure of deformation is based on the Lagrangian description of the motion of the body, and it automatically reduces to the Eulerian strain tensor $\boldsymbol{\varepsilon}$ when the displacement gradients are small.

In practical applications, the displacement vector $\boldsymbol{u}$ is typically used in the current configuration instead of the vector $\boldsymbol{x}$, which contains the spatial coordinates of the material points. However, it is clear that the two are related by $\mathbf{x} = \xi + \boldsymbol{u}$, thus we have

$$\boldsymbol{F} = \boldsymbol{I} + \frac{\partial \boldsymbol{u}}{\partial \boldsymbol{\xi}} \tag{8.13}$$

or expressed in component form as

$$F_{ij} = \frac{\partial(\xi_i + u_i)}{\partial \xi_j} = \delta_{ij} + \frac{\partial u_i}{\partial \xi_j} \tag{8.14}$$

Thus, the components of the Green–Lagrange strain tensor can be expressed as

$$\gamma_{ij} = \frac{1}{2}\left(F_{ki}F_{kj} - \delta_{ij}\right) \tag{8.15}$$

Substituting Eq. (8.14) into Eq. (8.15), we obtain

$$\gamma_{ij} = \frac{1}{2}\left(\frac{\partial u_i}{\partial \xi_j} + \frac{\partial u_j}{\partial \xi_i}\right) + \frac{1}{2}\frac{\partial u_k}{\partial \xi_i}\frac{\partial u_k}{\partial \xi_j} \tag{8.16}$$

When the displacement gradient $\boldsymbol{F}$ is small, the quadratic terms in the definition of the strain tensor (8.16) can be neglected, reducing it to the classical linear strain tensor expression. This is also the reason for introducing the factor 1/2 in the definition of the Green–Lagrange strain tensor: without this factor, omitting the quadratic terms would not yield the linear form of the strain tensor. For convenience, the incremental form of the Green–Lagrange strain tensor can be expressed as follows:

$$\Delta\boldsymbol{\gamma} = \frac{1}{2}\left(\boldsymbol{F}^{\mathrm{T}} \cdot \nabla_0(\Delta\boldsymbol{u}) + \nabla_0(\Delta\boldsymbol{u})^{\mathrm{T}} \cdot \boldsymbol{F}\right) + \frac{1}{2}(\nabla_0(\Delta\boldsymbol{u}))^{\mathrm{T}} \cdot \nabla_0(\Delta\boldsymbol{u}) \tag{8.17}$$

The incremental form of the Green–Lagrange strain tensor is obtained by taking the difference between $\boldsymbol{\gamma}^{t+\Delta t}$ and $\boldsymbol{\gamma}^{t}$. It should be noted that in the gradient notation Δ_0, the subscript 0 indicates that the gradient is taken with respect to the material coordinates $\boldsymbol{\xi}$, while the absence of a subscript indicates that the gradient is taken with respect to the spatial coordinates $\boldsymbol{x}$.

Stress is defined as the load per unit area. When the displacement gradient is relatively small compared to the whole, it does not matter whether the unit area bearing the load is measured in the reference configuration or the current configuration. However, in cases of large deformation, it is essential to specify in which configuration the stress is defined. In engineering practice, it is usually desirable to know the magnitude of the stress in the current configuration, that is, the current load divided by the current load-bearing area. This type of stress, also known as "true stress," is referred to as the Cauchy stress tensor.

In solid mechanics, the most common approach to describing the statics and kinematics of structures under large displacements is the Lagrangian description: all kinematic and static quantities must be referenced to a certain prior configuration. Since the Cauchy stress tensor is defined on the current, generally unknown, deformed configuration, it is necessary in the Lagrangian description to introduce an auxiliary stress measure defined with respect to the reference configuration. A commonly used auxiliary stress tensor in nonlinear analysis is the second Piola–Kirchhoff stress tensor $\boldsymbol{\tau}$, which is related to the Cauchy (true) stress tensor $\boldsymbol{\sigma}$ by:

$$\sigma = \frac{\rho}{\rho_0}\boldsymbol{F} \cdot \tau \cdot \boldsymbol{F}^{\mathrm{T}} \tag{8.18}$$

or, expressed in component form

$$\sigma_{ij} = \frac{\rho}{\rho_0} F_{ik}\tau_{kl}F_{jl} \tag{8.19}$$

where ρ and ρ_0 denote the mass density of the body in the current configuration and the reference configuration, respectively. The second Piola–Kirchhoff stress tensor does not have a direct physical interpretation. If the true magnitude of stress needs to be determined in nonlinear analysis, the Cauchy stress must be computed from the second Piola–Kirchhoff stress. For small displacement gradients $\rho \approx \rho_0$, $\boldsymbol{F} \approx \boldsymbol{I}$, so the Cauchy stress and the second Piola–Kirchhoff stress are consistent with each other.

The stress measures used in nonlinear finite element formulations must be defined in the reference configuration, and all quantities appearing in the principle of virtual work must also be based on the reference configuration. However, the principle of virtual work is often defined in the current configuration. Therefore, it is necessary to transform the virtual work equation into an expression that is integrated solely over the reference configuration. According to the law of mass conservation for an infinitesimal element, we have

$$\rho \mathrm{d}V = \rho_0 \mathrm{d}V_0 \tag{8.20}$$

Moreover

$$\det \boldsymbol{F} = \frac{\mathrm{d}x\mathrm{d}y\mathrm{d}z}{\mathrm{d}\xi\mathrm{d}\eta\mathrm{d}\zeta} = \frac{\mathrm{d}V}{\mathrm{d}V_0} \tag{8.21}$$

By comparing Eqs. (8.20) and (8.21), we obtain

$$\det \boldsymbol{F} = \frac{\rho_0}{\rho} \tag{8.22}$$

The variation of internal energy or internal virtual work can be expressed as

$$\delta W_{\text{int}} = \int_V \delta\boldsymbol{\varepsilon} : \boldsymbol{\sigma} \mathrm{d}V \tag{8.23}$$

By applying the law of mass conservation (Eq. (8.20)), and transforming the domain of integration, we obtain

$$\delta W_{\text{int}} = \int_{V_0} \frac{\rho_0}{\rho} \delta\boldsymbol{\varepsilon} : \boldsymbol{\sigma} \mathrm{d}V \tag{8.24}$$

Here, $\delta\boldsymbol{\varepsilon}$ and $\boldsymbol{\sigma}$ are functions defined on the current configuration. To express them in terms of quantities defined on the reference configuration, it is necessary to establish the relationship between the virtual strain field $\delta\boldsymbol{\varepsilon}$ and the virtual displacement field $\delta\boldsymbol{u}$. According to the definition of the Green–Lagrange strain tensor, its variation is given by

$$\delta\boldsymbol{\gamma} = \frac{1}{2}\delta\left(\boldsymbol{F}^{\mathrm{T}} \cdot \boldsymbol{F} - \boldsymbol{I}\right) = \frac{1}{2}\left(\delta\boldsymbol{F}^{\mathrm{T}} \cdot \boldsymbol{F} + \boldsymbol{F}^{\mathrm{T}} \cdot \delta\boldsymbol{F}\right) \tag{8.25}$$

or expressed in component form as

$$\begin{aligned}
\delta\gamma_{ij} &= \frac{1}{2}\left(\frac{\partial\delta u_k}{\partial\xi_i}\frac{\partial x_k}{\partial\xi_j} + \frac{\partial x_k}{\partial\xi_i}\frac{\partial\delta u_k}{\partial\xi_j}\right) \\
&= \frac{1}{2}\left(\frac{\partial\delta u_k}{\partial x_l}\frac{\partial x_l}{\partial\xi_i}\frac{\partial x_k}{\partial\xi_j} + \frac{\partial x_k}{\partial\xi_i}\frac{\partial\delta u_k}{\partial x_l}\frac{\partial x_l}{\partial\xi_j}\right)
\end{aligned}$$

$$= \frac{1}{2} F_{kj} \left(\frac{\partial \delta u_k}{\partial x_l} + \frac{\partial \delta u_l}{\partial x_k} \right) F_{li} \tag{8.26}$$

Therefore

$$\delta \boldsymbol{\gamma} = \boldsymbol{F}^{\mathrm{T}} \delta \boldsymbol{\varepsilon} \boldsymbol{F} \tag{8.27}$$

Equation (8.27) establishes the transformation relationship between the Green–Lagrange strain tensor defined in the reference configuration, and the strain tensor $\delta\varepsilon$, defined in the current configuration. Substituting the definition of the Green–Lagrange strain tensor yields

$$\delta \boldsymbol{\varepsilon} = \frac{1}{2} \left(\delta \boldsymbol{F}^{\mathrm{T}} + \delta \boldsymbol{F} \right) \tag{8.28}$$

or expressed in component form as

$$\delta \varepsilon_{ij} = \frac{1}{2} \left(\frac{\partial \delta u_j}{\partial x_i} + \frac{\partial \delta u_i}{\partial x_j} \right) \tag{8.29}$$

Substituting Eq. (8.28) into Eq. (8.24) yields

$$\delta W_{\text{int}} = \int_{V_0} \frac{\rho_0}{\rho} \nabla(\delta \boldsymbol{u}) : \boldsymbol{\sigma} \mathrm{d}V \tag{8.30}$$

Substituting the relationship between the Cauchy strain tensor and the second Piola–Kirchhoff strain tensor (Eq. (8.18)) into Eq. (8.30), we obtain

$$\int_{V_0} \frac{\rho_0}{\rho} \nabla(\delta \boldsymbol{u}) : \boldsymbol{\sigma} \mathrm{d}V = \int_{V_0} \mathrm{tr}\left(\nabla(\delta \boldsymbol{u}) \cdot \boldsymbol{F} \cdot \boldsymbol{\tau} \cdot \boldsymbol{F}^{\mathrm{T}} \right) \mathrm{d}V = \int_{V_0} \mathrm{tr}\left(\delta \boldsymbol{F} \cdot \boldsymbol{\tau} \cdot \boldsymbol{F}^{\mathrm{T}} \right) \mathrm{d}V \tag{8.31}$$

Using the variational definition of the Green–Lagrange strain in the reference configuration (Eq. (8.25)) and the symmetry of the second Piola–Kirchhoff stress tensor, Eq. (8.31) can be rewritten as

$$\delta W_{\text{int}} = \int_{V_0} \delta \boldsymbol{\gamma} : \boldsymbol{\tau} \mathrm{d}V \tag{8.32}$$

This result indicates that the second Piola–Kirchhoff stress tensor and the Green–Lagrange strain tensor are energetically conjugate. Moreover, since the former uses the configuration from a previous step as the reference configuration, it naturally becomes a more convenient stress measure within the Lagrangian description framework.

8.2 Formulation of Nonlinear Finite Elements in Continuum Geometry

In this section, the finite element formulation for continuum geometry nonlinear problems will be derived based on the total Lagrangian formulation. According to Eq. (8.32), the virtual work in the reference configuration, expressed using matrices and vectors, is given by

$$\int_{V_0} \delta\boldsymbol{\gamma}^{\mathrm{T}}\boldsymbol{\tau}^{t+\Delta t}\mathrm{d}V = \int_{S_0} \delta\boldsymbol{u}^{\mathrm{T}}\boldsymbol{t}_0\mathrm{d}S + \int_{V_0} \rho_0\delta\boldsymbol{u}^{\mathrm{T}}\boldsymbol{g}\mathrm{d}V \tag{8.33}$$

Here, t_0 denotes the nominal stress, which is defined as the ratio of the external load to the surface area in the undeformed configuration. The unknown stress $\boldsymbol{\tau}^{t+\Delta t}$ is decomposed into the known stress $\boldsymbol{\tau}^{t}$ from the previous loading step and the stress increment $\Delta\boldsymbol{\tau}$. Substituting this decomposition into the virtual work Eq. (8.33) (which is based on a previous configuration but remains valid at the current time $t + \Delta t$) yields

$$\int_{V_0} \delta\boldsymbol{\gamma}^{\mathrm{T}}\Delta\boldsymbol{\tau}\mathrm{d}V + \int_{V_0} \delta\boldsymbol{\gamma}^{\mathrm{T}}\boldsymbol{\tau}^{t}\mathrm{d}V = \int_{S_0} \delta\boldsymbol{u}^{\mathrm{T}}\boldsymbol{t}_0\mathrm{d}S + \int_{V_0} \rho_0\delta\boldsymbol{u}^{\mathrm{T}}\boldsymbol{g}\mathrm{d}V \tag{8.34}$$

For small strains, it can be assumed that there is a linear relationship between the stress increment $\Delta\boldsymbol{\tau}$ and the strain increment $\Delta\gamma$, that is,

$$\Delta\boldsymbol{\tau} = \boldsymbol{D}\Delta\boldsymbol{\gamma} \tag{8.35}$$

Here, the matrix $\boldsymbol{D}$ contains the instantaneous stiffness modulus of the material model. For linear elastic materials, Eq. (8.35) reduces to Hooke's law. Substituting into Eq. (8.34) yields

$$\int_{V_0} \delta\boldsymbol{\gamma}^{\mathrm{T}}\boldsymbol{D}\Delta\boldsymbol{\gamma}\mathrm{d}V + \int_{V_0} \delta\boldsymbol{\gamma}^{\mathrm{T}}\boldsymbol{\tau}^{t}\mathrm{d}V = \int_{S_0} \delta\boldsymbol{u}^{\mathrm{T}}\boldsymbol{t}_0\mathrm{d}S + \int_{V_0} \rho_0\delta\boldsymbol{u}^{\mathrm{T}}\boldsymbol{g}\mathrm{d}V \tag{8.36}$$

Unlike the constitutive relations of linear elastic materials, in large deformation problems, the strain increment $\Delta\gamma$ includes both linear and quadratic terms of the displacement increment $\Delta\boldsymbol{u}$ (see Eq. (8.17)), and can be expressed as follows:

$$\Delta\boldsymbol{\gamma} = \Delta\boldsymbol{e} + \Delta\boldsymbol{\eta} \tag{8.37}$$

Here, $\Delta\boldsymbol{e}$ is the linear component of the strain increment $\Delta\gamma$ with respect to the displacement increment $\Delta\boldsymbol{u}$, while $\Delta\boldsymbol{\eta}$ is the nonlinear component of the strain increment with respect to the displacement increment $\Delta\boldsymbol{u}$. Based on the variation $\delta\gamma^{t+\Delta t} = \delta\Delta\gamma$, substituting the above decomposition into the virtual work equation yields

$$\int_{V_0} \delta\Delta e^{\mathrm{T}} D \Delta e \mathrm{d}V + \int_{V_0} \delta\Delta e^{\mathrm{T}} D \Delta \eta \mathrm{d}V + \int_{V_0} \delta\Delta \eta^{\mathrm{T}} D \Delta e \mathrm{d}V + \int_{V_0} \delta\Delta \eta^{\mathrm{T}} D \Delta \eta \mathrm{d}V +$$
$$\int_{V_0} \delta\eta^{\mathrm{T}} \tau' \mathrm{d}V = \int_{S_0} \delta u^{\mathrm{T}} t_0 \mathrm{d}S + \int_{V_0} \rho_0 \delta u^{\mathrm{T}} g \mathrm{d}V - \int_{V_0} \delta e^{\mathrm{T}} \tau' \mathrm{d}V \tag{8.38}$$

Here, the term $\int (\delta\Delta \mathbf{e})^{\mathrm{T}} \tau' \mathrm{d}V$ is of zero order relative to the displacement increment Δu so it is moved to the right side of the equation to form the internal force vector. Since only the terms linear in the displacement increment Δu can be used to derive the tangent stiffness matrix, the terms that are nonlinear with respect to Δu (the second, third, and fourth terms on the left-hand side of the equation) must be linearized, that is,

$$\int_{V_0} \delta\Delta e^{\mathrm{T}} D \Delta e \mathrm{d}V + \int_{V_0} \delta\eta^{\mathrm{T}} \tau' \mathrm{d}V = \int_{S_0} \delta u^{\mathrm{T}} t_0 \mathrm{d}S + \int_{V_0} \rho_0 \delta u^{\mathrm{T}} g \mathrm{d}V - \int_{V_0} \delta e^{\mathrm{T}} \tau' \mathrm{d}V \tag{8.39}$$

The linear relationship between the linear part of the strain increment Δe and the displacement increment Δu is given by

$$\Delta \mathbf{e} = \mathbf{L} \Delta \mathbf{u} \tag{8.40}$$

The matrix L must be constructed according to the specific strain measure and strain type. For example, for the Green–Lagrange strain in a planar configuration, it is defined as follows:

$$\mathbf{L} = \begin{bmatrix} F_{11}\frac{\partial}{\partial \xi_1} & F_{21}\frac{\partial}{\partial \xi_1} \\ F_{12}\frac{\partial}{\partial \xi_2} & F_{22}\frac{\partial}{\partial \xi_2} \\ F_{11}\frac{\partial}{\partial \xi_2} + F_{12}\frac{\partial}{\partial \xi_1} & F_{21}\frac{\partial}{\partial \xi_2} + F_{22}\frac{\partial}{\partial \xi_1} \end{bmatrix} \tag{8.41}$$

The last row indicates that the matrix L uses engineering shear strains, which are twice the tensor form of Eq. (8.17). Additionally, for a three-dimensional configuration, the matrix L is expressed as:

$$L = \begin{bmatrix} F_{11}\frac{\partial}{\partial \xi_1} & F_{21}\frac{\partial}{\partial \xi_1} & F_{31}\frac{\partial}{\partial \xi_1} \\ F_{12}\frac{\partial}{\partial \xi_2} & F_{22}\frac{\partial}{\partial \xi_2} & F_{32}\frac{\partial}{\partial \xi_2} \\ F_{13}\frac{\partial}{\partial \xi_3} & F_{23}\frac{\partial}{\partial \xi_3} & F_{33}\frac{\partial}{\partial \xi_3} \\ F_{11}\frac{\partial}{\partial \xi_2} + F_{12}\frac{\partial}{\partial \xi_1} & F_{21}\frac{\partial}{\partial \xi_2} + F_{22}\frac{\partial}{\partial \xi_1} & F_{31}\frac{\partial}{\partial \xi_2} + F_{32}\frac{\partial}{\partial \xi_1} \\ F_{12}\frac{\partial}{\partial \xi_3} + F_{13}\frac{\partial}{\partial \xi_2} & F_{22}\frac{\partial}{\partial \xi_3} + F_{23}\frac{\partial}{\partial \xi_2} & F_{32}\frac{\partial}{\partial \xi_3} + F_{33}\frac{\partial}{\partial \xi_2} \\ F_{13}\frac{\partial}{\partial \xi_1} + F_{11}\frac{\partial}{\partial \xi_3} & F_{23}\frac{\partial}{\partial \xi_1} + F_{21}\frac{\partial}{\partial \xi_3} & F_{33}\frac{\partial}{\partial \xi_1} + F_{31}\frac{\partial}{\partial \xi_3} \end{bmatrix} \tag{8.42}$$

The continuum finite element method requires interpolation of the continuous displacement field. The matrix H, which contains the interpolation functions $h1$, $h2$,

$h3$, $h4$, ... is defined as follows:

$$\boldsymbol{H} = \begin{bmatrix} h_1 & 0 & 0 & h_2 & 0 & 0 & \cdots & h_n & 0 & 0 \\ 0 & h_1 & 0 & 0 & h_2 & 0 & \cdots & 0 & h_n & 0 \\ 0 & 0 & h_1 & 0 & 0 & h_2 & \cdots & 0 & 0 & h_n \end{bmatrix} \tag{8.43}$$

The relationship between the continuous displacement field and the discrete nodal displacements can be expressed as $\boldsymbol{u} = \boldsymbol{Ha}$ or, in incremental form, $\Delta\boldsymbol{u} = \boldsymbol{H}\Delta\boldsymbol{a}$. Based on $\boldsymbol{BL} = \boldsymbol{LH}$, the relationship between the linear strain increment $\Delta\boldsymbol{e}$ and the continuous displacement field $\boldsymbol{u}$ is

$$\Delta\mathbf{e} = \mathbf{B}_L\Delta\mathbf{a} \tag{8.44}$$

where, for the planar solids, we have

$$\mathbf{B}_L = \begin{bmatrix} F_{11}\frac{\partial h_1}{\partial \xi_1} & F_{21}\frac{\partial h_1}{\partial \xi_1} & \cdots\cdots \\ F_{12}\frac{\partial h_1}{\partial \xi_2} & F_{22}\frac{\partial h_1}{\partial \xi_2} & \cdots\cdots \\ F_{11}\frac{\partial h_1}{\partial \xi_2} + F_{12}\frac{\partial h_1}{\partial \xi_1} & F_{21}\frac{\partial h_1}{\partial \xi_2} + F_{22}\frac{\partial h_1}{\partial \xi_1} & \cdots\cdots \end{bmatrix} \tag{8.45}$$

For the three-dimensional solids, we have

$$\mathbf{B}_L = \begin{bmatrix} F_{11}\frac{\partial h_1}{\partial \xi_1} & F_{21}\frac{\partial h_1}{\partial \xi_1} & F_{31}\frac{\partial h_1}{\partial \xi_1} & \cdots\cdots \\ F_{12}\frac{\partial h_1}{\partial \xi_2} & F_{22}\frac{\partial h_1}{\partial \xi_2} & F_{32}\frac{\partial h_1}{\partial \xi_2} & \cdots\cdots \\ F_{13}\frac{\partial h_1}{\partial \xi_3} & F_{23}\frac{\partial h_1}{\partial \xi_3} & F_{33}\frac{\partial h_1}{\partial \xi_3} & \cdots\cdots \\ F_{11}\frac{\partial h_1}{\partial \xi_2} + F_{12}\frac{\partial h_1}{\partial \xi_1} & F_{21}\frac{\partial h_1}{\partial \xi_2} + F_{22}\frac{\partial h_1}{\partial \xi_1} & F_{31}\frac{\partial h_1}{\partial \xi_2} + F_{32}\frac{\partial h_1}{\partial \xi_1} & \cdots\cdots \\ F_{12}\frac{\partial h_1}{\partial \xi_3} + F_{13}\frac{\partial h_1}{\partial \xi_2} & F_{22}\frac{\partial h_1}{\partial \xi_3} + F_{23}\frac{\partial h_1}{\partial \xi_2} & F_{32}\frac{\partial h_1}{\partial \xi_3} + F_{33}\frac{\partial h_1}{\partial \xi_2} & \cdots\cdots \\ F_{13}\frac{\partial h_1}{\partial \xi_1} + F_{11}\frac{\partial h_1}{\partial \xi_3} & F_{23}\frac{\partial h_1}{\partial \xi_1} + F_{21}\frac{\partial h_1}{\partial \xi_3} & F_{33}\frac{\partial h_1}{\partial \xi_1} + F_{31}\frac{\partial h_1}{\partial \xi_3} & \cdots\cdots \end{bmatrix} \tag{8.46}$$

Define the first term of the tangent stiffness matrix as follows:

$$\boldsymbol{K}_L = \int_{V_0} \boldsymbol{B}_L^{\mathrm{T}}\boldsymbol{D}\boldsymbol{B}_L \mathrm{d}V \tag{8.47}$$

Its construction is based on the following expression:

$$\int_{V_0} (\delta\Delta\boldsymbol{e})^{\mathrm{T}}\boldsymbol{D}\Delta\boldsymbol{e}\mathrm{d}V = (\delta\Delta\boldsymbol{a})^{\mathrm{T}}\boldsymbol{K}_L\Delta\boldsymbol{a} \tag{8.48}$$

The contribution of the second term in the tangent stiffness matrix (Eq. (8.39)) can be expressed as

$$\int_{V_0} (\delta\Delta\boldsymbol{\eta})^{\mathrm{T}}\boldsymbol{\tau}^t\mathrm{d}V = (\delta\Delta\boldsymbol{a})^{\mathrm{T}}\boldsymbol{K}_{NL}\Delta\boldsymbol{a} \tag{8.49}$$

Thus, the geometric contribution to the tangent stiffness matrix can be constructed as

$$\boldsymbol{K}_{NL} = \int_{V_0} \boldsymbol{B}_{NL}^{\mathrm{T}} \mathbf{T}^t \boldsymbol{B}_{NL} \mathrm{d}V \tag{8.50}$$

in which, the second Piola–Kirchhoff stress is expressed in matrix form. For planar solids, the matrices in Eq. (8.50) are given by

$$\mathbf{T} = \begin{bmatrix} \tau_{xx} & \tau_{xy} & 0 & 0 \\ \tau_{xy} & \tau_{yy} & 0 & 0 \\ 0 & 0 & \tau_{xx} & \tau_{xy} \\ 0 & 0 & \tau_{xy} & \tau_{yy} \end{bmatrix} \tag{8.51}$$

$$\mathbf{B}_{NL} = \begin{bmatrix} \frac{\partial h_1}{\partial \xi_1} & 0 & \frac{\partial h_2}{\partial \xi_1} & 0 & \cdots\cdots \\ \frac{\partial h_1}{\partial \xi_2} & 0 & \frac{\partial h_2}{\partial \xi_2} & 0 & \cdots\cdots \\ 0 & \frac{\partial h_1}{\partial \xi_1} & 0 & \frac{\partial h_2}{\partial \xi_1} & \cdots\cdots \\ 0 & \frac{\partial h_1}{\partial \xi_2} & 0 & \frac{\partial h_2}{\partial \xi_2} & \cdots\cdots \end{bmatrix} \tag{8.52}$$

For the three-dimensional solids, we have

$$\mathbf{T} = \begin{bmatrix} \tau_{xx} & \tau_{xy} & \tau_{zx} & 0 & 0 & 0 & 0 & 0 & 0 \\ \tau_{xy} & \tau_{yy} & \tau_{yz} & 0 & 0 & 0 & 0 & 0 & 0 \\ \tau_{zx} & \tau_{yz} & \tau_{zz} & 0 & 0 & 0 & 0 & 0 & 0 \\ 0 & 0 & 0 & \tau_{xx} & \tau_{xy} & \tau_{zx} & 0 & 0 & 0 \\ 0 & 0 & 0 & \tau_{xy} & \tau_{yy} & \tau_{yz} & 0 & 0 & 0 \\ 0 & 0 & 0 & \tau_{zx} & \tau_{yz} & \tau_{zz} & 0 & 0 & 0 \\ 0 & 0 & 0 & 0 & 0 & 0 & \tau_{xx} & \tau_{xy} & \tau_{zx} \\ 0 & 0 & 0 & 0 & 0 & 0 & \tau_{xy} & \tau_{yy} & \tau_{yz} \\ 0 & 0 & 0 & 0 & 0 & 0 & \tau_{zx} & \tau_{yz} & \tau_{zz} \end{bmatrix} \tag{8.53}$$

$$\mathbf{B}_{NL} = \begin{bmatrix} \frac{\partial h_1}{\partial \xi_1} & 0 & 0 & \frac{\partial h_2}{\partial \xi_1} & 0 & 0 & \cdots\cdots \\ \frac{\partial h_1}{\partial \xi_2} & 0 & 0 & \frac{\partial h_2}{\partial \xi_2} & 0 & 0 & \cdots\cdots \\ \frac{\partial h_1}{\partial \xi_3} & 0 & 0 & \frac{\partial h_2}{\partial \xi_3} & 0 & 0 & \cdots\cdots \\ 0 & \frac{\partial h_1}{\partial \xi_1} & 0 & 0 & \frac{\partial h_2}{\partial \xi_1} & 0 & \cdots\cdots \\ 0 & \frac{\partial h_1}{\partial \xi_2} & 0 & 0 & \frac{\partial h_2}{\partial \xi_2} & 0 & \cdots\cdots \\ 0 & \frac{\partial h_1}{\partial \xi_3} & 0 & 0 & \frac{\partial h_2}{\partial \xi_3} & 0 & \cdots\cdots \\ 0 & 0 & \frac{\partial h_1}{\partial \xi_1} & 0 & 0 & \frac{\partial h_2}{\partial \xi_1} & \cdots\cdots \\ 0 & 0 & \frac{\partial h_1}{\partial \xi_2} & 0 & 0 & \frac{\partial h_2}{\partial \xi_2} & \cdots\cdots \\ 0 & 0 & \frac{\partial h_1}{\partial \xi_3} & 0 & 0 & \frac{\partial h_2}{\partial \xi_3} & \cdots\cdots \end{bmatrix} \tag{8.54}$$

Substituting Eqs. (8.48) and (8.49) into Eq. (8.39), we can obtain:

$$(\delta\Delta \boldsymbol{a})^{\mathrm{T}}(\boldsymbol{K}_L + \boldsymbol{K}_{NL})\Delta\, \boldsymbol{a} = (\delta\Delta \boldsymbol{a})^{\mathrm{T}}\left(\boldsymbol{f}_{\text{ext}}^{t+\Delta t} - \boldsymbol{f}_{\text{int}}^{t}\right) \tag{8.55}$$

Here, the load column vector and the internal force column vector are defined as follows:

$$\begin{cases} \boldsymbol{f}_{\text{ext}}^{t+\Delta t} = \int_{S_0} \boldsymbol{H}^{\mathrm{T}} \boldsymbol{t}_0 \mathrm{d}S + \int_{V_0} \rho_0 \boldsymbol{H}^{\mathrm{T}} \boldsymbol{g} \mathrm{d}V \\ \boldsymbol{f}_{\text{int}}^{t} = \int_{V_0} \boldsymbol{B}_L^{\mathrm{T}} \boldsymbol{\tau}^t \mathrm{d}V \end{cases} \tag{8.56}$$

Equation (8.55) holds for any virtual displacement increment $\delta\Delta\mathbf{a}$, therefore

$$(\boldsymbol{K}_L + \boldsymbol{K}_{NL})\Delta \boldsymbol{a} = \boldsymbol{f}_{\text{ext}}^{t+\Delta t} - \boldsymbol{f}_{\text{int}}^{t} \tag{8.57}$$

Note: The above derivation applies to both two-dimensional and three-dimensional continuum elements. Since the derivation is based on the total Lagrangian formulation, the initial, undeformed configuration is chosen as the reference configuration.

8.3 Hierarchical Quadrature Element Analysis of Mindlin Shallow Shells

8.3.1 Mindlin Shallow Shell Theory

The Mindlin shallow shell theory was initially proposed and applied to moderately thick plates to account for transverse shear deformation. Due to its consideration of shear deformation, the rotation of the mid-surface is no longer equal to the rotation of the normal, as shown in Fig. 8.3.

Assuming the cross-section remains planar, the stress state of a infinitesimal element located a distance l from the midsurface within the shallow shell structure is shown in Fig. 8.4.

The in-plane strain of the infinitesimal element, $\boldsymbol{\varepsilon}^{\mathrm{T}} = \left(\boldsymbol{\varepsilon}_{xx}, \boldsymbol{\varepsilon}_{yy}, \boldsymbol{\varepsilon}_{xy}\right)$, can be expressed as

$$\boldsymbol{\varepsilon} = \boldsymbol{\varepsilon}_l + Z_l \boldsymbol{\chi} \tag{8.58}$$

Here, the subscript l denotes the reference plane (in this case, the mid-surface of the shell, although this is not required).

Assuming that the in-plane strain is sufficiently small (i.e., $\left(\frac{\partial u}{\partial x}\right)^2 \leq \left(\frac{\partial w}{\partial x}\right)^2$) and the out-of-plane normal strain ε_{zz} is zero, the in-plane strain at the mid-surface is given by

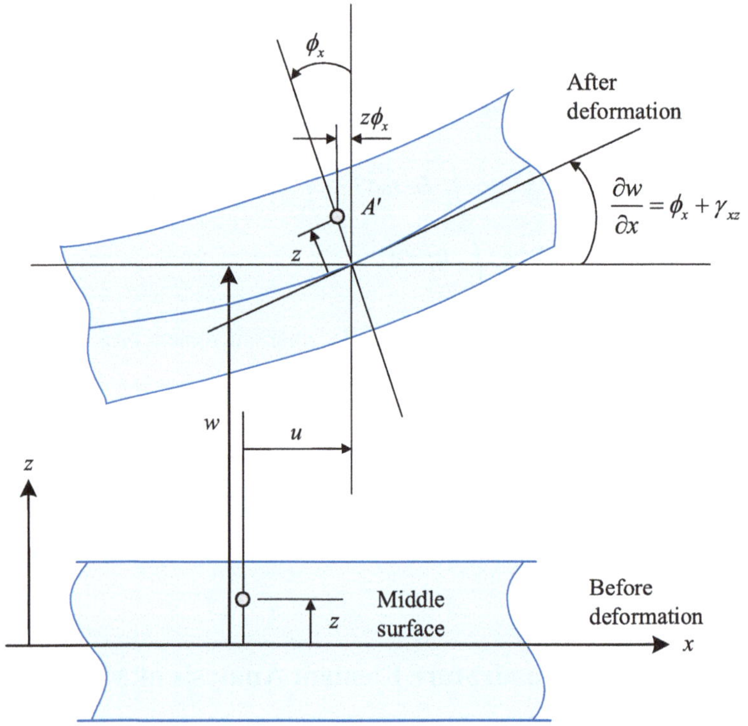

Fig. 8.3 In Mindlin's shallow shell theory, the rotation of the mid-surface differs from the rotation of the normal

$$\boldsymbol{\varepsilon}_l = \begin{pmatrix} \frac{\partial u_l}{\partial x} \\ \frac{\partial u_l}{\partial y} \\ \frac{\partial u_l}{\partial y} + \frac{\partial u_l}{\partial x} \end{pmatrix} + \begin{pmatrix} \frac{1}{2}\left(\frac{\partial w'}{\partial x}\right)^2 - \frac{1}{2}\left(\frac{\partial z}{\partial x}\right)^2 \\ \frac{1}{2}\left(\frac{\partial w'}{\partial y}\right)^2 - \frac{1}{2}\left(\frac{\partial z}{\partial x}\right)^2 \\ \frac{\partial w'}{\partial x}\frac{\partial w'}{\partial y} - \frac{\partial z}{\partial x}\frac{\partial z}{\partial y} \end{pmatrix} \tag{8.59}$$

Here, z is the initial vertical coordinate of the mid-surface of the shell; w is the vertical displacement relative to the initial configuration after shell deformation, and $w' = w + z$, as shown in Fig. 8.5.

In Mindlin's shallow shell theory, the bending deformation $\boldsymbol{\chi}$ is given by

$$\boldsymbol{\chi} = \begin{pmatrix} \chi_x \\ \chi_y \\ \chi_z \end{pmatrix} = \begin{pmatrix} \frac{\partial \theta_x}{\partial x} \\ \frac{\partial \theta_y}{\partial y} \\ \frac{\partial \theta_x}{\partial y} + \frac{\partial \theta_y}{\partial x} \end{pmatrix} \tag{8.60}$$

where θ_x and θ_y are the rotations of the normal to the mid-surface of the shell, and their directions are defined as shown in Fig. 8.6.

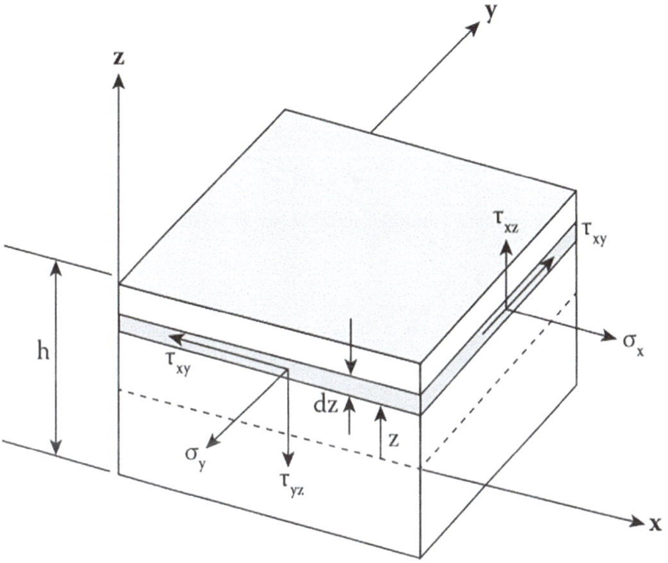

Fig. 8.4 Stress state of an infinitesimal element located a distance 1 from the mid-surface in a shallow shell

Fig. 8.5 Definition of the geometric configuration of a shallow shell

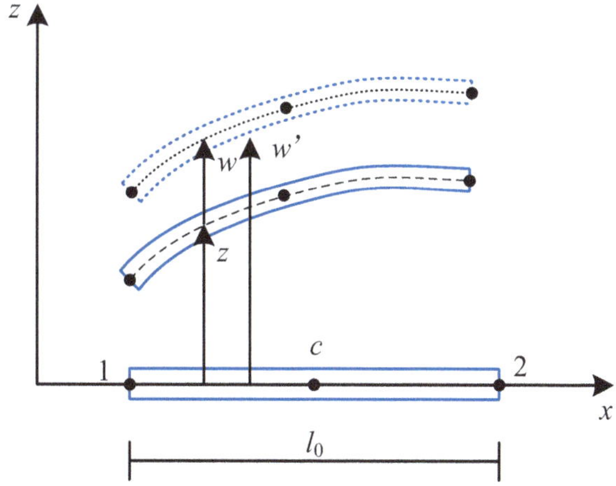

Since the cross-section is assumed to remain planar after shell deformation, the out-of-plane shear strain $\boldsymbol{\gamma}$ is defined as follows:

$$\boldsymbol{\gamma} = \begin{pmatrix} \gamma_{xz} \\ \gamma_{yz} \end{pmatrix} = \begin{pmatrix} \theta_x \\ \theta_y \end{pmatrix} + \begin{pmatrix} \frac{\partial w}{\partial x} \\ \frac{\partial w}{\partial y} \end{pmatrix} \tag{8.61}$$

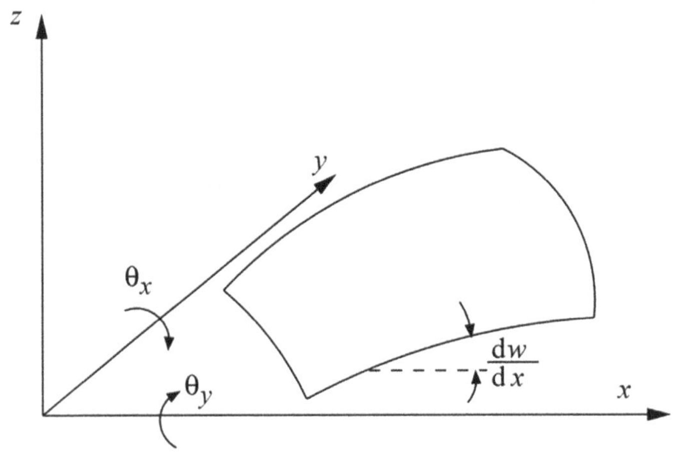

Fig. 8.6 Direction of the mid-surface normal rotation in shells

Equations (8.58) and (8.61) show that the stress state of any infinitesimal element within a shell can be completely determined by the stress state at the middle surface of the shell, which greatly facilitates the analysis of shell structure problems.

The variation of the above strain can be calculated according to the variational principle, where the variation of the in-plane strain $\boldsymbol{\varepsilon}$ is given by

$$\delta\boldsymbol{\varepsilon} = \delta\boldsymbol{\varepsilon}_l + Z_l\delta\boldsymbol{\chi} \tag{8.62}$$

where

$$\delta\boldsymbol{\varepsilon}_l = \begin{pmatrix} \frac{\partial \delta u_l}{\partial x} \\ \frac{\partial \delta u_l}{\partial y} \\ \frac{\partial \delta u_l}{\partial y} + \frac{\partial \delta u_l}{\partial x} \end{pmatrix} + \begin{pmatrix} \frac{\partial w'}{\partial x}\frac{\partial \delta w'}{\partial x} \\ \frac{\partial w'}{\partial y}\frac{\partial \delta w'}{\partial y} \\ \frac{\partial w'}{\partial x}\frac{\partial \delta w'}{\partial y} - \frac{\partial w'}{\partial y}\frac{\partial \delta w'}{\partial x} \end{pmatrix} \tag{8.63}$$

$$\delta\boldsymbol{\chi} = \begin{pmatrix} \frac{\partial \delta \theta_x}{\partial x} \\ \frac{\partial \delta \theta_y}{\partial y} \\ \frac{\partial \delta \theta_x}{\partial y} + \frac{\partial \delta \theta_y}{\partial x} \end{pmatrix} \tag{8.64}$$

The incremental form of the in-plane strain $\boldsymbol{\varepsilon}$ is given by

$$\Delta\boldsymbol{\varepsilon} = \Delta\boldsymbol{\varepsilon}_l + Z_l\Delta\boldsymbol{\chi} \tag{8.65}$$

where

$$\Delta\boldsymbol{\varepsilon}_l = \begin{pmatrix} \frac{\partial \Delta u_l}{\partial x} \\ \frac{\partial \Delta u_l}{\partial y} \\ \frac{\partial \Delta u_l}{\partial y} + \frac{\partial \Delta u_l}{\partial x} \end{pmatrix} + \begin{pmatrix} \frac{\partial w'}{\partial x}\frac{\partial \Delta w'}{\partial x} + \frac{1}{2}\left(\frac{\partial \Delta w}{\partial x}\right)^2 \\ \frac{\partial w'}{\partial y}\frac{\partial \Delta w'}{\partial y} + \frac{1}{2}\left(\frac{\partial \Delta w}{\partial y}\right)^2 \\ \frac{\partial w'}{\partial x}\frac{\partial \Delta w'}{\partial y} + \frac{\partial w'}{\partial y}\frac{\partial \Delta w'}{\partial x} - \frac{\partial \Delta w}{\partial x}\frac{\partial \Delta w}{\partial y} \end{pmatrix} \tag{8.66}$$

Here, the format of $\Delta\boldsymbol{\chi}$ is the same as that of $\boldsymbol{\delta\chi}$.

The variation of the out-of-plane shear strain $\boldsymbol{\gamma}$ is

$$\delta\boldsymbol{\gamma} = \begin{pmatrix} \delta\theta_x \\ \delta\theta_y \end{pmatrix} + \begin{pmatrix} \frac{\partial \delta w}{\partial x} \\ \frac{\partial \delta w}{\partial y} \end{pmatrix} \tag{8.67}$$

The format of $\Delta\boldsymbol{\gamma}$ is the same as that of $\delta\boldsymbol{\gamma}$.

In Mindlin's shallow shell theory, there are five nonzero stress components: in-plane normal stresses σ_{xx}, σ_{yy}, in-plane shear stress σ_{xy}, as well as out-of-plane shear stresses σ_{xz} and σ_{yz}. Integrating these stresses through the thickness yields the generalized forces, the following internal forces can be obtained:

Normal Force:

$$\mathbf{N} = \begin{pmatrix} N_x \\ N_y \\ N_{xy} \end{pmatrix} = \int_{-h/2}^{h/2} \begin{pmatrix} \sigma_{xx}(Z_l) \\ \sigma_{yy}(Z_l) \\ \sigma_{xy}(Z_l) \end{pmatrix} \mathrm{d}Z_l \tag{8.68}$$

Bending Moment:

$$\mathbf{M} = \begin{pmatrix} M_x \\ M_y \\ M_{xy} \end{pmatrix} = \int_{-h/2}^{h/2} \begin{pmatrix} \sigma_{xx}(Z_l) \\ \sigma_{yy}(Z_l) \\ \sigma_{xy}(Z_l) \end{pmatrix} Z_l \mathrm{d}Z_l \tag{8.69}$$

Shear Force:

$$\mathbf{Q} = \begin{pmatrix} Q_x \\ Q_y \end{pmatrix} = \int_{-h/2}^{h/2} \begin{pmatrix} \sigma_{xz}(Z_l) \\ \sigma_{yz}(Z_l) \end{pmatrix} \mathrm{d}Z_l \tag{8.70}$$

The directions of the normal force $\boldsymbol{N}$ and the bending moment $\boldsymbol{M}$ are shown in Fig. 8.7.

Based on the assumption that the strain remains small, Hooke's law can be used to relate the in-plane strain increment (which can be considered a degenerate form of the Green–Lagrange strain, see Sect. 8.1 for details) to the in-plane stress increment, namely,

$$\begin{pmatrix} \dot{\sigma}_{xx} \\ \dot{\sigma}_{yy} \\ \dot{\sigma}_{xy} \end{pmatrix} = \mathbf{D} \begin{pmatrix} \dot{\varepsilon}_{xx} \\ \dot{\varepsilon}_{yy} \\ \dot{\varepsilon}_{xy} \end{pmatrix} \tag{8.71}$$

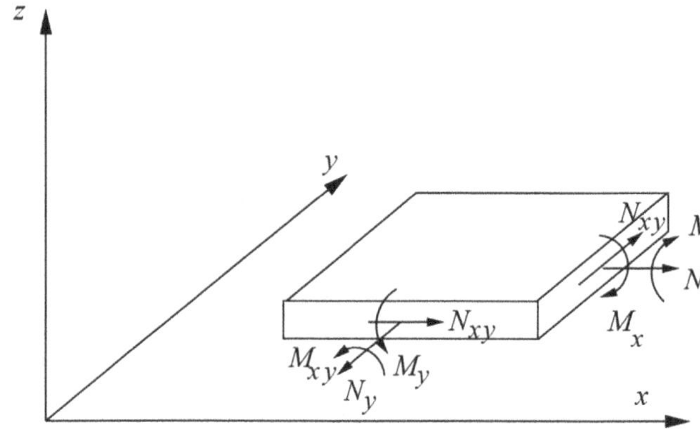

Fig. 8.7 Directions of the internal forces

Since the nonlinear analysis here does not involve material nonlinearity,

$$\mathbf{D} = \frac{E}{1-\upsilon^2}\begin{bmatrix} 1 & \upsilon & 0 \\ \upsilon & 1 & 0 \\ 0 & 0 & (1-\upsilon)/2 \end{bmatrix} \tag{8.72}$$

Here, E is the elastic modulus and υ is the Poisson's ratio.

Thus, the increments (or rates of change) of the in-plane generalized forces $\boldsymbol{M}$ and $\boldsymbol{N}$, as well as the out-of-plane shear force $\boldsymbol{Q}$, can be expressed as

$$\dot{\mathbf{N}} = \int_{-h/2}^{h/2} \mathbf{D}\varepsilon_l \mathrm{d}Z_l = \mathbf{D}h\boldsymbol{\varepsilon}_l = \mathbf{D}_{\mathrm{m}}\boldsymbol{\varepsilon}_l \tag{8.73}$$

$$\dot{\mathbf{M}} = \int_{-h/2}^{h/2} \mathbf{D}\dot{\boldsymbol{\chi}} Z_l^2 \mathrm{d}Z_l = \mathbf{D}\frac{h^3}{12}\dot{\chi} = \mathbf{D}_{\mathrm{b}}\dot{\boldsymbol{\chi}} \tag{8.74}$$

$$\dot{\mathbf{Q}} = \int_{-h/2}^{h.2} \mathbf{G}\dot{\boldsymbol{\gamma}} \mathrm{d}Z_l = \mathbf{G}h\dot{\boldsymbol{\gamma}} = \mathbf{D}_{\mathrm{s}}\dot{\boldsymbol{\gamma}} \tag{8.75}$$

where

$$\mathbf{G} = \frac{5}{6}\begin{bmatrix} u & 0 \\ 0 & u \end{bmatrix} \tag{8.76}$$

In the equation, u is the shear modulus of the material; $\boldsymbol{D}_m$ and $\boldsymbol{D}_s$ are the tangent constitutive matrices for the membrane and bending responses of the material, respectively; $\boldsymbol{D}_{\mathrm{s}}$ is the tangent constitutive matrix for the shear response of the material.

Therefore, by combining the above generalized stresses, the stress–strain relationship for all components can be obtained as follows:

$$\begin{pmatrix} \dot{\mathbf{N}} \\ \dot{\mathbf{M}} \\ \dot{\mathbf{Q}} \end{pmatrix} = \begin{bmatrix} \mathbf{D}_{\mathrm{m}} & \mathbf{0} & \mathbf{0} \\ \mathbf{0} & \mathbf{D}_{\mathrm{b}} & \mathbf{0} \\ \mathbf{0} & \mathbf{0} & \mathbf{D}_{\mathrm{s}} \end{bmatrix} \begin{pmatrix} \dot{\varepsilon}_l \\ \dot{\chi} \\ \dot{\gamma} \end{pmatrix} \tag{8.77}$$

8.3.2 Hierarchical Quadrature Element Formulation for Mindlin Shallow Shell

The Mindlin shallow shell theory reduces the three-dimensional shell structure to a reference surface for analysis, thus allowing the use of the two-dimensional hierarchical quadrature element shape functions introduced in Chap. 4 as its own shape functions. The following shape function matrices are defined:

$$\mathbf{H} = \begin{bmatrix} h_1 & 0 & h_2 & 0 & \cdots & \cdots & h_n & 0 \\ 0 & h_1 & 0 & h_2 & \cdots & \cdots & 0 & h_n \end{bmatrix} \tag{8.78}$$

$$\mathbf{h} = \begin{bmatrix} h_1 & h_2 & \cdots & h_n \end{bmatrix}^{\mathrm{T}} \tag{8.79}$$

The above shape functions can be either Serendipity or Lagrange interpolation shape functions, or hierarchical quadrature shape functions. Employing hierarchical quadrature shape functions can effectively mitigate the effects of shear locking.

In Mindlin's shallow shell theory, the primary unknowns are u_l, v_l, w, θ_x, θ_y. The nodal displacement vector and nodal rotation vector are defined as follows:

$$\mathbf{a} = \begin{pmatrix} (u_l)_1 \\ (v_l)_1 \\ \cdots \\ \cdots \\ (u_l)_n \\ (v_l)_n \end{pmatrix}, \quad \mathbf{w} = \begin{pmatrix} w_1 \\ w_2 \\ \cdots \\ \cdots \\ w_n \end{pmatrix}, \quad \boldsymbol{\theta} = \begin{pmatrix} (\theta_x)_1 \\ (\theta_y)_1 \\ \cdots \\ \cdots \\ (\theta_x)_n \\ (\theta_y)_n \end{pmatrix} \tag{8.80}$$

Thus, the discretized form of the displacement field is

$$\begin{pmatrix} \mathbf{u}_l \\ \mathbf{v}_l \end{pmatrix} = \mathbf{Ha}, \quad w = \mathbf{h}^{\mathrm{T}}\mathbf{w}, \quad \begin{pmatrix} \boldsymbol{\theta}_x \\ \boldsymbol{\theta}_y \end{pmatrix} = \mathbf{H}\theta \tag{8.81}$$

By discretizing the variational expression (8.62) for strain, we obtain

$$\delta\boldsymbol{\varepsilon}_l = \mathbf{B}\delta\mathbf{a} + (\mathbf{Bw}')\mathbf{B}_w\delta\mathbf{w} \tag{8.82}$$

where

$$\mathbf{B} = \begin{bmatrix} \frac{\partial h_1}{\partial x} & 0 & \cdots\cdots & \frac{\partial h_n}{\partial x} & 0 \\ 0 & \frac{\partial h_1}{\partial y} & \cdots\cdots & 0 & \frac{\partial h_n}{\partial y} \\ \frac{\partial h_1}{\partial y} & \frac{\partial h_1}{\partial x} & \cdots\cdots & \frac{\partial h_n}{\partial y} & \frac{\partial h_n}{\partial x} \end{bmatrix} \tag{8.83}$$

$$\mathbf{w}' = \begin{bmatrix} w_1' & 0 \\ 0 & w_1' \\ \cdots & \cdots \\ \cdots & \cdots \\ w_n' & 0 \\ 0 & w_n' \end{bmatrix} \tag{8.84}$$

$$\mathbf{B}_w = \begin{bmatrix} \frac{\partial h_1}{\partial x} & \cdots & \frac{\partial h_n}{\partial x} \\ \frac{\partial h_1}{\partial y} & \cdots & \frac{\partial h_n}{\partial y} \end{bmatrix} \tag{8.85}$$

By discretizing the variation of rotation $\delta\boldsymbol{\chi}$ and the variation of out-of-plane shear strain $\delta\boldsymbol{\gamma}$, we obtain

$$\delta\boldsymbol{\chi} = \mathbf{B}\delta\boldsymbol{\theta} \tag{8.86}$$

$$\delta\boldsymbol{\gamma} = \mathbf{H}\delta\boldsymbol{\theta} + \mathbf{B}_w\delta\mathbf{w} \tag{8.87}$$

Next, we derive the tangent stiffness matrix and the internal force vector. For shallow shell configurations, the virtual work equation at time $t + \Delta t$ is given by

$$\int_{\mathbf{V_0}} (\delta\boldsymbol{\varepsilon}^{\mathbf{t+\Delta t}})^{\mathrm{T}}\boldsymbol{\sigma}^{\mathbf{t+\Delta t}}\mathrm{d}\mathbf{V} = (\delta\boldsymbol{u}^{\mathbf{t+\Delta t}})^{\mathrm{T}}\mathrm{f}_{\mathrm{ext}}^{\mathbf{t+\Delta t}} \tag{8.88}$$

Assuming the surface area of the shell remains unchanged before and after deformation, $A = A_0$, and integrating along the thickness direction, we obtain

$$\int_A \left((\delta\boldsymbol{\varepsilon}_l^{t+\Delta t})^{\mathrm{T}}\mathbf{N}^{t+\Delta t} + (\delta\boldsymbol{\chi}_l^{t+\Delta t})^{\mathrm{T}}\mathbf{M}^{t+\Delta t} + (\delta\boldsymbol{\gamma}_l^{t+\Delta t})^{\mathrm{T}}\mathbf{Q}^{t+\Delta t}\right)\mathrm{d}A = (\delta\mathbf{u}^{t+\Delta t})^{\mathrm{T}}\mathbf{f}_{\mathrm{ext}}^{t+\Delta t} \tag{8.89}$$

Substituting it into the discretized strain variational Eqs. (8.82), (8.86), and (8.87), we obtain

$$\begin{aligned} &\int_A \left[\delta\mathbf{a}^{\mathrm{T}}\mathbf{B}^{\mathrm{T}}\mathrm{d}\mathbf{N} + \delta\boldsymbol{\theta}^{\mathrm{T}}\mathbf{B}^{\mathrm{T}}\mathrm{d}\mathbf{M} + (\delta\boldsymbol{\theta}^{\mathrm{T}}\mathbf{H}^{\mathrm{T}} + \delta\mathbf{w}^{\mathrm{T}}\mathbf{B}_w^{\mathrm{T}})d\mathbf{Q} + \delta\mathbf{w}^{\mathrm{T}}\mathbf{B}_w^{\mathrm{T}}(\mathbf{B}d\mathbf{w}')^{\mathrm{T}}\mathbf{N}\right]\mathrm{d}A \\ &= \delta\mathbf{a}^{\mathrm{T}}(\mathbf{f}_{\mathrm{ext}}^{\mathrm{a}} - \mathbf{f}_{\mathrm{int}}^{\mathrm{a}}) + \delta\mathbf{w}^{\mathrm{T}}(\mathbf{f}_{\mathrm{ext}}^{\mathrm{w}} - \mathbf{f}_{\mathrm{int}}^{\mathrm{w}}) + \delta\boldsymbol{\theta}^{\mathrm{T}}(\mathbf{f}_{\mathrm{ext}}^{\theta} - \mathbf{f}_{\mathrm{int}}^{\theta}) \end{aligned} \tag{8.90}$$

Here, the internal force column vector is defined as

$$\mathbf{f}_{\text{int}}^{\text{a}} = \int_A \mathbf{B}^{\text{T}}\mathbf{N}\text{d}A \tag{8.91}$$

$$\mathbf{f}_{\text{int}}^{\text{w}} = \int_A \mathbf{B}_w^{\text{T}}\left(\left(\mathbf{Bw}'\right)^{\text{T}}\mathbf{N} + \mathbf{Q}\right)\text{d}A \tag{8.92}$$

$$\mathbf{f}_{\text{int}}^{\theta} = \int_A (\mathbf{B}^{\text{T}}\mathbf{M} + \mathbf{H}^{\text{T}}\mathbf{Q})\text{d}A \tag{8.93}$$

Substituting the constitutive relation (8.77) for Mindlin plates and shells into Eq. (8.90), and noting that Eq. (8.90) holds for any virtual displacements ($\delta\boldsymbol{a}$, $\delta\boldsymbol{w}$, $\delta\boldsymbol{\theta}$), we obtain

$$\begin{bmatrix} \mathbf{K}_{\text{aa}} & \mathbf{K}_{\text{aw}} & \mathbf{K}_{\text{a}\theta} \\ \mathbf{K}_{\text{aw}}^{\text{T}} & \mathbf{K}_{\text{ww}} & \mathbf{K}_{\text{w}\theta} \\ \mathbf{K}_{\text{a}\theta}^{\text{T}} & \mathbf{K}_{\text{w}\theta}^{\text{T}} & \mathbf{K}_{\theta\theta} \end{bmatrix} \begin{pmatrix} \mathbf{da} \\ \mathbf{dw} \\ \mathbf{d\theta} \end{pmatrix} = \begin{pmatrix} \mathbf{f}_{\text{ext}}^{\text{a}} - \mathbf{f}_{\text{int}}^{\text{a}} \\ \mathbf{f}_{\text{ext}}^{\text{w}} - \mathbf{f}_{\text{int}}^{\text{w}} \\ \mathbf{f}_{\text{ext}}^{\theta} - \mathbf{f}_{\text{int}}^{\theta} \end{pmatrix} \tag{8.94}$$

The individual components of the tangent stiffness matrix are as follows:

$$\begin{aligned}
\mathbf{K}_{aa} &= \int_A \mathbf{B}^{\text{T}}\mathbf{D}_m\mathbf{B}\text{d}A \\
\mathbf{K}_{aw} &= \int_A \mathbf{B}^{\text{T}}\mathbf{D}_m(\mathbf{Bw}')\mathbf{B}_w\text{d}A \\
\mathbf{K}_{a\theta} &= \mathbf{0} \\
\mathbf{K}_{ww} &= \int_A \mathbf{B}_w^{\text{T}}\mathbf{D}_s\mathbf{B}_w\text{d}A + \int_A \mathbf{B}_w^{\text{T}}(\mathbf{Bw}')^{\text{T}}\mathbf{D}_m(\mathbf{Bw}')\mathbf{B}_w\text{d}A + \int_A \mathbf{B}_w^{\text{T}}\mathcal{N}\mathbf{B}_w\text{d}A \\
\mathbf{K}_{w\theta} &= \int_A \mathbf{B}_w^{\text{T}}\mathbf{D}_s\mathbf{H}\text{d}A + \int_A \mathbf{B}_w^{\text{T}}(\mathbf{Bw}')^{\text{T}}\mathbf{D}_c\mathbf{B}\text{d}A \\
\mathbf{K}_{\theta\theta} &= \int_A \mathbf{B}^{\text{T}}\mathbf{D}_b\mathbf{B}\text{d}A + \int_A \mathbf{H}^{\text{T}}\mathbf{D}_s\mathbf{H}\text{d}A
\end{aligned} \tag{8.95}$$

Here, the third term in $\boldsymbol{K}_{ww}$ is regarded as the contribution of geometric nonlinearity, with $\mathbf{N}$ being the matrix formed from the normal forces $\boldsymbol{N}$.

$$\mathcal{N} = \begin{pmatrix} N_x & N_{xy} \\ N_{xy} & N_y \end{pmatrix} \tag{8.96}$$

8.4 Numerical Examples of Geometric Nonlinear Analysis

The earlier sections of this chapter have provided a detailed theoretical foundation and operational procedure for geometric nonlinear finite element analysis. Based on the hierarchical differential quadrature element method (HQEM), this section presents the application of the HQEM to two-dimensional planar elements, three-dimensional elements, and shell elements based on Mindlin's shallow shell theory in the context of geometrically nonlinearity. The programs developed are based on the C++ platform and the Eigen matrix library, enabling quasi-static geometric nonlinear analysis of various planar and complex three-dimensional spatial structures. Visualization and post-processing are implemented using ParaView (5.2.2).

8.4.1 Geometric Nonlinear Analysis of a Two-Dimensional Cantilever Beam Structure Using the Hierarchical Quadrature Element Method

The clamped large-deformation cantilever beam problem is a classic geometric nonlinear problem. Consider the cantilever beam shown in Fig. 8.1, with specified geometric parameters: length $l = 8.0$ mm, height $h = 0.5$ mm, and a width corresponding to a plane stress thickness of 1mm. The material parameters are specified as: Young's modulus $E = 100$ N/mm^2, Poisson's ratio $\upsilon = 0.3$. A concentrated load $F = 0.2\,N$ at the free end of the cantilever beam, distributed over 20 loading steps: $\Delta F = 0.01N$.

To verify the correctness and effectiveness of the method presented in this section, the results computed by the C++ code are compared with those obtained from the commercial software Abaqus/CAE 2018 for the same structure, which is highly reputed in the field of nonlinear analysis. First, the convergence of the two-dimensional HQEM used in this chapter and the Abaqus CPS4 elements is investigated. As can be seen in Fig. 8.8, with a convergence criterion of "change less than 0.01," the two-dimensional HQEM achieves convergence with only tens of degrees of freedom (64 DOFs), whereas the Abaqus CPS4 elements require hundreds of degrees of freedom (800 DOFs) to achieve convergence. The solutions provided by the two-dimensional hierarchical quadrature elements at lower numbers of degrees of freedom exhibit good accuracy, making them highly suitable for nonlinear analyses that involve continuous iteration and updating of stiffness matrices, as the element matrices constructed by the two-dimensional hierarchical quadrature element method are significantly smaller than those of other low-order finite element methods.

In terms of nonlinear solution methods, the Newton–Raphson method and the arc-length method were respectively employed to solve the problem. The load–displacement curves obtained from the geometric nonlinear analysis of a cantilever beam structure using the two-dimensional HQEM and the ABAQUS (CPS4) element, employing the Newton–Raphson method and the arc-length method respectively, are

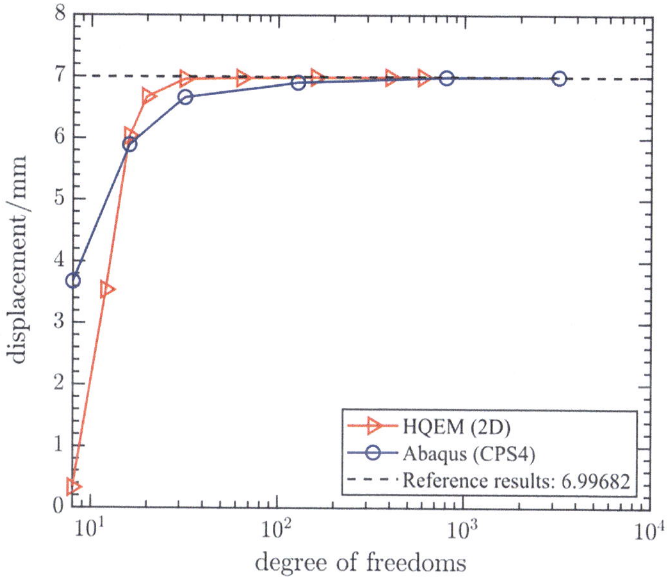

Fig. 8.8 Comparison of convergence rates between the two-dimensional hierarchical quadrature elements and the ABAQUS (CPS4) elements

shown in Fig. 8.9. It can be seen that, regardless of the element type or solution method used, the calculated load–displacement curves agree well with each other. Notably, when the two-dimensional hierarchical quadrature elements are combined with the arc-length method, only 6 data points are required (i.e., equilibrium is reached after 5 load steps), which demonstrates that this approach offers exceptionally high computational efficiency and accuracy.

The displacement at integration points, calculated using the two-dimensional hierarchical quadrature element, is exported to ParaView (5.5.2). Results for the displacement cloud diagram of the two-dimensional cantilever beam structure after deformation are shown in Fig. 8.10.

The two-dimensional hierarchical quadrature element method used in this section possesses C^0 continuity and employs interpolation transformation during its construction, which facilitates the element assembly process. To verify this, we conducted a numerical test of element assembly: a cantilever beam structure was divided axially into two two-dimensional hierarchical quadrature elements (see Fig. 8.11). The results, along with those obtained using a single two-dimensional hierarchical quadrature element and ABAQUS (CPS4) elements, are listed in Table 8.1. It can be seen that regardless of whether a single element or assembling two elements, the two-dimensional hierarchical quadrature elements can achieve results consistent with ABAQUS (CPS4) elements with significantly fewer degrees of freedom. In contrast, the number of degrees of freedom required for the latter to reach convergence is often 1 to 2 orders of magnitude higher than that of the former. The deformation

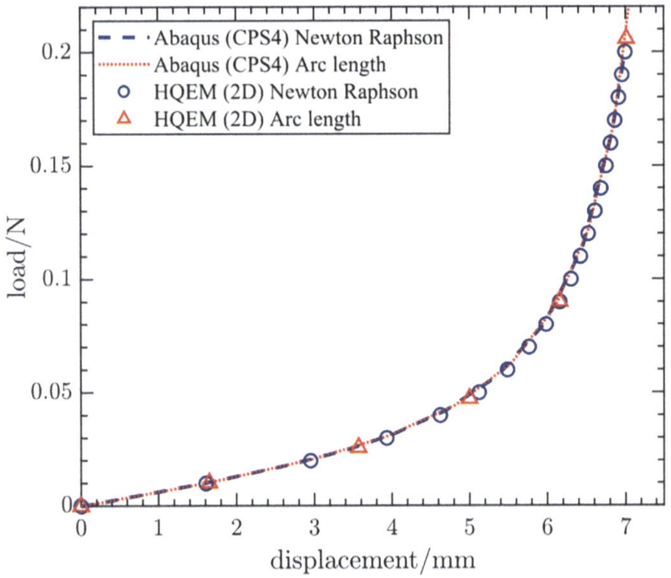

Fig. 8.9 Load–displacement curve of a two-dimensional cantilever beam structure

mode and displacement contour of the cantilever beam structure, calculated using two assembled hierarchical quadrature elements, are shown in Fig. 8.12. The data range of the displacement contour is set to 0 ~ 6.9967, which demonstrates good displacement continuity between the two hierarchical quadrature elements.

8.4.2 Geometric Nonlinear Analysis of Three-Dimensional Structures Using the Hierarchical Quadrature Element Method

Three-dimensional finite elements have always been a crucial component of finite element analysis, offering high accuracy, complete preservation of nodal information, and ease of modeling complex spatial structures. Although their large number of degrees of freedom makes them highly dependent on computational resources, the rapid advancements in computer processing power in recent years have made modern computational resources abundantly available. The disadvantages of 3D elements appear to be increasingly offset by these improvements in computational capability, while their advantages—high accuracy and the ability to easily model complex structures—have become especially important in contemporary engineering practice, where nonlinear behavior and structural complexity are receiving greater emphasis.

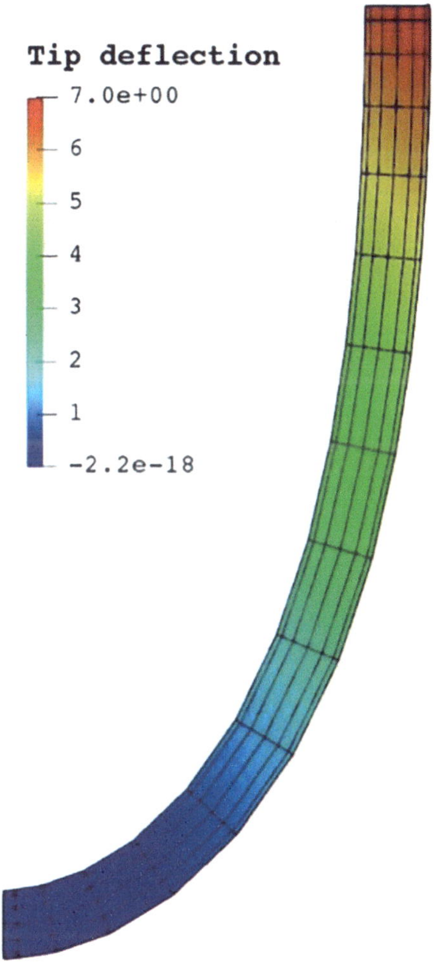

Fig. 8.10 Displacement contour plot of the deformed two-dimensional cantilever beam structure

This subsection first applies the three-dimensional hierarchical quadrature element method to the cantilever beam problem discussed in the previous subsection, and then extends its application to several complex spatial structures to assess its effectiveness in geometric nonlinear analysis of complex structures.

8.4.2.1 Three-Dimensional Cantilever Beam Structure

In this section, the cantilever beam structure shown in Fig. 8.8 is constructed as a three-dimensional configuration, as illustrated in Fig. 8.13. This structure shares the same geometric and material parameters as the two-dimensional cantilever beam, with a length $l = 8.0$ mm, height $h = 0.5$ mm and width $w = 1$ mm. The Young's

Fig. 8.11 Assembly pattern of two two-dimensional hierarchical quadrature elements

Table 8.1 Comparison of results between the two-dimensional hierarchical quadrature element method (HQEM) and ABAQUS (CPS4) element (two-dimensional cantilever beam structure)

Method	HQEM (2D)	HQEM (2D, Double element)	Abaqus (CPS4)
Displacement/mm	6.99682	6.9967	6.993

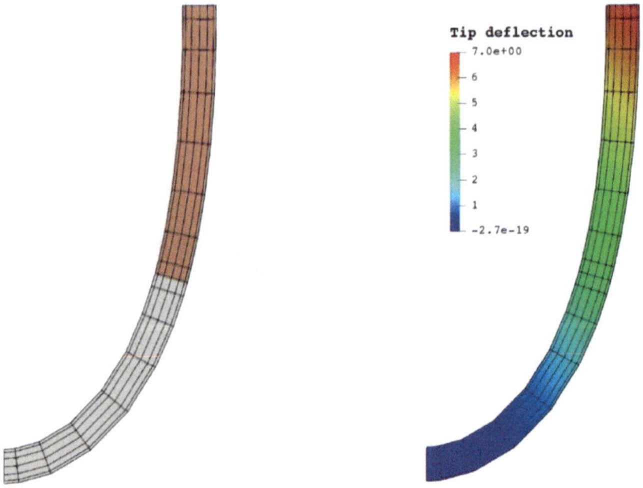

(a) Cantilever beam deformation mode (b) Displacement contour plot

Fig. 8.12 Deformation and displacement contour plots of a cantilever beam structure using two two-dimensional hierarchical quadrature elements

modulus is specified as $E = 100$ N/mm^2, and Poisson's ratio as $\upsilon = 0.3$. Unlike the two-dimensional cantilever beam structure, the load at the free end is applied as a uniformly distributed surface force on the free-end face in the three-dimensional configuration to better align with the characteristics of three-dimensional structures. The resultant of the uniformly distributed surface force is $F = 0.2\,N$, applied over 20 loading steps, i.e., $\Delta F = 0.01N$.

Similar to the previous section, the results obtained using the three-dimensional hierarchical quadrature elements are compared with those from ABAQUS (C3D8) elements. The C3D8 element is an 8-node hexahedral element. First, the convergence of the results calculated with both types of elements is compared. As shown in Fig. 8.14, similar to the convergence behavior in two-dimensional structures, the three-dimensional hierarchical quadrature element exhibits a significantly faster convergence rate than the ABAQUS (C3D8) element. When the convergence criterion is set as a tip displacement change of less than 0.01, the three-dimensional hierarchical quadrature element converges with 540 degrees of freedom, whereas the ABAQUS (C3D8) element requires 3,528 degrees of freedom to achieve convergence.

Next, the three-dimensional hierarchical quadrature elements and the ABAQUS (C3D8) elements were analyzed using two nonlinear solution methods: the Newton–Raphson method and the arc-length method. The resulting load–displacement curves are shown in Fig. 8.15. It can be observed that the four curves obtained from the two nonlinear solution methods and the two types of elements are in good agreement. Similar to the results for the two-dimensional hierarchical quadrature element, the three-dimensional hierarchical quadrature elements solved by the arc-length method reach equilibrium after only four load steps. This demonstrates that the three-dimensional hierarchical quadrature elements, when combined with the arc-length method, can achieve results for large geometric nonlinear deformations with fewer degrees of freedom and faster convergence.

The post-processing of the computational results from the three-dimensional hierarchical quadrature element and the ABAQUS (C3D8) element was performed, and displacement contour plots were generated for comparison, as shown in Fig. 8.16.

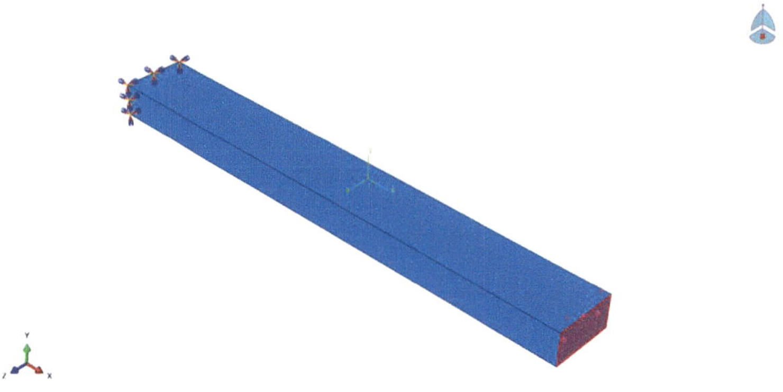

Fig. 8.13 A three-dimensional cantilever beam configuration

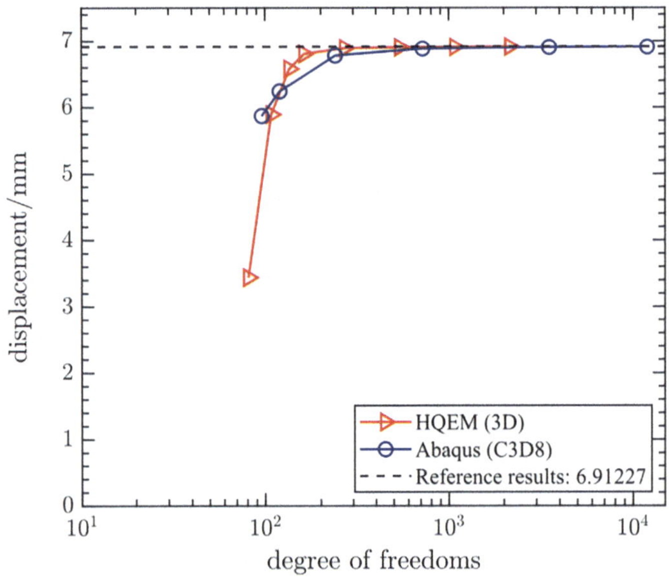

Fig. 8.14 Comparison of convergence rates between the three-dimensional hierarchical quadrature element and the ABAQUS (C3D8) element

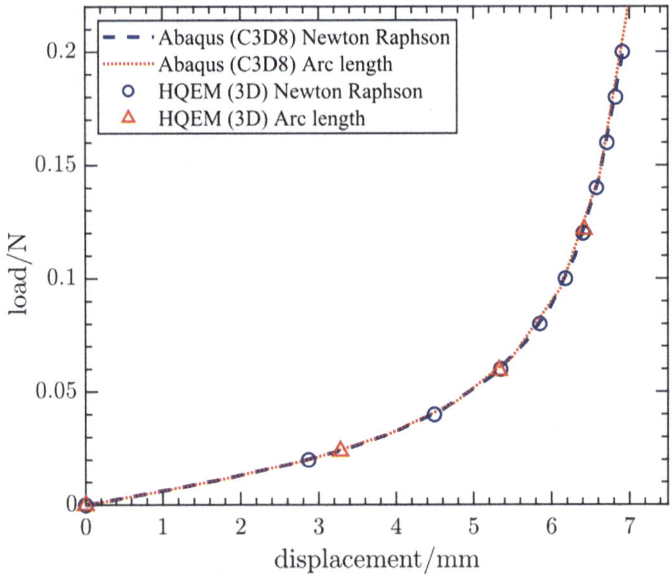

Fig. 8.15 Load–displacement curve of a three-dimensional cantilever beam structure

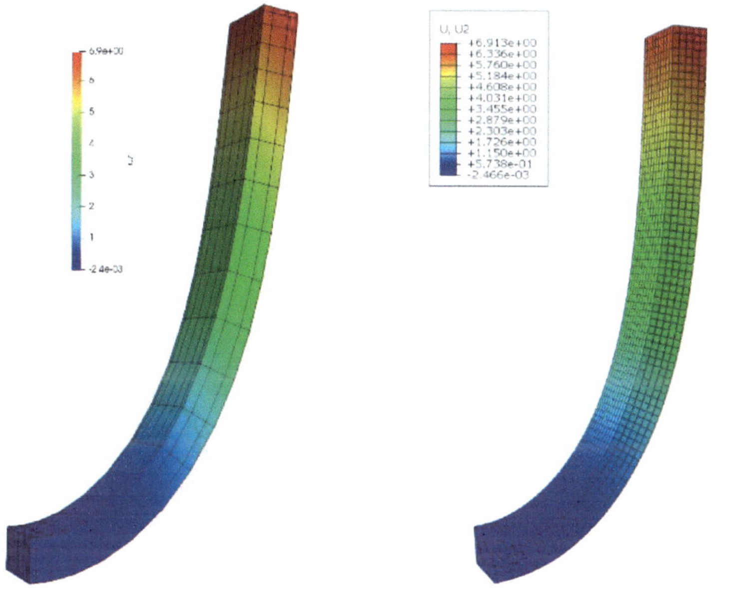

(a) HQEM 3D (b) Abaqus C3D8

Fig. 8.16 Displacement contour plot of the three-dimensional cantilever beam structure

Both contour plots are displayed using the Rainbow color scheme, with the data range set to 0–6.913. The results obtained from the hierarchical quadrature element integration are exported to ParaView (5.5.2) for visualization, while the results from the ABAQUS (C3D8) elements calculations are directly output and post-processed using ABAQUS CAE/2018.

The three-dimensional hierarchical elements can also be conveniently assembled. For example, assembling three three-dimensional hierarchical elements along the axis of the beam, the resulting deformation and displacement contour plots from the assembled element calculations are shown in Fig. 8.17. It can be observed that the displacement remains continuous at the element boundaries.

The computational results for a single three-dimensional hierarchical quadrature element, three three-dimensional higher-order hierarchical elements, and the ABAQUS (C3D8) element are listed in Table 8.2. It can be seen that the results obtained using the three-dimensional HQEM, whether calculated with a single element or through assembly, are in good agreement with those obtained using the ABAQUS (C3D8) elements.

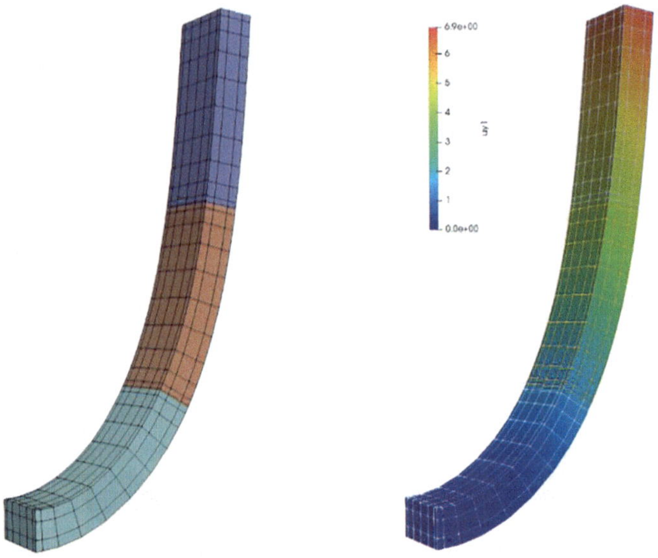

Fig. 8.17 Deformation and displacement contour plots of a cantilever beam structure using three three-dimensional hierarchical quadrature elements

Table 8.2 Comparison of results between three-dimensional hierarchical quadrature element method (HQEM) and ABAQUS (C3D8) elements (three-dimensional cantilever beam)

	HQEM (3D)	HQEM (3D, Triple element)	Abaqus (C3D8)
Displacement/mm	6.91227	6.91406	6.913

8.4.2.2 Three-Dimensional Tensile Spring Structure

Since the three-dimensional hierarchical quadrature elements can conveniently simulate various complex spatial structures, it is next used to model the stretching process of a spring. The geometric shape of the stretching spring is shown in Fig. 8.18, with parameters as follows: wire diameter $d = 1$ mm, center diameter $D = 10$ mm, number of turns $N_t = 2$, and height h = 10 mm. The material parameters are Young's modulus E = 100 N/mm^2 and Poisson's ratio $\upsilon = 0.3$. As illustrated in Fig. 8.18b, a fixed support constraint is applied to the bottom cross-section of the spring, while a uniformly distributed load of magnitude 0.01 N/mm^2, directed upward along the z-axis, is applied to the top cross-section of the spring.

The geometric nonlinear problem was analyzed using both the three-dimensional hierarchical quadrature elements and the ABAQUS (C3D8) elements. The tensile displacements of the spring are listed in Table 8.3, which shows that the two methods yield consistent numerical results. The displacement contour plots obtained from the three-dimensional hierarchical quadrature elements and the ABAQUS (C3D8)

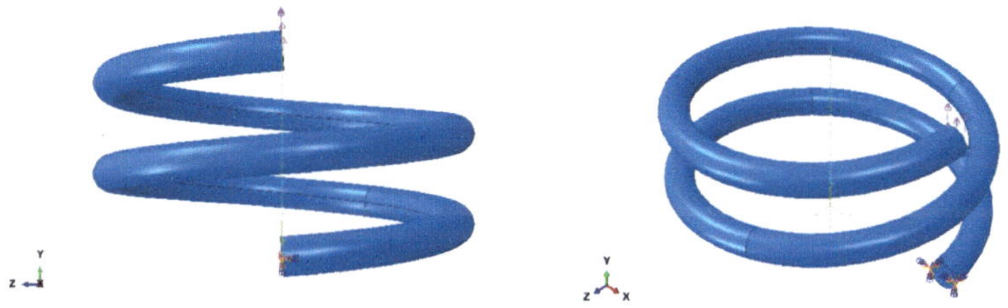

(a) Front view (b) Perspective view

Fig. 8.18 Geometric shape of the spring

Table 8.3 Comparison of three-dimensional hierarchical quadrature element method (HQEM) results and ABAQUS (C3D8) element results (three-dimensional tensile spring)

	HQEM (3D)	Abaqus (C3D8)
Displacement/mm	10.6139	10.65

elements are shown in Fig. 8.19, with both using the Rainbow color scheme. The style and data range are set to 0–10.65.

8.4.2.3 Three-Dimensional Cylindrical Shell Structures

Similar to the previous cases, consider a cylindrical shell shown in Fig. 8.20, with a length $a = 3$ mm, radius $R = 1$ mm, and thickness $h= 0.1$ mm. A fixed support constraint is applied along one straight edge, while a uniformly distributed load $p= 0.05$ N/mm^2 is applied along the opposite straight edge.

The end-face displacements were calculated using the three-dimensional hierarchical quadrature elements and the Abaqus C3D8 elements, respectively. As shown in Table 8.4, the results from both methods are in good agreement. The rapid convergence of the HQEM is once again demonstrated: the results for the three-dimensional HQEM in Table 8.4 were obtained with node arrangements of $\left[N_x, N_y, N_z\right] = [7, 7, 3]$, whereas the results for the Abaqus C3D8 elements require reducing the element size to $h = 0.05$ (with 11,712 nodes) to achieve comparable accuracy.

The displacement contour plots obtained using the three-dimensional hierarchical quadrature elements and the ABAQUS (C3D8) elements are shown in Fig. 8.21. For both methods, the style is set to Rainbow and the data range is set to 0 ~ 2.6247.

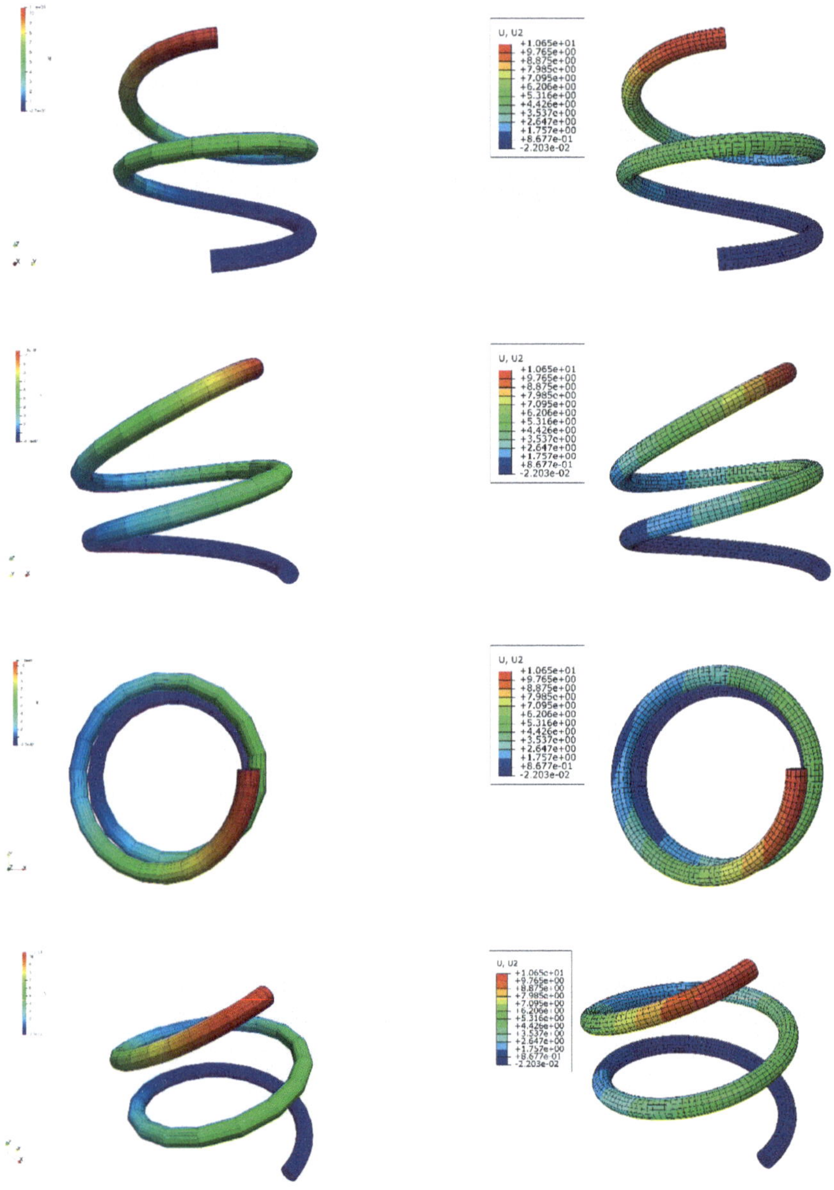

(a) HQEM 3D (b) Abaqus C3D8

Fig. 8.19 Displacement contour plot of the 3D tension spring structure

Fig. 8.20 Geometric configuration of a cylindrical shell

Table 8.4 Comparison between three-dimensional HQEM results and ABAQUS (C3D8) element results (three-dimensional cylindrical shell)

	HQEM (3D)	Abaqus (C3D8)
End face displacement/mm	2.6247	2.614

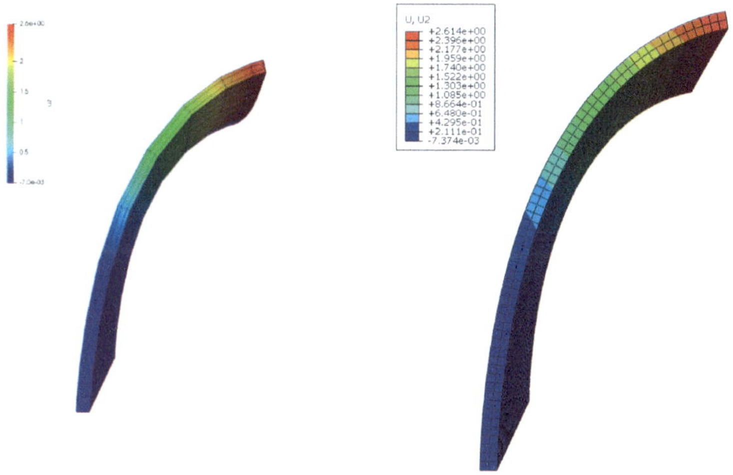

(a)HQEM 3D (b)Abaqus C3D8

Fig. 8.21 Displacement contour plot of the three-dimensional cylindrical shell

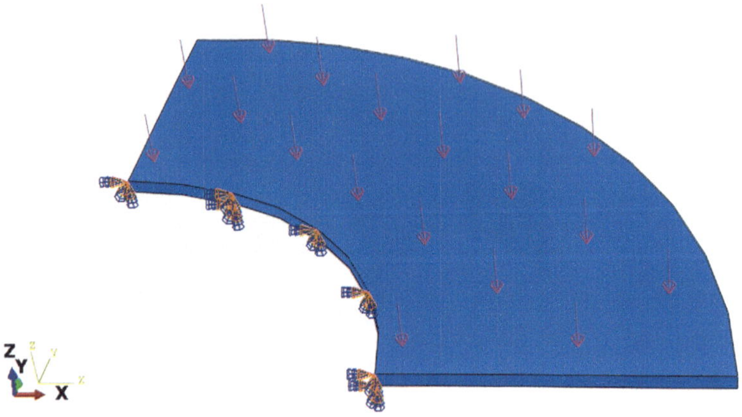

Fig. 8.22 Geometric shape of a sector plate

8.4.2.4 Three-Dimensional Sector Plate Structure

Next, consider the three-dimensional sector plate shown in Fig. 8.22, with geometric parameters as follows: inner radius $r = 5$ mm, outer radius $R = 10$ mm, and thickness $h= 0.2$ mm. A fixed support constraint is applied to the inner end face of the sector plate, while a uniformly distributed load of magnitude $p= 0.01$ N/mm^2, directed vertically downward, is applied to the top surface.

The outer end-face deflections calculated using the three-dimensional hierarchical quadrature element method and Abaqus C3D8 elements are shown in Table 8.5. The results from both methods agree well, with the three-dimensional HQEM employing a node arrangement of $[N_x,N_y,N_z]$=[8, 16, 2], while the Abaqus C3D8 elements used a mesh size of $h= 0.1$ mm (corresponding to 18,207 nodes).

The displacement contour plots obtained using the three-dimensional hierarchical quadrature elements and the ABAQUS (C3D8) elements are shown in Fig. 8.23.

Table 8.5 Comparison between three-dimensional HQEM results and ABAQUS (C3D8) element results (three-dimensional sector plate)

	HQEM (3D)	Abaqus (C3D8)
Outer end face deflection/mm	4.23252	4.260

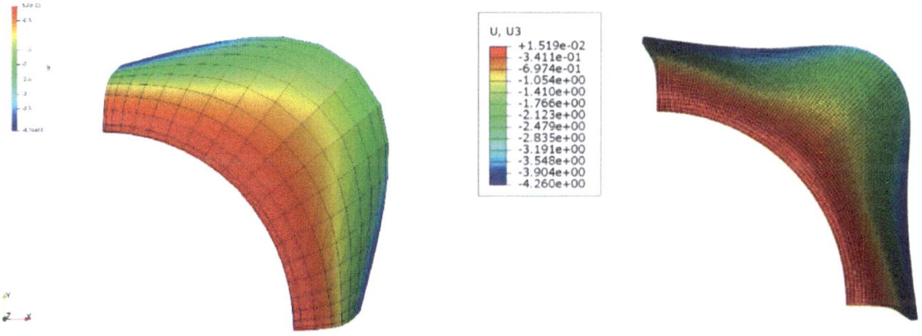

(a)HQEM 3D (b)Abaqus C3D8

Fig. 8.23 Displacement contour plot of the three-dimensional sector plate

8.4.3 Geometric Nonlinear Analysis of Mindlin Shallow Shell Structures Using the Hierarchical Quadrature Element Method

Consider the mid-surface configuration of the cylindrical shell shown in Fig. 8.24, with geometric parameters as follows: straight edge length $l = 10$ mm, curvature radius of the curved edge $r = 10$ mm, angle $\theta = \pi/3$, and shell thickness h = 1 mm. The material parameters are Young's modulus $E = 100$ N/mm^2, Poisson's ratio $\upsilon = 0.3$, and shear correction factor $\kappa = 5/6$. Fixed boundary conditions are applied on all four edges of the shell, and a uniformly distributed load p = 1N/mm^2 is applied to the top surface of the shell.

The geometric nonlinear finite element analysis of the above structure was conducted using the Mindlin hierarchical quadrature element method and the Abaqus

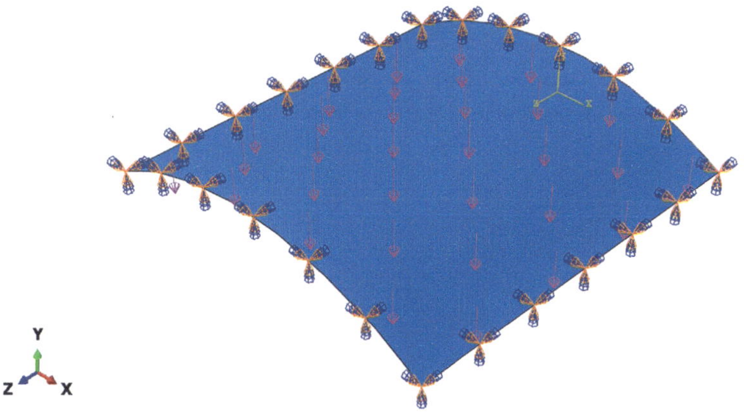

Fig. 8.24 Geometric shape of the middle surface of a cylindrical shell

Table 8.6 Comparison of Mindlin (Mindlin) hierarchical quadrature element method (HQEM) results and ABAQUS (S8R) element results

	HQEM (Mindlin)	Abaqus S8R
Maximum deflection/mm	1.32689	1.327

S8R elements, respectively. The maximum deflection at the center of the shell structure obtained from both methods is shown in Table 8.6, which indicates good agreement between the two sets of results. For the Mindlin hierarchical quadrature element method, the mesh arrangement was $[N_x, N_y]$=[10, 10], while for the Abaqus S8R elements, the mesh size was $h = 0.5$ mm (corresponding to 462 nodes).

The displacement contour plots obtained using the Mindlin hierarchical quadrature element and the ABAQUS (S8R) elements are shown in Fig. 8.25. In Fig. 8.25, the results for u_z, u_x and θ_x calculated with both elements are presented. It can be seen that the results of the geometric nonlinear analysis using the two types of elements are in good agreement.

8.5 Summary

This chapter provides a detailed introduction to the measures of stress and strain in the reference configuration under a fully Lagrangian description. Based on these measures, the virtual work equation in the reference configuration is presented, and the geometrically nonlinear finite element formulations for hierarchical quadrature two-dimensional and three-dimensional elements are derived. Based on the Mindlin theory, the geometrically nonlinear formulation for hierarchical quadrature shallow shell elements is also provided. The various algorithms for solving the resulting nonlinear algebraic systems follow the procedures introduced in Chap. 2. Finally, based on the constructed elements, numerous numerical examples are provided, including planar structures, complex spatial structures, and shallow shell structures. A comparison with corresponding ABAQUS results demonstrates that the hierarchical quadrature element method requires very low computational resources and shows promising application prospects in the field of geometric nonlinear analysis.

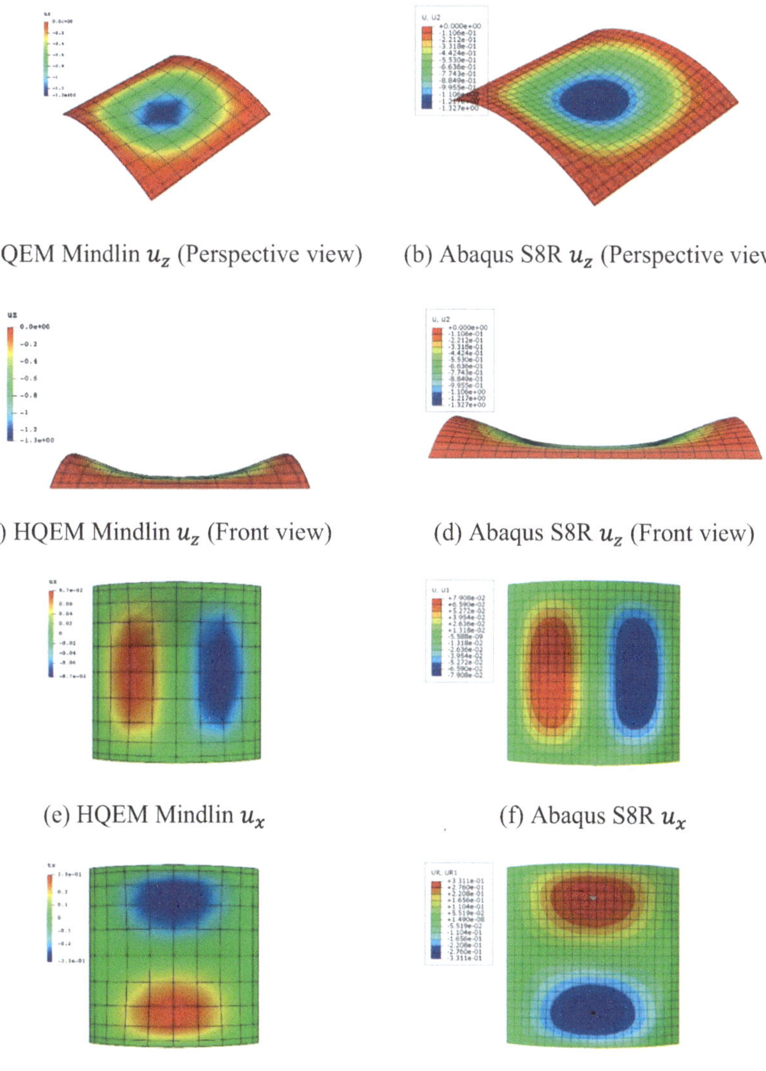

(a) HQEM Mindlin u_z (Perspective view) (b) Abaqus S8R u_z (Perspective view)

(c) HQEM Mindlin u_z (Front view) (d) Abaqus S8R u_z (Front view)

(e) HQEM Mindlin u_x (f) Abaqus S8R u_x

(g) HQEM Mindlin θ_x (h)Abaqus S8R θ_x

Fig. 8.25 Displacement contour plots generated by the Mindlin hierarchical quadrature element method (HQEM) and ABAQUS (S8R) elements

The manufacturer's authorised representative in the EU is Springer Nature Customer Service Centre GmbH, Europaplatz 3, 69115 Heidelberg, Germany. If you have any concerns regarding our products, please contact ProductSafety@springernature.com

Printed and bound by CPI Group (UK) Ltd, Croydon, CR0 4YY
07/10/2026
02229357-0014